T0226674

Protein Targeting, Transport & Translocation

PROTEIN TARGETING, TRANSPORT & TRANSLOCATION

Edited by

ROSS E. DALBEY
Ohio State University, Columbus, USA

GUNNAR VON HEIJNE
Stockholm University, Sweden

ACADEMIC PRESS
An Elsevier Science Imprint

Amsterdam   London   Oxford   New York   Boston
San Diego   San Francisco   Singapore   Sydney

Academic Press
An Elsevier Science Imprint
Harcourt Place, 32 Jamestown Road, London NW1 7BY, UK
http://www.academicpress.com

Academic Press
An Elsevier Science Imprint
525 B Street, Suite 1900, San Diego, California 92101-4495, USA
http://www.academicpress.com

ISBN 0-12-200731-X

Library of Congress Catalog Number: 2001094562

A catalogue record for this book is available from the British Library

Typeset by Charon Tec Pvt. Ltd, Chennai, India

Printed and bound by CPI Group (UK) Ltd, Croydon, CR0 4YY

Transferred to Digital Print 2011

Contents

Contributors   vii

Foreword   xi
Günter Blobel

Preface   xv
Ross E. Dalbey and Gunnar von Heijne

1 **Introduction/Overview**   1
Ross E. Dalbey and Gunnar von Heijne

2 **Methods in Protein Targeting, Translocation and Transport**   5
Ross E. Dalbey, Minyong Chen and Martin Wiedmann

3 **Targeting Sequences**   35
Gunnar von Heijne

4 **Protein Export in Bacteria**   47
Arnold J.M. Driessen and Chris van der Does

5 **Protein Sorting at the Membrane of the Endoplasmic Reticulum**   74
Nora G. Haigh and Arthur E. Johnson

6 **Membrane Protein Insertion into Bacterial Membranes and the Endoplasmic Reticulum**   107
Andreas Kuhn and Martin Spiess

7 **Disulfide Bond Formation in Prokaryotes and Eukaryotes**   131
James Regeimbal and James C.A. Bardwell

8 **The Unfolded Protein Response**   151
Carmela Sidrauski, Jason H. Brickner and Peter Walter

9 **Protein Quality Control in the Export Pathway: The Endoplasmic Reticulum and its Cytoplasmic Proteasome Connection**   180
Zlatka Kostova and Dieter H. Wolf

10 **Translocation of Proteins into Mitochondria**   214
Thorsten Prinz, Nikolaus Pfanner and Kaye N. Truscott

11 **The Import and Sorting of Protein into Chloroplasts**   240
   Jürgen Soll, Colin Robinson and Lisa Heins

12 **Import of Proteins into Peroxisomes**   268
   Suresh Subramani, Vincent Dammai, Partha Hazra,
   Ivet Suriapranata and Soojin Lee

13 **Nucleocytoplasmic Transport**   293
   Dirk Görlich and Stefan Jäkel

14 **Protein Transport to the Yeast Vacuole**   322
   Todd R. Graham and Steven F. Nothwehr

15 **The Secretory Pathway**   358
   Benjamin S. Glick

16 **Vesicular Transport**   377
   Joachim Ostermann, Tobias Stauber and Tommy Nilsson

17 **Conclusion/Perspective**   402
   Ross E. Dalbey and Gunnar von Heijne

Index   407

Contributors

James C.A. Bardwell, Department of Molecular, Cellular and Developmental Biology, University of Michigan, Ann Arbor, Michigan 48105-1048, USA

Jason H. Brickner, Department of Biochemistry and Biophysics, University of California, San Francisco, California 94143-0448, USA

Minyong Chen, Department of Chemistry, The Ohio State University, 100 West 18th Avenue, Columbus, Ohio 43210, USA

Ross E. Dalbey, Department of Chemistry, Ohio State University, 305 Johnston Laboratory, 176 W Nineteenth Avenue, Columbus, Ohio 43210, USA

Vincent Dammai, Section of Molecular Biology, Division of Biology, University of California, San Diego, La Jolla, California 92093-0322, USA

Arnold J.M. Driessen, Department of Microbiology, Groningen Biomolecular Sciences and Biotechnology Institute, University of Groningen, Kerklaan 30, 9751 NN Haren, PO Box 14, 9750 AA Haren, The Netherlands

Benjamin Glick, Department of Molecular Genetics and Cell Biology, Cummings Life Science Center, University of Chicago, 920 E 58th Street, Chicago, Illinois 60637, USA

Dirk Görlich, Zentrum für Molekulare Biologie Universität Heidelberg (ZMBH), Im Neuenheimer Feld 282, 69120 Heidelberg, Germany

Todd R. Graham, Department of Biological Sciences, Vanderbilt University, Nashville, Tennessee 37235-1634, USA

Nora G. Haigh, Department of Medical Biochemistry and Genetics, Texas A&M University System Health Science Center, 116 Reynolds Medical Building, 1114 TAMU, College Station, Texas 77843-1114, USA

Partha Hazra, Section of Molecular Biology, Division of Biology, University of California, San Diego, La Jolla, California 92093-0322, USA

Lisa Heins, Botanisches Institut der Christian Albrechts Universität, Am Botanischen Garten 1-9, Olshausenstrasse 40, 24098 Kiel, Germany

Stefan Jäkel, Zentrum für Molekulare Biologie, Universität Heidelberg (ZMBH), Im Neuenheimer Feld 282, 69120 Heidelberg, Germany

Arthur E. Johnson, Department of Medical Biochemistry and Genetics, Texas A&M University System Health Science Center, 116 Reynolds Medical Building, 1114 TAMU, College Station, Texas 77843-1114, USA

Zlatka Kostova, Institut für Biochemie der Universität Stuttgart, Pfaffenwaldring 55, 70569 Stuttgart, Germany

Andreas Kuhn, University of Hohenheim, Institute for Microbiology and Molecular Biology, Garbenstrasse 30, D-70599, Stuttgart, Germany

Soojin Lee, Section of Molecular Biology, Division of Biology, University of California, San Diego, La Jolla, California 92093-0322, USA

Tommy Nilsson, Cell Biology Programme, European Molecular Biology Laboratory, Meyerhofstrasse 1, D-69012, Heidelberg, Germany

Steven F. Nothwehr, Division of Biological Sciences, University of Missouri, Columbia, Missouri 65211, USA

Joachim Ostermann, McGill University, Department of Anatomy and Cell Biology, 3640 University Avenue, H3A 2B2, Montreal, Quebec, Canada

Nikolaus Pfanner, Institut für Biochemie und Molekularbiologie Universität Freiburg, Hermann-Herder-Strasse 7, 79104 Freiburg, Germany

Thorsten Prinz, Institut für Biochemie und Molekularbiologie Universität Freiburg, Hermann-Herder-Strasse 7, 79104 Freidburg, Germany

James Reigembal, Department of Molecular, Cellular and Developmental Biology, University of Michigan, Ann Arbor, Michigan 48105-1048, USA

Colin Robinson, University of Warwick, Department of Biological Sciences, Coventry CV4 7AL, UK

Carmela Sidrauski, Department of Biochemistry and Biophysics, University of California, San Francisco, California 94143-0448, USA

Jürgen Soll, Botanisches Institut der Universität Kiel, Am Botanischen Garten 1-9, 24118 Kiel, Germany

Martin Spiess, Department of Biochemistry, Biozentrum, University of Basel, Klingelbergstrasse 70, CH-4056 Basel, Switzerland

Tobias Stauber, Cell Biology Programme, European Molecular Biology Laboratory, Meyerhofstrasse 1, D-69012, Heidelberg, Germany

Ivet Suriapranata, Section of Molecular Biology, Division of Biology, University of California, San Diego, La Jolla, California 92093-0322, USA

Suresh Subramani, Department of Biology, University of California, San Diego, La Jolla, California 92093-0322, USA

Kaye N. Truscott, Institut für Biochemie und Molekularbiologie Universität Freiburg, Hermann-Herder-Strasse 7, 79104 Freiburg, Germany

Chris van der Does, Department of Microbiology, Groningen Biomolecular Sciences and Biotechnology Institute, University of Groningen, Kerklaan 30, 9751 NN Haren, PO Box 14, 9750 AA Haren, The Netherlands

Gunnar von Heijne, Department of Biochemistry, Arrhenius Laboratory, Stockholm University, S-106 91 Stockholm, Sweden

Peter Walter, Department of Biochemistry and Biophysics, University of California, San Francisco, California 94143-0448, USA

Martin Wiedmann, Cellular Biochemistry and Biophysics Program, Memorial Sloan-Kettering Cancer Center, 1275 York Avenue, New York, New York 10021, USA

Dieter H. Wolf, Institut für Biochemie, Universität Stuttgart, Pfaffenwaldring 55, 70569 Stuttgart, Germany

FOREWORD

Thirty years ago, in 1971, we published a hypothesis in which we suggested that secretory proteins contain a shared amino-terminal sequence element. A cytosolic binding factor was predicted not only to bind this sequence but also to mediate the attachment of the translating ribosome to the endoplasmic reticulum (ER) membrane. Following completion of translation, the ribosomal subunits were proposed to join the pool of free ribosomal subunits, ready to begin a new round of translation.

This hypothesis attempted to explain the observation that mRNAs for secretory proteins are translated on ER-bound ribosomes and not on free ribosomes. It emphasized the idea that all ribosomes are created equal and opposed a then popular notion that ribosomes might differ in their composition and in their ability to select various mRNAs for translation. The fact that a shared amino-terminal sequence element was not discernible among the few secretory proteins that had been sequenced at that time did not deter us from advancing our proposals. It seemed conceivable to us that such a shared sequence element might be transient in nature and be cleaved off before chain completion and hence be absent in the mature secretory protein.

Earlier it had been established that nascent chains of ER-bound ribosomes are 'vectorially' discharged to the *trans* side of the membrane (the lumen of microsomal vesicles) after incubation with puromycin. Vectorial discharge was thought to proceed through a 'discontinuity' in the membrane. This discontinuity, however, remained undefined until 1975. In what was then dubbed the signal hypothesis, the ideas proposed in 1971 were further amplified to include an ER embedded channel that consists of integral membrane proteins and that functions specifically to allow the passage of nascent secretory proteins to the *trans* side of the ER membrane. The amino-terminal sequence of the nascent secretory protein in concert with several sites on the large ribosomal subunit were envisaged to serve as ligands to assemble (or open) the protein-conducting channel. The concept of a protein-conducting channel made up of integral membrane proteins remained the most contentious aspect of the signal hypothesis for more than 15 years until definitive electrophysiological experiments in 1991 and 1992 established its existence.

The first evidence in support of a transient amino-terminal extension in secretory proteins was obtained in 1972 when mRNA for the light chain of

IgG was translated in a membrane-free translation system. However, it could still be argued that the detected amino-terminal sequence extension is not a signal for translocation, but serves other functions, e.g. it might facilitate folding of nascent secretory protein. It was only in 1975, when we succeeded in developing an *in vitro* coupled translation–translocation system that compelling evidence for the function of the amino-terminal extension as a signal for membrane translocation was obtained. The amino-terminal extension of the light chain of IgG was found to be cleaved only when translation occurred in the presence of added microsomal vesicles, but not when the microsomal vesicles were added after translation. This indicated that the microsomal membrane contained an embedded signal peptidase with its active site exposed on the *trans* side of the membrane. Most importantly, the signal-peptidase-processed nascent chains were found to be protected from externally added proteases, indicating that they were translocated into the lumen of the microsomal vesicles to which the added proteases had no access.

Once this coupled *in vitro* translation–translocation system was set up, it was only a matter of time to identify the cast of characters that are involved in translocation of secretory proteins across the ER. The first component to be isolated in 1978/1980 was the binding factor, whose existence was predicted in 1971. Unexpectedly this binding factor turned out to be a ribonucleoprotein particle, consisting of an RNA and six distinct proteins. As predicted in 1971, this binding factor, now termed signal recognition particle (SRP), recognized the signal sequence and bound the translating ribosome to the microsomal membrane. Thereafter, a heterodimeric membrane protein that is located only in the ER and that functions as an SRP receptor was isolated and characterized. Hence the components involved in signal sequence recognition and targeting to the ER were defined. Next, the enzyme that cleaves off the signal sequence was isolated and shown to consist of a complex of five distinct integral ER membrane proteins. A most important advance for the subsequent characterization of the protein-conducting channel was the demonstration in 1989 that protein translocation occurred faithfully in proteoliposomes that were reconstituted after detergent solubilization of microsomal membranes. Finally, the identification and characterization of the protein-conducting channel was accomplished by genetic and biochemical methods. Recent reconstitution of isolated protein conducting channels with RNCs (ribosome–nascent chain complexes) and subsequent analysis by cryo-electron microscopy and three-dimensional image reconstruction at 15.4 Å resolution revealed that the protein-conducting channel is aligned with the tunnel in the large ribosomal subunit and is a rather compact structure that is apparently in intimate contact with the translocating chain. At least four attachment sites to distinct segments of large ribosomal subunit RNA and proteins have been discerned.

Another proposal of the 1975 signal hypothesis was that a nascent integral membrane protein contains a signal sequence that is functionally identical to that of a secretory protein. This signal sequence was suggested to initiate translocation of the nascent membrane protein. An additional sequence element, termed stop-transfer sequence, was proposed to prevent further translocation of the nascent chain to the *trans* side by opening the protein-conducting channel laterally to the lipid bilayer, thereby allowing displacement of the stop-transfer sequence from the aqueous channel to the lipid bilayer. Data supporting these proposals were obtained in 1977/1978, when mRNA of the vesicular stomatitis virus (VSV) membrane glycoprotein (G) was translated in the coupled translation–translocation system. These experiments were paradigmatic as they showed that the asymmetric integration of a membrane protein into the lipid bilayer is not a spontaneous process, as was widely believed at the time, but is catalyzed.

Yet another proposal of the 1975 signal hypothesis was that proteins to be translocated across other intracellular membranes would possess signal sequences that are distinct from those addressed to the ER. Such signal sequences were indeed detected in the late 1970s and early 1980s for translocation across the bacterial plasma membrane, for protein import into mitochondria, chloroplasts and peroxisomes and finally for import and export across the nuclear pore complexes of the nuclear envelope. In many ways, the experiments of the ER translocation system were paradigmatic for the experiments in these other systems. Cell-free translocation systems were set up followed by genetic and biochemical experiments to identify the cast of characters involved in each of the cases. Similar strategies were also used to study intercompartmental transport. The various chapters of this book give us an account of these efforts and where we presently stand.

Although signal sequences, cognate recognition factors, targeting and passage through a membrane are common to all of the translocation systems, nature has created fascinating and ingenious variations of that general theme. Bacteria are clearly the masters of this game. A more recent example of their virtuosity is that practiced by pathogenic Gram-negative bacteria. These bacteria polymerize a needle-like structure from a single small protein to puncture the plasma membrane of a eukaryotic cell to transfer certain proteins through a very narrow gauge from the bacterial cytosol across three membranes into the eukaryotic cytosol.

Many important questions remain to be answered and several areas of intracellular macromolecular traffic are just in the beginning phases of exploration. It is clear, for example, that nuclear import or export does not end or begin, respectively, with transport across the nuclear pore complex. Export is preceded and import is followed by an intranuclear phase of transport. Another largely unexplored area is how various segments of nascent membrane proteins interact with protein-conducting channels to achieve the great variety of polytopic orientations in the membrane. The structural analysis of the various transport systems by X-ray crystallography and cryo-electron microscopy has just begun and should continue to provide major new insights into their function.

The field of macromolecular intracellular traffic is by no means in a stationary phase. To the contrary, it has barely entered the logarithmic phase. This book will be an important milestone and a guide to those who enter this exciting phase. No doubt, a deeper understanding of cellular macromolecular traffic systems will ultimately yield a broader understanding into how a cell, any cell, organizes itself.

Günter Blobel
New York, September 2001

PREFACE

Not since the *Protein Targeting Book* by Tony Pugsley in 1989 has the topic of protein localization been covered in depth in a textbook. We felt, therefore, the time was right to put together an up-to-date book that could be used both by scientists in general and in graduate level and/or advanced undergraduate courses. It is our strong belief that only when a book finds use in teaching is it really worthwhile.

In just the past ten years, there has been an explosion of activity in the protein targeting and transport area. Many major advances have been made just in this time period. For instance, most of the components that comprise that targeting factors and translocation systems have been identified and some of the protein structures have been solved to high resolution. It is now clear that there are diverse and extremely intricate machineries used to move proteins around within the cell.

In 1999, much attention was focused on the protein targeting area when Günter Blobel was awarded the Nobel Prize in Physiology or Medicine 'for the discovery that proteins have intrinsic signals that govern their transport and localization in the cell'. Now, in the early days of the 21st century, only thirty years after Blobel initiated his first ground-breaking experiments, the protein targeting area has proved to be of fundamental importance in areas ranging from biotechnology and molecular biology to apoptosis, immunology, signal transduction, and others. This book is intended to give some impression of this wide significance while at the same time not losing sight of the basic principles of protein targeting.

We would like to offer our sincerest thanks to all the contributors for their hard work and devotion to research which made this book both necessary and possible.

Ross E. Dalbey
Gunnar von Heijne

1

INTRODUCTION/OVERVIEW

ROSS E. DALBEY AND GUNNAR VON HEIJNE

All living cells contain proteins that carry out specialized functions within various subcellular membrane or aqueous spaces. Recent estimates suggest that approximately half of all the proteins of a typical cell are transported into or across a membrane. How are proteins synthesized in the cytoplasm of the cell, inserted into or across membranes, and how are they transported to their correct subcellular destinations? Questions such as these have been a central theme in cell biology for nearly four decades, starting with the pioneering work of George Palade (Nobel Laureate in Physiology or Medicine in 1974) that defined the basic structure of the secretory pathway in eukaryotic cells, and continued by among others Günter Blobel (Nobel Laureate in Physiology or Medicine in 1999) who discovered that proteins possess intrinsic signals that govern their localization in the cell.

Bacterial cells have at least one membrane that separates the inside of the cell from its environment. Gram-positive bacteria have only one membrane, and Gram-negative bacteria have an additional outer membrane. Therefore, in Gram-positive cells there are three compartments – the cytoplasm, the plasma membrane, and the extracellular medium – whereas in Gram-negative bacteria there are five – the cytoplasm, the plasma membrane, the periplasm, the outer membrane, and the extracellular medium.

In contrast to most bacterial cells, eukaryotic cells contain, in addition to the plasma membrane, internal membranes (Figure 1.1). These internal membranes are structural components of organelles and vesicles. Proteins embedded in membranes or localized in the aqueous spaces surrounded by membranes give rise to the specialized functions carried out in these compartments. Thus, the nucleus houses the machinery for DNA replication, transcription and RNA splicing; the mitochondrion specializes in respiration that produces adenosine triphosphate (ATP) for the cell; the chloroplast contains the proteins that are responsible for photosynthesis and the

Protein Targeting, Transport & Translocation
ISBN 0-12-200731-X

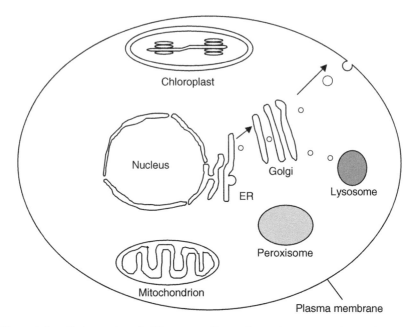

Figure 1.1   Eukaryotic cells. The organelles and membranes are shown for a typical eukaryotic cell. Each of the organelles has a specialized function. Most proteins are synthesized in the cytoplasm on ribosomes. Some proteins are directly targeted from the cytoplasm. This includes proteins directed to the ER, mitochondria, chloroplast, or peroxisome by intrinsic signals within their polypeptide chain. Some proteins that are targeted to the secretion pathway via the ER are further sorted to the Golgi, lysosome/vacuole, secretory vesicles, plasma membrane, or the extracellular medium. Chloroplast and mitochondrion have their own genomes that synthesize a small number of proteins. The chloroplast-synthesized proteins either remain in the stroma or are exported to the thylakoid membrane or thylakoid lumen. In mitochondria, the newly synthesized proteins remain in the matrix or are exported to the inner membrane in mitochondria.

synthesis of energy-rich compounds from carbon dioxide and water; the Golgi apparatus contains enzymes that modify sugars attached to exported proteins; the lysosome/vacuole contains digestive enzymes responsible for intracellular digestion, and the peroxisome houses enzymes for fatty acid oxidation and for producing and metabolizing hydrogen peroxide.

Most proteins are synthesized in the cytoplasm of the cell, except for a small number that are encoded in the mitochondrial and chloroplast genomes. This raises the question of how proteins are transported from the cytoplasm to other destinations within or outside of the cell. Approximately 20% of the proteins in a typical cell are located in the non-cytoplasmic aqueous spaces bounded by a membrane. An additional 25–30% of the proteins are located within a membrane.

Proteins are imported directly from the cytoplasm into the endoplasmic reticulum (ER), mitochondria, peroxisomes and chloroplasts by mechanisms that use a targeting sequence and a translocation machinery. Exported proteins are usually made in a precursor form with an amino-terminal signal peptide that directs the protein into the export pathway. Such amino-terminal signal peptides target proteins to the ER membrane where they are recognized by the translocation machinery (see Chapter 5). For peroxisomal proteins, there are two types of targeting sequences directing import: an amino-terminal signal or a carboxyl-terminal targeting sequence (see Chapter 12). Mitochondrial targeting sequences target proteins to the mitochondrial membrane by being recognized by surface exposed mitochondrial receptors (see Chapter 10). The chloroplast targeting signal directs chloroplast proteins to the chloroplast for import into the organelle (see Chapter 11).

In addition to importing proteins from the cytoplasm into the organelle, mitochondria and chloroplasts also export proteins from the mitochondrial matrix or the chloroplast stroma where proteins are encoded by their respective organellar genomes. It is not surprising that mitochondrial and chloroplast export machineries share some important features with those found in bacterial cells since these organelles descended from bacterial progenitors millions of years ago.

Proteins that are localized to the Golgi, lysosome/vacuole, and plasma membrane are first inserted into the ER. Within the ER, disulfide bonds are introduced into the proteins by a protein disulfide isomerase (see Chapter 7). Additionally, a 14-residue oligosaccharide core is attached to glycoproteins containing asparagine-linked sugars. The oligosaccharyl core is processed initially in the ER and then further trimmed and modified in the Golgi apparatus. Misfolded proteins in the ER are recognized and retrotranslocated out of the ER lumen into the cytoplasm where the protein is ubiquinated and degraded by the proteasome (see Chapter 9). Proteins that are folded correctly move from the ER to the *cis* Golgi and further along the secretory pathway by vesicular transport.

The details of how vesicles are formed at the ER and move through the Golgi stacks (from *cis* to *trans*) are being actively worked out. The SNARE hypothesis (see Chapter 16) proposes that vesicles mediate trafficking in the anterograde – forward – direction from the *cis* Golgi cisternae to the *medial* Golgi cisternae and then from the *medial* Golgi cisternae to the *trans* cisternae. The donor vesicle fuses with its target vesicle using a number of proteins (NSF, SNAP, SNARE, etc.). A competing hypothesis states that the vesicles do not mediate movement of cargo through the stacks. Rather, the *cis* Golgi cisternae mature into the *medial* cisternae; the *medial* cisternae then mature into the *trans* cisternae. In this model, the cisternae mature because certain Golgi components contained within the cisternae are removed by retrograde vesicle transport (see Chapters 15 and 16).

Transport into and out of the nucleus is unlike the mechanism for insertion into the ER, mitochondria, chloroplast and peroxisome in that it occurs via large aqueous pores that span both nuclear membranes. These nuclear pore complexes are huge structures that support two-way trafficking (Chapter 13). Proteins imported into the nucleus typically contain a positively charged nuclear localization signal. A number of soluble factors are also required for transport.

A good understanding of protein targeting and translocation is important for many areas in biology and medicine. It has applications in biotechnology, where growth hormones, insulin, interleukins and coagulation factor VIII, to name but a few, have been engineered to be secreted into the culture media. In immunology, knowledge of the secretion pathway has been very useful for the understanding of how peptides from antigens are displayed by the major histocompatibility complex proteins on the cell surface. In programmed cell death, protein translocation to and from the plasma membrane, mitochondria and the nucleus is critical for regulating apoptosis. Lastly, nuclear trafficking is very important for signal transduction and cell cycle regulation.

This book brings together a number of important topics in the protein localization field. First, we will describe some of the common techniques used to study protein translocation and transport (Chapter 2). Second, we review the targeting signals within exported proteins that direct the export of proteins to their subcellular compartment (Chapter 3). Third, we review how proteins cross and insert into membranes in bacteria and in the ER of a eukaryotic cell (see Chapters 4–6). Fourth, we will describe how disulfide bonds are introduced into exported or membrane proteins as they enter the ER lumen (Chapter 7). Fifth, we will report on the unfolded protein response where the cells can adapt to the condition where unfolded proteins accumulate in the lumen of the ER (Chapter 8) and quality control mechanisms allowing proteolysis of misfolded proteins (Chapter 9). Sixth, we will describe protein import into the mitochondria, chloroplast and peroxisome (Chapters 10–12). Seventh, we will review the import and export of nuclear proteins and regulation of this process (Chapter 13) and the movement of proteins along the secretion pathway (ER to Golgi to either the vacuole or the plasma membrane) (Chapters 14–16).

ACKNOWLEDGMENT

This work was supported by grant MCB-9808843 from NSF to R.E.D. and grants from the Swedish Research Council and the Swedish Cancer Foundation to G.V.H.

2

METHODS IN PROTEIN TARGETING, TRANSLOCATION AND TRANSPORT

ROSS E. DALBEY, MINYONG CHEN AND
MARTIN WIEDMANN

INTRODUCTION

Protein targeting, translocation and transport mechanisms have been studied extensively by scientists over the last 30 years using biochemical, genetic, cell biological, molecular biological, and electron microscopic techniques. In this chapter, we will cover some of the key techniques used to study protein export. We have divided them into four categories: *in vivo*, genetic, *in vitro*, and cell biology techniques. The *in vivo* techniques are necessary to examine the fate of a protein within an intact cell and often take advantage of mutants that were identified using genetics. The genetic section is separate from the *in vivo* section because of the premier importance it plays in the protein transport area. Genetics have unraveled most of the protein components that make up the translocation machinery involved in protein export. A powerful role is also played by the *in vitro* techniques where the functions of the purified proteins are defined and where the goal is to reconstitute transport events in a test tube. Finally, cell biology techniques exploiting electron microscopy and fluorescence light microscopy have allowed the researcher to follow the fate of a protein within a cell.

Protein Targeting, Transport & Translocation
ISBN 0-12-200731-X

IN VIVO STUDIES: PULSE-CHASE STUDIES WITH WHOLE CELLS AND SUBCELLULAR FRACTIONATION

Bacteria

Almost all proteins exported to the outer membrane and periplasmic space of *Escherichia coli* are made in a precursor form containing an amino-terminal extension peptide called a signal peptide. The export of these proteins requires the Sec machinery comprising SecA, SecY, SecE, SecG, SecD, SecF (Schatz and Beckwith, 1990; Wickner et al., 1991), and YajC (Duong and Wickner, 1997). Also needed for export is the electrochemical membrane potential (Geller et al., 1986) and ATP hydrolysis (Chen and Tai, 1985; Geller et al., 1986).

The use of drugs and Sec mutants to study protein export
To examine export *in vivo*, cells are typically labeled with [^{35}S]-methionine for a short time (15 s) and chased with non-radioactive methionine for various times. The labeled proteins are immunoprecipitated with antiserum to the respective protein, and analyzed by SDS–PAGE (sodium dodecyl sulfate-polyacrylamide gel electrophoresis) and phosphorimaging. In these pulse-chase experiments, preproteins are rapidly inserted into the membranes and processed by signal peptidase, an integral membrane protease that removes signal sequences. The addition to a bacterial culture of carbonyl cyanide *p*-chlorophenylhydrazone (CCCP), an uncoupler of the membrane electrochemical potential, causes accumulation of non-translocated prepro-teins at the membrane (Daniels et al., 1981; Date et al., 1980). The addition of azide, an inhibitor of SecA, causes Sec-dependent proteins to accumulate (Oliver et al., 1990). The effects of these drugs on the export of preproteins is tested by examining whether the precursor form of the exported protein accumulates. Usually, the precursor form that accumulates is easily detected by SDS–PAGE and fluorography.

Using Sec mutants is instrumental with *in vivo* studies for determining whether a protein is exported by the Sec machinery. For instance, ther-mosensitive (t.s.) mutations in SecA (Oliver and Beckwith, 1981) and SecY (Ito et al., 1983) and cold-sensitive (c.s.) mutations in SecE (Schatz et al., 1989), SecG (Nishiyama et al., 1994) and SecD (Gardel et al., 1987) have been isolated. T.s. and c.s. mutants are grown at the non-permissive temperature for certain times to deplete (synthetic mutants) or to inacti-vate (folding mutants) the Sec protein. When the cells are grown at the non-permissive temperature the kinetics of protein translocation can be investigated. If the newly synthesized preproteins accumulate at the non-permissive temperature, the protein is Sec-dependent. If no translocation defect is observed in these conditional mutants, it is useful to analyze a SecE depletion strain where the Sec machinery can be inactivated to a much greater degree by depletion. This strain has SecE under control of the *araBAD* promoter (Traxler and Murphy, 1996) and growth in the absence of arabinose can lead to very strong depletion in SecE as well as SecY, SecG and SecF (Yang et al., 1997).

Assay for membrane protein insertion

For bacterial inner membrane proteins, which do not contain cleavable signal peptides, it is necessary to determine directly the translocation of a membrane protein's hydrophilic domain across the membrane. Insertion can be determined by testing whether the membrane protein is accessible to protease in spheroplasts. Cells are converted to spheroplasts by a lysozyme and osmotic shock treatment, which causes the outer membrane to peel away from the inner membrane (Osborn et al., 1972). This enables the added protease access to the outer surface of the inner membrane. The spheroplasts are incubated on ice with or without protease. Digestion of a periplasmic domain within the membrane protein decreases the amount of the full-length protein and indicates that the membrane protein is inserted. For example, the full-length leader peptidase, which spans the membrane twice with a large C-terminal domain, is digested by protease, indicating that the C-terminal domain inserts across the plasma membrane (Dalbey and Wickner, 1986). Typically, one uses a cytoplasmic protein such as ribulokinase as a control to monitor the integrity of the spheroplasts and to show that there is no lysis. OmpA is routinely used in *E. coli* to monitor the efficiency of spheroplast formation. OmpA is not digested by protease in intact cells, whereas it is digested in spheroplasts.

To assay membrane protein insertion of polytopic membrane proteins, other methods must be used because polytopic membrane proteins may contain more than one hydrophilic loop exposed to the periplasmic side of the membrane. Multiple translocation events may occur in these proteins. Protease mapping studies in spheroplasts can reveal whether a polytopic membrane protein inserts across the membrane. However, determining which hydrophilic loop of the membrane protein is translocated across the membrane is not always so simple. To obtain this kind of information, three different types of approaches are used. In the first approach, alkaline phosphatase is fused to the different periplasmic loop regions of the membrane protein. Alkaline phosphatase is only active if it is exported to the periplasmic space (Manoil and Beckwith, 1986). Alkaline phosphatase activity is thus indicative of the topology of the particular protein domain. Alternatively, some investigators introduce short (less than 10 residues long) uncharged peptides into the periplasmic loops, which function as epitopes for antibodies (Konninger et al., 1999). Immunoprecipitation assays, together with protease-mapping studies, can then determine whether a given loop is accessible to protease and therefore membrane-translocated. If the epitope is translocated, then it is digested by protease and therefore the protein can no longer be immunoprecipitated with the antibody against the epitope. A third approach involves engineering a unique protease cleavage site within the membrane protein. The protease that recognizes the introduced protease cleavage site is then added to spheroplasts. If the cleavage site is exposed and located in the periplasmic space, the protease can lead to site-specific digestion of the membrane protein (Ehrmann et al., 1997).

Isolation of membrane and aqueous compartments in bacteria
Subcellular fractionation studies can determine whether a protein is localized to the correct cellular compartment. In *E. coli*, the periplasmic fraction is easily obtained by osmotically shocking bacterial cells (Neu and Heppel, 1965). After lysing the cells by sonication or using a French pressure cell, the membranes are separated from the cytosol by differential centrifugation. Inner and outer membranes are separated by isopycnic sucrose gradient centrifugation. Each of the four compartments of *E. coli* (cytoplasm, inner membrane, periplasm, outer membrane) can be isolated with minimum cross contamination. A valuable indicator for cross contamination and for the isolation of the correct compartment is to assay for a known protein found uniquely in the particular fraction. OmpA is a good marker for the outer membrane; precursor to maltose binding protein or alkaline phosphatase are good markers of the periplasmic fraction; NADH oxidase is an inner membrane marker; glucose-6-phosphate dehydrogenase is a good cytoplasmic marker.

Eukaryotes

Translocation of proteins into the ER lumen can be accessed *in vivo* in *Saccharomyces cerevisiae* by monitoring signal peptide cleavage and glycosylation. Signal peptidase removes the signal peptide of the preprotein after membrane translocation. In addition, many proteins are glycosylated during translocation where the ER located oligosaccharyl transferase adds the high mannose oligosaccharide onto Asn-X-Ser/Thr acceptor sites within the translocated preprotein (Silberstein and Gilmore, 1996). This modification results in an increased size of the exported proteins, which is visualized by SDS–PAGE.

In *S. cerevisiae*, two different export pathways have been characterized. In the first pathway, proteins are translocated co-translationally, requiring the Sec61p complex that functions as a protein-conducting channel (Simon and Blobel, 1991; Hanein et al., 1996). In the second pathway, proteins are translocated post-translationally after the protein is released from the ribosome. Post-translational translocation requires, in addition to the Sec61p complex, the protein components Sec62p, Sec63p, Sec71p and Sec72p (Corsi and Schekman, 1996).

Sec mutants to illuminate the secretory pathway
In yeast, as in bacteria, Sec mutants are very useful for defining the translocation/processing step. For example, genetic studies identified Sec61p as one of the candidates of the translocation channel (Deshaies and Schekman, 1987), whereas Sec11 was shown to be important for signal peptide processing of several exported proteins (Bohni et al., 1988).

In a temperature-sensitive Sec61α mutant, the secretory preproteins, like pre-pro-alpha factor and carboxypeptidase, accumulate in the cytosol of the cell (Deshaies and Schekman, 1987). This was shown by performing a protease experiment with a cytoplasmic extract under conditions where the ER vesicles, termed microsomes, were intact. In these experiments, proteins translocated into the ER lumen are protease resistant; cytoplasmic proteins are protease sensitive. The cell-free extract was prepared by pulse-labeling the Sec61α mutant at the non-permissive temperature and lysing the spheroplasts using a Potter–Elvehjem glass–Teflon type of homogenizer. Added protease digested the cytosolic pre-pro-alpha-factor and procarboxypeptidase.

In the temperature-sensitive mutant Sec11-7, the exported protein invertase accumulated at the restrictive temperature in the lumen of the ER where it was glycosylated (Bohni et al., 1988). The addition of tunicamycin, which inhibits glycosylation at Asn residues of invertase, prior to radioactive labeling, leads to the appearance of 58 kDa invertase precursor species. This species corresponds to the non-glycosylated mature protein with the leader sequence intact. Later studies showed that Sec11 codes for a subunit of the signal peptidase complex that is homologous to the bacterial, mitochondrial and chloroplast signal peptidases (Dalbey et al., 1997).

After a protein translocates into the ER, it can remain there, move to the Golgi, the plasma membrane, the vacuole, or be secreted. Not surprisingly, many different Sec mutations that affect the various steps in the secretion pathway have been isolated. One such mutant is in the *vps15* gene. A *vps15* mutant was isolated that cannot sort the soluble vacuolar hydrolases such as carboxypeptidase Y (CPY), proteinase A and proteinase B to the vacuole (Herman et al., 1991; Robinson et al., 1988). Consequently, these proteins are secreted as inactive proteins. In this mutant and many other *vps* mutants, protein glycosylation and secretion of other proteins is unaffected.

Protein modification with sugars in the ER and Golgi
In mammalian cells, the processing state of Asn-linked glycoproteins provides very useful information regarding their cellular location. As shown in Figure 2.1, the addition of the N-linked core oligosaccharide takes place

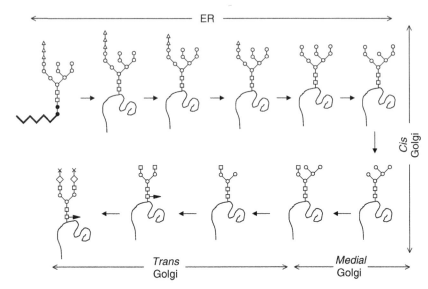

Figure 2.1   Biosynthesis and trimming of the core-glycosylated secretory proteins in the ER and Golgi. The core oligosaccharide, comprising 2 *N*-acetylglucosamines, 9 mannose and 3 glucose residues, is transferred from dolichol phosphate to the asparagine (Asn) residues of the growing polypeptide in the lumen of the ER. The glucosidase I and II, which remove the glucose residues of the N-linked core oligosaccharide, are in the ER. The first mannose residue is removed by an ER alpha-mannosidase I. Further mannose trimming by mannosidase I in the *cis* Golgi, produces a 5 mannose core. *N*-acetylglucosamine is introduced by *N*-acetylglu-cosamine-transferase I. Following further mannose trimming by mannosidase II and the addition of *N*-acetylglucosamine by *N*-acetylglucosamine-transferase II (in the *medial* Golgi), galactose, fucose, and sialic acid are added in the *trans* Golgi cister-nae. □, *N*-acetylglucosamine; ○, mannose; △, glucose; →, fucose; ◇, galactose; ×, sialic acid; ●, phosphate. Adapted from Goldberg and Kornfeld (1983) *J Biol Chem* **258**: 3160.

in the lumen of the ER. The oligosaccharide moiety is then trimmed by glu-cosidases I and II, which remove terminal glucose residues. The ER alpha-mannosidase I can then remove a mannose residue. Transport to the *cis* Golgi compartment results in further trimming and then *N*-acetylglucosamine-transferase I catalyzes the addition of *N*-acetylglucosamine in the *medial* Golgi compartment. After further trimming by mannosidases I and II in the *medial* Golgi compartment, *N*-acetylglucosamine residues are added by another *N*-acetylglucosamine-transferase, *N*-acetylglucosamine-transferase II. Finally, galactose residues and the sialic residues are introduced in the *trans* Golgi compartment.

These modifications are specifically traced by pulse-chase experiments where radioactive oligosaccharides or other groups are incorporated.

For instance, fucose, galactose and sialic acid are incorporated in the *trans* Golgi cisternae.

In addition, useful information about the location of a protein in the secretory pathway is provided by endoglycosidase H (endo H) sensitivity. The oligosaccharides of secretory proteins are sensitive to endo H before they reach the *medial* Golgi cisternae where the oligosaccharides are modified by the *N*-acetylglucosamine-transferases and mannosidases I and II. After reaching the *medial* Golgi cisternae the oligosaccharides are endo H-resistant.

Isolation of organelles and membranes in eukaryotic cells
Determining the localization of a protein to the correct subcellular membrane or organelle by biochemical methods requires cell disruption and a tedious membrane separation step since a eukaryotic cell contains many different membranes. These include the plasma membrane, endoplasmic reticulum, nuclear membrane, the membranes of the Golgi apparatus, peroxisomes, lysosomes, mitochondria, chloroplasts and the chloroplast thylakoids in plant cells. Providing evidence of membrane targeting is the co-localization of the protein to its target membrane.

To isolate organelles, the cells are first disrupted using the most gentle method possible. Typically, a Potter–Elvehjem glass–Teflon tissue homogenizer is used. This method is gentle enough to maintain the integrity of the organelles. Density gradient centrifugation is used to separate the various membranes and organelles. For example, chloroplasts and mitochondria can be purified from a crude lysate by density gradient centrifugation using either Percoll gradients for chloroplasts, or a sucrose gradient for mitochondria (Meisinger et al., 2000). ER membranes – also called microsomes – are usually isolated from dog pancreas where the pancreas is homogenized by a Potter–Elvehjem homogenizer. The microsomes are then sedimented at low speed and washed by running over a gel filtration column. Marker enzymes of the organelles and other membranes are used to determine the purity.

Summary: In vivo *studies*

- The use of drugs and Sec mutants can reveal whether membrane translocation requires the proton motive force and the function of the Sec proteins.
- Protease-accessibility and signal peptide processing studies can measure protein translocation across the bacterial and ER membrane.
- Modification of Asn-linked oligosaccharides can be used to trace the location of glycoproteins within the secretory pathway.
- Targeting to an organelle, intracellular membrane or aqueous compartment can be monitored by determining whether the exported protein fractionates with the appropriate isolated compartment.

GENETICS

Bacteria

The study of the genetics of protein translocation began in the late 1970s with the Beckwith group. These studies played a pivotal role in identifying the protein components that comprise the machinery and proving that the information for export is located within signal sequences.

Sec mutants

Beckwith and colleagues exploited fusions of the preprotein maltose binding protein (MBP) with the cytosolic reporter β-galactosidase encoded by the *lacZ* gene which, when active, allows the cell to use lactose as a carbon source. PreMBP encoded by the *malE* gene is a periplasmic protein and is needed for the cell to use maltose as a carbon source. The MalE-LacZ fusion protein is toxic when expressed by the addition of maltose in *E. coli*; therefore, the cells are maltose sensitive (Figure 2.2A). The toxicity is most likely due to jamming of the protein secretion machinery caused by the overexpressed LacZ fusion protein, which cannot be translocated and is stuck in the Sec machinery. Indeed, the fused β-galactosidase moiety of the fusion proteins cannot be fully exported across the membrane. The MalE-LacZ protein remains partly inserted into the membrane, thus interfering with the assembly of β-galacto-sidase into an active tetramer. Cells containing these fusion proteins are *lac⁻*, unable to grow on lactose. Therefore, the *lac⁺* revertants, which can grow on lactose, indicate that the hybrid protein relocalizes to the cytoplasm.

Maltose-resistant cells, selected from cells expressing a MalE-LacZ pro-tein, were then obtained and named *sec* mutants. These revertants were *lac⁺* and thus were able to grow on lactose. They had mutations in the *secA* (Oliver and Beckwith, 1981) or *secB* (Kumamoto and Beckwith, 1983) genes. Some of the *secA* mutants had a conditional lethal phenotype. At the non-permissive 42°C, the *secA* mutants accumulated the preprotein form of exported proteins (Oliver and Beckwith, 1981). The *secB* mutant did not show a temperature-sensitive phenotype and, in fact, a null *secB E. coli* mutant was viable in minimal media where protein synthesis is slower and therefore the need for SecB is reduced. Similar approaches using LacZ fusions of alkaline phosphatase were used to isolate cold-sensitive mutants in *secD* (Gardel et al., 1987). Finally, a different LacZ approach involving a SecA-LacZ fusion was used to isolate conditional-lethal mutations in SecE that increase SecA transcription by causing a global defect in protein export (Riggs et al., 1988).

Prl mutants

Silhavy, Bassford, Emr and colleagues used a different approach to identify components of the secretion pathway. Their approach was to isolate *protein*

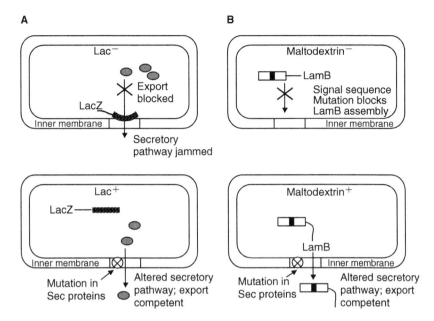

Figure 2.2 Genetic strategy for isolating mutants in the export pathway in bacteria. **A**, Gene fusion method. Maltose induces expression of the hybrid protein and causes jamming of the secretory pathway; the cell dies. Mutations inactivate either the signal sequence or a Sec component required for protein secretion such that jamming no longer occurs. Selected cells grow on lactose as a sugar source. **B**, The suppressor method takes advantage of a defective signal sequence attached to LamB, which is required for transport of maltodextrin across the outer membrane. Since there is an export defect with LamB, the cell cannot grow on maltodextrin as a carbon source. Mutations are selected which restore the signal sequence or occur within a Sec component that allows the mutant LamB to reach the outer membrane. These cells can grow on maltodextrin as a carbon source.

*l*ocalization (*prl*) mutants that acted as suppressors of signal sequence mutations. Silhavy and Emr studied LamB, an outer membrane protein, transports maltodextrins across the outer membrane. LamB containing the signal sequence mutation is not exported and remains in the cytoplasm. These cells are defective for maltodextrin uptake. Revertants that could grow on maltodextrins were then selected. Bassford's approach introduced the signal sequence mutations identified by the MalE-LacZ fusion method into the wild-type *malE* gene and then isolated revertants that could grow on maltose. Five *prl* mutants were isolated: *prlA*, *prlB*, *prlC*, *prlD* and *prlG* (Emr et al., 1981; Fikes and Bassford, 1989; Stader et al., 1989). *PrlA*, *prlD* and *prlG* are the most interesting. Mutations in *prlA*, *prlD* and *prlG* were identified with the identical genes of *secY*, *secA* and *secE*, respectively. Additional methods led to further mutants in the *sec* genes (see Table 2.1 for

Table 2.1 The *sec* and signal peptidase genes

Gene	Map position (min)	Protein	Mutant phenotype
secA (*prlD*)	2.5	ATPase, molecular motor; 102 kDa peripheral membrane protein	t.s. or c.s. lethal export defect; suppress signal sequence mutation
secB	81	Export chaperone 16.6 kDa	Non-essential on minimal media; export defect for certain proteins
secD	9.5	67 kDa inner membrane protein	c.s. lethal export defect
secE (*prlG*)	90	13.6 kDa inner membrane protein; component of the membrane channel	c.s. lethal export defect
secF	9.5	35 kDa inner membrane protein	c.s. lethal export defect
secG	69	11.4 kDa inner membrane protein	Null not lethal
secY (*prlA*)	72	49 kDa inner membrane protein; component of the membrane channel	t.s. or c.s. lethal export defect; suppress signal sequence mutations
lepB	55.5	Signal peptidase I; 36 kDa inner membrane protein	Lethal, signal peptide processing defect
lspA	0.5	Signal peptidase II; 18 kDa inner membrane protein	Lethal, signal peptide processing defect

a list of the *sec* as well as signal peptidase genes). Localized mutagenesis of the *E. coli* genome by Ito et al. (1983) resulted in a temperature-sensitive mutant in *secY* that accumulated precursor proteins at the non-permissive temperature (Ito et al., 1983).

Suppressors of suppressors
A third approach for isolating new components of the secretion pathway is to identify suppressors of suppressors within other genes. In this approach, suppressors are isolated in other mutant proteins that interact with a Sec protein. For example, using a temperature-sensitive *secA* mutant strain, suppressors in *secY* were isolated that allowed cell viability at the non-permissive temperature (Brickman et al., 1984). The identified suppressors of *secA* were found in *secY*, which was originally discovered as a suppressor

of signal peptide mutations (see the *prl* section above). That is why they are called suppressors of suppressors. However, other suppressor mutations gave false leads as many suppressor mutations were localized in ribosomal genes and were not involved in protein translocation (Ferro-Novick et al., 1984). These and other results (Lee et al., 1989; Oliver, 1985) suggested that simply slowing down protein synthesis could suppress the lethal phenotype of a temperature-sensitive *sec* mutant at the non-permissive temperature.

Selection of maltose-resistant or *lac*$^+$ cells, containing MalE-LacZ protein, also gave rise to mutants that had alterations in the hydrophobic core of the signal sequence within the MalE-LacZ protein (Bedouelle et al., 1980). Similar signal peptide mutations were found in pre-LamB using the LamB-LacZ fusion (Emr et al., 1980). Although the signal sequence mutations were isolated in the fusion proteins as 'gain of function' mutants, the mutations impaired export of MBP or LamB when combined within the intact gene. Conversely, starting with an export defective MalE mutant, researchers obtained revertants that could export MBP to the periplasm and therefore allow the cell to use maltodextrin as a carbon source (Bankaitis et al., 1984). Many of these mutants had mutations that increased the hydrophobicity of the signal sequence. The genetic studies were very important because they demonstrated that the information for protein export is mainly localized in the N-terminal region of the protein (for review see Michaelis and Beckwith (1982)).

Eukaryotes

Sec mutants

Temperature-sensitive mutants in the secretion pathway were isolated in yeast by Schekman and colleagues using a Ludox density enrichment strategy (Novick et al., 1980). Such mutants were identified by looking for cells with increased density due to the internalization of the normally secreted proteins. The *sec* mutants that accumulated large amounts of invertase at the non-permissive temperature, 37°C, were characterized. A number of mutants isolated by this technique have been characterized and are blocked at different steps in the secretion pathway (Figure 2.3).

In addition to this genetic strategy to isolate *sec* mutants, Deshaies and Schekman (1987) used a gene fusion approach (Deshaies and Schekman, 1987) similar to the one developed by Beckwith and coworkers for studying the bacterial system. In this approach, the cells are grown in the absence of histidine. The cytoplasmic enzyme histidinol dehydrogenase (HD) is required for the cell to grow in the absence of histidine. A fusion was made between the secreted invertase and HD and this was exported into the lumen of the ER in the process of secretion, rendering the cell unable to grow without histidine. Revertants are then selected that grow in minimal medium lacking histidine. These revertants retain some HD in the cytosol where it can function. This selection method yielded a temperature-sensitive mutation in *sec61* (Deshaies and Schekman, 1987), *sec62* and *sec63* (Rothblatt et al., 1989). The identities of the *sec* genes were determined by subcloning the cell with a yeast library, and isolating the transformed cell capable of growing at higher temperatures.

Vps and Vpl mutants
Mutants defective in the late stages of the secretion pathway that affect vacuolar sorting have been isolated using fusions between a vacuolar preprotein and the secreted protein invertase. Invertase is normally secreted from yeast and allows the cell to grow on sucrose as a carbon source. When the vacuolar invertase fusion protein is routed to the vacuole it renders the cell unable to grow in media containing sucrose. Exploiting this system, the Emr and Stevens groups isolated many mutants defective in sorting proteins to the vacuole (Bankaitis et al., 1986; Rothman and Stevens, 1986). Emr's laboratory called them *vacuolar protein sorting* (*vps*) mutants. The laboratory isolated mutants that secreted the invertase fusion protein into the extracellular medium allowing the cell to grow on sucrose media. A similar strategy using carboxypeptidase Y was used by Stevens' laboratory to isolate *vacuolar protein localization* (*vpl*) mutants. Some of the key *vps* mutants are indicated in Figure 2.3.

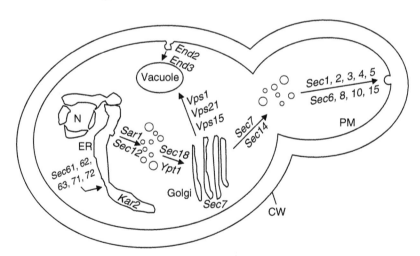

Figure 2.3   Genes involved in protein traffic in the yeast *Saccharomyces cerevisiae*. Only a subset of mutants are shown in the different pathways. The *sec* genes encode proteins involved in the secretory pathway, whereas *vps* encode vacuolar protein sorting proteins. The *end* mutants (not described in the text) are involved in the endocytosis pathway. ER, endoplasmic reticulum; N, nucleus; PM, plasma membrane; CW, cell wall. Adapted from Pugsley (1989) Protein Targeting, Academic Press.

Summary: *Genetics*

- Genetic techniques are very powerful for identifying components involved in protein export in both bacteria and eukaryotes.
- These techniques also define the region of the exported protein that is critical for export.

IN VITRO TRANSLOCATION AND TRANSPORT STUDIES

Translocation into membrane vesicles

Cell-free systems are quite useful in defining the function of the components of the protein translocation machinery. These systems can determine whether energy and chaperones are needed for translocation, and whether translocation can occur post-translationally. *In vitro* systems led the way to defining the export pathway in ER translocation, many years before extensive *in vivo* studies were conducted.

Endoplasmic reticulum
In the classic studies, Blobel and Dobberstein succeeded in co-translational translocation using pancreas microsomes (Blobel and Dobberstein, 1975a, 1975b). The exported protein is typically radiolabeled using a cell-free transcription/translation system (Figure 2.4). Radioactive amino acids such as [^{35}S]-methionine are often used currently to label the newly synthesized protein. Translocation into the microsomal vesicles are monitored by the removal of the signal sequence of the preprotein by signal peptidase that has its active side on the inside of the microsomes. Moreover, the translocated protein is resistant to digestion when proteases are added to the reaction mixture but is sensitive to digestion if a detergent is added to disrupt the membrane vesicle.

This ER microsomal system can be dissected by biochemical methods. Warren and Dobberstein (1978) found that microsomes washed with high salt buffer were incompetent for translocation; the addition of the salt wash fraction restored protein translocation. This activity was then purified by Walter and Blobel (1980) and the identified component, named signal recognition particle (SRP) (Walter and Blobel, 1981a, 1981b; Walter et al., 1981), is essential for protein translocation. Eukaryotic SRP consists of six polypeptides and a 7S RNA and was found to arrest protein synthesis. Using the *in vitro* system, the Blobel and Dobberstein groups also isolated the SRP receptor (docking protein). This was done as follows: Elastase treatment of the ER produced a proteolytic fragment of the SRP receptor

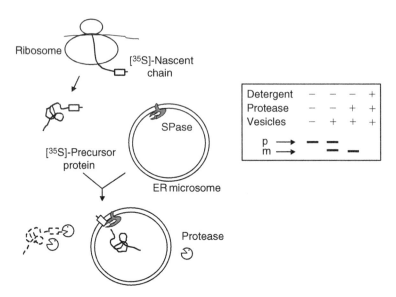

Figure 2.4   *In vitro* assay of the translocation of a preprotein. [^{35}S]-labeled prepro-tein can be synthesized using a cell-free system with membrane added either during or after synthesis. Translocation is monitored by the conversion of the preprotein to the mature protein by signal peptide cleavage by signal peptidase (SPase). Protease is added to determine whether the exported protein is translocated into the lumen of the vesicle. The [^{35}S]-labeled protein is separated by SDS–PAGE and visualized by autoradiography. As a control, protease is added to the membrane which is lysed by detergent to confirm that the labeled exported protein was protected from protease digestion by the membrane.

or docking protein. When this fragment was removed from the membranes, the microsomes were translocation incompetent (Meyer et al., 1982). The re-addition of the fragment led to restoration of the translocation activity. The fragment was then found to be part of the 72 kDa membrane protein called the α-subunit of the SRP receptor (Gilmore et al., 1982).

Information about membrane targeting is obtained by performing a flotation assay (Figure 2.5). In this approach, the nascent chain is radio-labeled using a transcription–translation system and incubated with mem-branes. The incubation mixture is then mixed with high concentrations of sucrose and placed at the bottom of a sucrose gradient. Centrifugation is performed using an ultracentrifuge and membranes will float to the lower densities of the sucrose gradient. The gradient fractions after centrifugation are collected and in each fraction the protein is trichloroacetic acid pre-cipitated and analyzed by SDS–PAGE to determine whether the ribosome nascent chain floats with the membranes. Targeting to the membrane occurs if the protein floats with the membranes. Alkali treatment of the membrane

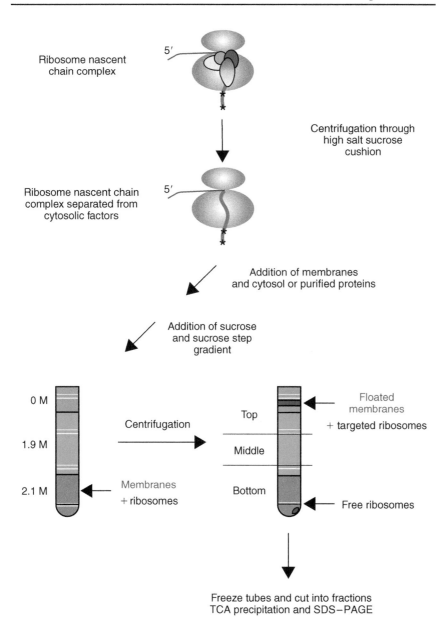

Figure 2.5 Flotation assay. Ribosome nascent chain complexes produced by translating truncated mRNA that lack a stop codon. After salt stripping, followed by centrifugation through a high salt sucrose cushion to remove the cytosol, membranes can be added as well as cytosol or purified proteins. The targeting of the ribosome nascent chain complexes to the membranes can be determined by sucrose gradient centrifugation and the fractions are analyzed by SDS–PAGE and autoradiography.

associated fractions is then used to determine whether the membrane protein integrated into the lipid bilayer, or is peripherally associated (Fujiki et al., 1982). Using this approach, Nicchitta et al. (1995) examined the targeting of the nascent chains of preprolactin to the ER membrane.

Bacteria

In vitro studies in bacteria began much later than the eukaryotic *in vitro* studies. Not until the 1980s were the first *in vitro* studies initiated (Chen and Tai, 1985; Geller et al., 1986; Muller and Blobel, 1984a, 1984b). The protein translocation systems require inverted membrane vesicles (INV or IMV) and preproteins as a substrate. As for the *in vitro* study with ER microsomes (Figure 2.4), translocation is monitored by signal peptide processing and protease protection. Using this *in vitro* system, one can determine whether cytosolic proteins, ATP, or the proton motive force (PMF) is required for the translocation process. Translocation of proteins into bacterial inverted membrane vesicles occurs after synthesis (Chen et al., 1985), requiring ATP and the PMF (Chen and Tai, 1985; Geller et al., 1986). In the cell-free system, purified SecB promotes translocation of the MBP (Weiss et al., 1988) and SecA is essential for *in vitro* protein translocation (Cabelli et al., 1988). Membranes depleted of SecA using a SecA amber mutant and the use of a SecA-depleted S100 *E. coli* extract lead to a severe block of protein translocation. Translocation can be restored by the re-addition of purified SecA.

A biochemical analysis of the bacterial cell-free system provided insights into the SecA and SecB functions (Hartl et al., 1990). [125I]-SecB binds with high affinity to inner membrane vesicles containing SecA; additionally, [125I]-proOmpA binds to vesicles with high affinity in a SecB-dependent manner. This shows that SecB plays a targeting role in protein export. Furthermore, preproteins are translocated across the membrane post-translationally in a series of growing loops with the N- and C-termini in the cytoplasm (Schiebel et al., 1991). One can observe these protease-resistant translocation intermediates when a lowered ATP concentration slows down translocation. In the translocation process, [125I]-SecA undergoes a conformational change upon binding ATP when inner membrane vesicles are present. This results in a 30 kDa protease-resistant band that likely occurs when a 30 kDa SecA domain inserts into the membrane and is protected from the protease by the membrane (Economou and Wickner, 1994). In addition to this study, the use of membrane impermeant reagents showed that the SecA protein is partly exposed to the periplasmic surface during translocation (Kim et al., 1994).

Although this section focused on the *in vitro* protein translocation in ER and bacteria, similar signal peptide processing, protease protection and cofractionation studies are used to study both protein targeting and translocation into the mitochondria and chloroplasts.

Photocrosslinking and fluorescence techniques to examine the translocation process

The protein components that interact with a nascent exported protein in the cytoplasm have been defined using crosslinking techniques. The exported protein in these studies is first radiolabeled and then trapped in the translocation process using the truncated mRNA technology (Gilmore et al., 1991) (Figure 2.6). Due to the absence of a stop codon in the truncated

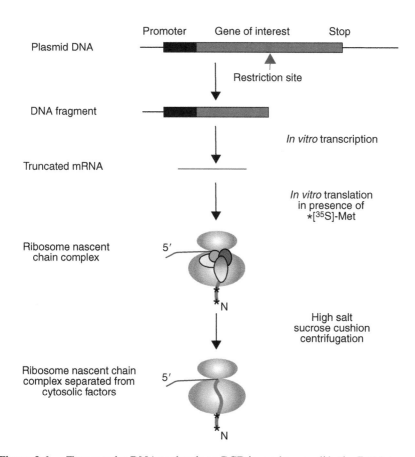

Figure 2.6 Truncated mRNA technology. PCR is used to amplify the DNA to produce PCR fragments containing the promoter region and the 5' end of the protein coding region of the gene of interest lacking a stop codon. This DNA fragment acts as a template for the synthesis of a truncated mRNA. The truncated mRNA is then added to a cell-free translation system from wheat germ, reticulocyte lysate, or *E. coli* lysate and the protein is labeled with [^{35}S]-methionine(*). Translation of the truncated mRNA produces ribosome-bound nascent chains. Finally, the ribosome nascent chain complexes are separated from cytosolic components using centrifugation through a high salt sucrose cushion.

mRNA the nascent protein remains bound to the ribosome. This results in nascent chains of equal length bound to the ribosome.

Photocrosslinking using translocation intermediates
To identify the proteins that interact with the nascent exported chain in the cytosol, a photoreactive probe is incorporated into the nascent chain. Shining UV light on the photoreactive probe containing sample generates a reactive species that can react with interacting partner proteins. SDS–PAGE and immunoprecipitations of the crosslinked product can identify the inter-acting proteins. If the signal sequence is to be crosslinked, the nascent chain has to be approximately 70 residues long since 35 to 40 residues are within the polypeptide channel of the ribosome. Such crosslinking experiments showed that the signal sequence interacts with the SRP 54 kDa subunit of SRP in the eukaryotic system (Krieg et al., 1986; Kurzchalia et al., 1986). The nascent chains can also be crosslinked to the nascent-polypeptide-associated complex (Nac) in the cytosol of eukaryotes (Wiedmann et al., 1994). The bacterial signal sequence interacts with trigger factor and in some cases, with SRP Ffh in bacteria (Beck et al., 2000; Valent et al., 1998).

Photocrosslinking is also used to identify factors involved in translocation. Translocation intermediates can be generated in the presence of membranes by using truncated nascent chains, as described above (Gilmore et al., 1991). The truncated nascent protein cannot be released from ribosomes, therefore the translocation process cannot finish. A photocrosslinking probe can be incorporated into the newly synthesized protein by means of a modified aminoacyl-tRNA (Brunner, 1996). Shown in Figure 2.7 is the Lys-tRNA charged with crosslinker, 4-(3-trifluoro-methyldiazirino) benzoyl-*N*-hydroxy succinimido ester. This photoreactive lysine analog is incorporated into the nascent polypeptide where lysines are normally located. After translocation intermediates are generated, irradiation results in the generation of a highly reactive carbene that reacts with neighboring proteins (Brunner, 1996).

Another application of the mRNA technology, in which the site-specific photocrosslinking approach is used, is to introduce an amber codon within the nascent polypeptide and then use an amber suppressor tRNA that is charged with a photoactivatable probe. The merits of this approach is that the probe can be placed at a specific site within the integrating membrane protein. The application of this approach demonstrated that Sec61α (Do et al., 1996; Mothes et al., 1997) and TRAM (Do et al., 1996) interact with polypeptides inserting into the ER membrane. Interaction of SecY and SecA (Houben et al., 2000) with exported proteins was also shown using this approach in the bacterial system.

Introduction of a fluorescence probe into exported proteins
Fluorescent reporter molecules may be introduced into nascent exported proteins using the mRNA technology and Lys-tRNA analogs to which

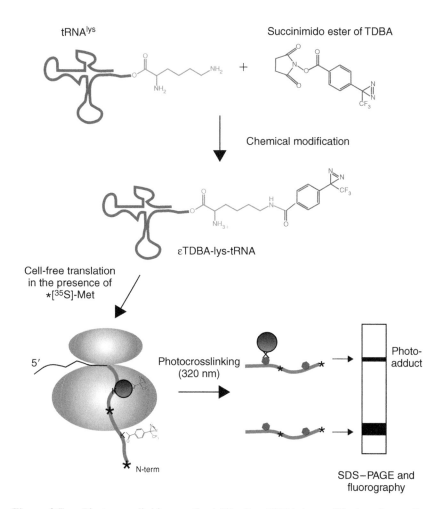

Figure 2.7 Photocrosslinking method. The Lys-tRNA is modified at the epsilon-amino group with 4-(3-trifluoro-methyldiazirino) benzoyl-*N*-hydroxy-succinimido ester. After the reaction with the succinimido ester of TDBA (4-(3-trifluorome-thyldiazirino) benzoic acid), one must purify the modified tRNA by chromatography, since the reaction is never complete. The truncated mRNA is translated in the cell-free system in the presence of [^{35}S]-methionine(*) to produce nascent chains attached to the ribosome. In the nascent chains, the photoactivatable lysine derivative is incorporated at positions where lysines normally occur. Irradiation at 320 nm produces a carbene that reacts with neighboring proteins. Thus, crosslinks are formed between the nascent polypeptide and neighboring proteins.

fluorescent probes are attached. Johnson and coworkers attached NBD (7-nitrobenz-2-oxa-1,3-diazole) to lysine and used the fluorescence properties of NBD to determine whether it is in a polar or apolar environment. The fluorescence lifetime of an NBD probe is much greater in a nonpolar environment than in an aqueous environment. Using this fluorescence approach, they were able to determine that the signal sequence moves through an aqueous ribosome tunnel (Crowley et al., 1993) as well as an aqueous environment during membrane translocation (Crowley et al., 1994). There appears to be a tight ribosome–membrane junction during translocation because the signal sequence containing the fluorescent probe is not accessible to the aqueous cytoplasm; the addition of the hydrophilic quencher iodide ions showed very little quenching of the NBD fluorescence.

Reconstitution of protein translocation

Bill Wickner's group was first to reconstitute protein translocation with proteoliposomes using purified components. First, bacterial inner membrane vesicles were solubilized in the detergent n-octyl-β-D-glucopyranoside and glycerol (Driessen and Wickner, 1990) and then the SecYE complex was purified (Brundage et al., 1990). The purified SecYE protein was reconstituted into proteoliposomes by the rapid detergent-dilution procedure (Racker, 1979). Briefly, purified SecYE in octyl-β-D-glucopyranoside is mixed with an *E. coli* phospholipid suspension. Samples were then diluted to lower the detergent below the critical micelle concentration and then the proteoliposomes were fused by adding $CaCl_2$. After isolating the large unilamelar proteoliposomes following EGTA (ethylenediamine tetraacetic acid) treatment to chelate the calcium, the proteoliposomes were sonicated to disrupt the aggregates within the sample. Using these proteoliposomes, proOmpA was translocated in a SecA and ATP-dependent manner. Wickner and colleagues found approximately twice as much SecYE was required for translocation in proteoliposomes as for inner membrane vesicles (Brundage et al., 1990). Further characterization of the purified SecYE revealed another polypeptide, Band I (SecG). Translocation of the preprotein proOmpA into proteoliposomes was assayed using accessibility to added protease. Soon after publication of this work, Akimara et al. (1991) published the successful reconstitution of protein translocation using only the integral membrane components SecY and E and the peripheral membrane protein SecA.

In conjunction with the bacterial reconstitution studies, scientists in the ER field were attempting to reconstitute microsomal protein translocation. In seminal studies, Nicchitta and Blobel solubilized ER membranes and made proteoliposomes that were active in protein translocation (Nicchitta and Blobel, 1990; Nicchitta et al., 1991). These studies were significant because they showed that protein translocation could be reconstituted from ER integral membrane protein components. Then, in a landmark study, Görlich and Rapoport (1993), using the purified Sec61p complex comprised of the Sec61 α, β and γ subunits, showed that these proteins, along with the SRP receptor, were sufficient for translocation. The Rapoport laboratory also went on to show that for a subset of proteins, an additional translocating chain associating membrane protein (TRAM) was needed for translocation.

The power of reconstitution studies is that they show which minimum components are necessary for protein translocation. In addition, components can be identified that stimulate protein translocation and these can be studied more easily because of the fewer protein components that may interfere and complicate the analysis.

Protein translocation into lipid vesicles

The reconstitution studies described above demonstrated that membrane proteinaceous components facilitate the translocation event. However, in certain cases, proteins were shown to insert directly into protein-free lipid vesicles. Geller and Wickner (1985) reported that the M13 procoat protein was capable of inserting *in vitro* into liposomes as procoat was degraded when chymotrypsin-encapsulated liposomes were added. This spontaneous insertion mechanism was verified by Kuhn and Vogel (Soekarjo et al., 1996). However, in both studies, the amount of the M13 procoat protein that inserted into the membrane was low. Recently, de Kruijff's laboratory used the Pf3 coat protein, a 44-residue single-transmembrane protein, to show that insertion into large unilamellar vesicles could occur spontaneously (Ridder et al., 2000). The generation of a proteinase resistant fragment provided evidence that a mutant Pf3 coat with a lengthened transmembrane segment with three leucine residues introduced, inserted directly into liposomes. For example, when the N-terminus of Pf3 coat inserts across the membrane to the inside of the lipid vesicle, proteinase K added to the outside of the vesicle causes a slight shift of the molecular weight by cleaving within the cytoplasmic tail. Therefore, at least *in vitro*, for some proteins with short translocated regions, membrane channels are apparently not absolutely required for translocation.

Nuclear transport assay using permeabilized cells

Permeabilized cells are very useful for studying the cytosolic factors and the nucleotide triphosphate requirements in nuclear transport. Cells are permeabilized with low concentrations of digitonin which perforates the plasma membrane causing the release of cytosolic components of the cell while simultaneously leaving the nuclear envelope and other organelles intact. A fluorescently labeled protein, which can enter the permeabilized cell, can then be examined for import into the nucleus using fluorescence microscopy. Using permeabilized Hela cells, Görlich and coworkers identified importin-α and importin-β, which promote import into the nucleus of the fluorescein-labeled nucleoplasmin substrate in a Ran-GTPase and energy dependent manner (Görlich et al., 1994, 1995).

Reconstitution of Golgi transport

Transport through the Golgi complex was reconstituted by studying the transport of the viral G-protein in Chinese hamster ovary (CHO) cells (Balch et al., 1984). The donor Golgi fraction came from a VSV-infected 15B mutant which lacked the enzyme UDP-GlcNAc glycosyltransferase I leaving the oligosaccharide chain incompletely processed (Figure 2.8). The acceptor Golgi fraction came from an uninfected mutant that contained UDP-GlcNAc glycosyltransferase I. In this *in vitro* system, attachment of [³H]-GlcNAc to the G-protein was observed and, therefore, must have resulted from the transport of components from the donor and acceptor Golgi cisternae. Rothman and colleagues interpreted these results as an indication of the transport of the incompletely glycosylated G-protein from the donor to the acceptor. Another possibility, however, is that there was transport of UDP-glycosyltransferase I from the *medial* cisternae of the

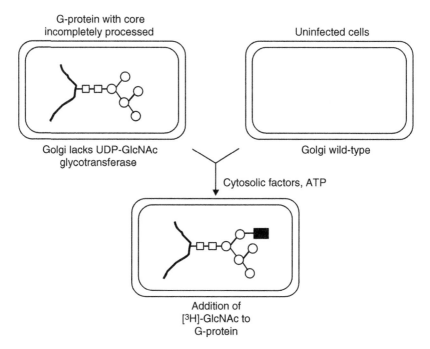

Figure 2.8   Cell-free system to monitor transport from *cis* to *medial* Golgi cisternae. The mutant CHO cells are infected with the vesicular stomatitis virus. The VSV-G proteins have an incompletely modified core as the CHO cells lack *N*-acetylglucosaminetransferase I and cannot attach [³H]-GlcNAc to the core oligosaccharide. The acceptor cisternal compartment is from wild-type CHO cells which are uninfected. Therefore, the attachment of [³H]-GlcNAc to the G-protein must result from the transport of components from the donor and acceptor cisternae. Adapted from Balch et al. (1984) *Cell* **39**: 405.

acceptor compartment to the donor compartment containing the G-protein with the incompletely modified oligosaccharide core. This work is significant because it showed it is possible to reconstitute biochemical vesicular traffic and enabled the purification of a number of proteins involved in intra Golgi transport.

Summary: In vitro *translocation and transport studies*

- The function of a Sec protein can be determined using *in vitro* systems.
- Photocrosslinking studies identify proteins that interact with the exported protein during targeting and membrane translocation.
- The minimum components necessary for membrane translocation can be determined by reconstituting protein translocation from purified components.
- Import of fluorescently labeled proteins into semi-permeabilized mammalian cells can be used to monitor nuclear import.
- The purification of proteins required for transport across the Golgi apparatus was possible by dissecting the reconstituted Golgi transport system.

CELL BIOLOGY TECHNIQUES

The study of protein export depends upon methods which examine the location of proteins within the cell. One very useful technique is to employ electron microscopy in which organelles and vesicles of eukaryotic cells are very clearly seen. In this approach, the specimen is sliced into thin sections, embedded in plastic at low temperatures to minimize structural perturbations, and stained with heavy metal salts. In addition, immunogold electron microscopy is used to determine the location of proteins within the cell (Clark, 1991). Here, electron micrograph samples are treated with antibodies to the desired proteins, which have small gold particles (typically 50 or 100 Å in diameter) attached to the antibody. Control antibodies that react against the resident organellar or vesicle protein in question can be added linked with different sized gold particles. The small and large black dots, which correspond to small and large gold particles on the electron micrograph, are compared to determine the cellular localization of the protein.

Examining the location of exported proteins in a living cell is also beneficial. For this purpose, fluorescence microscopy is used (Spector, 1998). The protein is visualized by adding and internalizing a fluorescently labeled antibody and, after several washing steps, the location is determined through fluorescence microscopy. Alternatively, an unlabeled primary antibody is first added to the protein under investigation and visualized with a secondary antibody that is labeled with fluorescein or rhodamine. The fluorescence image is then compared with a control antibody which localizes, for example, to the same compartment. Other applications exploit antibodies that recognize epitope tags (Harlow, 1999). Often a short epitope tag (comprising a short amino acid sequence) is fused to the protein of interest. Typically, the addition of the tag does not affect the structure/function of the protein. Immunofluorescence is then used to determine the protein's location by adding antibodies with a fluorescent label attached to either the primary or secondary antibody prepared against the tag.

In the last five years, the cell biology area exploded with the use of green fluorescence protein (GFP) as a tool to study protein trafficking in living cells (Lippincott-Schwartz et al., 2000). GFP is a protein that emits visible green light when excited with UV light. If GFP is fused to the exported protein, the chimeric protein is expressed in the cell and its localization can be monitored using fluorescence microscopy. Rarely, the attachment of GFP affects the intracellular function of the exported protein. Recently, GFP chimeras were used to unravel membrane traffic in living cells.

Summary: *Cell biology techniques*

- Gold-labeled antibodies bind to the exported proteins and their subcellular location is detected using electron microscopy.
- Fluorescently labeled antibodies that bind to the exported protein can be detected by light microscopy.
- Fusions of the GFP with the exported protein can be used to monitor the location of exported proteins in living cells.

ACKNOWLEDGMENT

This work was supported by grant MCB-9808843 from NSF to R.E.D.

REFERENCES

Akimaru, J., Matsuyama, S., Tokuda, H. and Mizushima, S. (1991) Reconstitution of a protein translocation system containing purified SecY, SecE and SecA from *Escherichia coli. Proc Natl Acad Sci USA* **88**: 6545–6549.

Balch, W.E., Dunphy, W.G., Braell, W.A. and Rothman, J.E. (1984) Reconstitution of the transport of protein between successive compartments of the Golgi measured by the coupled incorporation of N-acetylglucosamine. *Cell* **39**: 405–416.

Bankaitis, V.A., Rasmussen, B.A. and Bassford, P.J., Jr. (1984) Intragenic suppressor mutations that restore export of maltose binding protein with a truncated signal peptide. *Cell* **37**: 243–252.

Bankaitis, V.A., Johnson, L.M. and Emr, S.D. (1986) Isolation of yeast mutants defective in protein targeting to the vacuole. *Proc Natl Acad Sci USA* **83**: 9075–9079.

Beck, K., Wu, L.F., Brunner, J. and Muller, M. (2000) Discrimination between SRP- and SecA/SecB-dependent substrates involves selective recognition of nascent chains by SRP and trigger factor. *EMBO J* **19**: 134–143.

Bedouelle, H., Bassford, P.J., Jr., Fowler, A.V. et al. (1980) Mutations which alter the function of the signal sequence of the maltose binding protein of *Escherichia coli. Nature* **285**: 78–81.

Blobel, G. and Dobberstein, B. (1975a) Transfer of proteins across membranes. I. Presence of proteolytically processed and unprocessed nascent immunoglobulin light chains on membrane-bound ribosomes of murine myeloma. *J Cell Biol* **67**: 835–851.

Blobel, G. and Dobberstein, B. (1975b) Transfer to proteins across membranes. II. Reconstitution of functional rough microsomes from heterologous components. *J Cell Biol* **67**: 852–862.

Bohni, P.C., Deshaies, R.J. and Schekman, R.W. (1988) SEC11 is required for signal peptide processing and yeast cell growth. *J Cell Biol* **106**: 1035–1042.

Brickman, E.R., Oliver, D.B., Garwin, J.L., Kumamoto, C. and Beckwith, J. (1984) The use of extragenic suppressors to define genes involved in protein export in *Escherichia coli. Mol Gen Genet* **196**: 24–27.

Brundage, L., Hendrick, J.P., Schiebel, E., Driessen, A.J. and Wickner, W. (1990) The purified *E. coli* integral membrane protein SecY/E is sufficient for reconstitution of SecA-dependent precursor protein translocation. *Cell* **62**: 649–657.

Brunner, J. (1996) Use of photocrosslinkers in cell biology. *Trends Cell Biol* **6**: 154–157.

Cabelli, R.J., Chen, L., Tai, P.C. and Oliver, D.B. (1988) SecA protein is required for secretory protein translocation into *E. coli* membrane vesicles. *Cell* **55**: 683–692.

Chen, L., Rhoads, D. and Tai, P.C. (1985) Alkaline phosphatase and OmpA protein can be translocated post-translationally into membrane vesicles of *Escherichia coli. J Bacteriol* **161**: 973–980.

Chen, L. and Tai, P.C. (1985) ATP is essential for protein translocation into *Escherichia coli* membrane vesicles. *Proc Natl Acad Sci USA* **82**: 4384–4388.

Clark, M.W. (1991) Immunogold labeling of yeast ultrathin sections. *Methods Enzymol* **194**: 608–626.

Corsi, A.K. and Schekman, R. (1996) Mechanism of polypeptide translocation into the endoplasmic reticulum. *J Biol Chem* **271**: 30299–30302.

Crowley, K.S., Reinhart, G.D. and Johnson, A.E. (1993) The signal sequence moves through a ribosomal tunnel into a noncytoplasmic aqueous environment at the ER membrane early in translocation. *Cell* **73**: 1101–1115.

Crowley, K.S., Liao, S., Worrell, V.E., Reinhart, G.D. and Johnson, A.E. (1994) Secretory proteins move through the endoplasmic reticulum membrane via an aqueous, gated pore. *Cell* **78**: 461–471.

Dalbey, R.E. and Wickner, W. (1986) The role of the polar, carboxyl-terminal domain of *Escherichia coli* leader peptidase in its translocation across the plasma membrane. *J Biol Chem* **261**: 13844–13849.

Dalbey, R.E., Lively, M.O., Bron, S. and van Dijl, J.M. (1997) The chemistry and enzymology of the type I signal peptidases. *Protein Sci* **6**: 1129–1138.

Daniels, C.J., Bole, D.G., Quay, S.C. and Oxender, D.L. (1981) Role for membrane potential in the secretion of protein into the periplasm of *Escherichia coli. Proc Natl Acad Sci USA* **78**: 5396–5400.

Date, T., Goodman, J.M. and Wickner, W.T. (1980) Procoat, the precursor of M13 coat protein, requires an electrochemical potential for membrane insertion. *Proc Natl Acad Sci USA* **77**: 4669–4673.

Deshaies, R.J. and Schekman, R. (1987) A yeast mutant defective at an early stage in import of secretory protein precursors into the endoplasmic reticulum. *J Cell Biol* **105**: 633–645.

Do, H., Falcone, D., Lin, J., Andrews, D.W. and Johnson, A.E. (1996) The co-translational integration of membrane proteins into the phospholipid bilayer is a multistep process. *Cell* **85**: 369–378.

Driessen, A.J. and Wickner, W. (1990) Solubilization and functional reconstitution of the protein-translocation enzymes of *Escherichia coli. Proc Natl Acad Sci USA* **87**: 3107–3111.

Duong, F. and Wickner, W. (1997) The SecDFyajC domain of preprotein translocase controls preprotein movement by regulating SecA membrane cycling. *EMBO J* **16**: 4871–4879.

Economou, A. and Wickner, W. (1994) SecA promotes preprotein translocation by undergoing ATP-driven cycles of membrane insertion and deinsertion. *Cell* **78**: 835–843.

Ehrmann, M., Bolek, P., Mondigler, M., Boyd, D. and Lange, R. (1997) TnTIN and TnTAP: mini-transposons for site-specific proteolysis *in vivo. Proc Natl Acad Sci USA* **94**: 13111–13115.

Emr, S.D., Hedgpeth, J., Clement, J.M., Silhavy, T.J. and Hofnung, M. (1980) Sequence analysis of mutations that prevent export of lambda receptor, an *Escherichia coli* outer membrane protein. *Nature* **285**: 82–85.

Emr, S.D., Hanley-Way, S. and Silhavy, T.J. (1981) Suppressor mutations that restore export of a protein with a defective signal sequence. *Cell* **23**: 79–88.

Ferro-Novick, S., Honma, M. and Beckwith, J. (1984) The product of gene secC is involved in the synthesis of exported proteins in *E. coli. Cell* **38**: 211–217.

Fikes, J.D. and Bassford, P.J., Jr. (1989) Novel secA alleles improve export of maltose-binding protein synthesized with a defective signal peptide. *J Bacteriol* **171**: 402–409.

Fujiki, Y., Hubbard, A.L., Fowler, S. and Lazarow, P.B. (1982) Isolation of intracellular membranes by means of sodium carbonate treatment: application to endoplasmic reticulum. *J Cell Biol* **93**: 97–102.

Gardel, C., Benson, S., Hunt, J., Michaelis, S. and Beckwith, J. (1987) secD, a new gene involved in protein export in *Escherichia coli. J Bacteriol* **169**: 1286–1290.

Geller, B.L. and Wickner, W. (1985) M13 procoat inserts into liposomes in the absence of other membrane proteins. *J Biol Chem* **260**: 13281–13285.

Geller, B.L., Movva, N.R. and Wickner, W. (1986) Both ATP and the electrochemical potential are required for optimal assembly of pro-OmpA into *Escherichia coli* inner membrane vesicles. *Proc Natl Acad Sci USA* **83**: 4219–4222.

Gilmore, R., Blobel, G. and Walter, P. (1982) Protein translocation across the endoplasmic reticulum. I. Detection in the microsomal membrane of a receptor for the signal recognition particle. *J Cell Biol* **95**: 463–469.

Gilmore, R., Collins, P., Johnson, J., Kellaris, K. and Rapiejko, P. (1991) Transcription of full-length and truncated mRNA transcripts to study protein translocation across the endoplasmic reticulum. *Methods Cell Biol* **34**: 223–239.

Görlich, D. and Rapoport, T.A. (1993) Protein translocation into proteoliposomes reconstituted from purified components of the endoplasmic reticulum membrane. *Cell* **75**: 615–630.

Görlich, D., Prehn, S., Laskey, R.A. and Hartmann, E. (1994) Isolation of a protein that is essential for the first step of nuclear protein import. *Cell* **79**: 767–778.

Görlich, D., Vogel, F., Mills, A.D., Hartmann, E. and Laskey, R.A. (1995) Distinct functions for the two importin subunits in nuclear protein import. *Nature* **377**: 246–248.

Hanein, D., Matlack, K.E., Jungnickel, B. et al. (1996) Oligomeric rings of the Sec61p complex induced by ligands required for protein translocation. *Cell* **87**: 721–732.

Harlow, E. and Lane, D. (1999) *Using Antibodies: A Laboratory Manual*. Cold Spring Harbor, NY: Cold Spring Harbor Laboratory Press.

Hartl, F.U., Lecker, S., Schiebel, E., Hendrick, J.P. and Wickner, W. (1990) The binding cascade of SecB to SecA to SecY/E mediates preprotein targeting to the *E. coli* plasma membrane. *Cell* **63**: 269–279.

Herman, P.K., Stack, J.H., DeModena, J.A. and Emr, S.D. (1991) A novel protein kinase homolog essential for protein sorting to the yeast lysosome-like vacuole. *Cell* **64**: 425–437.

Houben, E.N., Scotti, P.A., Valent, Q.A. et al. (2000) Nascent Lep inserts into the *Escherichia coli* inner membrane in the vicinity of YidC, SecY and SecA. *FEBS Lett* **476**: 229–233.

Ito, K., Wittekind, M., Nomura, M., Shiba, K., Yura, T., Miura, A. and Nashimoto, H. (1983) A temperature-sensitive mutant of *E. coli* exhibiting slow processing of exported proteins. *Cell* **32**: 789–797.

Kim, Y.J., Rajapandi, T. and Oliver, D. (1994) SecA protein is exposed to the periplasmic surface of the *E. coli* inner membrane in its active state. *Cell* **78**: 845–853.

Konninger, U.W., Hobbie, S., Benz, R. and Braun, V. (1999) The haemolysin-secreting ShlB protein of the outer membrane of *Serratia marcescens*: determination of surface-exposed residues and formation of ion-permeable pores by ShlB mutants in artificial lipid bilayer membranes. *Mol Microbiol* **32**: 1212–1225.

Krieg, U.C., Walter, P. and Johnson, A.E. (1986) Photocrosslinking of the signal sequence of nascent preprolactin to the 54-kilodalton polypeptide of the signal recognition particle. *Proc Natl Acad Sci USA* **83**: 8604–8608.

Kumamoto, C.A. and Beckwith, J. (1983) Mutations in a new gene, secB, cause defective protein localization in *Escherichia coli*. *J Bacteriol* **154**: 253–260.

Kurzchalia, T.V., Wiedmann, M., Girshovich, A.S. et al. (1986) The signal sequence of nascent preprolactin interacts with the 54K polypeptide of the signal recognition particle. *Nature* **320**: 634–636.

Lee, C., Li, P., Inouye, H., Brickman, E.R. and Beckwith, J. (1989) Genetic studies on the inability of beta-galactosidase to be translocated across the *Escherichia coli* cytoplasmic membrane. *J Bacteriol* **171**: 4609–4616.

Lippincott-Schwartz, J., Roberts, T.H. and Hirschberg, K. (2000) Secretory protein trafficking and organelle dynamics in living cells. *Annu Rev Cell Dev Biol* **16**: 557–589.

Manoil, C. and Beckwith, J. (1986) A genetic approach to analyzing membrane protein topology. *Science* **233**: 1403–1408.

Meisinger, C., Sommer, T. and Pfanner, N. (2000) Purification of *Saccharomyces cerevisiae* mitochondria devoid of microsomal and cytosolic contaminations. *Anal Biochem* **287**: 339–342.

Meyer, D.I., Krause, E. and Dobberstein, B. (1982) Secretory protein translocation across membranes – the role of the 'docking protein'. *Nature* **297**: 647–650.

Michaelis, S. and Beckwith, J. (1982) Mechanism of incorporation of cell envelope proteins in *Escherichia coli*. *Annu Rev Microbiol* **36**: 435–465.

Mothes, W., Heinrich, S.U., Graf, R. et al. (1997) Molecular mechanism of membrane protein integration into the endoplasmic reticulum. *Cell* **89**: 523–533.

Muller, M. and Blobel, G. (1984a) *In vitro* translocation of bacterial proteins across the plasma membrane of *Escherichia coli*. *Proc Natl Acad Sci USA* **81**: 7421–7425.

Muller, M. and Blobel, G. (1984b) Protein export in *Escherichia coli* requires a soluble activity. *Proc Natl Acad Sci USA* **81**: 7737–7741.

Neu, H.C. and Heppel, L.A. (1965) The release of enzymes from *Escherichia coli* by osmotic shock and during the formation of spheroplasts. *J Biol Chem* **240**: 3685–3692.

Nicchitta, C.V. and Blobel, G. (1990) Assembly of translocation-Competent proteoliposomes from detergent-solubilized rough microsomes. *Cell* **60**: 259–269.

Nicchitta, C.V., Migliaccio, G. and Blobel, G. (1991) Biochemical fractionation and assembly of the membrane components that mediate nascent chain targeting and translocation. *Cell* **65**: 587–598.

Nicchitta, C.V., Murphy, E.C. 3rd, Haynes, R. and Shelness, G.S. (1995) Stage- and ribosome-specific alterations in nascent-chain-Sec61p interaction accompany translocation across the ER membrane. *J Cell Biol* **129**: 957–970.

Nishiyama, K., Hanada, M. and Tokuda, H. (1994) Disruption of the gene encoding p12 (SecG) reveals the direct involvement and important function of SecG in the protein translocation of *Escherichia coli* at low temperature. *EMBO J* **13**: 3272–3277.

Novick, P., Field, C. and Schekman, R. (1980) Identification of 23 complementation groups required for post-translational events in the yeast secretory pathway. *Cell* **21**: 205–215.

Oliver, D.B. (1985) Identification of five new essential genes involved in the synthesis of a secreted protein in *Escherichia coli*. *J Bacteriol* **161**: 285–291.

Oliver, D.B. and Beckwith, J. (1981) *E. coli* mutant pleiotropically defective in the export of secreted proteins. *Cell* **25**: 765–772.

Oliver, D.B., Cabelli, R.J., Dolan, K.M. and Jarosik, G.P. (1990) Azide-resistant mutants of *Escherichia coli* alter the SecA protein, an azide-sensitive component of the protein export machinery. *Proc Natl Acad Sci USA* **87**: 8227–8231.

Osborn, M.J., Gander, J.E., Parisi, E. and Carson, J. (1972) Mechanism of assembly of the outer membrane of *Salmonella typhimurium*. Isolation and characterization of cytoplasmic and outer membrane. *J Biol Chem* **247**: 3962–3972.

Racker, E. (1979) Reconstitution of membrane processes. *Methods Enzymol* **55**: 699–711.

Ridder, A.N., Morein, S., Stam, J.G. et al. (2000) Analysis of the role of interfacial tryptophan residues in controlling the topology of membrane proteins. *Biochemistry* **39**: 6521–6528.

Riggs, P.D., Derman, A.I. and Beckwith, J. (1988) A mutation affecting the regulation of a *sec-lacZ* fusion defines a new sec gene. *Genetics* **118**: 571–579.

Robinson, J.S., Klionsky, D.J., Banta, L.M. and Emr, S.D. (1988) Protein sorting in *Saccharomyces cerevisiae*: isolation of mutants defective in the delivery and processing of multiple vacuolar hydrolases. *Mol Cell Biol* **8**: 4936–4948.

Rothblatt, J.A., Deshaies, R.J., Sanders, S.L., Daum, G. and Schekman, R. (1989) Multiple genes are required for proper insertion of secretory proteins into the endoplasmic reticulum in yeast. *J Cell Biol* **109**: 2641–2652.

Rothman, J.H. and Stevens, T.H. (1986) Protein sorting in yeast: mutants defective in vacuole biogenesis mislocalize vacuolar proteins into the late secretory pathway. *Cell* **47**: 1041–1051.

Schatz, P.J. and Beckwith, J. (1990) Genetic analysis of protein export in *Escherichia coli. Annu Rev Genet* **24**: 215–248.

Schatz, P.J., Riggs, P.D., Jacq, A., Fath, M.J. and Beckwith, J. (1989) The secE gene encodes an integral membrane protein required for protein export in *Escherichia coli. Genes Dev* **3**: 1035–1044.

Schiebel, E., Driessen, A.J., Hartl, F.U. and Wickner, W. (1991) Delta mu H+ and ATP function at different steps of the catalytic cycle of preprotein translocase. *Cell* **64**: 927–939.

Silberstein, S. and Gilmore, R. (1996) Biochemistry, molecular biology, and genetics of the oligosaccharyltransferase. *Faseb J* **10**: 849–858.

Simon, S.M. and Blobel, G. (1991) A protein-conducting channel in the endoplasmic reticulum. *Cell* **65**: 371–380.

Soekarjo, M., Eisenhawer, M., Kuhn, A. and Vogel, H. (1996) Thermodynamics of the membrane insertion process of the M13 procoat protein, a lipid bilayer traversing protein containing a leader sequence. *Biochemistry* **35**: 1232–1241.

Spector, D.E.A. (1998) *Cells: A Laboratory Manual*, Vol. 2. Cold Spring Harbor, NY: Cold Spring Harbor Laboratory Press.

Stader, J., Gansheroff, L.J. and Silhavy, T.J. (1989) New suppressors of signal-sequence mutations, prlG, are linked tightly to the secE gene of *Escherichia coli. Genes Dev* **3**: 1045–1052.

Traxler, B. and Murphy, C. (1996) Insertion of the polytopic membrane protein MalF is dependent on the bacterial secretion machinery. *J Biol Chem* **271**: 12394–12400.

Valent, Q.A., Scotti, P.A., High, S. et al. (1998) The *Escherichia coli* SRP and SecB targeting pathways converge at the translocon. *EMBO J* **17**: 2504–2512.

Walter, P. and Blobel, G. (1980) Purification of a membrane-associated protein complex required for protein translocation across the endoplasmic reticulum. *Proc Natl Acad Sci USA* **77**: 7112–7116.

Walter, P. and Blobel, G. (1981a) Translocation of proteins across the endoplasmic reticulum. II. Signal recognition protein (SRP) mediates the selective binding to microsomal membranes of *in-vitro*-assembled polysomes synthesizing secretory protein. *J Cell Biol* **91**: 551–556.

Walter, P. and Blobel, G. (1981b) Translocation of proteins across the endoplasmic reticulum. III. Signal recognition protein (SRP) causes signal sequence-dependent and site-specific arrest of chain elongation that is released by microsomal membranes. *J Cell Biol* **91**: 557–561.

Walter, P., Ibrahimi, I. and Blobel, G. (1981) Translocation of proteins across the endoplasmic reticulum. I. Signal recognition protein (SRP) binds to *in-vitro*-assembled polysomes synthesizing secretory protein. *J Cell Biol* **91**: 545–550.

Warren, G. and Dobberstein, B. (1978) Protein transfer across microsomal membranes reassembled from separated membrane components. *Nature* **273**: 569–571.

Weiss, J.B., Ray, P.H. and Bassford, P.J., Jr. (1988) Purified secB protein of *Escherichia coli* retards folding and promotes membrane translocation of the maltose-binding protein *in vitro. Proc Natl Acad Sci USA* **85**: 8978–8982.

Wickner, W., Driessen, A.J. and Hartl, F.U. (1991) The enzymology of protein translocation across the *Escherichia coli* plasma membrane. *Annu Rev Biochem* **60**: 101–124.

Wiedmann, B., Sakai, H., Davis, T.A. and Wiedmann, M. (1994) A protein complex required for signal-sequence-specific sorting and translocation. *Nature* **370**: 434–440.

Yang, Y.B., Yu, N. and Tai, P.C. (1997) SecE-depleted membranes of *Escherichia coli* are active. SecE is not obligatorily required for the *in vitro* translocation of certain protein precursors. *J Biol Chem* **272**: 13660–13665.

SUGGESTED READING

Danese, P.N. and Silhavy, T.J. (1998) Targeting and assembly of periplasmic and outer membrane proteins in *Escherichia coli. Annu Rev Genet* **32**: 59–94.

Ellgaard, L., Molinari, M. and Helenius, A. (1999) Setting the standards: quality control in the secretory pathway. *Science* **286**: 1882–1888.

Franzusoff, A., Rothblatt, J. and Schekman, R. (1991) Analysis of polypeptide transit through yeast secretory pathway. *Methods Enzymol* **194**: 662–674.

Lippincott-Schwartz, J., Roberts, T. and Hirschberg, K. (2000) Secretory protein trafficking and organelle dynamics in living cells. *Annu Rev Cell Dev Biol* **16**: 557–589.

Tartakoff, A.M. (1991) Vectorial transport of proteins into and across membranes. *Methods Cell Biol* **34**: 1–426.

3

TARGETING SEQUENCES

GUNNAR VON HEIJNE

SUMMARY

Targeting sequences direct proteins to the correct cellular compartment. They are often but not always N-terminal extensions to the polypeptide chain, they are recognized by either cytoplasmic or organelle-bound receptors, and are in many cases removed once targeting has been achieved. Here, the main classes of 'primary' targeting signals – secretory signal peptides, nuclear localization signals, mitochondrial targeting peptides, peroxisomal targeting sequences, and chloroplast transit peptides – will be reviewed.

SIGNAL PEPTIDES WERE DISCOVERED IN THE EARLY 1970s

César Milstein and his collaborators (Milstein et al., 1972) were the first to experimentally identify a signal peptide. In a study of the biosynthesis of immunoglobulin light chains, they observed that a precursor protein of slightly higher molecular weight than the mature protein isolated from serum was formed when translation was carried out in a cell-free *in vitro* system devoid of ER-derived microsomal membranes. In contrast, when translation was carried out by ribosomes bound to microsomal vesicles, the product had the same molecular weight as the mature protein. From this, they concluded: 'In contrast to intracellular proteins, which are made on free polysomes, secretory proteins are generally thought to be synthesized on microsomes. The signaling device whereby this segregation is achieved is unknown. It seems to us that a short amino acid sequence at

Protein Targeting, Transport & Translocation
ISBN 0-12-200731-X

the N-terminus of a precursor protein would be a simple way to provide such a signal'. Milstein's work followed hard on the heels of the original formulation of the 'signal hypothesis' by Blobel and Sabatini (1971), where the existence of an N-terminal signal peptide had been postulated from indirect evidence.

The first indication that signal peptides are largely hydrophobic came a few years later when Schechter et al. used various radioactive amino acids in an *in vitro* synthesis of immunoglobulin light chains and then determined their positions in the precursor protein by Edman degradation (Schechter et al., 1975). It was only with the advent of cDNA sequencing, however, that it became possible to collect sufficient numbers of signal peptides to start defining their overall architecture (von Heijne, 1983).

The first mitochondrial targeting sequence was identified in 1979 (Maccecchini et al., 1979), and the first chloroplast transit peptide in 1977 (Dobberstein et al., 1977).

SIGNAL PEPTIDES TARGET PROTEINS FOR SECRETION

Protein secretion in eukaryotic, bacterial, and archaeal cells depends on N-terminal signal peptides. The signal peptide is usually cleaved off by a signal peptidase, although some proteins have non-cleaved signal peptides. There are at least two distinct secretory pathways that both depend on an N-terminal signal peptide: the Sec ('secretion') pathway (Chapter 4) and the Tat ('twin arginine translocation') pathway (Chapter 11). In bacteria, there is also a third kind of secretory mechanism called type III secretion or the 'general secretory pathway'. Type III secretion signals are located at the C-terminal end of proteins.

Sec signal peptides have three distinct regions

The first targeting signals to be studied were the Sec signal peptides. Although there are minor differences between Sec signal peptides from eukaryotic and prokaryotic organisms (von Heijne and Abrahmsén, 1989), they all share a common architecture with a short, positively charged N-terminal region (n-region), a central, hydrophobic region (h-region), and a slightly polar C-terminal region (c-region) (Figure 3.1, top).

The best-conserved part of Sec signal peptides is the c-region, which contains the signal peptidase cleavage site. Small residues are invariably found in positions −1 and −3 relative to the cleavage site, and they fit into shallow depressions near the active site of the enzyme (Paetzel et al., 1998). The c-region binds to the peptidase in an extended β-strand conformation,

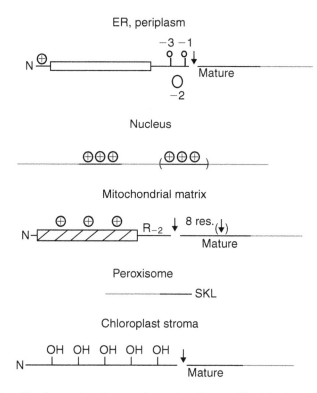

Figure 3.1   The five major classes of proteins discussed in this chapter.

while the h-region is believed to form an α-helical structure when bound to the so-called signal recognition particle (SRP; Chapter 4). Possibly, the positively charged n-region binds to negatively charged phosphate groups on the SRP 4.5S RNA (Batey et al., 2000). High-resolution structures of the signal-sequence binding domain of both human (Clemons et al., 1999) and a bacterial (Keenan et al., 1998) SRP are available.

Bacterial lipoproteins have a different c-region

Signal peptides from bacterial lipoproteins have the same kind of n- and h-regions as Sec signal peptides, but end with a so-called 'lipobox' with a consensus sequence L(A,S,I)(G,A)+C ('+' indicates the cleavage site) rather than a classical signal peptidase c-region (von Heijne, 1989). The lipobox is cleaved by lipoprotein signal peptidase in a reaction where a lipid anchor is simultaneously added to the N-terminal cysteine in the mature chain (Sankaran and Wu, 1994).

Tat signal peptides contain an RR motif in the n-region

The Tat pathway was discovered in thylakoids and was later found to exist in most bacteria (Berks et al., 2000). The name derives from the observation that a critical RR motif is invariably present in the n-region of the Tat signal peptides (Berks, 1996). The h-regions of Tat signal peptides have a lower average hydrophobicity than the h-regions in Sec signal peptides (Cristobal et al., 1999), and Tat signal peptides often have a positively charged residue in their c-region; a 'Sec-avoidance signal' (Bogsch et al., 1997).

Tat signal peptides are not recognized by the SRP or other components of the Sec machinery. It is not yet clear how they are recognized by the Tat machinery. Considering that there is no apparent relationship between the components of the Sec and Tat machineries, it is remarkable that the respective signal peptides are so similar in their overall design, and seem more like variations on a theme than two distinct kinds of targeting signals.

Signal peptides are degraded by various proteases

After the initial cleavage by signal peptidase, signal peptides are further degraded by both membrane-bound and cytosolic peptidases. In *Escherichia coli*, oligopeptidase IV and oligopeptidase A contribute to signal peptide degradation (Miller and Conlin, 1994). A subclass of *E. coli* signal peptides are cleaved in their n-region by prepilin peptidase (Lory, 1994).

In eukaryotic cells, unidentified ER peptidases make the initial cuts in the cleaved signal peptide (Weihofen et al., 2000). Hydrophobic fragments of signal peptides have been found bound to MHC complexes on the cell surface (O'Callaghan et al., 1998), and more polar n-terminal fragments have been found bound to cytosolic calmodulin, implying a possible signaling function (Martoglio et al., 1997).

SIGNALS FOR NUCLEAR IMPORT AND EXPORT CONTAIN CLUSTERS OF BASIC RESIDUES

Protein trafficking across nuclear pores is discussed in Chapter 13. The so-called nuclear localization signals (NLS) that target proteins for nuclear import by virtue of their affinity for a soluble cytosolic receptor (importin α) are generally composed of one (monopartite NLS) or two (bipartite NLS) short stretches of basic residues exposed on the surface of the folded protein (Hodel et al., 2001). The monopartite NLS has the core consensus sequence K(K/R)X(K/R), though NLS sequences tend to vary quite a lot (Cokol et al., 2000). Structures of importin α bound to two different NLSs have been determined recently (Conti and Kuriyan, 2000; Conti et al., 1998).

MITOCHONDRIAL IMPORT SIGNALS MEDIATE IMPORT ACROSS TWO MEMBRANES

Almost all nuclearly encoded mitochondrial proteins are imported through the so-called Tom and Tim complexes in the outer and inner mitochondrial membranes (see Chapter 10). The targeting signals seem to interact with the import machinery at multiple stages during import, yet their design is rather simple.

Matrix import signals form positively charged, amphiphilic α-helices

Most soluble matrix proteins have N-terminal mitochondrial targeting peptides that are removed by a matrix-localized protease upon import. Sequence analysis, mutational studies, and structure determination by nuclear magnetic resonance (NMR) have shown that an ability to form a positively charged, amphiphilic α-helix is critical for the import function, as seen most clearly in the recently determined structure of a targeting peptide bound to a domain of the outer membrane receptor Tom20 (Abe et al., 2000). The positively charged residues are thought to be important both for the interaction of the targeting peptide with 'acidic bristles' in proteins along the import pathway and for membrane-potential driven import across the inner membrane (Voos et al., 1999). Short mTP segments with a high propensity to bind the Tom20 and Tom70 receptors have been identified by biochemical and structural approaches (Brix et al., 1999; Muto et al., 2001).

The two-subunit matrix processing peptidase appears to cleave preferentially one residue C-terminal to an arginine, although only about two-thirds of all targeting peptides have Arg in positions -3 or -2 (Schneider et al., 1998). How the cleavage site is chosen in the remaining third is not known. The X-ray structure of the cytochrome bc_1 complex from bovine mitochondria has provided a first glimpse of how the matrix processing peptidase might recognize its substrate, since two subunits of this complex are homologous to the two subunits of the peptidase (Iwata et al., 1998).

About one-third of all imported matrix proteins are cleaved a second time by the so-called intermediate processing peptidase. This enzyme removes a further eight residues from the N-terminus of the mature protein, but its substrate specificity is not well understood (Chew et al., 2000).

Proteins belonging to the so-called carrier family in the inner membrane of mitochondria, as well as some unrelated inner membrane proteins, are imported via a distinct pathway that branches off from the matrix import pathway after passage through the Tom complex (see Chapter 10). The targeting information resides in positively charged loops between the transmembrane segments that appear to be recognized by the Tom20 receptor; these loops are then translocated across the inner membrane during the membrane insertion step (Brix et al., 1999; Endres et al., 1999; Koehler et al., 1999; Leuenberger et al., 1999).

Sorting to the intermembrane space depends on bipartite sorting signals

Many intermembrane space proteins have bipartite targeting signals, where a typical matrix-targeting peptide is followed by a second targeting peptide with similarities to Sec-type signal peptides. There has been a long-standing controversy regarding the mechanism by which these proteins are sorted. According to the 'conservative sorting' model, the first targeting peptide directs import into the matrix and is then cleaved off, exposing the second signal peptide that initiates re-export across the inner membrane to the intermembrane space. The 'stop-transfer' model, in contrast, postulates that the second signal peptide acts as a stop-transfer signal, leaving the prepro-tein spanning the inner membrane. It now seems that this controversy may be resolved by the realization that both mechanisms may be in operation, but for different proteins (Chauwin et al., 1998; Voos et al., 1999).

THERE ARE TWO KINDS OF PEROXISOMAL IMPORT SIGNALS

Peroxisomes import folded proteins in a post-translational manner (Chapter 12). A large number of cytosolic and peroxisomal membrane pro-teins have been implicated in this process (Fujiki, 2000). Two kinds of per-oxisome targeting signals (PTS) have been identified (Sacksteder and Gould, 2000): a ubiquitous C-terminal-Ser-Lys-Leu (SKL) motif (PTS1 signal), and a less common signal composed of a weakly conserved N-terminal exten-sion (PTS2 signal). A high-resolution structure of the PTS1 receptor with a bound SKL peptide was recently determined (Gatto et al., 2000).

CHLOROPLAST IMPORT SIGNALS FOR THE STROMAL AND THYLAKOID COMPARTMENTS

The chloroplast has three membranes: two envelope membranes and the internal thylakoid membrane (Chapter 11). Sorting signals thus need to be of different kinds, though only two classes are reasonably well understood at present: those that target proteins for import into the stromal compart-ment and those that mediate import into thylakoids. For other compart-ments such as the outer and inner envelope membranes, only a handful of proteins are known and no clear picture of the distinctive features of their targeting signals has emerged yet.

Stroma-targeting signals ('transit peptides') lack acidic but are rich in hydroxylated amino acids

N-terminal 'transit peptides' target proteins to the chloroplast import machinery and ensure translocation across the two envelope membranes into the stroma. They have a distinct amino acid composition, with few acidic but many hydroxylated residues (Emanuelsson et al., 1999). It has been suggested that phosphorylation of certain hydroxylated residues may be important for import (Waegemann and Soll, 1996), but beyond this it is unclear how transit peptides are recognized. The cleavage site between the transit peptide and the mature protein that is recognized by the stromal processing peptidase is likewise not well conserved, although certain weak positional amino acid preferences have been noted (Emanuelsson et al., 1999).

Targeting peptides for thylakoid import are similar to bacterial signal sequences

Once imported into the stroma, some chloroplast proteins need to be further sorted to the lumen of the thylakoids. At least three distinct pathways for thylakoid import have been found, and they are similar to the pathways identified in bacteria such as *E. coli*. Thus, thylakoids have a typical Sec machinery, an SRP-dependent import pathway, and a Tat machinery (Chapter 11). Not surprisingly, thylakoid import signals look much like their bacterial counterparts, with typical n-, h-, and c-regions (see above). Thylakoid proteins thus come with bipartite targeting signals composed of a transit peptide immediately followed by a thylakoid signal sequence (of either the Sec or Tat kind). The thylakoid processing peptidase is homologous to bacterial signal peptidases and has the same substrate specificity with small residues required in positions -1 and -3 relative to the cleavage site (Halpin et al., 1989).

PROTEIN LOCALIZATION CAN BE REASONABLY WELL PREDICTED

In these days of large-scale genome sequencing, prediction of protein localization has become an important aspect of functional prediction from sequence data. So far, three basic approaches for localization prediction have been proposed: (i) to search for targeting signals in the amino acid sequences, (ii) to search the mature parts of the protein for features known to correlate with the subcellular localization (overall amino acid composition, preponderance of glycosylation sites, etc.), and (iii) to look for prokaryotic homologs since many organellar proteins are more similar to their prokaryotic ancestors than to cytosolic eukaryotic proteins with corresponding functions.

Targeting signal predictors perform well but can only detect a subset of all signals

A priori, it would seem that prediction methods that 'imitate' the cellular recognition events discussed above should be able to discriminate between proteins destined for different subcellular locations. There is a long history of such prediction methods, where the earliest ones were designed for the detection of signal peptides on secretory proteins (McGeoch, 1985; von Heijne, 1986).

The most successful methods to date are based on 'machine learning' techniques such as neural networks and hidden Markov models (Baldi and Brunak, 1998). A good case in point is the so-called TargetP program (Emanuelsson et al., 2000). In TargetP, three different neural networks trained to recognize, respectively, signal peptides, mitochondrial targeting peptides, and chloroplast transit peptides have been integrated into a single predictor that takes the N-terminal 100 residues of a protein as input and produces an output where the most likely localization is given (including the possibility that the protein will not be targeted to either of the three compartments). TargetP works very well for secretory proteins (with both sensitivity and specificity around 95%), and works reasonably well for mitochondrial and chloroplast proteins (sensitivity and specificity around 85%). By adjusting the prediction thresholds one can increase the reliability of the predictions, but the method will then score an increasing fraction of all proteins as 'no prediction possible'. TargetP also predicts the most likely cleavage site in the targeting peptide, although these predictions are less reliable (signal peptide cleavage sites are correctly predicted in approximately 75% of all cases, while the performance is worse for mitochondrial targeting peptides and transit peptides).

An obvious drawback of TargetP and similar methods is that the targeting peptides that one tries to identify must be reasonably distinct in terms of amino acid sequence and composition, and that a fairly large number of experimentally verified examples must be available for training. Furthermore, signals such as nuclear localization signals or mannose-6-phosphate based lysosomal targeting signals that can only be recognized in the folded protein are presently out of reach.

'Global' prediction methods can handle many different compartments but with poorer overall performance

From a purely statistical point of view, one need not restrict oneself to look-ing for the same features that the cell uses for subcellular sorting when try-ing to predict protein localization. Thus if, say, nuclear proteins for reasons other than sorting per se tend to be different in, for example, overall amino acid composition from other proteins, this can be used as a basis for the prediction.

The best-known method based on this idea is PSORT (Nakai and Horton, 1999), although a couple of other similar methods have been presented recently (Chou and Elrod, 1999; Drawid and Gerstein, 2000). In PSORT, all kinds of relevant information can be incorporated: the existence of various targeting signals (as identified by TargetP or similar methods), predicted transmembrane segments, amino acid composition measures, frequency of predicted glycosylation sites, etc. The final prediction is made by evaluating all these features simultaneously, either using a rule-based approach ('if there is a predicted signal peptide, and if the overall amino acid composition of the protein is more similar to lysosomal than to secreted proteins, predict lysosome') or a purely statistical approach. Prediction performances vary for different localizations, and one needs to go through the published work care-fully to appreciate the difficulty of the prediction problem.

Protein localization can sometimes be predicted from 'phylogenetic profiles'

Since many organelles have evolved by endosymbiosis from once free-living bacteria engulfed by a eukaryotic cell, one often finds that organellar proteins have higher levels of sequence similarity to prokaryotic than to other eukaryotic proteins. This has recently been proposed as a basis for identifying probable locations of proteins in genome sequence data (Marcotte et al., 2000). In the present implementation, 'phylogenetic pro-files' (i.e., a list of the presence/absence of a given protein across all fully sequenced genomes) are used to decide whether the phylogenetic profile of a given protein is more similar to those of, for example, known mitochon-drial proteins than non-mitochondrial proteins. In this way, proteins can be predicted to be sorted to different compartments based on evolutionary relationships rather than sequence properties per se.

It appears that this method performs less well than, for example, TargetP for proteins that have classical targeting signals, though it has the distinct advantages that one needs no *a priori* knowledge about sorting mechanisms to make a prediction and that the method can be applied across all com-partments for which a sufficient number of known proteins and their corresponding phylogenetic profiles can be collected.

REFERENCES

Abe, Y., Shodai, T., Muto, T. et al. (2000) Structural basis of presequence recognition by the mitochondrial protein import receptor Tom20. *Cell* **100**: 551–560.

Baldi, P. and Brunak, S. (1998) *Bioinformatics: The Machine Learning Approach.* Boston, MA: MIT Press.

Batey, R.T., Rambo, R.P., Lucast, L., Rha, B. and Doudna, J.A. (2000) Crystal structure of the ribonucleoprotein core of the signal recognition particle. *Science* **287**: 1232–1239.

Berks, B.C. (1996) A common export pathway for proteins binding complex redox cofactors? *Mol Microbiol* **22**: 393–404.

Berks, B.C., Sargent, F. and Palmer, T. (2000) The Tat protein export pathway. *Mol Microbiol* **35**: 260–274.

Blobel, G. and Sabatini, D.D. (1971) Ribosome–membrane interaction in eukaryotic cells. *Biomembranes* **2**: 193–195.

Bogsch, E., Brink, S. and Robinson, C. (1997) Pathway specificity for a ΔpH-dependent precursor thylakoid lumen protein is governed by a 'Sec-avoidance' motif in the transfer peptide and a 'Sec-incompatible' mature protein. *EMBO J* **16**: 3851–3859.

Brix, J., Rudiger, S., Bukau, B., SchneiderMergener, J. and Pfanner, N. (1999) Distribution of binding sequences for the mitochondrial import receptors Tom20, Tom22, and Tom70 in a presequence-carrying preprotein and a non-cleavable preprotein. *J Biol Chem* **274**: 16522–16530.

Chauwin, J.F., Oster, G. and Glick, B.S. (1998) Strong precursor–pore interactions constrain models for mitochondrial protein import. *Biophys J* **74**: 1732–1743.

Chew, A., Sirugo, G., Alsobrook, J.P. and Isaya, G. (2000) Functional and genomic analysis of the human mitochondrial intermediate peptidase, a putative protein partner of frataxin. *Genomics* **65**: 104–112.

Chou, K.C. and Elrod, D.W. (1999) Protein subcellular location prediction. *Protein Eng* **12**: 107–118.

Clemons, W.M., Gowda, K., Black, S.D., Zwieb, C. and Ramakrishnan, V. (1999) Crystal structure of the conserved subdomain of human protein SRP54M at 2.1 Å resolution: evidence for the mechanism of signal peptide binding. *J Mol Biol* **292**: 697–705.

Cokol, M., Nair, R. and Rost, B. (2000) Finding nuclear localization signals. *EMBO Reports* **1**: 411–415.

Conti, E. and Kuriyan, J. (2000) Crystallographic analysis of the specific yet versatile recognition of distinct nuclear localization signals by karyopherin alpha. *Structure Fold Des* **8**: 329–338.

Conti, E., Uy, M., Leighton, L., Blobel, G. and Kuryian, J. (1998) Crystallographic analysis of the recognition of a nuclear localization signal by the nuclear import factor karyopherin α. *Cell* **94**: 192–204.

Cristobal, S., de Gier, J.W., Nielsen, H. and von Heijne, G. (1999) Competition between Sec- and TAT-dependent protein translocation in *Escherichia coli*. *EMBO J* **18**: 2982–2990.

Dobberstein, B., Blobel, G. and Chua, N.H. (1977) *In vitro* synthesis and processing of a putative precursor for the small subunit of ribulose-1,5-bisphosphate carboxylase of *Chlamydomonas reinhardtii*. *Proc Natl Acad Sci USA* **74**: 1082–1085.

Drawid, A. and Gerstein, M. (2000) A Bayesian system integrating expression data with sequence patterns for localizing proteins: comprehensive application to the yeast genome. *J Mol Biol* **301**: 1059–1075.

Emanuelsson, O., Nielsen, H. and von Heijne, G. (1999) ChloroP, a neural network-based method for predicting chloroplast transit peptides and their cleavage sites. *Protein Sci* **8**: 978–984.

Emanuelsson, O., Nielsen, H., Brunak, S. and von Heijne, G. (2000) Predicting subcellular localization of proteins based on their N-terminal amino acid sequence. *J Mol Biol* **300**: 1005–1016.

Endres, M., Neupert, W. and Brunner, M. (1999) Transport of the ADP-ATP carrier of mitochondria from the TOM complex to the TIM22.54 complex. *EMBO J* **18**: 3214–3221.

Fujiki, Y. (2000) Peroxisome biogenesis and peroxisome biogenesis disorders. *FEBS Lett* **476**: 42–46.

Gatto, G.J., Jr, Geisbrecht, B.V., Gould, S.J. and Berg, J.M. (2000) Peroxisomal targeting signal-1 recognition by the TPR domains of human PEX5. *Nat Struct Biol* **7**: 1091–1095.

Halpin, C., Elderfield, P.D., James, H.E. et al. (1989) The reaction specificities of the thylakoidal processing peptidase and *Escherichia coli* leader peptidase are identical. *EMBO J* **8**: 3917–3921.

Hodel, M., Corbett, A. and Hodel, A. (2001) Dissection of a nuclear localization signal. *J Biol Chem* **276**: 1317–1325.

Iwata, S., Lee, J.W., Okada, K. et al. (1998) Complete structure of the 11-subunit bovine mitochondrial cytochrome bc_1 complex. *Science* **281**: 64–71.

Keenan, R.J., Freymann, D.M., Walter, P. and Stroud, R.M. (1998) Crystal structure of the signal sequence binding subunit of the signal recognition particle. *Cell* **94**: 181–191.

Koehler, C.M., Merchant, S. and Schatz, G. (1999) How membrane proteins travel across the mitochondrial intermembrane space. *Trends Biochem Sci* **24**: 428–432.

Leuenberger, D., Bally, N.A., Schatz, G. and Koehler, C.M. (1999) Different import pathways through the mitochondrial intermembrane space for inner membrane proteins. *EMBO J* **18**: 4816–4822.

Lory, S. (1994) Leader peptidases of type IV prepilins and related proteins. In: von Heijne, G. (ed.) *Signal Peptidases*, pp. 31–48. Austin, TX: R.G. Landes Co.

Maccecchini, M.L., Rudin, Y., Blobel, G. and Schatz, G. (1979) Import of proteins into mitochondria: precursor forms of the extramitochondrially made F1-ATPase subunits in yeast. *Proc Natl Acad Sci USA* **76**: 343–347.

Marcotte, E., Xenarios, I., van Der Bliek, A. and Eisenberg, D. (2000) Localizing proteins in the cell from their phylogenetic profiles. *Proc Natl Acad Sci USA* **97**: 12115–12120.

Martoglio, B., Graf, R. and Dobberstein, B. (1997) Signal peptide fragments of preprolactin and HIV-1 p-gp160 interact with calmodulin. *EMBO J* **16**: 6636–6645.

McGeoch, D.J. (1985) On the predictive recognition of signal peptide sequences. *Virus Res* **3**: 271–286.

Miller, C. and Conlin, C. (1994) Signal peptide hydrolases. In: von Heijne, G. (ed.) *Signal Peptidases*, pp. 49–57. Austin, TX: R.G. Landes Co.

Milstein, C., Brownlee, G.G., Harrison, T.M. and Mathews, M.B. (1972) A possible precursor of immunoglobulin light chains. *Nature New Biol* **239**: 117–120.

Muto, T., Obita, T., Abe, Y. et al. (2001) NMR Identification of the Tom20 binding segment in mitochondrial presequences. *J Mol Biol* **306**: 137–143.

Nakai, K. and Horton, P. (1999) PSORT: a program for detecting sorting signals in proteins and predicting their subcellular localization. *Trends Biochem Sci* **24**: 34–35.

O'Callaghan, C., Tormo, J., Willcox, B. et al. (1998) Structural features impose tight peptide binding specificity in the nonclassical MHC molecule HLA-E. *Molec Cell* **1**: 531–541.

Paetzel, M., Dalbey, R.E. and Strynadka, N.C.J. (1998) Crystal structure of a bacterial signal peptidase in complex with a β-lactam inhibitor. *Nature* **396**: 186–190.

Sacksteder, K.A. and Gould, S.J. (2000) The genetics of peroxisome biogenesis. *Annu Rev Genet* **34**: 623–652.

Sankaran, K. and Wu, H. (1994) Signal peptidase II – Specific signal peptidase for bacterial lipoproteins. In: von Heijne, G. (ed.) *Signal Peptidases*, pp. 17–29. Austin, TX: R.G. Landes Co.

Schechter, I., McKean, D.J., Guyer, R. and Terry, W. (1975) Partial amino acid sequence of the precursor of immunoglobulin light chain programmed by messenger RNA *in vitro*. *Science* **188**: 160–162.

Schneider, G., Sjöling, S., Wallin, E. et al. (1998) Feature-extraction from endopeptidase cleavage sites in mitochondrial targeting peptides. *Protein Struct Funct Genet* **30**: 49–60.

von Heijne, G. (1983) Patterns of amino acids near signal-sequence cleavage sites. *Eur J Biochem* **133**: 17–21.

von Heijne, G. (1986) A new method for predicting signal sequence cleavage sites. *Nucleic Acids Res* **14**: 4683–4690.

von Heijne, G. (1989) The structure of signal peptides from bacterial lipoproteins. *Protein Eng* **2**: 531–534.

von Heijne, G. and Abrahmsén, L. (1989) Species-specific variation in signal peptide design: implications for protein secretion in foreign hosts. *FEBS Lett* **244**: 439–446.

Voos, W., Martin, H., Krimmer, T. and Pfanner, N. (1999) Mechanisms of protein translocation into mitochondria. *BBA Rev Biomembranes* **1422**: 235–254.

Waegemann, K. and Soll, J. (1996) Phosphorylation of the transit sequence of chloroplast precursor proteins. *J Biol Chem* **271**: 6545–6554.

Weihofen, A., Lemberg, M.K., Ploegh, H.L., Bogyo, M. and Martoglio, B. (2000) Release of signal peptide fragments into the cytosol requires cleavage in the transmembrane region by a protease activity that is specifically blocked by a novel cysteine protease inhibitor. *J Biol Chem* **275**: 30951–30956.

4

Protein Export in Bacteria

Arnold J.M. Driessen and Chris van der Does

INTRODUCTION

In Gram-negative bacteria such as *Escherichia coli*, various compartments can be identified, i.e., the *cytosol*, which is separated by the cytoplasmic membrane from the *periplasm*, which in turn is separated from the *external milieu* by the outer membrane. Periplasmic and outer membrane proteins are synthesized in the cytosol and therefore need to be transported across the cytoplasmic membrane to reach their final destination. Protein secretion involves another transport step across the outer membrane and is not further discussed here (for a review, see Thanassi and Hultgren, 2000). Gram-positive bacteria lack an outer membrane, and instead are surrounded by a thick cell wall. In these organisms, proteins only need to cross one membrane in order to be secreted. Analyses revealed that up to 10% of the cellular proteome of *Bacillus subtilis* comprises secretory proteins (Tjalsma et al., 2000).

The cytoplasmic membrane of bacteria is the site for energy-transducing processes such as respiration, ATP synthesis, solute transport and flagellar movement. The electrochemical gradient of protons across the membrane (or proton motive force, PMF) is a central energy intermediate in energy transduction. This implies that protein export across the cytoplasmic membrane must occur without compromising the proton and ion barrier function. The molecular mechanism of bacterial protein export has been studied in great detail during the last two decades using the powerful genetic and biochemical tools that are available for *E. coli*. These studies have revealed some striking mechanistic similarities with the translocation of proteins across the endoplasmic reticulum (ER) of eukaryotes (see Chapter 5) and

Protein Targeting, Transport & Translocation
ISBN 0-12-200731-X

thylakoid membrane of chloroplasts (see Chapter 11). The components involved in bacterial protein export are also needed for the integration of newly synthesized membrane proteins into the lipid bilayer (see Chapter 6).

In bacteria, the major route of protein translocation is the so-called secretion pathway abbreviated as 'Sec pathway' (Figure 4.1). Secretory proteins are synthesized at the ribosome as precursors with an amino-terminal extension, the signal peptide. These precursor proteins (prepro-teins) are either targeted directly or via molecular chaperones such as SecB or signal recognition particle (SRP) and its receptor (see also Chapter 6)

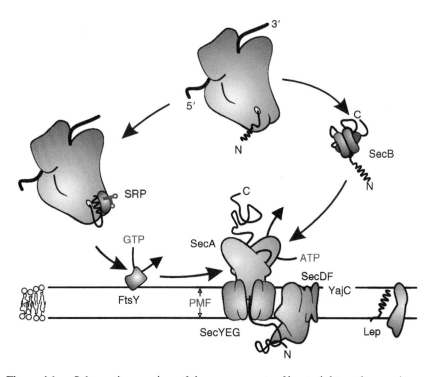

Figure 4.1 Schematic overview of the components of bacterial translocase. A protein with a signal sequence emerging from the ribosome can be targeted to the translocase by at least three different routes. Post-translationally, the preprotein may either directly associate with the translocase or use SecB as a targeting factor. Co-translationally, preproteins will associate with signal recognition particle (SRP), and via FtsY, associate with the translocase. This latter process requires the hydrolysis of GTP by both SRP and FtsY. Proteins are translocated across the membrane through a protein-conducting channel that consists of an oligomeric assembly of the SecYEG complex. Protein translocation is driven by the hydrolysis of ATP by SecA. Once translocation has been initiated at the expense of ATP, the PMF can also drive translocation. During translocation the signal sequence is removed by signal peptidase (Lep). SecD, SecF and YajC are membrane proteins involved in protein translocation, but their exact function is not yet known.

to a membrane-bound complex that combines the function of both a molecular motor and protein-conducting channel. This system is collectively termed 'translocase' and consists of a preprotein-stimulated ATPase, SecA, and a large integral membrane domain with the SecY and SecE polypeptides at its core, and SecG, SecD, SecF and YajC as additional subunits. Preproteins are translocated across the membrane in an unfolded state requiring energy input in the form of ATP and the PMF. Specific phospholipids are necessary for optimal functioning of the translocation reaction (van Voorst and de Kruijff, 2000). Finally, at the periplasmic face of the membrane, the signal sequence of the preprotein is removed by signal peptidase (see Chapter 3). The released mature domain will fold into its native conformation assisted by periplasmatic chaperones (see Chapter 7), and some proteins will subsequently integrate into the outer membrane. The role of the translocase in the integration of membrane proteins into the membrane is discussed in Chapter 6. Here we will describe the mechanism of bacterial protein export.

PROTEIN TARGETING TO THE BACTERIAL TRANSLOCASE

Secretory proteins are synthesized in the cytosol as precursors mostly with a cleavable amino-terminal signal sequence (see Chapter 3) (von Heijne, 1998). Signal sequences have a length that ranges from 18 up to 30 amino acids and show no conservation in amino acid sequence. Signal sequences are, however, equipped with the same physical properties and have a tripartite structure:

1. N-domain: The amino-terminal domain, with a size of 1 up to 5 amino acids, contains a net positive charge. Preproteins that do not have this positive charge are still recognized by the translocase but are translocated slowly. The N-domain interacts with the translocation ATPase SecA and with negatively charged phospholipids (van Voorst and de Kruijff, 2000). Due to the positive charge, translocation of the N-domain is prevented by the transmembrane electrical potential, $\Delta\psi$, which in *E. coli* is negative inside and positive outside. This results in the orientation of the signal sequence with the N-terminus in the cytosol.

2. H-domain: The hydrophobic core of the signal sequence consists of a stretch of 7 to 15 hydrophobic residues that may fold into an α-helical conformation. This domain is thought to insert into the lipid bilayer. Frequently, glycine and proline residues are found in the middle of this domain. These residues may act as helix breakers to form a hairpin-like structure that facilitates insertion of the signal sequence into the membrane or translocation channel. Inside the lipid bilayer,

the glycine and proline residues adopt an α-helical conformation, causing the 'unlooping' of the hairpin and further insertion of the signal sequence (van Dalen et al., 1999). The N- and H-domain have overlapping functions in protein export and both are needed for recognition of the signal sequence by SecA (Akita et al., 1990). The α-helical conformation of the H-domain is promoted by interaction of positively charged residues of the N-domain with anionic phospholipids.

3. C-domain: The polar C-domain stretches from 3 to 7 residues, and contains the cleavage site for signal peptidase. *E. coli* contains a signal peptidase for regular signal sequences, and another enzyme that cleaves the signal sequence of lipid-modified precursors of lipoproteins (Dalbey et al., 1997). In some bacteria, multiple signal peptidases are found with overlapping substrate specificities. Signal peptidase recognizes residues at the -1 and -3 positions relative to the cleavage site. These positions contain amino acids with small neutral side chains. The C-domain is oriented by the H-domain to be in the correct position for cleavage by signal peptidase. Signal peptides from bacterial secretory proteins, thylakoid luminal proteins and proteins that are translocated across the ER membrane can be functionally exchanged (von Heijne, 1998), suggesting similar mechanisms of protein translocation in these organelles. The signal peptide targets the protein to the translocase and is removed during or after translocation of the protein. This event is required for the release of the mature domain from the membrane. Released signal peptides are degraded by various peptidases and removed from the membrane.

Co- and post-translational export

When a protein with an amino-terminal signal sequence is translated by the ribosome, it can be targeted either co- or post-translationally to the translocase. During co-translational targeting the preprotein is directed to the translocase as ribosome-bound nascent chain, while in the post-translational modus, the preprotein is first synthesized to its full-length prior to targeting and translocation. The following paragraphs discuss these pathways in bacteria in detail.

Co-translational protein targeting

Co-translational protein targeting has first been discovered in the ER of mammals (see Chapter 5). The two key components in this targeting route are the *signal recognition particle* (SRP) and the *SRP receptor*. SRP is a ribonucleoprotein composed of six proteins (SRP72, 68, 54, 19, 14 and 9) assembled on a 300-nucleotide RNA scaffold. It binds to the signal sequence and the large subunit of the ribosome, and arrests further translation.

This *ribosome–nascent chain complex* then binds via SRP to the SRP receptor, which is composed of a peripheral (SRα) and a transmembrane (SRβ) GTPase. Upon GTP binding, the SRP receptor releases SRP from the RNC complex followed by binding of the signal sequence and the ribosome to Sec61p, the eukaryotic homolog of the bacterial protein-conducting pore. GTP hydrolysis subsequently releases SRP from the SRP receptor so that it recycles to the cytosol.

SRP and the SRP receptor are also present in bacteria. The *E. coli* SRP is much smaller than its eukaryotic homolog and consists only of a 48 kDa GTPase called Ffh (for *fifty-four-h*omolog) and a 4.5S RNA which is similar to the eukaryotic 7S RNA (for review see Herskovits et al., 2000). Both components are essential for viability and needed for the translocation of a subset of proteins. The SRP pathway in bacteria, however, is mostly involved in the co-translational targeting of α-helical membrane proteins to the translocase for membrane integration (see Chapter 6). Ffh and 4.5S RNA form a complex that interacts specifically with the signal sequence of nascent preproteins (Luirink et al., 1992). The bacterial homolog of the mammalian SRα is termed FtsY (Gill and Salmond, 1990). FtsY is a soluble GTPase that can bind to the membrane surface. It is essential for viability. A homolog of the mammalian SRβ has not been found in bacteria, but FtsY may fulfill the functions of both the SRα and SRβ subunits as it can bind to the membrane without the involvement of a receptor. In some bacteria, FtsY is anchored to the membrane by means of an α-helical transmembrane domain. The *E. coli* SRP binds tightly to FtsY in a GTP-dependent manner. Binding of 4.5S RNA to Ffh controls the association and dissociation of SRP and FtsY (Peluso et al., 2000).

Ffh consists of three regions: an amino-terminal N-domain, a middle G-domain, and a carboxyl-terminal M-domain. The G-domain is closely related to the p21^{Ras} GTPase family and contains the GTP binding site. The N-domain senses or controls the nucleotide occupancy of the G-domain through interfacial contacts. The GTPase cycle of Ffh, however, differs from those of other GTPases. The active site side chains of the nucleotide-free form of the G-domain are effectively sequestered and provide a relatively stable non-nucleotide bound state (Freymann et al., 1999). Ffh can be chemically crosslinked to hydrophobic signal sequences. The signal sequence-binding groove in the M-domain of Ffh is lined up with a large number of clustered methionine residues whose hydrophobic side chains (Freymann et al., 1999), together with the conserved domain IV of the RNA, form a surface that by a combination of hydrophobic and electrostatic interactions recognizes the signal sequence (Batey et al., 2000). The signal sequence recognition surface thus consists of both protein and RNA.

The SRP RNA from Gram-negative bacteria is much shorter than the eukaryotic homolog and does not contain the 5′ and 3′ RNA Alu domain. In eukaryotes, this region associates with the SRP9/14 heterodimer that retards the ribosomal elongation of preproteins before the engagement with the translocon. Translation arrest may not be essential in bacteria due to the shorter traffic distances and the faster translocation rates. However, in archaea and Gram-positive bacteria like *Bacillus subtilis*, the SRP RNA has a more complex structure and includes the 5' and 3' RNA Alu domain. These organisms lack homologs of SRP9/14, but in *B. subtilis*, a histone-like protein, HBsu, exists that interacts with the Alu domain of the SRP-RNA. Translocation arrest may thus be a feature in Gram-positive bacteria (Herskovits et al., 2000).

As we have learned from the mammalian system (see Chapters 5 and 8), the co-translational protein targeting serves to direct the ribosome–nascent chain complex to the translocon for subsequent insertion of the signal sequence and translocation of the growing polypeptide chain across the cytoplasmic membrane. A schematic overview for co-translational targeting of nascent preproteins to the translocase is shown in Figure 4.2.

Post-translational protein targeting

In *E. coli*, most preproteins are translocated post-translationally (Randall, 1983). These proteins are retained in the cytosol as full-length unfolded precursors with the help of molecular chaperones. SecB is a chaperone with a function dedicated to protein translocation (for review see Fekkes and Driessen, 1999). It is present in most Gram-negative bacteria. SecB is a highly acidic homotetramic protein, and is needed for the efficient translocation of preproteins (Kumamoto and Beckwith, 1983). Its function is, however, not required for the viability of *E. coli*. *In vivo*, SecB is very selective and found to interact with only a subset of preproteins, mainly outer membrane proteins that are rich in β-sheet structure and that are prone to aggregation (Kumamoto and Francetiç, 1993). These proteins are kept in a translocation-competent state, i.e., a loosely folded, non-aggregated state. *In vitro*, SecB is rather unselective and interacts with any protein that is in a non-native conformation (Randall and Hardy, 1986).

How does SecB differentiate between secretory and cytosolic proteins? The signal sequence does not provide a SecB binding site of detectable affinity but retards the folding of the mature region of proteins. SecB preferentially binds the unfolded conformation of the mature part of preproteins. A typical SecB-binding motif is about nine amino acid residues long and enriched in aromatic and basic residues, while acidic residues are strongly disfavored (Knoblauch et al., 1999). The slow folding dictated by the signal sequence assists SecB in recognizing such peptide sequences that typically occur within internal regions of folded proteins (Liu et al., 1989). SecB associates with ribosome-bound nascent chains after they have reached a length of about 150 residues. These long binding regions simultaneously occupy multiple binding sites on the tetrameric SecB to allow a high affinity of interaction (K_d 5–50 nM). The three-dimensional structure of

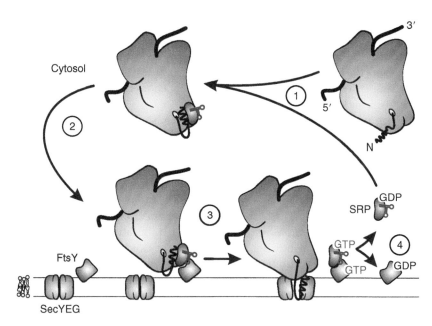

Figure 4.2   Schematic overview of SRP-mediated co-translational targeting of nascent preproteins to the translocase. (1) The synthesis of the preprotein begins with an unattached ribosome in the cytosol. The SRP complex recognizes the ribosome-emerged signal sequence when about 70 amino acids have been synthesized. (2) The ribosome–nascent chain complex is subsequently targeted by SRP to the membrane-bound FtsY. (3) This binding event increases the GTP-binding affinity of Ffh and FtsY, and in an as yet unspecified order, binding of GTP to the membrane-bound FtsY and/or SRP dissociates the ribosome–nascent chain complex from SRP and releases it to the SecYEG complex. The large ribosomal subunit also binds directly to the SecYEG complex via the 23S rRNA. (4) Subsequent hydrolysis of GTP dissociates SRP from FtsY allowing it to recycle into the cytosol while FtsY remains bound to the membrane. At this stage the ribosome and SecA collaborate for the translocation of the preprotein across the membrane. The ribosome continues to elongate the polypeptide chain until translation is completed, while SecA drives the translocation of the polypeptide segments across the membrane, as will be discussed later. For simplicity SecA is not shown.

SecB has been determined without a peptide substrate in its binding site. The structure reveals a long surface-exposed channel on each side of the SecB tetramer that has all the characteristics of a peptide-binding site (Xu et al., 2000). To occupy the peptide-binding grooves on both sites, long unstructured polypeptide segments presumably wrap around the chaperone.

General chaperones such as GroEL and DnaK can substitute for SecB in stabilizing preproteins in a translocation competent state, but they cannot substitute for the specific targeting function of SecB in protein export. This in particular concerns the ability of SecB to associate with SecA

(Hartl et al., 1990). In the cytosol, this interaction is of low affinity, but at the membrane, SecB binds with high affinity to SecA which itself is bound to the preprotein-conducting SecYEG channel. A negatively charged solvent-exposed surface on each of the sides of the SecB tetramer (Xu et al., 2000) electrostatically associates with the positively charged carboxyl-termini of the SecA dimer (Fekkes et al., 1997).

The preprotein transfer activity and the release of SecB from the membrane is not contained in the SecB structure. These events depend on the catalytic activity of SecA. The SecB–preprotein complex docks at the SecYEG-bound SecA, and subsequent binding of the exposed signal sequence region of the preprotein to SecA results in the tightening of the SecB–SecA interaction (Fekkes et al., 1997). The latter causes the release of the preprotein from SecB and its transfer to SecA. SecB is released from this ternary complex when SecA binds ATP to initiate translocation. The released SecB recycles into the cytosol where it can bind a newly synthesized preprotein. The catalytic cycle of SecB mediated preprotein targeting is schematically shown in Figure 4.3.

The SRP and SecB pathways converge at the translocase

How do preproteins decide between the SRP- and SecB-dependent targeting pathways? The choice which targeting route is used occurs immediately after the signal sequence protrudes from the ribosome. SRP binds specifically to long hydrophobic signal sequences. Recognition of less hydrophobic signal sequences by SRP is prevented by the ribosome-associated chaperone *trigger factor*, a cytosolic peptidyl-prolyl-*cis/trans*-isomerase capable of catalyzing protein folding *in vitro* (Beck et al., 2000). Preproteins that escape SRP recognition are either bound by SecB that interacts only with long nascent chains or targeted directly to the translocase by the signal sequence. Subtle differences in the hydrophobicity of signal peptides thus define the pathway that will be followed for targeting. Indeed, the SRP dependency of preprotein translocation increases with the hydrophobicity of the signal sequence, while the requirement for SecB is reduced. Both targeting pathways converge at the translocase (Valent et al., 1998) (Figure 4.1).

Summary: *Protein targeting to the bacterial translocase*

- Secretory proteins destined for the periplasm and outer membrane are synthesized as precursors in the cytosol mostly with a cleavable N-terminal signal sequence.
- Hydrophobic signal sequences of nascent preproteins are recognized and bound by a signal recognition particle (SRP), which in turn is bound by a SRP receptor at the cytoplasmic membrane.

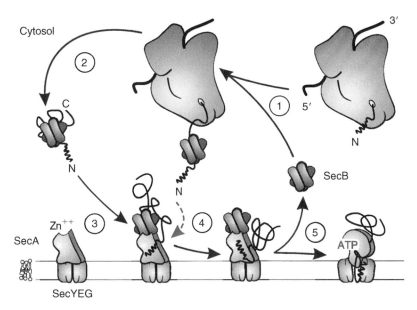

Figure 4.3   Schematic overview of SecB-dependent targeting of preproteins to the translocase. (1) Cytosolic SecB binds to the mature domain of a nascent preprotein, and (2) stabilizes its unfolded state. (3) The binary SecB–preprotein complex is targeted to the SecYEG-bound SecA where SecB binds with a high affinity binding to a carboxyl-terminal zinc-binding domain of SecA. Alternatively, it may first associate with low affinity with the cytosolic SecA and stay in the cytosol until translocation sites at the membrane become available. (4) Binding of the signal sequence to SecA tightens the SecB–SecA interaction and elicits the release of the preprotein from SecB-bound state with the concomitant transfer to SecA. (5) The release of SecB from the membrane is coupled to the binding of ATP to SecA. Under these conditions, the carboxyl-termini of SecA are no longer available for SecB binding.

- GTP hydrolysis dissociates the nascent preprotein from its association with SRP with the concomitant transfer to the translocase.
- Preproteins that escape binding by SRP are synthesized completely in the cytosol and bound by the molecular chaperone SecB. SecB stabilizes the preprotein in a translocation competent state and targets it to SecA, the peripheral subunit of the translocase.

TRANSLOCASE, A MULTI-SUBUNIT INTEGRAL MEMBRANE PROTEIN COMPLEX

The bacterial translocase (Figure 4.1) can be dissected into two modules: a protein-conducting pore formed by a set of transmembrane proteins termed the SecYEG complex and a unit that directs the movement of a translocating polypeptide chain. For post-translational translocation, SecYEG associates with SecA (Manting and Driessen, 2000), while for co-translational translocation and membrane insertion, it can also bind to the ribosome (Prinz et al., 2000). Unlike co-translational translocation in eukaryotes, protein translocation in bacteria is not coupled to chain elongation at the ribosome (Neumann-Haefelin et al., 2000) and requires the activity of SecA (Lill et al., 1989). SecY and SecE are the central components of the translocation pore. They normally associate with a third membrane protein, SecG, to form the heterotrimeric SecYEG complex (Brundage et al., 1990) and with another trimeric complex consisting of the SecD, SecF and YajC polypeptides to form a large supramolecular translocase complex (Duong and Wickner, 1997b). In *E. coli*, the SecYEG complex is the most abundant form of the integral membrane domain of translocase, ranging from 100 to 500 copies per cell. SecD and SecF are present in substoichiometric amounts, while SecA is present in large excess.

ATP hydrolysis by SecA powers protein export in bacteria

The *secA* gene was discovered in a screen for proteins that are pleiotropically defective in secretion (Oliver and Beckwith, 1982). It is an essential gene and encodes a soluble protein of 901 amino acids. SecA is found in all bacteria, and represents the most conserved Sec protein. SecA performs a central role in preprotein translocation (for review see Manting and Driessen, 2000). In the cell, SecA can cycle between the cytosol and the cytoplasmic membrane where it binds either with low affinity to phospholipids or with high affinity to the SecYEG complex. The latter represents the functional state of the SecA protein that drives preprotein translocation at the expense of ATP. As discussed previously, SecA binds to SecB and accepts the preprotein in the subsequent transfer reaction that involves recognition of the signal sequence. In addition, SecA can accept nascent preproteins from SRP or bind to preproteins directly without the involvement of chaperones.

SecA is the only ATPase involved in preprotein translocation. Its role is to couple the hydrolysis of ATP to the stepwise translocation of the preprotein across the membrane (Schiebel et al., 1991). SecA functions as a homodimer. Each of the monomers contains two domains: a 68 kDa N-terminal domain and a 34 kDa C-terminal domain. The N-domain contains a high affinity nucleotide binding site (NBD1) with the typical Walker A and B regions. NBD1 is essential for the SecA function, and is responsible for the ATPase activity of SecA (Mitchell and Oliver, 1993). The C-domain contains another nucleotide binding site (NBD2) that does not function as a hydrolysis site but instead acts as a regulatory domain (Nakatogawa et al., 2000).

The C-domain of SecA is also involved in dimerization. In the cytoplasm, the N-terminal ATPase domain is repressed by interactions with the C-terminal domain (Karamanou et al., 1999). Upon the interaction of SecA with the SecYEG complex and preprotein, the repression by the C-domain is relieved and the N-domain is activated for ATPase activity. This ATPase activity is termed 'SecA translocation ATPase' as it is coupled to the translocation of the preprotein across the membrane (Lill et al., 1989).

The interaction between SecA and the major channel subunit SecY involves multiple regions of both proteins. Whereas SecA is a cytosolic protein, parts of the N- and C-domain are accessible to membrane-impermeable reagents and proteases added from the periplasmic face of the membrane (van der Does et al., 1996; Ramamurthy and Oliver, 1997). This suggests a complex membrane topology of the SecYEG-bound SecA. SecA may be accessible from the periplasmic side of the membrane via the proteinaceous pore that would build the translocation channel (see next section). When SecYEG-bound SecA is activated by ATP in the presence of a preprotein, or in the presence of non-hydrolyzable ATP analog alone, SecA becomes highly protease-resistant (Economou and Wickner, 1994). The proteolytic fragments correspond to the N- and C-domains, and signify a conformational change of SecA associated with preprotein translocation. The protease resistance of these fragments is lost upon membrane disruption with the detergent triton X-100, and this finding has been taken to suggest that SecA inserts with nearly its entire mass into the membrane during its catalytic cycle (Economou and Wickner, 1994). According to this model, cycles of SecA insertion and de-insertion will result in the translocation of bound polypeptide segments across the membrane. A major complication with this hypothesis is that the stable SecA fragments can also be formed in the absence of membranes when a detergent is used that does not interfere with the SecA–SecYEG interaction (van der Does et al., 1998). Another problem is that during catalysis, SecA is not in contact with phospholipids (van Voorst et al., 1998) while the molecule is too large (rectangular protein of about 10 nm long and 9 nm wide) to fit in the phospholipid membrane. Other models of SecA-mediated protein translocation involve nucleotide-dependent hinge-like movements of the C- and N-domain like a kind of wrench that pushes the bound preprotein into the translocation channel (see section on 'Energetics of protein export', below).

The heterotrimeric SecYEG complex forms the core of the protein conducting channel

The gene encoding SecY was discovered in a screen for mutations that could specifically suppress the secretion defect of signal sequence mutations (Emr et al., 1981). This genetic locus was termed *prlA* for *protein localization* (see Historical Note 1). SecE was originally found in a screen for mutations affecting the expression of the *secA* gene (Schatz et al., 1989). Mutations that cause protein secretion defects normally lead to elevated levels of SecA in the cell. SecY and SecE mutants cause pleiotropic protein export defects and are often cold-sensitive for growth. SecY and SecE are both essential for viability and protein export (Bieker and Silhavy, 1989, 1990). SecY has 10 membrane-spanning α-helices, while SecE is much smaller (Figure 4.4). In *E. coli*, SecE contains three membrane-spanning α-helices but in most bacteria, SecE contains only one transmembrane domain. SecY and SecE form a stable complex, but in the absence of SecE, SecY is degraded by FtsH, an ATP-dependent membrane-bound protease (Kihara et al., 1995). Subsequent biochemical experiments showed that liposomes reconstituted with only the SecYE complex and supplemented with purified SecA are functional in the ATP-dependent translocation of preproteins (Historical Note 2) (Brundage et al., 1990). SecY and SecE are thus the only bacterial membrane proteins required for translocation. In the cell, SecY is bound not only to SecE but also to another small membrane protein, SecG. SecG contains two membrane-spanning α-helices (Figure 4.4). It was originally not identified through a genetic screen, but found to stimulate SecYE-mediated translocation (Nishiyama et al., 1993) and was co-purified with SecYE (Brundage et al., 1990). SecG is not essential for viability but needed for efficient protein export, in particular at lower temperature. SecG undergoes a remarkable topology inversion of its two α-helical membrane-spanning domains when SecA is activated by translocation ligands (Nishiyama et al., 1996).

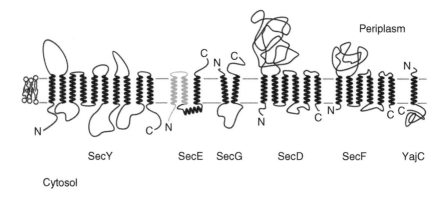

Figure 4.4 Topology of the integral membrane subunits of the bacterial translocase. SecY, SecE and SecG form a stable heterotrimeric complex that can be associated with another heterotrimeric complex composed of the SecD, SecF and YajC polypeptides. SecE in *Enterobacteriaceae* consists of three α-helical transmembrane segments, but in most bacteria it comprises only one α-helical transmembrane segment that is homologous to the third membrane domain of the *E. coli* SecE. The first two α-helical transmembrane segments of the *E. coli* SecE (indicated in gray) can be removed without loss of activity and are needed only for stability.

Several lines of evidence indicate that the SecYEG complex forms an aqueous protein-conducting pore. Addition of preproteins to isolated *E. coli* membranes opens aqueous pores that allow the conductance of ions across the membrane. The conductance relates to the expression levels of the SecYE complex (Kawasaki et al., 1993). Both SecY and SecA can be chemically crosslinked to preproteins that are 'stuck' in the translocation channel (Joly and Wickner, 1993). The translocating preprotein is largely shielded from crosslinking to phospholipids, indicating that it is protected from the lipid phase by a proteinaeous environment (Eichler et al., 1997). The translocase also allows the passage of large substrates across the membrane as preproteins with a stable disulfide bonded tertiary loop of up to 20 amino acids can be transported across the membrane (Tani et al., 1990). It thus appears as if the translocase is involved in creating a protein-conducting pore across the membrane that is aligned by protein, not phospholipid. An arrested state in the preprotein translocation reaction induced by a non-hydrolyzable ATP analog adenosine 5′-(β,γ-imidotriphosphate) (AMP-PNP) enforces the crosslinking of neighboring SecE molecules in translocase (Kaufmann et al., 1999). When assuming a heterotrimeric stoichiometry of the SecYEG complex, the latter experiment indicates the presence of more than one SecYEG complex per translocase, a postulation confirmed by structural analyses.

The actual protein-conducting channel is lined up by four SecYEG complexes. Images of the channel have been generated by negative stain high-resolution electron microscopy of single particles and computer averaging (Manting et al., 2000). These show that the channel is a conical shaped protein structure, 10.5–12 nm in diameter with a 5 nm wide, stain-filled central pore or indentation. This channel-like structure resembles the ribosome-association induced ring-like structures found with the eukaryotic Sec61p complex that are thought to contain three to four Sec61 heterotrimers (Hanein et al., 1996). That the central pore in Sec61p could be part of a protein-conducting channel has been enforced by the three-dimensional reconstruction of the ribosome–Sec61p complex structures from electron microscopical images. The exit channel of the large ribosomal subunit aligned with the putative protein-conducting pore of Sec61p complex to form a continuous protein conduit (Beckmann et al., 1997). The purified SecYEG complex forms much smaller structures that fit either one or two heterotrimeric complexes. *In vitro*, the large channel-like structure is formed only in the presence of SecA and AMP-PNP, or in the presence of SecA and ATP and a preprotein 'stuck' in the SecYEG complex (Manting et al., 2000). Preprotein translocation normally initiates upon the binding of ATP to the SecYEG-bound SecA, but *in vitro* the binding of AMP-PNP circumvents the requirement for a preprotein to induce the SecA conformational change. This nucleotide-induced conformational change of SecA thus creates one large translocation pore across the membrane that seems to consist of multiple copies of the smaller structures and is stable only during translocation conditions.

The presence of a large pore in the translocase may be deleterious to the bacterial cells as for its size one would predict that it allows passage of small solutes and protons. Since the oligomerization of the SecYEG complex requires an active state SecA, it is likely that the large channel structure accommodates part of the SecA protein.

A second heterotrimeric integral membrane protein complex is needed for efficient protein export

The *secD* gene was identified in a screen for mutants altered in export of preproteins fused to the rapidly folding protein β-galactosidase (Gardel et al., 1990). Such fusion proteins are toxic to the cells as β-galactosidase cannot be translocated, but instead 'jams' the translocase. The mutants are cold-sensitive for growth, and accumulate preproteins in the cell (Historical Note 3). A further analysis of the genomic region around the *secD* gene identified a second gene, *secF*, of which the deletion caused the same cold-sensitive phenotype as of the *secD* strain.

SecD and SecF are integral membrane proteins with six membrane-spanning α-helices and a large periplasmic domain (Figure 4.4). SecD and SecF homologs are found in nearly all prokaryotes, including archaea, and sometimes they are fused as one polypeptide. SecD and SecF show some structural similarity to transport systems of the RND- (resistance/nodulation/cell division) family (Tseng et al., 1999). *YajC* is the third open reading frame of the *secDF* operon, and encodes a small membrane protein with a single membrane spanning segment and a large cytosolic domain (Figure 4.4). YajC is neither essential for growth nor involved in protein export, but it co-purifies with SecD and SecF as a heterotrimeric integral membrane protein complex. The SecDFYajC complex co-purifies with the SecYEG complex when mild detergents are used (Duong and Wickner, 1997b), but in biochemical reconstitution experiments, no catalytic activity could be assigned to SecDFYajC. On the other hand, cells that are depleted for SecD and SecF are greatly deficient in protein export and barely viable, which points at an essential function (Pogliano and Beckwith, 1994). The essential function could be an activity that is normally not detected with *in vitro* assays that monitor the translocation of a preprotein into membrane vesicles. For instance, SecD has been implicated in protein release at the periplasmic side of the membrane (Matsuyama et al., 1993). In another study the SecDFYajC complex has been proposed to control the ATP-driven catalytic cycle of SecA and to slow down reverse and forward movement of preprotein (Duong and Wickner, 1997a). This effect on the SecA catalytic cycle must be indirect, as archaea do not contain SecA whereas SecD and SecF are present (Pohlschröder et al., 1997). SecD and SecF have also been implicated in the maintenance of the PMF. The homology of SecD and SecF with the RND

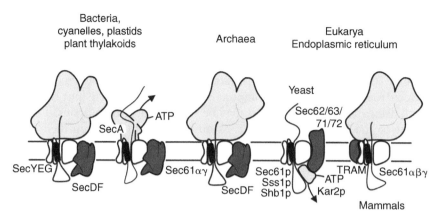

Figure 4.5   The protein-conducting channel is conserved in all kingdoms of life. The bacterial SecY and SecE proteins are homologous to the Sec61α (Sec61p) and Sec61γ (Sss1p) proteins of the protein-conducting channel of the ER of eukarya and archaea. SecG and Sec61β (Sbh1p) are found in bacteria and eukarya, respectively, but bear no significant sequence similarity. Although the heterotrimeric organization of the protein-conducting channel is conserved, post-translational protein translocation in bacteria and the yeast ER is driven by different mechanisms. Hydrolysis of ATP by SecA (bacteria, cyanelles, plastids and plant thylakoids) pushes the preprotein in a step-wise fashion through the pore, while hydrolysis of ATP by BiP/Kar2p pulls the preprotein through the channel, a process that is assisted by Sec62, Sec63, Sec71 and Sec72. SecD and SecF are accessory proteins that are found only in eubacteria and archaea, while the TRAM protein is found only in eukarya. During co-translational translocation of proteins across the bacterial and ER membrane, the ribosome directly associates with the translocation channel. Unlike co-translational translocation in eukarya, chain elongation at the ribosome is not sufficient to drive complete translocation of proteins across the bacterial cytoplasmic membrane but requires the catalytic activity of SecA. For simplicity SecA is not shown in the diagram that depicts the ribosome–SecYEG complex association. The model of the translocase of archaea is speculative and based only on the identification of homologous proteins found in the genetic databases.

transporter family suggests another function, for instance, the removal of the signal peptide or phospholipids from the aqueous protein-conducting pore formed by the translocase.

Conservation of the protein translocation mechanism in bacteria, archaea and eukaryotes

The heterotrimeric organization of the protein-conducting channel is highly conserved in all three kingdoms of life (Pohlschröder et al., 1997) (see Chapters 5 and 11) (Figure 4.5). Homologs of SecY and SecE have also been identified in archaea and eukaryotes. The first eukaryotic homolog of

SecY was identified in the yeast *Saccharomyces cerevisiae* via a genetic screening for translocation defects. This gene, *SEC61*, is essential for viability and needed for the translocation of proteins across the ER. The SecE homolog Ssslp was identified as a suppressor of a *SEC61* temperature-sensitive strain. The *S. cerevisiae* translocon was purified and shown to consist of Sec61p, Ssslp and a third component, Sbh1p. The latter is not essential for translocation, and although it may be functionally homologous, it is not similar to the bacterial SecG. The mammalian Sec61p was purified from canine pancreatic microsomes on the basis of its ribosome association. This complex consists of the SecY and SecE homologs Sec61α and Sec61γ, and also Sec61β, which is similar to the yeast Sbh1p. The ER of *S. cerevisiae* contains a second, but non-essential translocon, the Ssh1p complex. The α- and β-subunits, Ssh1p and Sbh2p, are close homologs of Sec61α and Sec61β, while the γ-subunit, Ssslp, is common to both trimeric complexes.

SecY and SecE homologs have also been identified in chloroplasts of higher plants, cyanobacteria and algae. The plant *Arabidopsis thaliana* contains both a Sec61α and a SecY homolog. Sec61α is present in the ER, while SecY functions in the thylakoid membrane of the chloroplast (see Chapter 11). In archaea, the translocon components are more closely related to the eukaryotic Sec61α and Sec61γ protein as compared to the bacterial SecY and SecE subunits.

The mechanism of protein translocation in the various systems is vastly different (Figure 4.5). In mammals, protein translocation occurs mainly co-translationally and is driven by chain-elongation at the ribosomes. In yeast, proteins can also be translocated post-translationally and this process involves in addition to Sec61p, Sec62p and Sec63p the ER lumenal ATPase Kar2p/BiP, which *pulls* the proteins across the membrane (see Chapter 5). Also in mammals, homologs of Sec62p and Sec63p exist that associate with the Sec61 complex. Translocation in bacteria can be both co- and post-translational, but both modes of protein export involve SecA to drive polypeptide segments across the membrane. In contrast to Kar2p/BiP, SecA *pushes* preprotein across the membrane. In chloroplasts, homologs of SecA and SRP have been identified, and in these organelles, a bacterial-like translocase is involved in the translocation of stromal proteins into the lumen of the thylakoid. Finally, it is not known how proteins are exported in archaea. Genome analysis so far failed to identify homologs of SecA or BiP/Kar2p. Since SRP and its receptor are present in all archaea, protein export in these organisms may be co-translational.

Translocation of folded redox proteins

In addition to the Sec pathway, another translocation pathway exists in bacteria, the twin arginine translocation (Tat) pathway (Berks et al., 2000).

The Tat pathway is a specialized translocation route mostly for fully folded redox proteins often with the bound co-factor. The Tat signal peptide resembles the typical signal peptides but includes a conserved double arginine motif [S/T]RRxFLK and appears to have less hydrophobic H-regions. This conserved motif may function as a Sec-avoidance motif, but other signals may contribute as well to the selectivity. The *tatABCD* and *tatE* loci encode the components of the Tat protein export pathway (Sargent et al., 1998; Weiner et al., 1998). The integral membrane proteins TatA, TatB and TatC presumably form the Tat translocase. The energy requirement of Tat translocase mediated translocation is unknown, but homologous components mediate the ΔpH dependent import of folded proteins in thylakoid (see Chapter 11).

Summary: *Translocase, a multi-subunit integral membrane protein complex*

- Translocase is a multi-subunit complex composed of two modules, a protein-conducting pore formed by the SecYEG complex and a unit that directs the movement of a translocating polypeptide chain, the SecA ATPase.
- For co-translational translocation, the translocase associates with the ribosome.
- Preproteins cross the membrane in an unfolded state through a protein-lined channel that consists of an oligomeric assembly of SecYEG complexes.
- The protein-conducting channels of bacteria, archaea, the plant thylakoid membrane and the ER of eukaryotes are highly conserved, while the energy coupling mechanisms for protein translocation differ.
- Folded redox proteins are translocated across the cytoplasmic membrane by a specialized export system.

ENERGETICS OF PROTEIN EXPORT

Protein export is not a spontaneous process but requires the input of energy. Proteins are translocated across the membrane in an unfolded state, and this process is driven by ATP hydrolysis and the PMF. These energy sources function at different stages of the protein translocation reaction. ATP is needed for the initiation of translocation, while the PMF can drive translocation at later stages.

Mechanism of ATP driven protein translocation

ATP is an essential energy source for protein translocation. How is ATP used to drive the movement of a protein across the membrane? The cycle of events during ATP-dependent protein export has been partially resolved in the last decade by biochemical studies using isolated membrane vesicles of *E. coli* (Schiebel et al., 1991; van der Wolk et al., 1997). SecA functions as an ATP-dependent stepping motor protein that drives unfolded proteins across the membrane (Figure 4.6). In the absence of any translocation ligands, the SecYEG-bound SecA is in an ADP-bound state. Subsequent binding of a preprotein to SecA promotes the exchange of ADP for ATP and this triggers a conformational change (Economou and Wickner, 1994) that converts the SecA protein from the compact ADP-bound state into the more extended ATP-bound state (den Blaauwen et al., 1996). This process is coupled to the transmembrane translocation of a loop of the signal

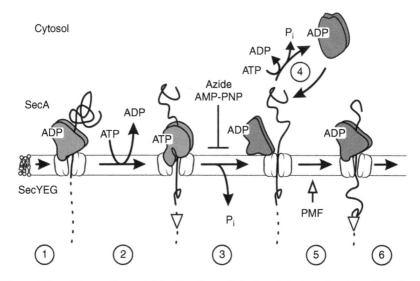

Figure 4.6 Schematic model for SecA-mediated protein translocation. Translocation of a polypeptide segment occurs as a step-wise process, related to the SecA reaction cycle. (1) A translocation intermediate is associated with ADP-bound SecA. (2) In a rate-limiting step ADP is exchanged for ATP, resulting in a conformational change of SecA with the concomitant translocation of the bound preprotein segment of about 2.5 kDa (or a stretch of about 20 amino acids). This stage of the translocation reaction is short-lived and only results in the formation of translocation intermediates when trapped with AMP-PNP or sodium azide. (3) Upon the hydrolysis of ATP, SecA reverses its conformational change and releases the preprotein. (4) The SecYEG-bound SecA can now be exchanged with cytosolic SecA in a process that requires the hydrolysis of another ATP molecule. (5) SecA can bind to a translocation intermediate, and this binding event results in the translocation of another 2.5 kDa. The total step-size resulting from the SecA reaction cycle is about 5 kDa. (6) Rebinding of ATP and subsequent repeats of the ATP-dependent catalytic cycle of SecA permit the step-wise translocation of the preprotein. The PMF can drive translocation when SecA is not associated with the preprotein, but the step-size is not known.

sequence and early mature domain to the extent that the signal sequence can be processed by signal peptidase which has its catalytic site exposed to the periplasm. Hydrolysis of ATP reverses the conformational change of SecA. The protein returns to its compact, membrane-surface bound state and at the same time releases the preprotein to the SecYEG complex. SecA may then rebind to the partially translocated preprotein. Without the input of nucleotide, binding alone is already sufficient to drive the translocation of 2–2.5 kDa of polypeptide mass across the membrane. In the next step, SecA-bound ADP is exchanged for ATP and this results in a conformational change as discussed before which is accompanied by the translocation of another of 2–2.5 kDa of polypeptide mass. A complete catalytic cycle of SecA thus permits the step-wise translocation of 5 kDa polypeptide segments by two consecutive events, i.e. about 2.5 kDa upon binding of the polypeptide by SecA, and another 2.5 kDa upon binding of ATP to SecA. Azide is an inhibitor of the growth of *E. coli* cells. Mutations that render cells resistant to azide, map in the *secA* gene (Oliver et al., 1990). Azide inhibits protein export by blocking the SecA translocation ATPase (Oliver et al., 1990).

SecA is a dissociable subunit of the translocase. It can be released from the membrane, cycle into the cytosol and associate with preproteins and subsequently rebind to the SecYEG complex. It is not known if release from the membrane is an essential feature of ATP-driven translocation. The event requires the hydrolysis of ATP (Economou et al., 1995), which indicates a very tight interaction with the SecYEG complex. SecY (PrlA) mutants that translocate preproteins with defective signal sequences also translocate normal preproteins with a greater efficiency. Such mutants bind SecA even tighter than the wild-type (van der Wolk et al., 1998), which lends support to the suggestion that SecA release is a factor that slows down translocation. Along these lines is the finding that a significant fraction of the membrane-bound SecA resists extraction with high concentrations of the chaotropic agent urea or with carbonate (Chen et al., 1996). SecA remains catalytically active after this treatment.

Mechanism of proton motive force driven translocation

After the initiation of translocation by SecA, the PMF can drive further protein translocation. Both components of the PMF, the $\Delta\psi$ and ΔpH (the transmembrane gradient of protons) can function as a driving force. In most bacteria, the $\Delta\psi$ and ΔpH are outside negative and acidic, respectively. In liposomes reconstituted with the purified translocase, this polarity of the PMF was found to stimulate protein translocation whereas the opposite polarity was inhibitory (Driessen, 1992). This suggests that the PMF provides directionality to the ATP-driven protein translocation reaction.

Other *in vitro* experiments show that when SecA is inactivated, translocation can be completed by the PMF in the absence of ATP hydrolysis (Schiebel et al., 1991). This has led to the hypothesis that the PMF functions as a driving force once SecA has initiated preprotein translocation at the expense of ATP.

Models of PMF-driven protein export involve an electrophoretic mechanism by which the $\Delta\psi$ would drive the translocation of negatively charged residues across the membrane. For instance, the $\Delta\psi$ facilitates the proper orientation of the signal sequence into the membrane by promoting the stretching of the 'looped' signal sequence as discussed before (van Dalen et al., 1999). The charge distribution of the mature domain of the preprotein may also affect the PMF requirement for translocation, but a preprotein with an uncharged mature domain also requires a PMF for translocation. Once translocation has been initiated at the expense of ATP hydrolysis, and SecA is no longer associated with the translocating preprotein, the PMF can drive rapid translocation. It is not known whether PMF-driven translocation is a step-wise or continuous event. In the presence of a PMF, less ATP is required to translocate a preprotein. Mechanistically, PMF-driven translocation is SecA-independent, while ATP-driven translocation is SecA-dependent. These two modes of translocation are, however, strongly interrelated. For instance, in the presence of a high concentration of SecA, ATP-driven translocation is favored and the PMF requirement for translocation is reduced (Yamada et al., 1989). The PMF lowers the apparent K_m value of the translocation reaction for ATP, which suggests that the PMF directly intervenes with the SecA catalytic cycle. The PMF stimulates the release of ADP from SecA (Shiozuka et al., 1990), a step that may be rate determining in the SecA catalytic cycle, and thereby promotes the release of SecA from the membrane. SecD, SecF and SecG also influence PMF-driven translocation. In more general terms, it appears that SecG and SecDF direct translocation into the ATP-dependent pathway by stabilizing SecA in its active membrane-bound state, while the PMF disengages SecA from the translocating polypeptide chain and the SecYEG complex to allow an efficient PMF-driven translocation.

Both the PMF and ATP are essential to drive the translocation of stable loop-like structures in preproteins such as an intramolecular disulfide bridge (Tani et al., 1990). Strikingly, in the earlier mentioned PrlA strains, the translocation of stabilized loop structures no longer requires the PMF (Nouwen et al., 1996). In the PrlA strains preprotein translocation is in general less PMF-dependent, which likely relates to the increased SecA binding affinity (van der Wolk et al., 1998) that directs translocation into the ATP-dependent modus (Yamada et al., 1989). However, for the translocation of looped polypeptide structures other aspects need to be considered as well. The PMF may modulate the opening or even the formation of the translocation channel, and possibly exclude the SecA from the channel mouth to allow a large opening. In *Prl* strains, the translocation pore may be in a 'more-or-less' open state due to a looser interaction between the SecY and SecE subunits (Duong and Wickner, 1999).

Summary: *Energetics of protein export*

- SecA is an ATPase that couples the hydrolysis of ATP to the step-wise translocation of the preprotein across the membrane.
- The proton motive force drives translocation when SecA has released the preprotein upon the hydrolysis of ATP.

HISTORICAL NOTES

Historical Note 1

In the early 1980s, sophisticated genetic screens were developed to select mutants in the genes for both exported proteins and the proteins that facilitate export. As many of the genes that encode components of the translocase are essential for viability, conditional lethal mutants were selected. In one of the genetic selection protocols, a strain was constructed that carried a small deletion in the *lamB* signal sequence by which the cells were unable to export the LamB protein to the outer membrane (Emr et al., 1981). Consequently, such cells are unable to transport maltodextrins across the outer membrane for use as a carbon source. Strains containing suppressor mutations that restored secretion of this export-defective LamB protein were selected for by growth on maltodextrin. Such suppressor mutations map at various loci, one of which was designated as *prlA*. The *prlA* mutations can suppress signal sequence mutations in the hydrophobic core region of a range of secretory proteins, resulting in export and processing of the normally export-defective protein to the correct cellular location. The *prlA* mutations do not cause any detectable growth defect nor do they interfere with normal protein export and processing. The *prlA* mutation was mapped to the *secY* gene that encodes the major and essential subunit of the integral membrane domain of the translocase.

Historical Note 2

In the early 1990s, the bacterial SecYEG complex was purified and functionally reconstituted into lipid vesicles. The level of SecYEG protein in the native membrane is very low and less than 0.1% of total membrane protein. Therefore, for the purification, a sensitive biochemical assay was used that monitored the preprotein-stimulated ATPase activity of SecA (translocation ATPase) (Lill et al., 1989). This activity requires a functional association of SecA with the SecYEG complex. Membranes of *E. coli* were solubilized with a detergent, fractionated by column chromatography techniques, and the fractions were reconstituted into liposomes (Brundage et al., 1990). These liposomes were supplemented with SecA and assayed for translocation ATPase activity. This led to the purification of an integral membrane protein complex that consisted of SecY and SecE, whose genes were previously identified in a genetic screen for secretion factors, and a third component, SecG. The major outcome of this study was the finding that the SecYEG complex together with the SecA ATPase represents the minimal constituents that allow the reconstitution of an authentic preprotein translocation reaction *in vitro*.

Historical Note 3

Many mutations in the membrane domain of the translocase result in a cold-sensitive growth phenotype. This may reflect a common cold-sensitive step in protein export (Pogliano and Beckwith, 1993). The cold sensitivity would become exacerbated by any mutations that interfere with the activity of the translocase. The mechanism underlying thermal sensitivity of protein export is not known, but potential cold-sensitive steps could be insertion of the signal sequence and/or SecA into the membrane, or the oligomerization of the SecYEG complex into a protein-conducting channel.

REFERENCES

Akita, M., Sasaki, S., Matsuyama, S. et al. (1990) SecA interacts with secretory proteins by recognizing the positive charge at the amino terminus of the signal peptide in *Escherichia coli*. *J Biol Chem* **265**: 8164–8169.

Batey, R.T., Rambo, R.P., Lucast, L. et al. (2000) Crystal structure of the ribonucleoprotein core of the signal recognition particle. *Science* **87**: 1232–1239.

Beck, K., Wu, L.F., Brunner, J. et al. (2000) Discrimination between SRP- and SecA/SecB-dependent substrates involves selective recognition of nascent chains by SRP and trigger factor. *EMBO J* **19**: 134–143.

Beckmann, R., Bubeck, D., Grassucci, R. et al. (1997) Alignment of conduits for the nascent polypeptide chain in the ribosome–Sec61 complex. *Science* **278**: 2123–2126.

Berks, B.C., Sargent, F. and Palmer, T. (2000) The Tat protein export pathway. *Mol Microbiol* **35**: 260–274.

Bieker, K.L. and Silhavy, T.J. (1989) PrlA is important for the translocation of exported proteins across the cytoplasmic membrane of *Escherichia coli*. *Proc Natl Acad Sci USA* **86**: 968–972.

Bieker, K.L. and Silhavy, T.J. (1990) PrlA (SecY) and PrlG (SecE) interact directly and function sequentially during protein translocation in *E. coli. Cell* **61**: 833–842.

Brundage, L., Hendrick, J.P., Schiebel, E. et al. (1990) The purified *E. coli* integral membrane protein SecY/E is sufficient for reconstitution of SecA-dependent precursor protein translocation. *Cell* **62**: 649–657.

Chen, X., Xu, H. and Tai, P.C. (1996) A significant fraction of functional SecA is permanently embedded in the membrane. SecA cycling on and off the membrane is not essential during protein translocation. *J Biol Chem* **271**: 29698–29706.

Dalbey, R.E., Lively, M.O., Bron, S. et al. (1997) The chemistry and enzymology of the type I signal peptidases. *Protein Sci* **6**: 1129–1138.

den Blaauwen, T., Fekkes, P., de Wit, J.G. et al. (1996) Domain interactions of the peripheral preprotein translocase subunit SecA. *Biochemistry* **35**: 11994–12004.

Driessen, A.J.M. (1992) Precursor protein translocation by the *Escherichia coli* translocase is directed by the protonmotive force. *EMBO J* **11**: 847–853.

Duong, F. and Wickner, W. (1997a) Distinct catalytic roles of the SecYE, SecG and SecDFyajC subunits of preprotein translocase holoenzyme. *EMBO J* **16**: 2756–2768.

Duong, F. and Wickner, W. (1997b) The SecDFyajC domain of preprotein translocase controls preprotein movement by regulating SecA membrane cycling. *EMBO J* **16**: 4871–4879.

Duong, F. and Wickner, W. (1999) The PrlA and PrlG phenotypes are caused by a loosened association among the translocase SecYEG subunits. *EMBO J* **18**: 3263–3270.

Economou, A. and Wickner, W. (1994) SecA promotes preprotein translocation by undergoing ATP-driven cycles of membrane insertion and deinsertion. *Cell* **78**: 835–843.

Economou, A., Pogliano, J.A., Beckwith, J. et al. (1995) SecA membrane cycling at SecYEG is driven by distinct ATP binding and hydrolysis events and is regulated by SecD and SecF. *Cell* **83**: 1171–1181.

Eichler, J., Brunner, J. and Wickner, W. (1997) The protease-protected 30 kDa domain of SecA is largely inaccessible to the membrane lipid phase. *EMBO J* **16**: 2188–2196.

Emr, S.D., Hanley-Way, S. and Silhavy, T.J. (1981) Suppressor mutations that restore export of a protein with a defective signal sequence. *Cell* **23**: 79–88.

Fekkes, P. and Driessen, A.J.M. (1999) Protein targeting to the bacterial cytoplasmic membrane. *Microbiol Mol Biol Rev* **63**: 161–173.

Fekkes, P., van der Does, C. and Driessen, A.J.M. (1997) The molecular chaperone SecB is released from the carboxy-terminus of SecA during initiation of precursor protein translocation. *EMBO J* **16**: 6105–6113.

Freymann, D.M., Keenan, R.J., Stroud, R.M. et al. (1999) Functional changes in the structure of the SRP GTPase on binding GDP and Mg^{2+}GDP. *Nat Struct Biol* **6**: 793–801.

Gardel, C., Johnson, K., Jacq, A. and Beckwith, J. (1990) The *secD* locus of *E. coli* codes for two membrane proteins required for protein export. *EMBO J* **9**: 4205–4206.

Gill, D.R. and Salmond, G.P. (1990) The identification of the *Escherichia coli ftsY* gene product: an unusual protein. *Mol Microbiol* **4**: 575–583.

Hanein, D., Matlack, K.E., Jungnickel, B. et al. (1996) Oligomeric rings of the Sec61p complex induced by ligands required for protein translocation. *Cell* **87**: 721–732.

Hartl, F.U., Lecker, S., Schiebel, E. et al. (1990) The binding cascade of SecB to SecA to SecY/E mediates preprotein targeting to the *E. coli* plasma membrane. *Cell* **63**: 269–279.

Herskovits, A.A., Bochkareva, E.S. and Bibi, E. (2000) New prospects in studying the bacterial signal recognition particle pathway. *Mol Microbiol* **38**: 927–939.

Joly, J.C. and Wickner, W. (1993) The SecA and SecY subunits of translocase are the nearest neighbors of a translocating preprotein, shielding it from phospholipids. *EMBO J* **12**: 255–263.

Karamanou, S., Vrontou, E., Sianidis, G. et al. (1999) A molecular switch in SecA protein couples ATP hydrolysis to protein translocation. *Mol Microbiol* **34**: 1133–1145.

Kaufmann, A., Manting, E.H., Veenendaal, A.K. et al. (1999) Cysteine-directed cross-linking demonstrates that helix 3 of SecE is close to helix 2 of SecY and helix 3 of a neighboring SecE. *Biochemistry* **38**: 9115–9125.

Kawasaki, S., Mizushima, S. and Tokuda, H. (1993) Membrane vesicles containing overproduced SecY and SecE exhibit high translocation ATPase activity and countermovement of protons in a SecA- and presecretory protein-dependent manner. *J Biol Chem* **268**: 8193–8198.

Kihara, A., Akiyama, Y. and Ito, K. (1995) FtsH is required for proteolytic elimination of uncomplexed forms of SecY, an essential protein translocase subunit. *Proc Natl Acad Sci USA* **92**: 4532–4536.

Knoblauch, N.T., Rudiger, S., Schonfeld, H.J. et al. (1999) Substrate specificity of the SecB chaperone. *J Biol Chem* **274**: 34219–34225.

Kumamoto, C.A. and Beckwith, J. (1983) Mutations in a new gene, *secB*, cause defective protein localization in *Escherichia coli*. *J Bacteriol* **154**: 253–260.

Kumamoto, C.A. and Francetiç, O. (1993) Highly selective binding of nascent polypeptides by an *Escherichia coli* chaperone protein *in vivo*. *J Bacteriol* **175**: 2184–2188.

Lill, R., Cunningham, K., Brundage, L.A. et al. (1989) SecA protein hydrolyzes ATP and is an essential component of the protein translocation ATPase of *Escherichia coli*. *EMBO J* **8**: 961–966.

Liu, G., Topping, T.B. and Randall, L.L. (1989) Physiological role during export for the retardation of folding by the leader peptide of maltose-binding protein. *Proc Natl Acad Sci USA* **86**: 9213–9217.

Luirink, J., High, S., Wood, H. et al. (1992) Signal-sequence recognition by an *Escherichia coli* ribonucleoprotein complex. *Nature* **359**: 741–743.

Manting, E.H. and Driessen, A.J.M. (2000) *Escherichia coli* translocase: the unravelling of a molecular machine. *Mol Microbiol* **37**: 226–238.

Manting, E.H., van der Does, C., Remigy, H. et al. (2000) SecYEG assembles into a tetramer to form the active protein translocation channel. *EMBO J* **19**: 852–861.

Matsuyama, S., Fujita, Y. and Mizushima, S. (1993) SecD is involved in the release of translocated secretory proteins from the cytoplasmic membrane of *Escherichia coli*. *EMBO J* **12**: 265–270.

Mitchell, C. and Oliver, D. (1993) Two distinct ATP-binding domains are needed to promote protein export by *Escherichia coli* SecA ATPase. *Mol Microbiol* **10**: 483–497.

Nakatogawa, H., Mori, H. and Ito, K. (2000) Two independent mechanisms down-regulate the intrinsic SecA ATPase activity. *J Biol Chem* **275**: 33209–33212.

Neumann-Haefelin, C., Schafer, U., Müller, M. and Koch, H.G. (2000) SRP-dependent co-translational targeting and SecA-dependent translocation analyzed as individual steps in the export of a bacterial protein. *EMBO J* **19**: 6419–6426.

Nishiyama, K., Mizushima, S. and Tokuda, H. (1993) A novel membrane protein involved in protein translocation across the cytoplasmic membrane of *Escherichia coli*. *EMBO J* **12**: 3409–3415.

Nishiyama, K., Suzuki, T. and Tokuda, H. (1996) Inversion of the membrane topology of SecG coupled with SecA- dependent preprotein translocation. *Cell* **85**: 71–81.

Nouwen, N., de Kruijff, B. and Tommassen, J. (1996) PrlA suppressors in *Escherichia coli* relieve the proton electrochemical gradient dependency of translocation of wild-type precursors. *Proc Natl Acad Sci USA* **93**: 5953–5957.

Oliver, D.B. and Beckwith, J. (1982) Identification of a new gene (*secA*) and gene product involved in the secretion of envelope proteins in *Escherichia coli*. *J Bacteriol* **150**: 686–691.

Oliver, D.B., Cabelli, R.J., Dolan, K.M. et al. (1990) Azide-resistant mutants of *Escherichia coli* alter the SecA protein, an azide-sensitive component of the protein export machinery. *Proc Natl Acad Sci USA* **87**: 8227–8231.

Peluso, P., Herschlag, D., Nock, S. et al. (2000) Role of 4.5S RNA in assembly of the bacterial signal recognition particle with its receptor. *Science* **288**: 1640–1643.

Pogliano, K.J. and Beckwith, J. (1993) The Cs sec mutants of *Escherichia coli* reflect the cold sensitivity of protein export itself. *Genetics* **133**: 763–773.

Pogliano, J.A. and Beckwith, J. (1994) SecD and SecF facilitate protein export in *Escherichia coli*. *EMBO J* **13**: 554–561.

Pohlschröder, M., Prinz, W.A., Hartmann, E. et al. (1997) Protein translocation in the three domains of life: variations on a theme. *Cell* **91**: 563–566.

Prinz, A., Behrens, C., Rapoport, T.A. et al. (2000) Evolutionarily conserved binding of ribosomes to the translocation channel via the large ribosomal RNA. *EMBO J* **19**: 1900–1906.

Ramamurthy, V. and Oliver, D. (1997) Topology of the integral membrane form of *Escherichia coli* SecA protein reveals multiple periplasmically exposed regions and modulation by ATP binding. *J Biol Chem* **272**: 23239–23246.

Randall, L.L. (1983) Translocation of domains of nascent periplasmic proteins across the cytoplasmic membrane is independent of elongation. *Cell* **33**: 231–240.

Randall, L.L. and Hardy, S.J. (1986) Correlation of competence for export with lack of tertiary structure of the mature species: a study *in vivo* of maltose-binding protein in *E. coli*. *Cell* **46**: 921–928.

Sargent, F., Bogsch, E.G., Stanley, N.R. et al. (1998) Overlapping functions of components of a bacterial Sec-independent protein export pathway. *EMBO J* **17**: 3640–3650.

Schatz, P.J., Riggs, P.D., Jacq, A., Fath, M.J. and Beckwith, J. (1989) The *secE* gene encodes an integral membrane protein required for protein export in *Escherichia coli*. *Genes Dev* **3**: 1035–1044.

Schiebel, E., Driessen, A.J., Hartl, F.U. et al. (1991) $\Delta\mu$ H+ and ATP function at different steps of the catalytic cycle of preprotein translocase. *Cell* **64**: 927–939.

Shiozuka, K., Tani, K., Mizushima, S. and Tokuda, H. (1990) The proton motive force lowers the level of ATP required for the *in vitro* translocation of a secretory protein in *Escherichia coli*. *J Biol Chem* **265**: 18843–18847.

Tani, K., Tokuda, H. and Mizushima, S. (1990) Translocation of ProOmpA possessing an intramolecular disulfide bridge into membrane vesicles of *Escherichia coli*. Effect of membrane energization. *J Biol Chem* **265**: 17341–17347.

Thanassi, D.G. and Hultgren, S.J. (2000) Multiple pathways allow protein secretion across the bacterial outer membrane. *Curr Opin Cell Biol* **12**: 420–430.

Tjalsma, H., Bolhuis, A., Jongbloed, J.D. et al. (2000) Signal peptide-dependent protein transport in *Bacillus subtilis*: a genome-based survey of the secretome. *Microbiol Mol Biol Rev* **64**: 515–547.

Tseng, T.T., Gratwick, K.S., Kollman, J. et al. (1999) The RND permease superfamily: An ancient, ubiquitous and diverse family that includes human disease and development proteins. *J Mol Microbiol Biotechnol* **1**: 107–125.

Valent, Q.A., Scotti, P.A., High, S. et al. (1998) The *Escherichia coli* SRP and SecB targeting pathways converge at the translocon. *EMBO J* **17**: 2504–2512.

van Dalen, A., Killian, A. and de Kruijff, B. (1999) Δψ stimulates membrane translocation of the C-terminal part of a signal sequence. *J Biol Chem* **274**: 19913–19918.

van der Does, C., den Blaauwen, T., de Wit, J.G. et al. (1996) SecA is an intrinsic subunit of the *Escherichia coli* preprotein translocase and exposes its carboxyl terminus to the periplasm. *Mol Microbiol* **22**: 619–629.

van der Does, C., Manting, E.H., Kaufmann, A. et al. (1998) Interaction between SecA and SecYEG in micellar solution and formation of the membrane-inserted state. *Biochemistry* **37**: 201–210.

van der Wolk, J.P., de Wit, J.G. and Driessen, A.J.M. (1997) The catalytic cycle of the *Escherichia coli* SecA ATPase comprises two distinct preprotein translocation events. *EMBO J* **16**: 7297–7304.

van der Wolk, J.P., Fekkes, P., Boorsma, A. et al. (1998) PrlA4 prevents the rejection of signal sequence defective preproteins by stabilizing the SecA–SecY interaction during the initiation of translocation. *EMBO J* **17**: 3631–3639.

van Voorst, F. and de Kruijf, B. (2000) Role of lipids in the translocation of proteins across membranes. *Biochem J* **347**: 601–612.

van Voorst, F., van der Does, C., Brunner, J. et al. (1998) Translocase-bound SecA is largely shielded from the phospholipid acyl chains. *Biochemistry* **37**: 12261–12268.

von Heijne, G. (1998) Life and death of a signal peptide. *Nature* **396**: 111–113.

Weiner, J.H., Bilous, P.T., Shaw, G.M. et al. (1998) A novel and ubiquitous system for membrane targeting and secretion of cofactor-containing proteins. *Cell* **93**: 93–101.

Xu, Z., Knafels, J.D. and Yoshino, K. (2000) Crystal structure of the bacterial protein export chaperone SecB. *Nat Struct Biol* **7**: 1172–1177.

Yamada, H., Matsuyama, S., Tokuda, H. et al. (1989) A high concentration of SecA allows proton motive force-independent translocation of a model secretory protein into *Escherichia coli* membrane vesicles. *J Biol Chem* **264**: 18577–18581.

SUGGESTED READING

Brundage, L., Hendrick, J.P., Schiebel, E. et al. (1990) The purified *Escherichia coli* integral membrane protein SecY/E is sufficient for reconstitution of SecA-dependent precursor protein translocation. *Cell* **62**: 649–657.

Manting, E.H., van der Does, C., Remigy, H. et al. (2000) SecYEG assembles into a tetramer to form the active protein translocase channel. *EMBO J* **19**: 852–861.

Pohlschröder, M., Prinz, W.A., Hartmann, E. et al. (1997) Protein translocation in the three domains of life: variations on a theme. *Cell* **91**: 563–566.

Schiebel, E., Driessen, A.J.M., Hartl, F.-U. et al. (1991) $\Delta\mu_{H+}$ and ATP function at different steps of the catalytic cycle of preprotein translocase. *Cell* **64**: 927–939.

Wickner, W.T. (1994) How ATP drives proteins across membranes. *Science* **266**: 1197–1198.

5

PROTEIN SORTING AT THE MEMBRANE OF THE ENDOPLASMIC RETICULUM

NORA G. HAIGH AND ARTHUR E. JOHNSON

INTRODUCTION

Nearly all proteins in the eukaryotic cell are synthesized by ribosomes in the cytoplasm, but many proteins ultimately perform their functions in other locations such as the various organelles or compartments inside the cell or, in the case of secreted proteins, outside the cell. The process of delivering proteins to their final destinations is termed protein trafficking or protein sorting. This sorting of proteins requires a means to identify and transport cellular proteins to various locations, and often involves the translocation of a protein through a membrane bilayer. Cells have therefore evolved specialized mechanisms to deliver individual proteins across the appropriate membranes and to the proper locations to perform their functions.

GENERAL PRINCIPLES

A few fundamental principles control all protein sorting in the cell. First, there must be a systematic method for identifying those proteins that are to be sorted. This information is encoded in the primary sequence of the protein, often near the amino-terminus of the protein where the sorting signal can be recognized and decoded while the nascent protein chain is still bound to the ribosome. Following the identification of specific proteins as substrates for protein sorting, a targeting mechanism is required to deliver each individual protein substrate to the appropriate cellular membrane. In some

Protein Targeting, Transport & Translocation
ISBN 0-12-200731-X

cases, specialized molecules in the cytoplasm are required for efficient delivery of cytoplasmic-encoded proteins to a particular membrane. Each cellular destination or organelle must have a unique receptor on the cytoplasmic side of the membrane to recognize protein substrates and ensure that they are delivered to the correct organelle.

After targeting, the cell often must direct the facilitated transport of polypeptides across a membrane bilayer. Since the primary function of a membrane is to create a barrier and separate two aqueous compartments, the movement of a macromolecule from the cytoplasm into an organelle, i.e. from one side of a membrane to the other, raises a number of fundamental mechanistic issues. Does the protein move through the membrane spontaneously, or are specialized sites or machinery required to facilitate the transport? In either case, what type of environment does the protein substrate pass through, a hydrophobic one, an aqueous one, or something in between? Depending on the precise mechanism, it may be difficult to transfer a protein once it is already folded, so it might be necessary to keep the protein in an unfolded state prior to and during transport across the membrane bilayer. And the transfer of a polypeptide across the membrane will almost certainly require some type of energy input to provide the power necessary for the process. These mechanistic requirements must all be met while maintaining a permeability barrier across the membrane, so that the various cellular membranes can still function efficiently to separate cellular compartments.

The focus of this chapter will be protein translocation across the membrane of the endoplasmic reticulum (ER), the first step in the cellular secretory pathway in eukaryotic cells. Proteins destined to reside in the ER, Golgi and other organelles, as well as secretory proteins and many membrane proteins, are directed first to the ER membrane. Early investigations of the mammalian ER using the electron microscope (EM) revealed the presence of ribosomes positioned along the cytoplasmic side of the rough ER membrane (Palade, 1975). This observation suggested that polypeptides might be threaded directly across the ER membrane and into the lumen of the ER as they are being synthesized by the ribosomes (co-translational translocation). Yet the precise mechanism by which a polypeptide chain is transported across the ER membrane bilayer remained a mystery for many years.

TARGETING TO THE ER MEMBRANE

As discussed in Chapter 3, cells have developed a specialized system for the identification and targeting of those proteins that are destined to cross the ER membrane. In the case of co-translational translocation, the process of targeting to the ER can be divided into three stages: signal sequence recognition,

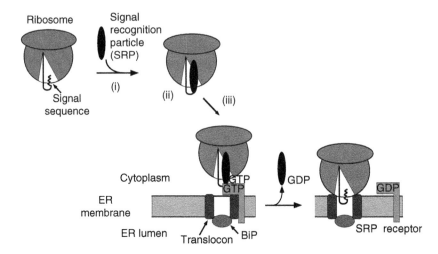

Figure 5.1 SRP-dependent targeting of RNC complexes to the ER membrane. SRP-dependent targeting of ribosomes containing nascent chains with exposed signal sequences to the ER membrane can be divided into three stages: (i) signal sequence recognition, (ii) elongation arrest, and (iii) ribosome binding to the ER membrane. After the signal sequence is synthesized and emerges from the ribosome in the cytoplasm, the SRP recognizes and binds the signal sequence. Elongation of translation is arrested in the resulting ribosome–nascent–SRP chain complex. This complex is then targeted to the ER membrane via a GTP-dependent interaction between SRP and the SRP receptor.

elongation arrest, and binding to the ER membrane (Figure 5.1; reviewed in Walter and Johnson, 1994; Stroud and Walter, 1999).

Targeting substrates are identified by the presence of a particular series of amino acid residues at their amino-terminus called a signal sequence. Signal sequences show no obvious sequence conservation, but instead share a common pattern of charged and nonpolar amino acids. Signal sequences are 20 to 30 amino acids in length, with a central hydrophobic core of 8 to 12 amino acid residues. Positively charged residues are often found amino-terminal to the hydrophobic core, while uncharged polar amino acids are often found on the carboxyl-terminal side of the hydrophobic core (see Chapter 3).

The signal recognition particle (SRP) is a ribonucleoprotein complex that is responsible for signal sequence recognition. SRP is composed of a single 300-nucleotide 7S RNA molecule and six different proteins with molecular masses of 9, 14, 19, 54, 68 and 72 kDa (termed SRP9, SRP14, SRP19, SRP54, SRP68 and SRP72) (Walter and Blobel, 1980, 1982). The overall shape of SRP is that of an elongated rod, with the RNA spanning almost the entire length and the protein subunits associating with various RNA domains (Walter and Johnson, 1994). The most highly conserved of the protein subunits is SRP54, which has a homolog in *Escherichia coli* called Ffh (fifty-four homolog) that is the sole protein subunit of bacterial SRP.

Within the context of the SRP complex, the SRP54 protein is bound to one end of the 7S RNA and has three domains. The central G-domain is a GTPase domain that is structurally similar to the GTPase domains in the SRP receptor and the Ras protein. The amino-terminal N-domain is a four-helix bundle and the M-domain, which contains multiple methionine residues, comprises the carboxyl-terminus of the SRP54 protein. The M-domain is connected to the N- and G-domains by a flexible hinge region and is responsible for both the RNA-binding and signal sequence-binding activities of SRP54 (reviewed in Stroud and Walter, 1999).

How does SRP faithfully recognize signal sequences that vary either in length of hydrophobic sequence and/or in specific amino acid sequence? Recent structural determinations of a bacterial Ffh protein provide insights into how the M-domain recognizes and binds signal sequences (Keenan et al., 1998; Batey et al., 2000). The M-domain structure contains a deep groove that is lined with hydrophobic side chains that associate with the hydrophobic residues in the signal sequence.

Because it usually resides at the amino-terminus of the nascent polypeptide, the signal sequence is the first part of the protein to be synthesized by the ribosome and to emerge from the ribosome during the translation process (Figure 5.1). The affinity of SRP for ribosomes is increased by nearly three orders of magnitude when a signal sequence is exposed during translation of a nascent chain. Thus, the exposure of the signal sequence results in the tight binding of SRP to the ribosome–nascent chain complex. Since it has been estimated that there is approximately one SRP molecule for every ten ribosomes in the cell, it is likely that the SRP cycles between ribosomes until it locates an exposed signal sequence that stabilizes SRP binding to a particular ribosome (reviewed in Walter and Johnson, 1994).

Upon binding of SRP to the signal sequence and the ribosome, further translation is inhibited (Figure 5.1) (Walter and Blobel, 1981). This 'elongation arrest' function that blocks protein synthesis is effected by the SRP9 and SRP14 proteins that are bound to the opposite end of the elongated SRP RNA structure from SRP54. This arrangement allows the SRP molecule to bind the signal sequence near the ribosome exit tunnel and simultaneously affect elongation at the distant peptidyltransferase center of the ribosome. Elongation arrest is not absolutely required for protein translocation, but it is thought to help maintain the targeting and translocation competence of the nascent chains. Thus, elongation arrest allows sufficient time to target ribosomes to the ER membrane before the length of the nascent chain becomes too long to support efficient targeting (reviewed in Walter and Johnson, 1994).

The ribosome–nascent chain–SRP (RNC–SRP) complex is targeted to the cytoplasmic side of the ER membrane via a specific interaction between SRP and the SRP receptor (SR), a heterodimeric membrane protein found only in the rough ER membrane and composed of peripheral and transmembrane subunits called SRα and SRβ, respectively. The interaction between SR and RNC–SRP initiates a series of steps that includes the binding of the ribosome to the membrane, the release of the signal sequence from SRP, the release of SR and SRP from the ribosome, and the initiation of nascent chain translocation (Figure 5.1) (Walter and Johnson, 1994). This targeting process also requires the hydrolysis of GTP. SRP, SRα and SRβ each contain a GTP binding site that together accomplish and regulate the targeting process. This may involve SRP54 and SRα acting to stimulate each other's GTPase activities (Powers and Walter, 1995; Rapiejko and Gilmore, 1997; Millman and Andrews, 1997), while GTP binding to SRβ appears to mediate the translocon-dependent release of the signal sequence from the SRP (Fulga et al., 2001). In the presence of non-hydrolyzable GTP analogs, targeting proceeds, but SRP and the SRP receptor remain tightly locked together, suggesting that GTP hydrolysis is required to release the proteins and regenerate free SRP and SR (Walter and Johnson, 1994; Rapiejko and Gilmore, 1997). The GTP-dependent dissociation of SRP from SR completes the targeting process as those proteins are released and the RNC complex binds to the ER membrane to begin co-translational translocation.

Summary: *Targeting to the ER membrane*

- Targeting to the ER can be divided into three stages: signal sequence recognition, elongation arrest, and binding to the ER membrane.
- An amino-terminal signal sequence identifies substrates for ER translocation. The signal sequence is 20 to 30 amino acids long and contains a core of 8–12 hydrophobic amino acid residues.
- A ribonucleoprotein called the SRP is responsible for signal sequence recognition and binding, and for arresting elongation.
- A GTP-dependent interaction between SRP and the SRP receptor is responsible for the targeting of RNC complexes to the ER membrane.

TRANSLOCATION ACROSS THE ER MEMBRANE

In mammalian cells, proteins are translocated across the ER membrane in a co-translational manner. This means that the polypeptide chain is translocated across the membrane bilayer at the same time that it is being synthesized by the ribosome. Translocation begins after the RNC complex binds to the ER membrane. Protein synthesis resumes after the departure of the SRP, and the nascent chain proceeds to move across the membrane bilayer and into the ER lumen. As protein synthesis continues, the nascent chain is transferred across the membrane, presumably in an extended conformation without defined secondary structure. While translocation proceeds, the signal sequence is cleaved and the nascent polypeptide chain is modified, folded, and processed in a co-translational manner by enzymes and chaperones located in the lumen of the ER.

Although translocation at the ER membrane occurs co-translationally in mammals, translocation at the ER membrane in yeast can occur by either a co-translational or post-translational mechanism. In post-translational translocation, a secretory protein is completely synthesized by the ribosome in the cytoplasm before the translocation process begins. Because transfer of the polypeptide across the ER membrane presumably occurs while the polypeptide is in an extended conformation, the nascent chain must be kept unfolded or loosely folded in the cytoplasm prior to translocation. Cytoplasmic chaperones perform this function and may also participate in targeting the polypeptide to the translocon. Substrates for post-translational translocation contain signal sequences, but the targeting process is completely independent of SRP or a similar factor (reviewed in Corsi and Schekman, 1996). Once targeting is complete, post-translational translocation substrates are transferred across the ER membrane and into the ER lumen, where a different set of chaperone proteins reside. Following translocation into the ER lumen, lumenal enzymes and chaperones act to cleave the signal sequence and correctly modify and fold the translocated proteins.

Long after the basic steps involved in targeting had been identified (Sanders and Schekman, 1992), the mechanism of passage of proteins across the ER membrane remained a 'black box' because of the experimental difficulties of working with membrane-associated events. How exactly are polypeptides transferred from one side of a membrane bilayer to the other without compromising the permeability barrier (Figure 5.2)? Is the process spontaneous, or is some type of protein machinery required? Through what type of environment does the nascent chain travel?

The nature of the translocation site is revealed: the translocon

In 1975, Blobel and Dobberstein hypothesized that secretory proteins are translocated through the ER membrane via aqueous channels formed by integral ER membrane proteins (Figure 5.2C) (Blobel and Dobberstein, 1975). However, for many years, the existence of aqueous channels and proteinaceous translocation sites was doubted by most workers in the field. A competing hypothesis was that the signal sequence directs the spontaneous insertion of the polypeptide into the membrane and the rest of the

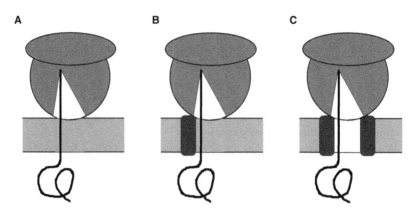

Figure 5.2   Possible nascent chain environments during translocation. Models for the co-translational translocation of a nascent chain across the ER membrane. **A**, The nascent chain may be translocated across the membrane through its hydrophobic interior in a 'spontaneous' manner, completely unassisted by ER proteins. **B**, The nascent chain may be translocated through the nonpolar core of the bilayer with the assistance of ER proteins. **C**, The nascent chain may be translocated through an aqueous pore that is formed by ER proteins in the membrane.

polypeptide is translocated through the nonpolar core of the membrane bilayer without the assistance of specific translocation channels or proteins (Figure 5.2A) (Engelman and Steitz, 1981). The mechanism of protein transfer across the ER membrane was vigorously debated for many years, but not resolved because of the absence of convincing experimental data in support of a single hypothesis. New experimental approaches ultimately provided such data and led to a general agreement in the early 1990s that proteinaceous sites on the ER membrane were involved in translocation.

While many experiments in the 1980s showed that peptides could spontaneously insert into membrane bilayers, consistent with the second model above, nascent chains bound to ribosomes appeared to act differently. Ribosome-bound nascent secretory proteins targeted to the ER membrane could be extracted with urea under conditions that would not extract membrane proteins from the membranes, suggesting that the ribosome-bound nascent chains were in an environment accessible to aqueous perturbants (Gilmore and Blobel, 1985).

The first strong indication that membrane proteins other than SR might be involved in the translocation of proteins across the ER membrane came from photocrosslinking experiments. In these studies, probes were incorporated directly into the nascent chain so that they could report on the local environment of the nascent chain during translocation. When translocation intermediates containing photoreactive moieties in the nascent chain were formed and then photolyzed, the nascent chain was crosslinked to membrane proteins throughout the duration of the translocation process, suggesting that these proteins might be involved in facilitating polypeptide movement across the ER membrane (Wiedmann et al., 1987; Krieg et al., 1989; Wiedmann et al., 1989; Thrift et al., 1991; High et al., 1991).

The presence of transmembrane aqueous channels in the ER membrane was detected by ion conductivity measurements in electrophysiological studies of the ER membrane (Simon and Blobel, 1991). This ion conductivity was observed after treatment of the membranes with puromycin, an antibiotic that causes the ribosome to release the nascent chain. This result suggested that nascent chain release was required before transmembrane ion flow could occur. Since the channels closed when ribosomes were washed off the membrane, the existence of these ion-conducting channels was both nascent chain- and ribosome-dependent.

Final proof that nascent proteins are in an aqueous milieu as they traverse the membrane bilayer was provided by a series of experiments in which water-sensitive fluorescent probes were incorporated directly into the nascent chains of translocation intermediates. The probes covalently attached to the nascent chains were then positioned at sites within the ER membrane to detect whether the nascent chain was in an aqueous or a nonpolar milieu inside the bilayer. The fluorescence lifetimes of the resulting samples revealed that all of the probes inside the membrane-bound ribosome and the membrane itself were in an aqueous environment (Crowley et al., 1994).

Independent confirmation that the nascent chain occupies an aqueous pore through the membrane during translocation was obtained using iodide ions as hydrophilic collisional quenchers of fluorescence. Iodide ions on the lumenal side of the ER membrane, but not the cytoplasmic side, were able to collide with and quench fluorescent dyes located inside the ribosome on the cytoplasmic side of the membrane. This result showed that there was an aqueous pathway for the lumenal iodide ions that completely spanned the ER membrane and extended into the ribosome (Crowley et al., 1994). The collisional quenching experiments also demonstrated that the aqueous nascent chain tunnel in the ribosome and the aqueous pore in the ER membrane are sealed off from the cytoplasm by the binding of the ribosome to the membrane (Crowley et al., 1993, 1994). Together, the fluorescence experiments showed that the nascent chain was in an aqueous environment during translocation, and that this aqueous space was contiguous only with the ER lumen, the destination of the nascent secretory proteins (Figure 5.2C).

Recently, electron microscopy of purified membrane proteins has provided an image of an aqueous translocation pore. Some of the membrane proteins that were implicated in translocation during photocrosslinking studies were purified and found to form a ring-like oligomer that resembles structures observed in electron micrographs of native ER membranes (Hanein et al., 1996). Subsequent cryo-electron microscopy reconstructions

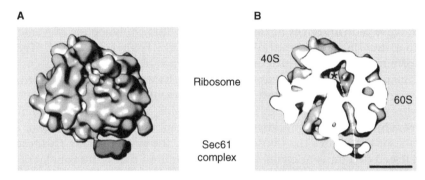

Figure 5.3 Ribosome–Translocon alignment. Three-dimensional reconstruction of cryo-electron microscopic images of a complex formed by non-translating ribosomes and purified Sec61 heterotrimers. This surface representation shows the alignment of the nascent chain tunnel in the ribosome with the pore formed by the Sec61 complex. **A**, Side view, with the ribosome shown above the Sec61 complex pore. **B**, Same orientation as **A**, but cut along a plane that cross sections the pore of the Sec61 oligomer and the ribosome tunnel. The ribosomal tunnel and its alignment with the Sec61 pore are indicated by a broken line. The star indicates the location of the peptidyltransferase center. Scale bar, 100 Å. Excerpted with permission from Beckmann, R., Bubek, D., Grassucci, R., Penczek, P., Verschoor, A., Blobel, G. and Frank, J. (1997) Alignment of conduits for the nascent polypeptide chain in the ribosome-Sec61 complex. *Science* **278**: 2123–2126. Copyright 1997 American Association for the Advancement of Science.

of the complex in combination with ribosomes demonstrated that the aqueous pore through which the nascent chain passes aligns with the ribosome tunnel (Figure 5.3) (Beckmann et al., 1997). This arrangement indicates, consistent with the fluorescence data, that the ribosome and translocation pore combine to form a single sealed conduit that runs from the site of protein synthesis in the ribosome to the destination of the secretory protein, the ER lumen (Figure 5.3B).

Thus, in co-translational translocation, the ribosome binds to specific sites on the ER membrane where aqueous pores traverse the membrane bilayer. As protein synthesis proceeds, the nascent polypeptide chain is translocated across the ER membrane and into the ER lumen through these pores. Even though the existence of these sites was controversial in 1986, Walter and Lingappa coined the term 'translocon' to identify sites in the ER membrane at which secretory protein translocation and membrane protein integration occur (Walter and Lingappa, 1986).

Summary: *Translocation across the ER membrane*

• Proteins move across the ER membrane through an aqueous pore at sites called translocons.

- Ribosome binding to the translocon prevents ion and small molecule movement through the aqueous pore.
- In co-translational translocation, the ribosome binds to the translocon to form a single aqueous conduit that stretches across the ER membrane and is sealed off from the cytoplasm.

TRANSLOCON COMPOSITION AND STRUCTURE

Even before it was firmly established that protein translocation occurs through an aqueous pore in the ER membrane, efforts were made to establish the protein composition and structure of the putative translocon. We now know, based on work from many different laboratories, that the translocon 'core' required for translocation consists of four ER membrane proteins. In addition, several other accessory proteins act on the nascent chain while it is engaged with the translocon (reviewed in Johnson and van Waes, 1999; Rapoport et al., 1996).

Primary components of the translocon

The translocon components that form the protein-conducting channel in the ER membrane were first identified by photocrosslinking, using an approach that incorporated photoreactive probes directly into the nascent chain that was being translocated. When these photoprobes were positioned within the ER membrane in translocation intermediates and then photolyzed, the nascent chain was crosslinked to specific ER membrane proteins that were adjacent to the nascent chain throughout its translocation or during its integration (Krieg et al., 1989; Wiedmann et al., 1989; Thrift et al., 1991; High et al., 1991).

The translocon proteins that formed photoadducts with the nascent chain were then purified, reconstituted into proteoliposomes, and found to be sufficient to carry out translocation and integration (Görlich et al., 1992a, 1992b; Görlich and Rapoport, 1993). One component of the mammalian translocon, the translocation-associated membrane protein or TRAM (Görlich et al., 1992a), was found to be required for the translocation or integration of most, but not all, proteins (Görlich and Rapoport, 1993; Oliver et al., 1995; Voigt et al., 1996). TRAM is a glycoprotein with 8 transmembrane (TM) segments and a mass of 36 kDa. Another component, Sec61α (Görlich et al., 1992b), was so named because it was homologous to the previously identified yeast protein Sec61p (Deshaies and Schekman, 1987; Stirling et al., 1992). Sec61α is 476 amino acids in length and spans the membrane ten times. Two smaller polypeptides, single-spanning proteins of 8 and 14 kDa, were purified as components of a heterotrimer with Sec61α and were called Sec61β and Sec61γ, respectively (Görlich and Rapoport,

1993). The heterotrimeric Sec61 complex and TRAM are considered to be the core components of the mammalian translocon since translocation and integration activity can be successfully reconstituted with only these proteins (Table 5.1).

The post-translational translocon in yeast consists of similar core components, specifically Sec61p, Sbh1p and Sss1p, the yeast homologs of Sec61α, Sec61β and Sec61γ. This heterotrimeric Sec complex associates in yeast with three integral membrane proteins (Sec62p, Sec63p and Sec71p) and a cytoplasmic peripheral membrane protein, Sec72p, to make up the heptameric post-translational translocon (Corsi and Schekman, 1996; Rapoport et al., 1996). These seven membrane proteins plus a soluble lumenal protein, Kar2p, are required to reconstitute post-translational translocation *in vitro* (Table 5.1) (Panzner et al., 1995).

The photocrosslinking and reconstitution studies identified a minimum set of ER membrane proteins that are sufficient for translocation and membrane protein integration at the ER membrane. Translocation in the reconstituted system, however, is far less efficient than in intact ER membranes, so it is probable that other ER proteins are involved in translocation *in vivo*. In addition, a number of other activities, such as signal sequence cleavage, glycosylation and protein folding, occur during co-translational translocation, so the proteins that perform these functions are located near the translocon and may also be considered legitimate structural components of the translocon machinery *in vivo*.

Accessory components of the translocon and its stoichiometry

Signal peptidase (SP), which cleaves signal sequences, and oligosaccharyltransferase (OST), which glycosylates proteins, are enzymes that act on a nascent chain as it is being translocated and hence are located close to the translocon. In fact, the 25 kDa subunit of SP has been chemically crosslinked to Sec61β and is therefore adjacent to the translocon. One of the subunits of the OST crosslinks to ribosomes, and the active site of the OST is close to the translocon. Furthermore, the OST likely remains adjacent to the translocon throughout translocation since glycosylation occurs at any glycosylation site along the entire length of a nascent protein during translocation (reviewed in Johnson and van Waes, 1999).

Calnexin is an ER membrane protein that acts as a chaperone during the folding of nascent membrane proteins and has been crosslinked to nascent chains. Calnexin therefore appears to be positioned adjacent or proximal to the translocon, but there is currently no evidence that calnexin forms a specific complex with translocon proteins. Soluble lumenal proteins such as calreticulin, protein disulfide isomerase, BiP and ERp57 also interact with the nascent chain co-translationally (Table 5.1, reviewed in Johnson and van Waes, 1999).

Table 5.1   Translocon components

Component	Location	Function
Primary components		
Co-translational		
Sec61α	ER membrane ⎱	Multiple copies (2–4) of these three proteins
Sec61β	ER membrane ⎬	associate as heterotrimers to create the
Sec61γ	ER membrane ⎰	translocon pore in the ER membrane of mammalian cells; transmembrane helices of Sec61α form most or all of the inner surface of the pore
TRAM	ER membrane	Component of the co-translational translocon in mammals; may play a role in membrane protein integration and/or signal sequence recognition
Post-translational (*yeast*)		
Sec61p	ER membrane	Yeast homolog of Sec61α
Sbh1p	ER membrane	Yeast homolog of Sec61β
Sss1p	ER membrane	Yeast homolog of Sec61γ
Sec62p	ER membrane	Thought to be involved in signal sequence interaction
Sec63p	ER membrane	Serves as binding site for BiP at the lumenal side of the translocon
Sec71p	ER membrane	Thought to be involved in signal sequence interaction
Sec72p	Peripheral ER membrane	Cytoplasmic localization; thought to be involved in signal sequence interaction
Accessory components		
Signal peptidase (SP)	ER membrane	Multi-subunit complex; cleaves signal sequences from nascent chains in the ER lumen
Oligosaccharyl-transferase (OST)	ER membrane	Multi-subunit complex (includes ribophorin I and II); glycosylates proteins in the ER lumen not far from the translocon pore
BiP (Kar2p)	ER lumen	Chaperone activity; Hsp70 family member; required to power post-translational translocation; seals the lumenal side of the co-translational translocon
Calnexin	ER membrane	Chaperone activity; binds to glycoproteins via their covalently-attached carbohydrate moieties
Calreticulin	ER lumen	Chaperone activity; binds to glycoproteins via their covalently-attached carbohydrate moieties
Protein disulfide isomerase (PDI)	ER lumen	Chaperone activity; promotes the formation of disulfide bonds
ERp57	ER lumen	Chaperone activity; thiol-dependent reductase

The minimal components of the translocon have been identified, but there is still debate about the numbers of each component protein required to assemble a single translocon. There are likely equal numbers of Sec61α, Sec61β and Sec61γ in each co-translational translocon because the mammalian Sec61 complex is purified as a heterotrimer (Görlich and Rapoport, 1993). Based on EM, the number of Sec61α proteins per translocon has been estimated to be either 3–4 (Hanein et al., 1996) or 2 (Beckmann et al., 1997). TRAM has been estimated to be present at a level of 1–2 copies per ribosome (Görlich et al., 1992a). Rough ER microsomes contain approximately equimolar numbers of ribosomes and SP, and of ribosomes and ribophorin I, a subunit of OST. It is therefore likely that one SP and one OST are associated with each translocon (reviewed in Johnson and van Waes, 1999).

Properties of the aqueous translocon pore

EM data reveal that purified Sec61 complexes form ring-like structures (Hanein et al., 1996; Beckmann et al., 1997) and hence suggest that Sec61 polypeptides line the aqueous pore. Since the nascent chain passes through the aqueous pore during translocation, photocrosslinking of the nascent chain to translocon components indicates which polypeptides line the inside of the pore. In the intact translocon, Sec61α, Sec61β and TRAM have each been crosslinked to nascent chains (reviewed in Johnson and van Waes, 1999), but Sec61α is the primary target within the pore (High et al., 1993; Mothes et al., 1994). These results strongly indicate that the walls facing the aqueous interior of the translocon pore are formed largely by the α-helices of Sec61α, a conclusion that is consistent with the unusually limited hydrophobicity of some of the Sec61α TM segments (Wilkinson et al., 1996).

Various experimental approaches have been used to estimate the size of the translocon pore. Using a fluorescence approach, the aqueous pore in the actively translocating translocon was estimated to be 40–60 Å in diameter, much larger than might be expected for an aqueous pore in a membrane that maintains a permeability barrier. In these experiments, hydrophilic quenching agents of different sizes were added to the lumenal side of intact, fully-assembled translocation intermediates with fluorescent-labeled nascent chains to determine at what point the quenchers were too large to enter the aqueous translocon pore (Hamman et al., 1997). Although it has been speculated that the large diameter may be important during membrane protein integration (Hamman et al., 1997; Johnson and van Waes, 1999) or retrotranslocation (Johnson and Haigh, 2000), the functional significance of the large pore size of the ribosome-bound translocon remains unknown.

EM has also been used to estimate the size of the translocon pore. Images of detergent-solubilized Sec61 heterotrimers reveal a ring with an outer diameter near 110 Å (Hanein et al., 1996; Beckmann et al., 1997). However, the inner diameter of the pore in these images is estimated to be near 20 Å, in stark contrast to the 40–60 Å diameter hole determined using fluorescence. Fluorescence measurements of the ribosome-free, non-translocating translocon reveal a pore with an internal diameter of 9–15 Å (Hamman et al., 1998), so the dissociation of the ribosome causes a dramatic contraction of the pore. Since the rings observed in EM studies have a similar diameter, it seems likely that the translocons observed using EM are also in the ribosome-free conformation. A recent study (Ménétret et al., 2000) of translocation intermediates by EM has yielded small pore sizes and a gap between the ribosome and translocon similar to that observed with purified ribosomes and Sec61 heterotrimers (Beckmann et al., 1997). Since the translocation intermediates examined by Ménétret et al. lacked TRAM, translocon accessory proteins, phospholipids, and other membrane components following detergent extraction of samples, it may be that an intact translocon in a functional membrane is required to maintain the large diameter and tight seal detected in the fluorescence studies.

Summary: *Translocon composition and structure*

- The heterotrimeric Sec61 complex (Sec61α, Sec61β and Sec61γ) and TRAM are the core components of the co-translational translocon and are sufficient to reconstitute translocation activity *in vitro*.
- Post-translational translocation in yeast requires the yeast homologs of the Sec61 complex (Sec61p, Sbh1p and Sss1p), as well as Sec62p, Sec63p, Sec71p, Sec72p, and Kar2p, a lumenal chaperone.
- Accessory proteins such as OST, SP and other proteins function at the translocon to modify, fold, and process the nascent chain.
- The ribosome-free translocon has a pore diameter of 9–15 Å, while a ribosome-bound translocon engaged in translocation has a much larger pore diameter of 40–60 Å.

FUNCTIONS OF THE TRANSLOCON AT THE ER MEMBRANE

It is now firmly established that translocation across the ER membrane occurs through the aqueous pores of ER translocons. Furthermore, the major components of the translocon machinery have been identified and characterized. In addition, it is becoming increasingly clear that the translocon is a multifunctional protein complex (reviewed in Rapoport et al.,

1996; Matlack et al., 1998; Johnson and van Waes, 1999). In addition to co-translational and post-translational translocation of proteins across the ER membrane, the translocon is also the site of membrane protein integration into the ER membrane. Protein folding and assembly in the ER lumen are also translocon-associated processes. And intriguing new evidence is mounting for the retrotranslocation of proteins through the translocon from the ER lumen back into the cytoplasm for the purpose of degradation by the cytoplasmic proteasome. In this section, each of the functions of the translocon and their interrelationships will be discussed (Figure 5.4).

Co-translational translocation in mammalian cells

Prior to targeting and translocation, the translocon pore at rest has a diameter of 9–15 Å and is sealed on the lumenal side by BiP (Figure 5.4A) (Hamman et al., 1998). BiP is a soluble lumenal protein, and a member of the Hsp70 family of chaperones that is involved in protein folding and assembly (Haas and Wabl, 1983; Gething and Sambrook, 1992). The mechanism by which BiP closes the ribosome-free pore is not known, nor is (are) the translocon protein(s) with which BiP interacts known. BiP may bind directly to the lumenal end of the translocon pore so as to plug the hole, or alternatively, the pore may be sealed indirectly via an interaction between BiP and the recently-identified mammalian Sec63p homolog that has been found associated with the Sec61 complex (Skowronek et al., 1999; Meyer et al., 2000; Tyedmers et al., 2000).

Following SRP-dependent targeting (see 'Targeting to the ER Membrane', above), translocation begins as the RNC complex is transferred to the translocon machinery and a tight seal is formed between the ribosome and the translocon pore (Crowley et al., 1993). Protein synthesis is free to resume after the departure of SRP, and the growing nascent chain then proceeds to move through the translocon. Interestingly, BiP is still bound to the lumenal end of the pore at this point and hence the nascent chain is in a sealed aqueous compartment that is not contiguous with either the cytoplasm or the lumen (Crowley et al., 1994; Hamman et al., 1998). The delayed release of BiP and opening of the pore must constitute a safety mechanism for maintaining the permeability barrier of the ER membrane. This evolutionary design ensures that one end of the pore is not opened before the other end is firmly sealed.

After the nascent chain reaches a length of ~70 residues (Crowley et al., 1994), the translocon pore is opened to the ER lumen. The mechanistic details of the release of BiP and the opening of the translocon pore to the lumen are not yet understood. Following the departure of BiP, the ribosome-bound, functioning translocon pore has a diameter of 40–60 Å (Hamman et al., 1997). Translation of mRNA by the ribosome proceeds, and because the tight ribosome–translocon junction blocks any movement in the direction

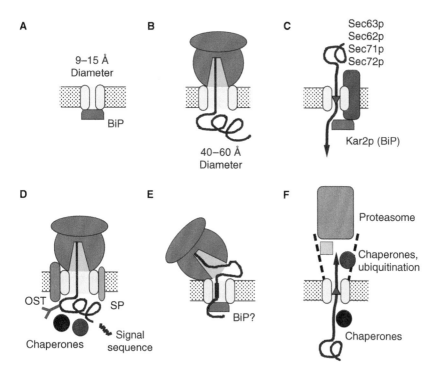

Figure 5.4   Functions of the translocon at the ER membrane. Various functional modes of the translocon (yellow) at the ER membrane (grey) are shown. The cytoplasmic and lumenal sides are above and below the membrane, respectively. For clarity, OST and SP are included only in **D**. **A**, Resting state. The ribosome-free translocon pore has an inner diameter of 9–15 Å and is sealed, either directly or indirectly, on the lumenal side by BiP. **B**, Co-translational nascent chain transloca-tion. The ribosome forms a tight seal with the translocon on its cytoplasmic side. The pore of a translocon functioning in translocation has an inner diameter of 40–60 Å. **C**, Post-translational protein translocation. Post-translational translocation requires the heterotrimeric Sec complex, as well as four additional translocon components and Kar2p (yeast BiP). The red triangle indicates the direction of protein move-ment. **D**, Nascent chain processing and folding. Co-translational translocation is shown. Chaperones act on the nascent chain to assist in protein folding, while SP acts to remove the signal sequence and OST acts to glycosylate the nascent chain at specific sites. Similar processing occurs during co-translational membrane protein integration and post-translational protein translocation. **E**, Co-translational membrane protein integration. When a TM domain is synthesized, the ribosome and translocon recognize the TM, orient it properly, and insert it laterally into the membrane bilayer. Only one of the intermediate states is depicted here. For a more detailed identification of some intermediate stages during membrane protein integration, see Figure 5.5. **F**, Retrotranslocation and protein degradation. Misfolded or misassembled proteins from the ER lumen are transported through the translocon into the cytoplasm for degradation by the proteasome. The red triangle indicates the direction of protein movement. Adapted from Johnson and Haigh (2000).

of the cytoplasm, the growing nascent chain is vectorially translocated across the membrane (Figure 5.4B). The tight ribosome–translocon seal is present throughout the process to maintain the permeability barrier during co-translational translocation.

Upon termination of translation, the release of the ribosome apparently causes the translocon to contract back to an inner diameter of 9–15 Å and to become sealed on the lumenal side via a BiP-mediated mechanism (Hamman et al., 1998). This contraction and pore closure presumably does not occur before the newly synthesized protein has moved entirely through the translocon and into the lumen. It seems likely that the nascent chain must be completely translocated before the ribosome is released from the translocon if the system is to preserve the permeability barrier of the ER membrane.

Co- and post-translational translocation in yeast

Yeasts utilize both co-translational and post-translational translocation mechanisms (reviewed in Corsi and Schekman, 1996; Rapoport et al., 1996; Kalies and Hartmann, 1998). However, all essential proteins must be capable of being translocated in a post-translational way since yeasts lacking SRP are still viable (Mutka and Walter, 2001). Co-translational translocation in yeast is thought to proceed in much the same way as described above for mammals, but in post-translational translocation, the protein is completely synthesized before it is targeted to the translocon. SRP-independent targeting of post-translational translocation substrates is poorly understood, but cytoplasmic chaperones are proposed to keep the protein in a loosely folded state and target it to the translocon. In general, proteins with more hydrophobic signal sequences are targeted in an SRP-dependent manner, while proteins with less hydrophobic signal sequences are translocated via the post-translational pathway (Ng et al., 1996).

Following targeting, the mechanism of post-translational translocation must differ substantially from that of co-translational translocation (Figure 5.4C). There is no ribosome present, and hence the ribosome cannot seal the membrane and prevent movement of the substrate polypeptide back into the cytoplasm. In the absence of the ribosome, it is unclear how the permeability barrier might be maintained across the membrane during the process, but it seems unlikely that the post-translational translocon pore expands to a diameter as large as 40–60 Å. Some energy source other than translation must also be provided to move the polypeptide across the ER membrane in the absence of a ribosomal tunnel stacked on the translocon.

Reconstitution of post-translational translocation in proteoliposomes requires Sec61p, Sbh1p and Sss1p, the yeast homologs of Sec61α, Sec61β and Sec61γ. In addition, Sec62p, Sec63p, Sec71p, Sec72p and the soluble lumenal Kar2p protein, the yeast homolog of the mammalian BiP, are required (Table 5.1) (Panzner et al., 1995). Kar2p is thought to act as a molecular ratchet to move the polypeptide through the translocon, across the membrane and into the lumen (Matlack et al., 1999). BiP is not required for co-translational translocation in the mammalian reconstituted system (Görlich and Rapoport, 1993), but Kar2p has been shown to be important for both co-translational (Brodsky, 1996; Young et al., 2001) and post-translational translocation in yeast (reviewed in Brodsky, 1996; Corsi and Schekman, 1996).

The photocrosslinking of Sec61p to post-translational translocation substrates indicates that Sec61p (Müsch et al., 1992; Sanders et al., 1992) forms the protein-conducting channel in the post-translational translocon, just as Sec61α is involved in the formation of the co-translational channel. The Sec62p, Sec63p, Sec71p and Sec72p proteins perform post-translational pathway-specific roles in targeting and transporting the secretory protein across the bilayer (Corsi and Schekman, 1996; Rapoport et al., 1996), and may also be involved in maintaining the permeability barrier.

The Sec63p integral membrane protein has a lumenal DnaJ domain that has been shown to interact with Kar2p, a member of the Hsp70 family of ATPases. This interaction stimulates Kar2p activity, and is required for post-translational translocation (reviewed in Brodsky, 1996; Johnson and van Waes, 1999). Kar2p binds to Sec63p in the yeast post-translational translocon to facilitate protein translocation, presumably by acting as the molecular motor that pulls the translocating polypeptide across the membrane (Brodsky, 1996; Matlack et al., 1999). The Sec63p protein and yeast Kar2p were also recently shown to be required for co-translational translocation in yeast *in vivo*, while the Sec62p protein was not required (Young et al., 2001). It is unclear whether Kar2p also serves to gate the yeast co- or post-translational translocons in a manner analogous to the role of BiP in co-translational translocation in mammals.

Nascent chain processing and folding

During translocation, while the nascent chain is still associated with the translocon, a variety of processing, modification, folding, and even assembly reactions can take place (Figure 5.4D) (reviewed in Johnson and van Waes, 1999). Signal sequence cleavage by SP occurs during translocation when the nascent chain reaches a length of 130–150 residues. OST covalently modifies the nascent chain by glycosylation with an oligosaccharide at the asparagine in Asn-X-Thr/Ser sequences. In addition to covalent modifications of the nascent chain, some of its folding also occurs co-translationally, assisted by resident chaperones of the ER lumen such as BiP, calnexin, calreticulin, protein disulfide isomerase, and others. Assembly of multicomponent complexes can also occur co-translationally. For example, the association of triglycerides and cholesteryl esters with nascent ApoB occurs co-translationally during the formation of lipoprotein particles (reviewed in Kang and Davis, 2000).

The regulation of these events is not well understood, but there must be some mechanism to ensure that each process occurs faithfully at the proper site and at the proper time. The extent to which these enzymes coordinate directly with the translocon machinery is also unclear. Membrane proteins such as the SP and the OST complex appear to be associated with the translocon (reviewed in Johnson and van Waes, 1999), while modification and folding enzymes may be recruited only when they are needed for a particular translocation substrate.

Membrane protein integration

In addition to secretory protein translocation across the ER, the translocon is also the site of integration of membrane proteins into the membrane bilayer (see also Chapter 6 and Johnson and van Waes, 1999). During co-translational translocation, when a transmembrane (TM) sequence of a nascent membrane protein is synthesized, its appearance in the translocon will prevent further translocation into the ER lumen. Instead, each TM sequence of a nascent membrane protein is moved laterally out of the translocon and into the lipid bilayer (Figure 5.4E). Since the translocon serves as the entry point for the integration of TM sequences into the lipid bilayer, the translocon is directly involved in the recognition, orientation, lateral movement, and insertion of TM sequences. Thus, in contrast to its relatively passive role in translocation, the translocon appears to be an active participant in the integration process.

Membrane protein substrates are identified at their amino-termini by the presence of either a typical cleavable signal sequence or an uncleaved signal sequence, termed a signal-anchor sequence, that will eventually serve as a TM sequence in the mature protein. In either instance, SRP recognizes the stretch of hydrophobic amino acid residues when it emerges from the ribosome and then targets the RNC complex to the translocon (see 'Targeting to the ER Membrane', above). Following targeting with a cleavable signal sequence, translation resumes as in secretory protein translocation until a TM sequence is synthesized and the nascent protein is recognized as a substrate for membrane protein integration (Figure 5.5).

A TM sequence in a nascent chain might be expected to be recognized only after it reaches the membrane, yet it was recently discovered that the TM sequence in a nascent chain with a cleavable signal sequence is first detected by the ribosome (Liao et al., 1997). The recognition occurs while the TM sequence is still located in the ribosomal tunnel close to the peptidyltransferase center. The appearance of the TM sequence initiates a series of events that prepares the translocon for membrane protein integration rather than translocation, including the closing of the lumenal end of

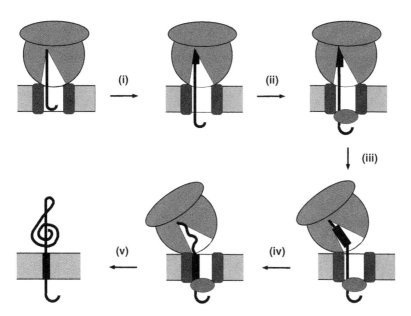

Figure 5.5   Co-translational membrane protein integration of a single-spanning membrane protein. The structural changes necessary to maintain the permeability barrier of the ER membrane during the co-translational integration of a single-spanning membrane protein into the ER membrane. (i) The RNC complex begins to synthesize a TM sequence. (ii) After the TM sequence has been synthesized and is still inside the ribosome, the translocon pore is sealed at both its cytoplasmic and lumenal sides. (iii) The ribosome–translocon junction is broken to allow nascent chain access to the cytoplasm. (iv) The TM is recognized by the translocon machinery and moves laterally into the translocon. (v) After completion of translation, the membrane protein is released into the ER membrane. Adapted from Liao et al. (1997).

the pore and the subsequent opening of the tight ribosome–translocon seal (see below and Figure 5.5) (Liao et al., 1997).

Even though a TM sequence is detected first by the ribosome, the translocon must still independently recognize the TM sequence and orient it properly for its lateral move into the bilayer. Once oriented properly, how does a TM sequence move laterally through the translocon and into the bilayer? In the case of a single-spanning membrane protein with a cleavable signal sequence, it appears that the TM sequence moves through the translocon via a multistep pathway that is regulated by protein–protein interactions until the TM sequence is released into the membrane bilayer after translation terminates (Do et al., 1996). In contrast, an uncleaved signal-anchor sequence appears to be surrounded by phospholipid molecules shortly after entering the translocon, suggesting that the signal-anchor sequence moves more quickly into the nonpolar core of the bilayer (Martoglio et al., 1995;

Mothes et al., 1997; Heinrich et al., 2000). It is therefore possible that signal sequence-bearing membrane proteins and signal-anchor membrane proteins are integrated via separate mechanisms, though this issue remains unresolved at present.

An integral membrane protein has one or more domains or loops on the cytoplasmic side of the membrane. How are domains of a nascent membrane protein introduced into the cytoplasm when such movement would disrupt the ribosome–translocon seal and therefore the permeability barrier of the ER membrane? In the case of a single-spanning membrane protein, the ribosome and translocon interact in a well-orchestrated series of events in order to localize the cytoplasmic domain correctly while maintaining the permeability barrier (Liao et al., 1997). Shortly after the ribosome recognizes the TM sequence in the ribosome tunnel, and before the ribosomal seal is broken, the lumenal side of the translocon pore is first closed by an unknown mechanism (Figure 5.5, ii). The lumenal seal during integration may be provided by BiP, by a conformational change in the translocon, or by some other mechanism. Subsequently, the ribosome–translocon seal is broken to allow movement of the cytoplasmic domain of the nascent membrane protein into the cytoplasm (Figure 5.5, iii).

In the case of multispanning membrane proteins, the situation is even more complex. Multiple TM sequences must be recognized, oriented correctly, assembled, and inserted properly into the membrane bilayer, all while the permeability barrier is maintained across the membrane. How are multiple TM sequences released laterally from the translocon into the membrane bilayer? Some data suggest that multiple TM sequences in a multispanning membrane protein are all present in the aqueous pore of the translocon until they are integrated into the bilayer at the same time (Borel and Simon, 1996). Other studies suggest that TM sequences are integrated into the bilayer in pairs (Skach and Lingappa, 1993; Lin and Addison, 1995), perhaps as helical hairpins (Engelman and Steitz, 1981). These and other issues still need to be resolved experimentally. There is much yet to learn about how nascent chains and TM sequences interact with the translocon, and about whether there are multiple modes of translocon-mediated integration.

Retrotranslocation and protein degradation

Each of the translocon functions discussed so far involves the transfer of proteins from the site of their synthesis in the cytoplasm across or into the ER membrane. However, recent data indicate that the translocon is also involved in the transport of proteins in the reverse direction, from the ER lumen to the cytoplasm, in a process termed retrotranslocation (Figure 5.4F). The translocon appears to be involved in the removal of misfolded or misassembled proteins from the ER lumen and their subsequent degradation by the cytoplasmic proteasome (Brodsky and McCracken, 1997; Kopito,

1997; Sommer and Wolf, 1997; Johnson and Haigh, 2000). By translocating proteins from the lumen into the cytoplasm for degradation by the ubiquitin–proteasome pathway, the cell is able to clear the ER lumen of non-functional proteins. Retrotranslocation appears to be part of a general quality control mechanism (see Chapter 9) and is linked to the unfolded protein response (see Chapter 8), where many lumenal chaperones and ER housekeeping enzymes are upregulated in response to excess unfolded proteins in the ER lumen (Travers et al., 2000).

The mechanistic details of translocon involvement in retrotranslocation are still very unclear. Both soluble lumenal proteins and membrane proteins are substrates for retrotranslocation and degradation by the cytoplasmic proteasome. Since these protein substrates have already had their signal sequences cleaved, there must be a separate and unique method for substrate identification and targeting during retrotranslocation. In the case of membrane protein substrates, there may be lateral reopening of the translocon for entry of the proteins back into the translocon pore. Other mechanistic issues such as energy requirements, maintenance of the permeability barrier, and coordination between the translocation and degradation machinery remain poorly understood for the process of retrotranslocation.

Summary: *Functions of the translocon at the ER membrane*

- In mammalian cells, secretory proteins are translocated across the ER membrane in a co-translational manner and a tight seal is formed between the ribosome and the translocon.
- In yeast, proteins are translocated by either a co-translational or a post-translational mechanism.
- The nascent chain is folded and processed in the lumen of the ER during and following translocation.
- The translocon is also the site of membrane protein integration into the ER membrane. The ribosome and translocon work together to recognize transmembrane segments, orient them properly, and insert them into the membrane bilayer, all while maintaining a permeability barrier at the ER membrane.
- Proteins may also move in the opposite direction through the translocon in a process called retrotranslocation, whereby misfolded or misassembled proteins can be delivered to the cytoplasm for degradation by the proteasome.

REGULATION OF TRANSLOCATION

Given the wide variety of functions performed by the translocon at the ER membrane, it is difficult to view the translocon as simply a passive pore through the membrane. Instead, the translocon is a complex and sophisticated molecular machine with various structural and functional modes (reviewed in Johnson and van Waes, 1999). The translocon is structurally and functionally coupled with the ribosome during co-translational translocation and integration. However, it is also capable of moving full-length polypeptide substrates either from the cytoplasm to the ER lumen during post-translational translocation or in the opposite direction during retro-translocation. How does the translocon participate in such a wide variety of functions? What is the driving force behind each of the polypeptide movements? Is the translocon a dynamic structure? With so many possible functions for the translocon machinery, the cell must have a method of regulating translocon activity, both at the level of individual translocon function and also on the cellular level to control the overall synthesis and movement of proteins within the cell.

Ribosome–Translocon alignment and coordination

EM images of a ribosome–Sec61 complex reveal that the RNC tunnel is co-axially aligned above the translocon pore (Figure 5.3) (Beckmann et al., 1997; Ménétret et al., 2000). This image highlights the fact that the ribosome and translocon are oriented relative to each other, presumably to effect a functional coupling during co-translational translocation and integration. Structurally, the ribosome and translocon associate in a functional membrane to form a tight seal that is impermeable to small ions during translocation (Crowley et al., 1993, 1994). The ribosome also shows a tight or high-affinity binding to the Sec61 complex (Kalies et al., 1994; Prinz et al., 2000). These data suggest that the binding surfaces on the ribosome and translocon must complement each other very well.

Together, the ribosome and translocon control nascent chain synthesis and movement at the ER membrane. This functional coupling is especially evident during the integration process, when the detection of a TM sequence inside the ribosomal tunnel leads to major structural changes at each end of the translocon pore: the lumenal end of the pore is first sealed, and then the cytoplasmic end is opened (Figure 5.5) (Liao et al., 1997). Thus, the ribosome and translocon are intimately merged during co-translational translocation and integration, and work together to accomplish each function. It is not clear what specific component–component interactions effect the long-range, transmembrane communications that mediate various stages of co-translational targeting, translocation and integration.

Dynamics

The translocon is not a simple conduit in the membrane, but instead is dynamic and can undergo changes in conformation required to perform a particular function. For example, the translocon changes its internal diameter from 9–15 Å to 40–60 Å upon binding to a ribosome and engaging in translocation (Figure 5.4A, B) (Hamman et al., 1997). The translocon also changes conformation as TM sequences pass laterally through the translocon during integration (Do et al., 1996). Conformational changes also accompany the sequential sealing of the lumenal end of the translocon pore and the opening of its cytoplasmic end during integration (Liao et al., 1997). In addition, the composition of the translocon changes as its functional state changes, as evidenced by the transient association of the translocon with BiP (Brodsky, 1996; Hamman et al., 1998) and the SRP receptor (Walter and Johnson, 1994).

In each of these cases, the regulation of the translocon conformational changes may not be well understood, but the observations demonstrate that the translocon should not be seen as a static pore. Instead, the translocon should be viewed as a multifunctional molecular assembly apparatus, capable of the structural and conformational changes necessary to accomplish many functions.

Directionality and energy requirements

During co-translational translocation, is the translocon directly involved in nascent chain movement across the bilayer? In the case of co-translational translocation, an active role for the translocon machinery does not appear to be necessary because the ribosome–translocon seal prevents nascent chain movement into the cytoplasm. Therefore, as the nascent chain grows in length, the polypeptide can only move towards the ER lumen as the newly-added amino acids enter the RNC tunnel. The tight association of the ribosome and translocon therefore appears to eliminate the need to actively push or pull the nascent chain across the ER membrane.

During post-translational translocation, the polypeptide must be actively transported from one side of the membrane to the other without the assistance of a ribosome. In this case, BiP acts as a molecular ratchet to pull the polypeptide into the ER lumen (Matlack et al., 1999). Does such a mechanism also occur during co-translational translocation? The reconstitution of co-translational translocation in the absence of soluble lumenal proteins suggests that BiP may not be required for co-translational translocation (Görlich and Rapoport, 1993). However, in another study, soluble lumenal proteins were required to complete the translocation of some nascent chains into the lumen (Nicchitta and Blobel, 1993). Furthermore, several studies suggest that a functional BiP and Sec63p are required for optimal co-translational translocation *in vitro* and *in vivo* (Brodsky, 1996; Boisramé et al., 1998; Young et al., 2001). Thus, the possible active involvement of BiP or similar proteins in facilitating co-translational nascent chain translocation still needs to be clarified.

Since the translocon has now been implicated in the retrotranslocation of polypeptides from the ER lumen back into the cytoplasm for degradation, the question of directionality becomes even more relevant. How does the translocon know in which direction to transport a particular polypeptide? And how does the translocon power the movement of the polypeptide in the appropriate direction? There may be completely separate populations of translocons for movement in each direction that differ in their accessory proteins. Alternatively, a common translocon may be used, but substrates for each process may be identified by interactions with particular chaperones or other targeting mechanisms that recruit the necessary accessory proteins for that particular function. In any case, the processes must be closely regulated in order to avoid confusion or a 'tug-of-war' scenario at the ER membrane.

Translocon assembly, modification and turnover

As discussed above in 'Functions of the Translocon at the ER Membrane', the translocon participates in a wide variety of functional operations at the ER membrane, each with its own requirements for accessory proteins (Figure 5.4). Do translocons assemble and disassemble between periods of activity? How does the cell regulate individual translocons to perform a given function? Are separate populations of translocons dedicated to each function, or are accessory proteins recruited to a translocon as needed for each task? If 'sub-populations' of translocons exist, are they interchangeable so that the cell can regulate overall translocation function?

The minimal translocon unit required to form an aqueous pore is the heterotrimeric Sec61 complex (Hanein et al., 1996; Beckmann et al., 1997). It is clear that translocons retain this minimal pore structure when not in use because after ribosomes are released by puromycin and high salt treatment, the translocons are still permeable to ions when their lumenal seals are absent (Hamman et al., 1998). Therefore translocon components remain assembled and form a pore even in the absence of a ribosome. Hence, it is unlikely that translocons completely assemble and disassemble between periods of activity. It is possible that a common minimal translocon unit may recruit accessory proteins as necessary for a given function. Alternatively, there may be distinct populations of translocons dedicated to a particular function.

In a recent study, TRAM, Sec61β and one of the SRP receptor subunits were found to be modified by phosphorylation (Gruss et al., 1999). Such modification may allow the cell to regulate translocons on an individual level. The localization of translocon proteins may also provide insight into the regulation of translocon machinery on a more global level. Translocons are found in the rough ER of the eukaryotic cell, but none of the translocon components contain the usual ER retention signal. Sec61 complexes have been observed in the ER–Golgi intermediate compartment, suggesting that there may be a retrieval mechanism at work (Greenfield and High, 1999). These results hint at the mechanisms that the cell uses to regulate translocon function.

Summary: *Regulation of translocation*

- The ribosome and the translocon are intimately coupled and coordinated during co-translational translocation and integration.
- The translocon is a dynamic structural entity that undergoes conformational and probably compositional changes to carry out its various functions.
- Protein synthesis by the ribosome may be sufficient to drive co-translational translocation, while BiP acts as a molecular ratchet to power post-translational translocation from the lumenal side of the ER membrane.
- The mechanisms that regulate translocon number, localization, and heterogeneity, if any, have not yet been identified.

CONCLUSIONS

The process of translocation across the ER membrane was once seen as a 'black box', and it was not clear whether polypeptide substrates were translocated directly through the lipid portion of the membrane bilayer, through a hydrophobic protein or lipid environment, or through an aqueous pore (Figure 5.2). It is now clear that translocation across the ER membrane occurs through a large aqueous pore formed by a minimal set of ER membrane protein components that consist of the heterotrimeric Sec61 complex and the TRAM protein. A variety of accessory proteins are associated with the translocon, either transiently or constantly. For co-translational translocation, these proteins include the SRP and SRP receptor that are required for targeting, the SP that cleaves signal sequences, the OST that glycosylates nascent chains, and chaperones that facilitate folding in the ER lumen (Table 5.1).

Although much is now known about the translocon machinery, there are still many interesting questions for future study. The structure of the translocon in terms of exact composition and stoichiometry is still largely an open issue. The recent discovery of mammalian homologs of Sec62p and Sec63p raises the possibility that there are components and aspects of translocon structure that are yet to be identified (Skowronek et al., 1999; Meyer et al., 2000; Tyedmers et al., 2000). The role of these proteins in co-translational translocation or other translocon-associated functions is unclear. There have been several EM studies that have provided a satisfying overall picture of the translocon machinery (Figure 5.3) (Beckmann et al., 1997; Hanein et al., 1996; Ménétret et al., 2000), but higher resolution structures would obviously provide a more complete picture. X-ray crystal structures of individual translocon components may be available in the near future, but obtaining atomic structures of the entire membrane-bound translocon, or even the core translocon, will be a substantial challenge. Such structural biology experiments will be difficult both to perform and to interpret in view of the dynamic nature of translocon conformation and composition, and in the absence of the membrane in the otherwise high-resolution structures.

The translocon is a sophisticated, dynamic molecular machine that performs a variety of functions at the ER membrane. Given the wide variety of functions performed by the translocon machinery (Figure 5.4), issues pertaining to translocon function and mechanism are still not very well understood and will be interesting topics for further experimentation. In terms of mechanism, membrane protein integration is perhaps the most complex of the tasks undertaken by the translocon machinery. It involves recognition of TM segments, their reorientation if necessary, and their lateral release into the bilayer. The demonstration that Kar2p acts as a molecular ratchet to power post-translational translocation has clarified one aspect of this process, but there are many other open questions, including events at the initial stages of post-translational translocation.

As a more complete picture of the various functions performed by the translocon emerges, global issues of regulation and cross-talk between functional pathways that use the translocon machinery will be of interest. These interactions have implications for overall cellular regulation because the translocon machinery is important both for protein sorting and assembly, and for protein degradation and quality control. The structure, function and regulation of translocon-mediated events at the ER membrane are the subject of much ongoing scientific investigation that will undoubtedly lead to many more important discoveries in the future.

HISTORICAL NOTES

Historical Note 1

In 1975, Günter Blobel and Bernhard Dobberstein made the bold proposal that secretory protein movement across the ER membrane occurred through an aqueous pore that was formed by membrane proteins of the ER. This hypothesis elicited alternative proposals in response, many of which argued that secretory protein translocation occurred directly through the hydrophobic interior of the ER membrane instead of through an aqueous pathway (Figure 5.2). In the absence of pertinent experimental data, the environment of the secretory protein during translocation across the ER membrane was the primary topic of vigorous and healthy debate at scientific meetings that often took on the character of a religious argument. The introduction of new experimental approaches and data ultimately resolved this issue in favor of an aqueous pore in the early 1990s, thereby clarifying our understanding of the mechanism of translocation and simultaneously reducing blood pressure levels at meetings. What is perhaps most remarkable about Blobel and Dobberstein's hypothesis is the extent to which the picture of the ribosome and pore in the original 1975 paper so closely resembles the current pictures of the ribosome and translocon (Figure 5.3).

Historical Note 2

The signal hypothesis put forth in 1971 by Günter Blobel and David Sabatini predicted that each secretory protein would be identified by an N-terminal signal sequence, and that this sequence would target the nascent secretory protein and ribosome to the ER membrane. This hypothesis ultimately proved to be astonishingly prescient and accurate, but did not anticipate one important feature of the targeting process. The simplest targeting mechanism would involve the direct binding of the signal sequence to a receptor found solely in the rough ER membrane. However, Peter Walter discovered in 1980 that a large cytoplasmic complex was required to target RNC complexes to the ER membrane. He termed this complex the SRP. This discovery was important because it allowed investigators to use *in vitro* experiments to examine the requirements for, and the mechanisms of, targeting and translocation with purified systems. Within a short time, the work of Walter and others demonstrated that the mechanism of targeting was actually much more complicated than originally envisioned. Among other things, SRP-dependent targeting was found to be a GTP-regulated process. Because of the availability of SRP, our understanding of mammalian protein translocation advanced more rapidly than that of prokaryotic or yeast translocation. This circumstance was unusual because the better-established genetic and biochemical techniques of bacterial and fungal systems typically result in a better understanding of fundamental processes in prokaryotes and yeast than in eukaryotes.

Historical Note 3

While cytoplasmic ribosomes were extensively investigated during the late 1950s, 1960s and 1970s, very little attention was paid to membrane-bound ribosomes. This lack of interest was due partly to the experimental difficulties in working with membrane-bound ribosomes, and partly to the presumption that ribosomal structure and function would be the same for cytoplasmic and membrane-bound ribosomes. Yet important features of translocation and integration were likely to be revealed if one could specifically characterize the environment and interactions of the nascent chain while it was being synthesized by a membrane-bound ribosome. Because of the presence of ribosomal proteins (and also ER membrane proteins for membrane-bound ribosomes), covalent attachment of reporter groups solely to the nascent chain in a ribosomal complex could only be achieved by incorporating probes into the nascent chain as it is being made by the ribosome. Thus, in the early 1970s Art Johnson created a Lys-tRNA analog with a probe covalently attached to the lysine side chain, and then showed that the ribosome would accept the modified Lys-tRNAs during *in vitro* protein synthesis and incorporate their probes into nascent chains.

After the discovery of SRP in 1980 and the subsequent development of *in vitro* systems to study protein sorting, modified Lys-tRNAs were used in a series of studies by different groups to incorporate photoreactive (and later fluorescent) probes into nascent secretory and membrane proteins to examine various aspects of co-translational targeting, translocation, and integration. In the late 1980s, Josef Brunner and several others extended this approach by synthesizing modified aminoacyl-tRNAs that recognized an amber stop codon during translation, thereby providing a means to position a single photoreactive probe at a specific site in a nascent chain. By allowing investigators to examine protein sorting from the point of view of the nascent chain substrate, aminoacyl-tRNA analogs have provided unique and extremely valuable insights into the mechanisms of targeting, translocation and integration.

REFERENCES

Batey, R.T., Rambo, R.P., Lucast, L., Rha, B. and Doudna, J.A. (2000) Crystal structure of the ribonucleoprotein core of the signal recognition particle. *Science* **287**: 1232–1238.

Beckmann, R., Bubeck, D., Grassucci, R. et al. (1997) Alignment of conduits for the nascent polypeptide chain in the ribosome-Sec61 complex. *Science* **278**: 2123–2126.

Blobel, G. and Dobberstein, B. (1975) Transfer of proteins across membranes. I. Presence of proteolytically processed and unprocessed nascent immunoglobulin light chains on membrane-bound ribosomes of murine myeloma. *J Cell Biol* **67**: 835–851.

Boisramé, A., Kabani, M., Beckerich, J.-M., Hartmann, E. and Gaillardin, C. (1998) Interaction of Kar2p and Sls1p is required for efficient co-translational translocation of secreted proteins in the yeast *Yarrowia lipolytica*. *J Biol Chem* **273**: 30903–30908.

Borel, A.C. and Simon, S.M. (1996) Biogenesis of polytopic membrane proteins: membrane segments assemble within translocation channels prior to membrane integration. *Cell* **85**: 379–389.

Brodsky, J.L. (1996) Post-translational protein translocation: not all hsc70s are created equal. *TiBS* **21**: 122–126.

Brodsky, J.L. and McCracken, A.A. (1997) ER-associated and proteasome-mediated protein degradation: how two topologically restricted events came together. *Trends Cell Biol* **7**: 151–156.

Corsi, A.K. and Schekman, R. (1996) Mechanism of polypeptide translocation into the endoplasmic reticulum. *J Biol Chem* **271**: 30299–30302.

Crowley, K.S., Reinhart, G.D. and Johnson, A.E. (1993) The signal sequence moves through a ribosomal tunnel into a noncytoplasmic aqueous environment at the ER membrane early in translocation. *Cell* **73**: 1101–1115.

Crowley, K.S., Liao, S., Worrell, V.E., Reinhart, G.D. and Johnson, A.E. (1994) Secretory proteins move through the endoplasmic reticulum membrane via an aqueous, gated pore. *Cell* **78**: 461–471.

Deshaies, R.J. and Schekman, R. (1987) A yeast mutant defective at an early stage in import of secretory protein precursors into the endoplasmic reticulum. *J Cell Biol* **105**: 633–645.

Do, H., Falcone, D., Lin, J., Andrews, D.W. and Johnson, A.E. (1996) The cotranslational integration of membrane proteins into the phospholipid bilayer is a multi-step process. *Cell* **85**: 369–378.

Engelman, D.M. and Steitz, T.A. (1981) The spontaneous insertion of proteins into and across membranes: the helical hairpin hypothesis. *Cell* **23**: 411–422.

Fulga, T.A., Sinning, I., Dobberstein, B. and Pool, M.R. (2001) SRβ coordinates signal sequence release from SRP with ribosome binding to the translocon. *EMBO J* **20**: 2338–2347.

Gething, M.J. and Sambrook, J. (1992) Protein folding in the cell. *Nature* **355**: 33–45.

Gilmore, R. and Blobel, G. (1985) Translocation of secretory proteins across the microsomal membrane occurs through an environment accessible to aqueous perturbants. *Cell* **42**: 497–505.

Görlich, D. and Rapoport, T.A. (1993) Protein translocation into proteoliposomes reconstituted from purified components of the endoplasmic reticulum membrane. *Cell* **75**: 615–630.

Görlich, D., Hartmann, E., Prehn, S. and Rapoport, T.A. (1992a) A protein of the endoplasmic reticulum involved early in polypeptide translocation. *Nature* **357**: 47–52.

Görlich, D., Prehn, S., Hartmann, E., Kalies, K.-U. and Rapoport, T.A. (1992b) A mammalian homolog of SEC61p and SECYp is associated with ribosomes and nascent polypeptides during translocation. *Cell* **71**: 489–503.

Greenfield, J.J.A. and High, S. (1999) The sec61 complex is located in both the ER and the ER-Golgi intermediate compartment. *J Cell Sci* **112**: 1477–1486.

Gruss, O.J., Feick, P., Frank, R. and Dobberstein, B. (1999) Phosphorylation of components of the ER translocation site. *Eur J Biochem* **260**: 785–793.

Haas, I.G. and Wabl, M. (1983) Immunoglobulin heavy chain binding protein. *Nature* **306**: 387–389.

Hamman, B.D., Chen, J.-C., Johnson, E.E. and Johnson, A.E. (1997) The aqueous pore through the translocon has a diameter of 40–60 Å during cotranslational protein translocation at the ER membrane. *Cell* **89**: 535–544.

Hamman, B.D., Hendershot, L.M. and Johnson, A.E. (1998) BiP maintains the permeability barrier of the ER membrane by sealing the lumenal end of the translocon pore before and early in translocation. *Cell* **92**: 747–758.

Hanein, D., Matlack, K.E.S., Jungnickel, B. et al. (1996) Oligomeric rings of the Sec61p complex induced by ligands required for protein translocation. *Cell* **87**: 721–732.

Heinrich, S.H., Mothes, W., Brunner, J. and Rapoport, T.A. (2000) The Sec61p complex mediates the integration of a membrane protein by allowing lipid partitioning of the transmembrane domain. *Cell* **102**: 233–244.

High, S., Görlich, D., Wiedmann, M., Rapoport, T.A. and Dobberstein, B. (1991) The identification of proteins in the proximity of signal-anchor sequences during their targeting to and insertion into the membrane of the ER. *J Cell Biol* **113**: 35–44.

High, S., Martoglio, B., Görlich, D. et al. (1993) Site-specific photocross-linking reveals that Sec61p and TRAM contact different regions of a membrane-inserted signal sequence. *J Biol Chem* **268**: 26745–26751.

Johnson, A.E. and Haigh, N.G. (2000) The ER translocon and retrotranslocation: is the shift into reverse manual or automatic. *Cell* **102**: 709–712.

Johnson, A.E. and van Waes, M.A. (1999) The translocon: a dynamic gateway at the ER membrane. *Annu Rev Cell Dev Biol* **15**: 799–842.

Kalies, K.-U. and Hartmann, E. (1998) Protein translocation into the endoplasmic reticulum (ER). Two similar routes with different modes. *Eur J Biochem* **254**: 1–5.

Kalies, K.-U., Görlich, D. and Rapoport, T.A. (1994) Binding of ribosomes to the rough endoplasmic reticulum mediated by the Sec61p-complex. *J Cell Biol* **126**: 925–934.

Kang, S. and Davis, R.A. (2000) Cholesterol and hepatic lipoprotein assembly and secretion. *Biochim Biophys Acta* **1529**: 223–230.

Keenan, R.J., Freymann, D.M., Walter, P. and Stroud, R.M. (1998) Crystal structure of the signal sequence binding subunit of the signal recognition particle. *Cell* **94**: 181–191.

Kopito, R.R. (1997) ER quality control: the cytoplasmic connection. *Cell* **88**: 427–430.

Krieg, U.C., Johnson, A.E. and Walter, P. (1989) Protein translocation across the endoplasmic reticulum membrane: identification by photocross-linking of a 39 kD integral membrane glycoprotein as part of a putative translocation tunnel. *J Cell Biol* **109**: 2033–2043.

Liao, S., Lin, J., Do, H. and Johnson, A.E. (1997) Both lumenal and cytosolic gating of the aqueous ER translocon pore is regulated from inside the ribosome during membrane protein integration. *Cell* **90**: 31–41.

Lin, J. and Addison, R. (1995) A novel integration signal that is composed of two transmembrane segments is required to integrate the neurospora plasma membrane H$^+$-ATPase into microsomes. *J Biol Chem* **270**: 6935–6941.

Martoglio, B., Hofmann, M.W., Brunner, J. and Dobberstein, B. (1995) The protein-conducting channel in the membrane of the endoplasmic reticulum is open laterally toward the lipid bilayer. *Cell* **81**: 207–214.

Matlack, K.E.S., Mothes, W. and Rapoport, T.A. (1998) Protein translocation: tunnel vision. *Cell* **92**: 381–390.

Matlack, K.E.S., Misselwitz, B., Plath, K. and Rapoport, T.A. (1999) BiP acts as a molecular ratchet during posttranslational transport of prepro-α factor across the ER membrane. *Cell* **97**: 553–564.

Ménétret, J.-F., Neuhof, A., Morgan, D.G. et al. (2000) The structure of ribosome-channel complexes engaged in protein translocation. *Mol Cell* **6**: 1219–1232.

Meyer, H.-A., Grau, H., Kraft, R. et al. (2000) Mammalian Sec61 is associated with Sec62 and Sec63. *J Biol Chem* **275**: 14550–14557.

Millman, J.S. and Andrews, D.W. (1997) Switching the model: a concerted mechanism for GTPases in protein targeting. *Cell* **89**: 673–676.

Mothes, W., Prehn, S. and Rapoport, T.A. (1994) Systematic probing of the environment of a translocating secretory protein during translocation through the ER membrane. *EMBO J* **13**: 3973–3982.

Mothes, W., Heinrich, S.U., Graf, R. et al. (1997) Molecular mechanism of membrane protein integration into the endoplasmic reticulum. *Cell* **89**: 523–533.

Müsch, A., Wiedmann, M. and Rapoport, T.A. (1992) Yeast Sec proteins interact with polypeptides traversing the endoplasmic reticulum membrane. *Cell* **69**: 343–352.

Mutka, S.C. and Walter, P. (2001) Multifaceted physiological response allows yeast to adapt to the loss of the signal recognition particle-dependent protein-targeting pathway. *Mol Biol Cell* **12**: 577–588.

Ng, D.T.W., Brown, J.D. and Walter, P. (1996) Signal sequences specify the targeting route to the endoplasmic reticulum membrane. *J Cell Biol* **134**: 269–278.

Nicchitta, C.V. and Blobel, G. (1993) Lumenal proteins of the mammalian endoplasmic reticulum are required to complete protein translocation. *Cell* **73**: 989–998.

Oliver, J., Jungnickel, B., Görlich, D., Rapoport, T. and High, S. (1995) The Sec61 complex is essential for the insertion of proteins into the membrane of the endoplasmic reticulum. *FEBS Lett* **362**: 126–130.

Palade, G. (1975) Intracellular aspects of the process of protein synthesis. *Science* **189**: 347–358.

Panzner, S., Dreier, L., Hartmann, E., Kostka, S. and Rapoport, T.A. (1995) Posttranslational protein transport in yeast reconstituted with a purified complex of Sec proteins and Kar2p. *Cell* **81**: 561–570.

Powers, T. and Walter, P. (1995) Reciprocal stimulation of GTP hydrolysis by two directly interacting GTPases. *Science* **269**: 1422–1424.

Prinz, A., Behrens, C., Rapoport, T.A., Hartmann, E. and Kalies, K.-U. (2000) Evolutionarily conserved binding of ribosomes to the translocation channel via the large ribosomal RNA. *EMBO J* **19**: 1900–1906.

Rapiejko, P.J. and Gilmore, R. (1997) Empty site forms of the SRP54 and SRα GTPases mediate targeting of ribosome–nascent chain complexes to the endoplasmic reticulum. *Cell* **89**: 703–713.

Rapoport, T.A., Jungnickel, B. and Kutay, U. (1996) Protein transport across the eukaryotic endoplasmic reticulum and bacterial inner membranes. *Annu Rev Biochem* **65**: 271–303.

Sanders, S.L. and Schekman, R. (1992) Polypeptide translocation across the endoplasmic reticulum membrane. *J Biol Chem* **267**: 13791–13794.

Sanders, S.L., Whitfield, K.M., Vogel, J.P., Rose, M.D. and Schekman, R.W. (1992) Sec61p and BiP directly facilitate polypeptide translocation into the ER. *Cell* **69**: 353–365.

Simon, S.M. and Blobel, G. (1991) A protein-conducting channel in the endoplasmic reticulum. *Cell* **65**: 371–380.

Skach, W.R. and Lingappa, V.R. (1993) Amino-terminal assembly of human P-glycoprotein at the endoplasmic reticulum is directed by cooperative actions of two internal sequences. *J Biol Chem* **268**: 23552–23561.

Skowronek, M.K., Rotter, M. and Haas, I.G. (1999) Molecular characterization of a novel mammalian DnaJ-like Sec63p homolog. *Biol Chem* **380**: 1133–1138.

Sommer, T. and Wolf, D.H. (1997) Endoplasmic reticulum degradation: reverse protein flow of no return. *FASEB J* **11**: 1227–1233.

Stirling, C.J., Rothblatt, J., Hosobuchi, M., Deshaies, R. and Schekman, R. (1992) Protein translocation mutants defective in the insertion of integral membrane proteins into the endoplasmic reticulum. *Mol Biol Cell* **3**: 129–142.

Stroud, R.M. and Walter, P. (1999) Signal sequence recognition and protein targeting. *Curr Opin Struct Biol* **9**: 754–759.

Thrift, R.N., Andrews, D.W., Walter, P. and Johnson, A.E. (1991) A nascent membrane protein is located adjacent to ER membrane proteins throughout its integration and translation. *J Cell Biol* **112**: 809–821.

Travers, K.J., Patil, C.K., Wodicka, L. et al. (2000) Functional and genomic analyses reveal an essential coordination between the unfolded protein response and ER-associated degradation. *Cell* **101**: 249–258.

Tyedmers, J., Lerner, M., Bies, C. et al. (2000) Homologs of the yeast Sec complex subunits Sec62p and Sec63p are abundant proteins in dog pancreas microsomes. *Proc Natl Acad Sci USA* **97**: 7214–7219.

Voigt, S., Jungnickel, B., Hartmann, E. and Rapoport, T.A. (1996) Signal sequence-dependent function of the TRAM protein during early phases of protein transport across the endoplasmic reticulum membrane. *J Cell Biol* **134**: 25–35.

Walter, P. and Blobel, G. (1980) Purification of a membrane-associated protein complex required for protein translocation across the endoplasmic reticulum. *Proc Natl Acad Sci USA* **77**: 7112–7116.

Walter, P. and Blobel, G. (1981) Translocation of proteins across the endoplasmic reticulum. III. Signal recognition protein (SRP) causes signal sequence-dependent and site-specific arrest of chain elongation that is released by microsomal membranes. *J Cell Biol* **91**: 557–561.

Walter, P. and Blobel, G. (1982) Signal recognition particle contains a 7S RNA essential for protein translocation across the endoplasmic reticulum. *Nature* **299**: 691–698.

Walter, P. and Johnson, A.E. (1994) Signal sequence recognition and protein targeting to the endoplasmic reticulum membrane. *Annu Rev Cell Biol* **10**: 87–119.

Walter, P. and Lingappa, V.R. (1986) Mechanism of protein translocation across the endoplasmic reticulum membrane. *Annu Rev Cell Biol* **2**: 499–516.

Wiedmann, M., Kurzchalia, T.V., Hartmann, E. and Rappoport, T.A. (1987) A signal sequence receptor in the endoplasmic reticulum membrane. *Nature* **328**: 830–833.

Wiedmann, M., Görlich, D., Hartmann, E., Kurzchalia, T.V. and Rapoport, T.A. (1989) Photocrosslinking demonstrates proximity of a 34 kDa membrane protein to different portions of preprolactin during translocation through the endoplasmic reticulum. *FEBS Lett* **257**: 263–268.

Wilkinson, B.M., Critchley, A.J. and Stirling, C.J. (1996) Determination of the transmembrane topology of yeast Sec61p, an essential component of the endoplasmic reticulum translocation complex. *J Biol Chem* **271**: 25590–25597.

Young, B.P., Craven, R.A., Reid, P.J., Willer, M. and Stirling, C.J. (2001) Sec63p and Kar2p are required for the translocation of SRP-dependent precursors into the yeast endoplasmic reticulum *in vivo*. *EMBO J* **20**: 262–271.

6

MEMBRANE PROTEIN INSERTION INTO BACTERIAL MEMBRANES AND THE ENDOPLASMIC RETICULUM

ANDREAS KUHN AND MARTIN SPIESS

MEMBRANE PROTEINS

Membranes separate two aqueous compartments by a thin, two-dimensional, lipid phase. Membrane proteins generally span this lipid phase and therefore need to accommodate to the hydrophilic milieu on both sides of the membrane as well as to the hydrophobic environment in the core of the lipid bilayer. The structure of the membrane-embedded portions consists either of transmembrane α-helices, often assembled into helix bundles, or of antiparallel β-sheets forming barrel-shaped pores. β-barrel structures are found in the outer membrane proteins of bacteria, mitochondria and chloroplasts. Proteins of all other membranes are of the helical type. Considering the many functions these proteins perform, the transmembrane helix is an astonishingly versatile structural element. To span the hydrocarbon core of the membrane of ~3 nm requires an α-helix of ~20 uncharged, predominantly apolar residues. Therefore, most membrane-spanning helical segments can be identified by a hydropathy analysis that plots the average hydrophobicity of amino acid side chains for a window of residues along the sequence (Kyte and Doolittle, 1982; Eisenberg et al., 1984). Based on the known structures of membrane proteins, most transmembrane helices are about 25 residues in length (Bowie, 1997; Wallin et al., 1997). The length

Protein Targeting, Transport & Translocation
ISBN 0-12-200731-X

of a transmembrane helix may even correlate with the thickness of its membrane. Single-spanning membrane proteins of the endoplasmic reticulum (ER) and Golgi generally have shorter transmembrane domains than plasma membrane proteins (Bretscher and Munro, 1993). Since cholesterol content and thus the thickness of the lipid bilayer also increases along the secretory pathway, this might reflect a role of the transmembrane segments and lipids in protein sorting. In some cases, manipulating the length of the transmembrane domain of Golgi or plasma membrane proteins indeed affected protein localization (Munro, 1995).

In multi-spanning membrane proteins, the transmembrane helices are tightly bundled to compact, globular structures from which lipids are excluded (Figure 6.1). Contacting helices are slightly tilted to each other

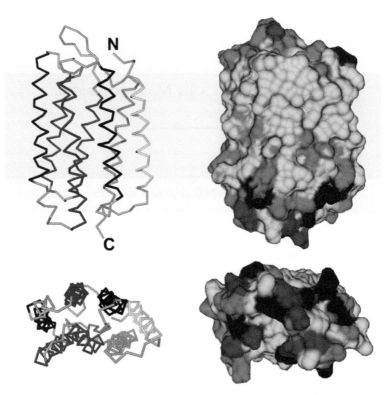

Figure 6.1 Typical membrane proteins form compact helical bundles. As an example, the backbone structure and a space-filling representation of bacteriorhodopsin is shown seen from the side (i.e. from within the membrane separating the periplasm above and the cytoplasm below) and from the cytoplasmic face. The backbones of the seven transmembrane helices are shown in different colors. The surface of basic, acidic, polar and apolar amino acids are colored in blue, red, green and yellow, respectively. (Courtesy of Markus Meier, Biozentrum, University of Basel.)

and with respect to the bilayer normal. Although there are a few examples where two helices are connected by a charge pair (Sahin-Toth et al., 1992), the helix–helix contacts are mainly based on hydrophobic interactions, not unlike that found in coiled-coil structures. The transmembrane segments of multi-spanning proteins are therefore hydrophobic not only towards the lipid phase of the bilayer, but also towards the polypeptide core of the protein. This is even the case for transport proteins, where only a narrow path for the substrate molecules is lined by individual polar side chains and hydrogen bonding groups of the peptide backbone (Kolbe et al., 2000).

For most amino acids, there is no obvious statistical preference for a particular position within transmembrane sequences. Exceptions are aromatic amino acids, particularly tyrosine and tryptophan, which are preferentially found at the ends of transmembrane segments near the membrane interface, the phospholipid head group regions of the bilayer (Wallin et al., 1997; Ridder et al., 2000). Prolines are rarely found in the transmembrane segments of single-spanning proteins, but are more frequently found in those of multi-spanning proteins (Deber et al., 1986). A proline located near the center of a transmembrane helix induces a slight kink in the α-helix.

Summary: *Membrane proteins*

- Most membrane proteins span the lipid bilayer with α-helical segments of ~25 hydrophobic amino acids.
- Many membrane proteins fold into globular structures by tight bundling of the transmembrane helices excluding lipid molecules.
- Like soluble proteins, membrane proteins are held together mainly by hydrophobic forces.

FROM THE CYTOSOL TO THE MEMBRANE

Upon synthesis on cytosolic ribosomes, secretory and membrane proteins are specifically targeted to the translocation/insertion machinery in the correct membrane by *signal sequences* (see Chapter 3). The main feature of a signal sequence is a hydrophobic stretch of uncharged, mainly apolar residues. In the classical case, as in most secretory and many eukaryotic membrane proteins, the signal sequence constitutes the most N-terminal segment of the polypeptide (in bacteria often called the *leader sequence*). It is required for the translocation of the C-terminal sequence across the membrane and is cleaved off by signal peptidase (or *leader peptidase*) on the *trans* side of the membrane. Cleaved signal sequences are generally short with a hydrophobic core of at least ~7, typically 10–15 residues.

Most bacterial inner membrane and many eukaryotic membrane proteins (particularly the multi-spanning ones) do not contain a cleaved signal sequence. For example, only about one-sixth of the seven-transmembrane receptor family members have an N-terminal, cleaved signal sequence. Interestingly, it has been shown for many hydrophobic transmembrane sequences that they are able to function as a signal for targeting to the membrane and for translocation of the flanking sequence (Friedlander and Blobel, 1985; Spiess and Lodish, 1986; Zerial et al., 1986). With ~18–25 apolar residues, such *uncleaved signals* are longer than cleavable signals and anchor the protein in the lipid bilayer. They are indistinguishable from other transmembrane segments that are not involved in targeting to the membrane.

In the mammalian system, both cleaved and uncleaved amino-terminal signals are recognized in the context of the nascent chain–ribosome complex by signal recognition particle (SRP) and are targeted co-translationally to the translocation machinery by interaction with the SRP receptor (see Chapter 5; and Historical Note 1). SRP binding slows or arrests translation (Wolin and Walter, 1989) preventing any problems caused by folding, misfolding or aggregation of the protein before integration or translocation has started.

In yeast, subsets of secretory proteins are targeted in a post-translational, SRP-independent manner, or are capable of using either pathway (Ng et al., 1996). SRP-independent membrane proteins are not known (except for a class of C-terminally anchored proteins mentioned below). The situation in bacteria is similar. Exported proteins contain a cleavable leader sequence and are generally not recognized by the bacterial SRP, but associate with chaperones like SecB, trigger factor, and GroEL/GroES (see Chapter 4). Membrane proteins, in contrast, are recognized by the bacterial homolog of SRP as demonstrated by chemical crosslinking studies (Valent et al., 1998). Why do signal sequences of eukaryotes interact with SRP, but not prokaryotic leader sequences? A plausible explanation is that bacterial leader sequences are less hydrophobic than cleaved signal sequences of eukaryotes or transmembrane segments (de Gier et al., 1998).

Summary: *From the cytosol to the membrane*

- Most proteins of the ER are synthesized with a signal sequence, whereas bacterial inner membrane proteins generally do not contain a leader sequence.
- Newly synthesized bacterial membrane proteins and proteins of the ER are recognized by the components of the signal regonition particle (SRP).
- Bacterial preproteins do not interact with SRP and are post-translationally translocated.

DIRECT VERSUS TRANSLOCASE-MEDIATED MEMBRANE INSERTION

Initially, it was a biochemically attractive hypothesis that predominantly hydrophobic membrane proteins might insert spontaneously into a lipid bilayer, driven by a conformational change from a somewhat water-soluble to a membrane-embedded conformation (see Historical Note 2). In experiments performed with liposomes, some hydrophobic proteins were indeed found to insert spontaneously into the membrane (Soekarjo et al., 1996). The driving force for this process is the hydrophobic effect. To achieve a transmembrane configuration, the protein entering the membrane from one face has to translocate one of its two hydrophilic ends. Remarkably, the proteins that successfully inserted *in vitro* did not insert into the bacterial membrane by themselves *in vivo*: they required at least one additional membrane protein, YidC (Samuelson et al., 2000). Why membrane insertion *in vivo* differs from that *in vitro* is most likely due to the high curvature of the liposome membrane and the increased distance between the lipid molecules. This may allow the protein to translocate a hydrophilic flanking sequence more easily.

It is an open question whether any natural proteins (besides pore-forming toxins) insert spontaneously into the cellular bilayer without the help of a proteinaceous machinery. To date only some mutant proteins of the single-spanning Pf3 coat protein with an extended hydrophobic region are known to translocate across the *Escherichia coli* membrane without a proteinaceous machinery. The protein inserts without a defined orientation: equal amounts of the 3L-mutant protein were found in N-out or C-out orientation (Kiefer and Kuhn, 1999). It seems that cells assure a defined orientation of a membrane protein by enzymatically controlling the insertion process.

The translocation machineries of prokaryotes and eukaryotes have been identified in recent years (see Chapters 4 and 5). Their sequences revealed evolutionary conservation of the major components from archaea to human (including mitochondria and chloroplasts; Dalbey and Kuhn, 2000). Most of the membrane proteins analyzed so far require the Sec translocase SecYEG/SecA in bacteria and Sec61αβγ in eukaryotes, which are also used for the export of secretory proteins (Figure 6.2A). One example of a substrate is the *E. coli* membrane protein leader peptidase. Translocation of the 250-amino acid C-terminal tail depends on SecA and SecYEG (Wolfe et al., 1985). SecA is the energy providing component that pushes the large, hydrophilic portions of the polypeptide through the translocon (Economou et al., 1995). Other *E. coli* membrane proteins lacking large soluble domains, however, only require SecYE, but not SecA and SecG (Kuhn, 1988; Andersson and von Heijne, 1994; Koch and Muller, 2000). This also supports the concept that SecA and SecG work together in the ATP-driven process.

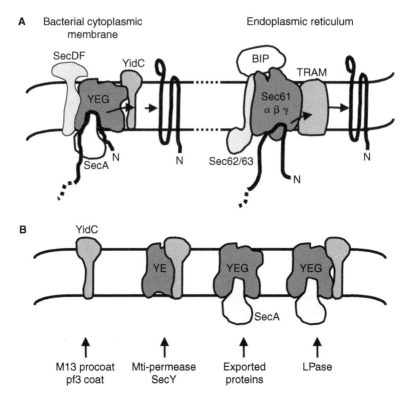

Figure 6.2 Membrane translocation and insertion devices in bacteria and in the endoplasmic reticulum. **A,** Components of the bacterial membrane translocase SecYEG with SecA, SecDFYajC and YidC and of the ER translocase Sec61αβγ with Sec62Sec63, BIP and TRAM are shown. Membrane inserting proteins are first contacting SecYEG or Sec61 and, in a later stage, YidC or TRAM, respectively. **B,** Distinct modules are used for the translocation or membrane insertion of pre-proteins and various membrane proteins, respectively.

The energetics of translocation without SecA is unknown. Possibly, the ribosome pushes the polypeptide chain (Herskovits and Bibi, 2000). This would require that translocation is coupled to ongoing translation, which has not been observed in *E. coli* so far.

The recently identified YidC protein has been found to catalyze the insertion of proteins into the membrane of *E. coli* (Samuelson et al., 2000). Photoactivatable analogs in the transmembrane region of leader peptidase nascent chains were crosslinked to YidC suggesting a direct role for YidC in this process. Presumably, YidC recognizes non-translocated hydrophobic regions and catalyzes their insertion to a transmembrane topology. Inter-estingly, YidC has been found to support the membrane insertion of Sec-independent as well as Sec-dependent proteins. For Sec-dependent proteins it was shown that a depletion of YidC causes jamming of the Sec translocase (Samuelson et al., 2001). This supports the idea that YidC can work in close association with SecYE.

The available data suggest that the Sec components might assemble dynamically into different subcomplexes (Figure 6.2B). The individual modules may be devoted to specific functions, whereby SecYE resembles the basic translocase unit. The addition of SecA/G is required to export proteins into the periplasm. SecDF/YajC is added to fold complex domains of translocated proteins in the periplasm for translocation of secretory proteins (Duong and Wickner, 1997). Finally, YidC is devoted to the integration of membrane proteins either in a complex with SecYE (Scotti et al., 2000), or as an independent enzyme (Samuelson et al., 2001). Similar to bacterial translocases, subcomplexes of the Sec61 ER translocation machinery may exist, e.g. with or without the Sec62/63/71/72 complex in post- or co-translational translocation (Panzner et al., 1995). Taken together, it seems that different substrates use specific sets of modular devices for membrane insertion and/or translocation.

Summary: *Direct versus translocase-mediated membrane insertion*

- *In vivo* insertion of membrane proteins requires a proteinaceous machinery.
- The major constituents of the translocation and insertion machineries are evolutionarily conserved.
- Membrane proteins are inserted by specific functional modules of a machinery that may also translocate secretory proteins.

MEMBRANE ANCHORING AND ORIENTING TRANSMEMBRANE HELICES

Single-spanning membrane proteins

For single-spanning membrane proteins, two final topologies exist: N_{cyt}/C_{exo} (cytoplasmic N- and exoplasmic C-terminus) or N_{exo}/C_{cyt}. However, if the topogenic determinants involved are taken into consideration, three types of single-spanning membrane proteins can be distinguished (Table 6.1). Type I membrane proteins are initially targeted to the ER by a cleavable, N-terminal signal sequence and are then anchored in the membrane by a subsequent *stop-transfer sequence*, a stretch of nonpolar residues that halts the further translocation of the polypeptide and acts as a transmembrane anchor. It can be as short as 11 residues, if the sequence is sufficiently hydrophobic (Davis et al., 1985; Chen and Kendall, 1995).

Table 6.1   Topogenic determinants of single-spanning membrane proteins

Signal type:	C-Terminus-translocating signals			N-Terminus-translocating signals
Topogenic determinants:	C-terminal signal	Cleaved signal + stop transfer sequence	Signal-anchor	Reverse signal-anchor
Machinery:	Unknown	SecYEG or YidC/Sec61 + signal peptidase	SecYEG/Sec61	SecYEG or YidC/Sec61
Final topology: exoplasmic / cytoplasmic				
Examples Bacteria:	Light harvesting proteins α and β	M13 coat protein; CSGG precursor	Penicillin-binding protein Ib; proline isomerase D	Bacteriaphage Pf3; DnaJ-like protein DnaJ
Eukaryotes:	Synaptobrevin; cytochrome b_5	Glycophorin; LDL receptor	Transferrin receptor; galactosyltransferase	Synaptotagmin; neuregulin; cytochromes P-450; Apy1p

In type II proteins, a *signal-anchor sequence* is responsible for both insertion and anchoring. It differs from cleaved signals in that it (1) can be positioned internally within the polypeptide chain, (2) lacks a signal-cleavage site, and (3) functions as a transmembrane anchor. However, like cleaved signals it initiates the translocation of its C-terminal end across the membrane. The opposite is the case for *reverse signal-anchors* of type III proteins: they translocate their N-terminal tail.

In eukaryotes, there is an additional class of proteins composed of an N-terminal cytoplasmic domain anchored in the membrane by a very C-terminal signal sequence. Examples are cytochrome b_5 and the SNARE proteins such as synaptobrevin. Targeting and insertion was shown in yeast to be independent of SRP and Sec61p and to use an as yet unknown ATP-requiring mechanism (Kutay et al., 1993, 1995; Whitley et al., 1996).

Flanking charges
In bacteria there is a strong correlation between positively charged amino acids flanking transmembrane segments and the membrane topology (von Heijne, 1986). Basic amino acids were four times more likely to be within cytoplasmic than periplasmic loops. An opposite, but weaker correlation was found for acidic residues. A similar charge bias was also observed for membrane proteins in the ER, chloroplasts and mitochondria, suggesting that the 'positive-inside' rule is quite general in nature (von Heijne, 1992; Wallin and von Heijne, 1998). The idea that positively charged residues were determinants of the membrane protein topology in bacteria was supported by site-directed mutagenesis (Laws and Dalbey, 1989; von Heijne, 1989; Nilsson and von Heijne, 1990). However, basic residues only revealed topological effects within a short distance of the hydrophobic sequence. This is explained by electrostatic binding to the acidic phospholipid head groups, which hinders translocation of the positively charged region (Gallusser and Kuhn, 1990). In a recent study, the topology of leader peptidase was affected by manipulating the amount of acidic phospholipids within *E. coli* (van Klompenburg et al., 1997). An interesting investigation was recently reported using membrane proteins of *Sulfolobus acidocaldarius*. This archaeal organism lives in a very acidic environment (pH 0.5–2.5) and has a positive-inside electrical potential maintaining an internal neutral pH and a $\Delta\mu$ of -200 eV (Moll and Schäfer, 1988; Michels and Bakker, 1985). Analysis of SecY and subunit I of cytochrome c oxidase showed that the charge bias in these proteins is also positive inside (van de Vossenberg et al., 1998). This implies that the 'positive-inside' rule is mainly due to electrostatic interactions with the phospholipid head groups.

Membrane orientation of a bacterial type III protein has been extensively analyzed regarding the charged residues flanking the hydrophobic region. The Sec-independent Pf3 coat protein has two negatively charged residues in the N-terminal region and two positively charged residues in the C-terminal tail. Reversal of all the charges by amino acid substitutions completely inverted the orientation of the protein. Surprisingly, if only negatively charged residues were present in both flanking regions, both orientations were found, whereas no membrane insertion was observed, if both tails were positively charged or uncharged. This suggests that the negatively charged residues determine the orientation of this protein (Kiefer and Kuhn, 1999). Most likely, the charge distribution matches with the transmembrane electrochemical potential, which favors the translocation of negatively charged residues into the periplasm.

For proteins inserted into the ER, the charge rule holds less strictly than for bacterial proteins. In addition, the charge difference between the two flanking sequences of an uncleaved signal's hydrophobic core, rather than the positive charge per se, correlates with the orientation: the cytoplasmic sequence generally carries a more positive charge than the exoplasmic one (Hartmann et al., 1989), although there are numerous exceptions. Experimentally, the type III protein cytochrome P-450 could be converted to a type II protein by insertion of positively charged residues into its short N-terminal domain (Monier et al., 1988; Szczesna-Skorupa et al., 1988). Mutation of flanking charges in the asialoglycoprotein receptor H1 and in the paramyxovirus hemagglutinin-neuraminidase, two type II proteins, caused a fraction of the polypeptides to insert with the opposite type III (N_{exo}/C_{cyt}) topology (Beltzer et al., 1991; Parks and Lamb, 1991, 1993). However, in these and other studies the asymmetric distribution of flanking charges in mutant proteins was not sufficient to generate a unique topology, indicating that additional factors contribute to efficient topogenesis in the ER.

Folding

In post-translational translocation, the polypeptide is kept translocation-competent either by preprotein-specific chaperones such as bacterial SecB (Randall and Hardy, 1986) or by general chaperones like Hsp70 (Chirico et al., 1988; Deshaies et al., 1988), which prevent aggregation and folding of the protein in the cytosol. In co-translational insertion, SRP and the ribosome protect the nascent polypeptide from premature folding. Sequences N-terminal of a (reverse) signal-anchor, however, are exposed to the cytosol before targeting is initiated. Their folding behavior may influence translocation competence and thus protein topology. This was experimentally tested in transfected mammalian cells (COS cells) using a diagnostic charge mutant of the asialoglycoprotein receptor H1 which inserted equally with N_{exo}/C_{cyt} and N_{cyt}/C_{exo} orientation. Extension of the N-terminus by a rapidly folding zinc-finger domain of 30 residues or dihydrofolate reductase of 237 amino acids, completely prevented N-terminal translocation (Denzer et al., 1995). These findings suggest that the insertion process is sufficiently flexible to allow signals to reorient in the translocon, when N-terminal translocation is blocked (as illustrated in Figure 6.3). Disruption of the

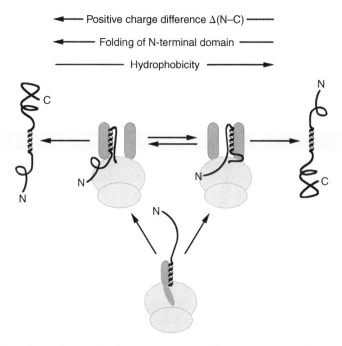

Figure 6.3 Orienting a signal sequence in the ER membrane. A signal sequence translocates either the C-terminus or the N-terminus across the membrane resulting in a type II (left) or a type III membrane protein (right). Cleavage of the signal by signal peptidase in the left branch would generate a secretory protein (not shown for simplicity). An N-terminal positive charge difference across the signal favors C-terminal translocation, as does a folded N-terminal domain. Longer and more hydrophobic signals favor N-terminal translocation.

N-terminal structures by destabilizing point mutations largely recovered N-terminal translocation across the membrane.

Rapidly folding N-terminal domains thus lock a signal-anchor into a type II orientation. In contrast, type III proteins require N-terminal regions that do not fold up in the cytosol. An example of a natural type III protein for which this is relevant is the neuregulin precursor with a 241-residue N-terminal domain containing an immunoglobulin-like and an epidermal growth factor (EGF)-like domain. Only in the ER lumen can disulfide formation and thus stable folding occur. Chaperones are likely to prevent misfolding and aggregation in the cytosol until the reverse signal-anchor emerges and targeting and translocation is initiated.

Hydrophobicity
The apolar core of the signal sequence also makes a significant contribution to orienting itself within the ER membrane. With diagnostic mutant proteins, an increased fraction of N-terminal translocation was observed with increasing length and hydrophobicity of this segment (Sato et al., 1990; Sakaguchi

et al., 1992; Wahlberg and Spiess, 1997; Harley et al., 1998). For example, an N-terminal type II signal-anchor with a positively charged N- and a negative C-terminal region was inserted with both orientations when the natural 19-residues apolar core was replaced by 19 consecutive leucines (Wahlberg and Spiess, 1997). Different oligo-leucine sequences covered the entire spectrum from almost complete C-terminal translocation to exclusive N-terminal translocation. The influence of the hydrophobic sequence on topology was additive with the effects of flanking charges and of an N-terminal hydrophilic extension (Wahlberg and Spiess, 1997; Harley et al., 1998). In addition, Harley et al. (1998) observed a correlation between a hydropathy gradient along the apolar sequence and signal orientation, with the more hydrophobic end more efficiently translocated across the membrane. This could be explained by a similar gradient in the signal binding site of the translocon. By photocrosslinking, the hydrophobic segment of the signal was found to specifically contact the transmembrane helices 2 and 7 of Sec61p in yeast (Plath et al., 1998). How the longer and/or more hydrophobic homo-oligomers induce increased N-terminal translocation is not obvious.

Homo-oligomers of other apolar sequences showed a similar bias for increasing N-terminal translocation with increasing length as oligo-leucines (Rösch et al., 2000). The ability to promote N-terminal translocation decreased in this order: $I > L > V \approx W > Y > F > M$. Except for oligo-alanine, which was not functional as a signal sequence, all homo-oligomers tested were efficient in targeting and insertion. Sequences as different in shape and volume as Val_{19} and Trp_{19} behaved identically, highlighting the ability of the translocon to accommodate an extremely broad spectrum of signal sequences. All uncharged amino acids have also been tested for the effect on signal orientation when inserted at identical positions in a Leu_{16} or Leu_{19} sequence (Rösch et al., 2000). The ranking order of residues with respect to promoting N-terminal translocation resembled a hydrophobicity scale:

$$I > V > L \approx W > F > Y > C > M > A > T > S > G > N > Q > H > P.$$

The topogenic contribution of the hydrophobic sequence was also shown to be important for natural proteins. The correct and unique insertion of the cleavable signal of the vasopressin precursor and of the reverse signal-anchor of microsomal epoxide hydrolase, both of which violate the charge distribution rule, was compromised upon extending or shortening the apolar sequence, respectively (Eusebio et al., 1998).

Summary: *Single-spanning membrane proteins*

- Positive charges near signal and transmembrane segments preferentially remain cytosolic ('positive-inside' rule in bacteria; charge difference rule in ER proteins).

- In bacteria, negatively charged residues can drive translocation across the membrane.
- A polypeptide must be unfolded for translocation. Folding of domains N-terminal to a signal sequence prevents N-terminal translocation.
- The longer and the more hydrophobic the apolar core of a signal, the higher its tendency to translocate the N-terminal region upstream of the ER signal sequences.

Multi-spanning membrane proteins

Multi-spanning membrane proteins can be considered to consist of a series of alternating signal and stop-transfer sequences which are linearly inserted into the bilayer. According to the simplest model, the initial signal defines its own orientation as well as the orientations of all subsequent transmembrane segments. The latter do not require any additional information, but will simply follow the lead of the first signal. Evidence for this 'linear insertion model' (initially proposed by Blobel, 1980) has been provided using chimeric proteins with two to four transmembrane segments separated by approximately 50–200 residues from each other (Wessels and Spiess, 1988; Lipp et al., 1989). The results showed that signal-anchor sequences (normally N_{cyt}/C_{exo}) can indeed be forced to insert as stop-transfer sequences (N_{exo}/C_{cyt}) depending on their relative positions in the polypeptide.

However, there is also strong evidence against a dominant role of 'a first signal sequence' in many membrane proteins. Statistics show that internal transmembrane domains also follow the charge rules, although in eukaryotic proteins less stringently than the most N-terminal signal (von Heijne, 1986, 1989). Experimentally, insertion of clusters of positive charges into short exoplasmic loops of model proteins caused individual hydrophobic domains not to insert at all ('frustrated' topologies; Gafvelin and von Heijne, 1994; Gafvelin et al., 1997). Deletion of individual membrane-spanning segments in bacterial proteins did not necessarily affect the topology of the downstream transmembrane domains (Figure 6.4A; Bibi et al., 1991; McGovern et al., 1991). Similarly, inversion of the charge difference of the first signal of the glucose transporter Glut1 did not affect the topology of the rest of the molecule, but just prevented insertion of the first signal (Sato et al., 1998). These studies showed that multi-spanning proteins contain topogenic information throughout their sequence and that insertion is not always strictly linear.

Successive determinants in a polypeptide influence each other during topogenesis, as was shown using simple chimeric proteins with two conflicting signal sequences, a cleavable signal and a signal-anchor (Figure 6.4B; Goder et al., 1999). When the signals were separated by more than 60 residues, linear insertion overriding the topological preference of the second signal was observed (Figure 6.4B, left topology). With shorter spacers, however, an increasing fraction of proteins inserted with a translocated C-terminus as dictated by the second signal (Figure 6.4B, right topology). The second signal thus co-determined the insertion process. This was similarly observed also in bacteria (Coleman et al., 1985). Topogenic competition between successive signals also indicates that the nascent polypeptide can reorient within the translocation machinery to explore its most 'comfortable' conformation. This was confirmed by insertion of a glycosylation site into the spacer sequence: glycosylation affected topogenesis by sterically trapping the modified spacer on the lumenal side (Goder et al., 1999). Potentially other exoplasmic modifications (e.g. signal cleavage) or protein folding could similarly stabilize one of the topologies.

Topogenic determinants throughout the membrane protein will ensure a defined and unique way of insertion into the membrane. Two transmembrane segments with a very short spacer sequence obviously cannot orient themselves independently of each other: they will insert together as a

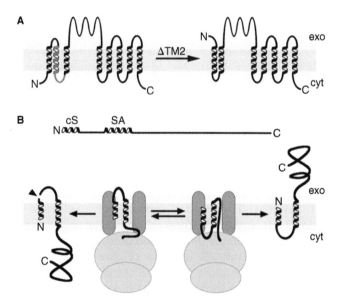

Figure 6.4 Multiple topogenic sequences in multi-spanning membrane proteins. **A**, Deletion of the second transmembrane segment of the bacterial MalF protein affects only the topology of the first transmembrane, but not of the downstream transmembrane domains (McGovern et al., 1991). Topology is thus defined by determinants throughout the sequence and not simply by the first transmembrane segment. **B**, Two conflicting signals, a cleavable N-terminal signal (cS) and an internal signal-anchor sequence (SA), generate different topologies depending on the length of the spacer sequence in between: the longer the spacer the larger the fraction of polypeptides with the left topology (Goder et al., 1999). Glycosylation in the spacer sequence shifts the ratio of topologies to the left (see text). The arrowhead indicates signal cleavage.

hairpin. With an increasing distance the cooperativity (or the competition) of successive transmembrane elements decreases and finally disappears (in the constructs of Goder et al., 1999, with spacers longer than 60 residues). Long spacer segments may be sterically unable to reorient themselves or may be trapped by binding of lumenal chaperones like BiP. The yeast homolog of BiP, Kar2p, has been shown to promote post-translational translocation (Brodsky et al., 1995; Panzner et al., 1995; Zimmermann, 1998; Matlack et al., 1999), but may also act in co-translational transport.

Another mechanism that might be relevant is the exit of transmembrane segments out of the translocon into the lipid environment. Studies in the bacterial system using a proOmpA mutant with artificial stop-transfer sequences indicated that more hydrophobicity is required for membrane partitioning than to stop SecA-driven translocation (Duong and Wickner, 1998). Photocrosslinking experiments showed that in the ER a stop-transfer sequence initially contacts Sec61α and TRAM, then only TRAM; contact to the translocation machinery was terminated only after release of the polypeptide from the tRNA (Do et al., 1996). Analysis of lipid contact, however, revealed that signals and signal-anchors could be crosslinked to lipids very early after entering the translocon (Martoglio et al., 1995; Mothes et al., 1997), suggesting that the translocon may be partially open to the lipid phase of the membrane or that signals may easily leave – and maybe reenter – the translocon (Figure 6.5). An alternative model suggests that the transmembrane segments of a nascent protein remain within the

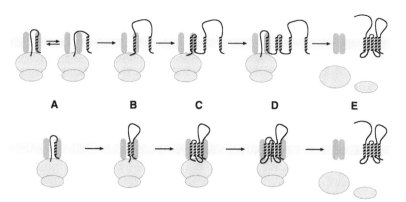

Figure 6.5 Two models for the assembly of multi-spanning membrane proteins. Transmembrane segments may be released sequentially into the lipid bilayer (upper panel, **A–D**) to be bundled to the final compact structure in a separate step. Partitioning between the translocon and the lipid may be a reversible process (**A**). Closely spaced transmembrane segments (hairpins) behave as a unit (**C–D**). Alternatively, it has been proposed that helix assembly takes place within the translocon before release of the final bundle upon completion of translation (bottom panel).

translocation complex and assemble in this 'protected' environment until protein completion (Borel and Simon, 1996). It seems also likely that a transmembrane helix with a central charged residue would stay associated with the pore complex until it can associate with its partner helix by complementary charge.

Summary: *Multi-spanning membrane proteins*

- Multi-spanning membrane proteins may contain topogenic determinants throughout the polypeptide.
- Topogenesis may involve the reversible movement of polypeptide segments within the translocon.
- Glycosylation (and potentially other exoplasmic modifications) can contribute to topogenesis.
- Photocrosslinking experiments suggest that transmembrane segments sequentially contact first Sec61α and then TRAM.

INTRA- AND INTERMOLECULAR BUNDLING OF TRANSMEMBRANE HELICES

In most cases, following insertion into the membrane, individual transmembrane helices recognize each other to form helix bundles. As an *intramolecular* process, this is part of the folding of a multi-spanning protein into a compact functional structure (Figure 6.1). As an *intermolecular* process, it is the basis for the formation of defined oligomeric membrane protein complexes. Some of these complexes have been isolated and crystallized as a whole. The cytochrome oxidase complex, ATPase, the light harvesting complexes, and the reaction center are the prominent examples illustrating the basic structural principles. In addition to such stable complexes, structurally more dynamic complexes might exist. It is even discussed that within the two-dimensional bilayer membrane different subcomplexes can form and separate with time (Leuenberger et al., 1999).

The main driving force of helix bundling is hydrophobic interaction, which requires the matching of helix surfaces. Helix–helix interaction has been thoroughly investigated for glycophorin A. Its transmembrane region spans the membrane with a stretch of 23 amino acids and forms stable dimers even in sodium dodecyl sulfate (SDS) micelles (Lemmon et al., 1992). Mutagenesis experiments identified the critical residues which are facing the contact surface of the transmembrane helix (Fleming et al., 1997). The dimerization motif is LIxxGVxxGVxxT, where x denotes amino acids that could be exchanged with little effect on dimerization. Using this information, a computer-simulated three-dimensional model of the right-handed dimer was built where the helix face, which is critical for packing, is formed by the residues sensitive to mutagenesis. The nuclear magnetic resonance (NMR) structure of a glycophorin A transmembrane dimer in dodecylphosphocholine confirmed this model (MacKenzie et al., 1997). A general motif GxxG emerged as a high-affinity homo-oligomerization motif from a randomized sequence library based on the glycophorin sequence (Russ and Engelman, 2000). In addition to hydrophobic interaction, inter-helical hydrogen bonding may contribute significantly (Zhou et al., 2000). Two helices can also be connected by a pair of charged residues. Lactose permease is the most prominent example that shows that charge interaction is used to tie two transmembrane helices together (Kaback et al., 1995).

Based on refolding experiments and expression of fragments, a two-step model for membrane protein folding and oligomerization has been proposed (Popot and Engelman, 1990). First, the hydrophobic helices are oriented and integrated into the lipid bilayer as a rather loose structure, before they assemble to the compact structure. In the case of oligomeric membrane proteins like glycophorin, it is obvious that the insertion of the transmembrane helices into the bilayer precedes bundling, since each polypeptide is synthesized and inserted on a separate ribosome and translocon. By analogy, the individual transmembrane segments of multi-spanning proteins may be inserted and released into the lipid bilayer in a first step, and assemble to the functional bundle in a second step. Alternatively, bundling may also occur within the translocon (Figure 6.5, bottom). Evidence that insertion and assembly can indeed take place separately was obtained by coexpressing or mixing complementary fragments of bacteriorhodopsin or Lac permease which yielded functional protein (Huang et al., 1981; Wrubel et al., 1990; Bibi and Kaback, 1990). Such experiments also showed that the connecting loops contribute rather little to bundling (Marti, 1998), although they appeared to play subtle roles for both stability and specificity of assembly.

Recently, the assembly of the tetrameric KcsA potassium channel from individual single-spanning subunits was studied. Subunit insertion and channel assembly was clearly separate, since insertion was independent of the membrane potential (and of SecA), whereas assembly of the complex required the electrical component of the membrane potential (van Dalen et al., 2000).

Summary: *Intra- and intermolecular bundling of transmembrane helices*

- Membrane protein folding and oligomerization may be separated into two steps: integration of transmembrane helices into the bilayer (including topogenesis) and helix bundling to a compact structure.
- Transmembrane helices are held together mainly by hydrophobic interactions, but in some cases also by ion pairs and hydrogen bonds.

HISTORICAL NOTES

Historical Note 1

In the eukaryotic system, it was proposed early on by Günter Blobel that translation of secretory and membrane proteins is arrested to synchronize translation with translocation. In this model, SRP serves as a device to target the ribosomes that synthesize proteins with signal sequences to the ER, leading to a synchronous process of protein translation and translocation. This proposal led to intense controversy as to whether protein translocation occurs co- or post-translationally. Since then it has become clear that, at least in yeast, both pathways exist. In all systems, multi-spanning membrane proteins appear to use the SRP-dependent pathway. Very hydrophobic sequences are likely to be incompatible with a post-translational mechanism.

Historical Note 2

Experiments with small bacterial membrane proteins and liposomes led to the understanding that the process of membrane insertion is a folding pathway. The possibility that this pathway is thermodynamically driven was summarized in the trigger hypothesis of Bill Wickner. The best examples in which this pathway is used in nature are the membrane toxins. However, it seems that the insertion of most membrane proteins is enzymatically controlled. Although folding pathways for membrane inserting proteins exist, these processes are far more complex and involve more factors than initially assumed.

Historical Note 3

It was a surprise to the membrane biochemists when Gobind Khorana and collegues showed in 1981 that two proteolytic fragments of bacteriorhodopsin assemble into a functional proton pump within liposomes. Later, Ruth Ehring, Ron Kaback and collegues found in 1990 that fragments of lactose permease that were expressed in *E. coli* were inserted into the membrane and oligomerized to a functional protein. This nicely showed that the membrane insertion process and the assembly into a folded complex are two separate steps. It also made evident that mechanisms exist for two fragments of a protein to recognize each other.

REFERENCES

Andersson, H. and von Heijne G. (1994) Positively charged residues influence the degree of SecA dependence in protein translocation across the *E. coli* inner membrane. *FEBS Lett* **347**: 169–172.

Beltzer, J.P., Fiedler, K., Fuhrer, C. et al. (1991) Charged residues are major determinants of the transmembrane orientation of a signal-anchor sequence. *J Biol Chem* **266**: 973–978.

Bibi, E. and Kaback, H.R. (1990) *In vivo* expression of the lacY gene in two segments lead to functional Lac permease. *Proc Natl Acad Sci USA* **87**: 4325–4329.

Bibi, E., Verner, G., Chang, C.Y. et al. (1991) Organization and stability of a polytopic membrane protein: deletion analysis of the lactose permease of *Escherichia coli*. *Proc Natl Acad Sci USA* **88**: 7271–7275.

Blobel, G. (1980) Intracellular protein topogenesis. *Proc Natl Acad Sci USA* **77**: 1496–1500.

Borel, A.C. and Simon, S.M. (1996) Biogenesis of polytopic membrane proteins: membrane segments assemble within translocation channels prior to membrane integration. *Cell* **85**: 379–389.

Bowie, J.U. (1997) Helix packing in membrane proteins. *J Mol Biol* **272**: 780–789.

Bretscher, M.S. and Munro, S. (1993) Cholesterol and the Golgi apparatus. *Science* **261**: 1280–1281.

Brodsky, J.L. and Schekman, R. (1993) A sec63p-BiP complex from yeast is required for protein translocation in a reconstituted proteoliposome. *J Cell Biol* **123**: 1355–1363.

Brodsky, J.L., Goeckeler, J. and Schekman, R. (1995) BiP and Sec63p are required for both co- and posttranslational protein translocation into the yeast endoplasmic reticulum. *Proc Natl Acad Sci USA* **92**: 9643–9646.

Chen, H. and Kendall, D.A. (1995) Artificial transmembrane segments. Requirements for stop transfer and polypeptide orientation. *J Biol Chem* **270**: 14115–14122.

Chirico, W.J., Waters, M.G. and Blobel, G. (1988) 70k heat shock related proteins stimulate protein translocation into microsomes. *Nature* **332**: 805–810.

Coleman, J., Inukai, M. and Inouye, M. (1985) Dual functions of the signal peptide in protein transfer across the membrane. *Cell* **43**: 351–360.

Dalbey, R.E. and Kuhn, A. (2000) Evolutionarily related insertion pathways of bacterial, mitochondrial, and thylakoid membrane proteins. *Annu Rev Cell Dev Biol* **16**: 51–87.

Davis, N.G., Boeke, J.D. and Model, P. (1985) Fine structure of a membrane anchor domain. *J Mol Biol* **181**: 111–121.

Deber, C.M., Brandl, C.J., Deber, R.B. et al. (1986) Amino acid composition of the membrane and aqueous domains of integral membrane proteins. *Arch Biochem Biophys* **251**: 68–76.

de Gier, J.W., Scotti, P.A., Saaf, A. et al. (1998) Differential use of the signal recognition particle translocase targeting pathway for inner membrane protein assembly in *Escherichia coli*. *Proc Natl Acad Sci USA* **95**: 14646–14651.

Denzer, A.J., Nabholz, C.E. and Spiess, M. (1995) Transmembrane orientation of signal-anchor proteins is affected by the folding state but not the size of the aminoterminal domain. *EMBO J* **14**: 6311–6317.

Deshaies, R.J., Koch, B.D., Werner-Washbourne, M. et al. (1988) A subfamily of stress proteins facilitates translocation of secretory and mitochondrial precursor polypeptides. *Nature* **332**: 800–805.

Do, H., Falcone, D., Lin, J. et al. (1996) The cotranslational integration of membrane proteins into the phospholipid bilayer is a multistep process. *Cell* **85**: 369–378.

Duong, F. and Wickner, W. (1997) Distinct catalytic roles of the SecYE, SecG and SecDFyajC subunits of preprotein translocase holoenzyme. *EMBO J* **16**: 2756–2768.

Duong, F. and Wickner, W. (1998) Sec-dependent membrane protein biogenesis: SecYEG, preprotein hydrophobicity and translocation kinetics control the stop-transfer function. *EMBO J* **17**: 696–705.

Economou, A., Pogliano, J.A., Beckwith J. et al. (1995) SecA membrane cycling at SecYEG is driven by distinct ATP binding and hydrolysis events and is regulated by SecD and SecF. *Cell* **83**: 1171–1181.

Eisenberg, D., Schwarz, E., Komaromy, M. et al. (1984) Analysis of membrane and surface protein sequences with the hydrophobic moment plot. *J Mol Biol* **179**: 125–142.

Eusebio, A., Friedberg, T. and Spiess, M. (1998) The role of the hydrophobic domain in orienting natural signal sequences within the ER membrane. *Exp Cell Res* **241**: 181–185.

Fleming, K.G., Ackerman, A.L. and Engelman, D.M. (1997) The effect of point mutations on the free energy of transmembrane alpha-helix dimerization. *J Mol Biol* **272**: 266–275.

Friedlander, M. and Blobel, G. (1985) Bovine opsin has more than one signal sequence. *Nature* **318**: 338–343.

Gafvelin, G. and von Heijne, G. (1994) Topological frustration in multispanning *E. coli* inner membrane proteins. *Cell* **77**: 401–412.

Gafvelin, G., Sakaguchi, M., Andersson, H. et al. (1997) Topological rules for membrane protein assembly in eukaryotic cells. *J Biol Chem* **272**: 6119–6127.

Gallusser, A. and Kuhn, A. (1990) Initial steps in protein insertion. Bacteriophage M13 procoat binds to the membrane surface by electrostatic interaction. *EMBO J* **9**: 2723–2729.

Goder, V., Bieri, C. and Spiess, M. (1999) Glycosylation can influence topogenesis of membrane proteins and reveals dynamic reorientation of nascent polypeptides within the translocon. *J Cell Biol* **147**: 257–266.

Harley, C.A., Holt, J.A., Turner, R. et al. (1998) Transmembrane protein insertion orientation in yeast depends on the charge difference across transmembrane segments, their total hydrophobicity, and its distribution. *J Biol Chem* **273**: 24963–24971.

Hartmann, E., Rapoport, T.A. and Lodish, H.F. (1989) Predicting the orientation of eukaryotic membrane-spanning proteins. *Proc Natl Acad Sci USA* **86**: 5786–5790.

Herskovitz, A.A. and Bibi, E. (2000) Association of *Escherichia coli* ribosomes with the inner membrane requires the signal recognition particle receptor but is independent of the signal recognition particle. *Proc Natl Acad Sci USA* **97**: 4621–4626.

Huang, K.S., Bayley, H., Liao, M.J. et al. (1981) Refolding of an integral membrane protein. Denaturation, renaturation, and reconstitution of intact bacteri-orhodopsin and two proteolytic fragments. *J Biol Chem* **256**: 3802–3809.

Kaback, H.R., Jung, K., Jung, H. et al. (1995) Helix packing in the c-terminal half of lactose permease. *Adv Cell Mol Biol Mem Org* **4**: 129–144.

Kiefer, D. and Kuhn, A. (1999) Hydrophobic forces drive the spontaneous membrane insertion of the bacteriophage Pf3 coat protein without topological control. *EMBO J* **18**: 6299–6306.

Koch, H.G. and Muller, M. (2000) Dissecting the translocase and integrase functions of the *Escherichia coli* SecYEG translocon. *J Cell Biol* **150**: 689–894.

Kolbe, M., Besir, H., Essen, L.O. et al. (2000) Structure of the light-driven chloride pump halorhodopsin at 1.8 Å resolution. *Science* **288**: 1390–1396.

Kuhn A. (1988) Alterations in the extracellular domain of M13 procoat protein make its membrane insertion dependent on secA and secY. *Eur J Biochem* **177**: 267–271.

Kutay, U., Hartmann, E. and Rapoport, T.A. (1993) A class of membrane proteins with a C-terminal anchor. *Trends Cell Biol* **3**: 72–75.

Kutay, U., Ahnert-Hilger, G., Hartmann, E. et al. (1995) Transport route for synaptobrevin via a novel pathway of insertion into the endoplasmic reticulum membrane. *EMBO J* **14**: 217–223.

Kyte, J. and Doolittle, R.F. (1982) A simple method for displaying the hydrophobic character of a protein. *J Mol Biol* **157**: 105–132.

Laws, J.K. and Dalbey, R.E. (1989) Positive charges in the cytoplasmic domain of *Escherichia coli* leader peptidase prevent an apolar domain from functioning as a signal. *EMBO J* **8**: 2095–2099.

Lemmon, M.A., Flanagan, J.M., Hunt, J.F. et al. (1992) Glycophorin A dimerization is driven by specific interactions between transmembrane alpha-helices. *J Biol Chem* **267**: 7683–7689.

Leuenberger, D., Bally, N.A., Schatz, G. et al. (1999) Different import pathways through the mitochondrial intermembrane space for inner membrane proteins. *EMBO J* **18**: 4816–4822.

Lipp, J., Flint, N., Haeuptle, M.T. et al. (1989) Structural requirements for membrane assembly of proteins spanning the membrane several times. *J Cell Biol* **109**: 2013–2022.

MacKenzie, K.R., Prestegard, J.H. and Engelman, D.M. (1997) A transmembrane helix dimer: structure and implication. *Science* **276**: 131–133.

Marti, T. (1998) Refolding of bacteriorhodopsin from expressed polypeptide fragments. *J Biol Chem* **273**: 9312–9322.

Martoglio, B., Hofmann, M.W., Brunner, J. et al. (1995) The protein-conducting channel in the membrane of the endoplasmic reticulum is open laterally toward the lipid bilayer. *Cell* **81**: 207–214.

Matlack, K.E.S., Misselwitz, B., Plath, K. et al. (1999) BiP acts as a molecular ratchet during posttranslational transport of prepro-αfactor across the ER membrane. *Cell* **97**: 553–564.

McGovern, K., Ehrmann, M. and Beckwith, J. (1991) Decoding signals for membrane protein assembly using alkaline phosphatase fusions. *EMBO J* **10**: 2773–2782.

Michels, M. and Bakker, E.P. (1985) Generation of a large, protonophore-sensitive proton motive force and pH difference in the acidophilic bacteria *Thermoplasma acidophilum* and *Bacillus acidocaldarius*. *J Bacteriol* **161**: 231–237.

Moll, R. and Schäfer, G. (1988) Chemiosmotic H+ cycling across the plasma membrane of the thermoacidophilic archaebacterium *Sulfolobus acidocaldarius*. *FEBS Lett* **232**: 359–363.

Monier, S., Van Luc, P., Kreibich, G. et al. (1988) Signals for the incorporation and orientation of cytochrome P450 in the endoplasmic reticulum membrane. *J Cell Biol* **107**: 457–470.

Mothes, W., Heinrich, S.U., Graf, R. et al. (1997) Molecular mechanism of membrane protein integration into the endoplasmic reticulum. *Cell* **89**: 523–533.

Munro, S. (1995) An investigation of the role of transmembrane domains in Golgi protein retention. *EMBO J* **14**: 4695–4704.

Ng, D.T., Brown, J.D. and Walter, P. (1996) Signal sequences specify the targeting route to the endoplasmic reticulum membrane. *J Cell Biol* **134**: 269–278.

Nilsson, I.M. and von Heijne, G. (1990) Fine-tuning the topology of a polytopic membrane protein – role of positively and negatively charged amino acids. *Cell* **62**: 1135–1141.

Panzner, S., Dreier, L., Hartmann, E. et al. (1995) Posttranslational protein transport in Yeast reconstituted with a purified complex of Sec-proteins and Kar2p. *Cell* **81**: 561–570.

Parks, G.D. and Lamb, R.A. (1991) Topology of eukaryotic type-II membrane proteins – Importance of N-terminal positively charged residues flanking the hydrophobic domain. *Cell* **64**: 777–787.

Parks, G.D. and Lamb, R.A. (1993) Role of NH_2-terminal positively charged residues in establishing membrane protein topology. *J Biol Chem* **268**: 19101–19109.

Plath, K., Mothes, W., Wilkinson, B.M. et al. (1998) Signal sequence recognition in posttranslational protein transport across the yeast ER membrane. *Cell* **94**: 795–807.

Popot, J.L. and Engelman, D.M. (1990) Membrane protein folding and oligomerization: the two-stage model. *Biochemistry* **29**: 4031–4037.

Randall, L.L. and Hardy, S.J. (1986) Correlation of competence for export with lack of tertiary structure of the mature species: a study *in vivo* of maltose-binding protein in *E. coli*. *Cell* **46**: 921–928.

Ridder, A.N., Morein, S., Stam, J.G. et al. (2000) Analysis of the role of interfacial tryptophan residues in controlling the topology of membrane proteins. *Biochemistry* **39**: 6521–6528.

Rösch, K., Naeher, D., Laird, V. et al. (2000) The topogenic contribution of uncharged amino acids on signal sequence orientation in the endoplasmic reticulum. *J Biol Chem* **275**: 14916–14922.

Russ, W.P. and Engelman, D.M. (2000) The GxxxG motif: a framework for transmembrane helix–helix association. *J Mol Biol* **296**: 911–919.

Sahin-Toth, M., Dunten, R.L., Gonzalez, A. et al. (1992) Functional interactions between putative intramembrane charged residues in the lactose permease of *Escherichia coli*. *Proc Natl Acad Sci USA* **89**: 10547–10551.

Sakaguchi, M., Tomiyoshi, R., Kuroiwa, T. et al. (1992) Functions of signal and signal-anchor sequences are determined by the balance between the hydrophobic segment and the N-terminal charge. *Proc Natl Acad Sci USA* **89**: 16–19.

Samuelson, J.C., Chen, M., Jiang, F. et al. (2000) YidC mediates both Sec-dependent and Sec-independent membrane protein insertion. *Nature* **406**: 637–641.

Samuelson, J.C., Jiang, F., Yi, L. et al. (2001) Function of YidC for the insertion of M13 procoat protein in *E. coli*: translocation of mutants that show differences in their membrane potential dependence and Sec-requirement. *J Biol Chem* **278**: 34847–34852.

Sato, M., Hresko, R. and Mueckler, M. (1998) Testing the charge difference hypothesis for the assembly of a eucaryotic multispanning membrane protein. *J Biol Chem* **273**: 25203–25208.

Sato, T., Sakaguchi, M., Mihara, K. et al. (1990) The amino-terminal structures that determine topological orientation of cytochrome-P-450 in microsomal membrane. *EMBO J* **9**: 2391–2397.

Scotti, P.A., Urbanus, M.L., Brunner, J. et al. (2000) YidC, the *Escherichia coli* homologue of mitochondrial Oxa1p, is a component of the Sec translocase. *EMBO J* **19**: 542–549.

Soekarjo, M., Eisenhawer, M., Kuhn, A. et al. (1996) Thermodynamics of the membrane insertion process of the M13 procoat protein, a lipid bilayer traversing protein containing a leader sequence. *Biochemistry* **35**: 1232–1241.

Spiess, M. and Lodish, H.F. (1986) An internal signal sequence: the asialogycoprotein receptor membrane anchor. *Cell* **44**: 177–185.

Szczesna-Skorupa, E., Browne, N., Mead, D.A. et al. (1988) Positive charges at the NH_2 terminus convert the membrane-anchor signal peptide of cytochrome P-450 to a secretory peptide. *Proc Natl Acad Sci USA* **85**: 738–742.

Valent, Q.A., Scotti, P.A., High, S. et al. (1998) The *Escherichia coli* SRP and SecB targeting pathways converge at the translocon. *EMBO J* **17**: 2504–2512.

van Dalen, A., Schrempf, H., Killian, J.A. et al. (2000) Efficient membrane assembly of the KcsA potassium channel in *Escherichia coli* requires the protonmotive force. *EMBO Rep* **1**: 340–346.

van de Vossenberg, J.L., Albers, S.V., van der Does, C. et al. (1998) The positive inside rule is not determined by the polarity of the delta psi. *Mol Microbiol* **29**: 1125–1127.

van Klompenburg, W., Nilsson, I., von Heijne G. et al. (1997) Anionic phospholipids are determinants of membrane protein topology. *EMBO J* **16**: 4261–4266.

von Heijne, G. (1986) The distribution of positively charged residues in bacterial inner membrane proteins correlates with the trans-membrane topology. *EMBO J* **5**: 3021–3027.

von Heijne, G. (1989) Control of topology and mode of assembly of a polytopic membrane protein by positively charged residues. *Nature* **341**: 456–458.

von Heijne, G. (1992) Membrane protein structure prediction. Hydrophobicity analysis and the positive-inside rule. *J Mol Biol* **225**: 487–494.

Wahlberg, J.M. and Spiess, M. (1997) Multiple determinants direct the orientation of signal-anchor proteins: The topogenic role of the hydrophobic signal domain. *J Cell Biol* **137**: 555–562.

Wallin, E. and von Heijne, G. (1998) Genome-wide analysis of integral membrane proteins from eubacterial, archaean, and eukaryotic organisms. *Protein Sci* **7**: 1029–1038.

Wallin, E., Tsukihara, T., Yoshikawa, S. et al. (1997) Architecture of helix bundle membrane proteins: an analysis of cytochrome *c* oxidase from bovine mitochondria. *Protein Sci* **6**: 808–815.

Wessels, H.P. and Spiess, M. (1988) Insertion of a multispanning membrane protein occurs sequentially and requires only one signal sequence. *Cell* **55**: 61–70.

Whitley, P., Grahn, E., Kutay, U. et al. (1996) A 12-residue-long polyleucine tail is sufficient to anchor synaptobrevin to the endoplasmic reticulum membrane. *J Biol Chem* **271**: 7583–7586.

Wolfe, P.B., Rice, M. and Wickner, W. (1985) Effects of two sec genes on protein assembly into the plasma membrane of *Escherichia coli*. *J Biol Chem* **260**: 1836–1841.

Wolin, S.L. and Walter, P. (1989) Signal recognition particle mediates a transient elongation arrest of preprolactin in reticulocyte lysate. *J Cell Biol* **109**: 2617–2622.

Wrubel, W., Stochaj, U., Sonnewald, U. et al. (1990) Reconstitution of an active lactose carrier *in vivo* by simultaneous syntheses of two complementary protein fragments. *J Bacteriol* **172**: 5374–5381.

Zerial, M., Melancon, P., Schneider, C. et al. (1986) The transmembrane segment of the human transferrin receptor functions as a signal peptide. *EMBO J* **5**: 1545–1550.

Zhou, F.X., Cocco, M.J., Russ, W.P. et al. (2000) Interhelical hydrogen bonding drives strong interactions in membrane proteins. *Nat Struct Biol* **7**: 154–160.

Zimmermann, R. (1998) The role of molecular chaperones in protein transport into the mammalian endoplasmic reticulum. *Biol Chem* **379**: 275–282.

SUGGESTED READING

Dalbey, R.E. and Kuhn, A. (2000) Evolutionarily related insertion pathways of bacterial, mitochondrial, and thylakoid membrane proteins. *Annu Rev Cell Dev Biol* **16**: 51–87.

High, S., Andersen, S.S., Görlich, D. et al. (1993) Sec61p is adjacent to nascent type I and type II signal-anchor proteins during their membrane insertion. *J Cell Biol* **121**: 743–750.

Popot, J.L. and Engelman, D.M. (2000) Helical membrane protein folding, stability, and evolution. *Annu Rev Biochem* **69**: 881–922.

Wahlberg, J.M. and Spiess, M. (1997) Multiple determinants direct the orientation of signal-anchor proteins: The topogenic role of the hydrophobic signal domain. *J Cell Biol* **137**: 555–562.

White, S.H. and Wimley, W.C. (1999) Membrane protein folding and stability: physical principles. *Annu Rev Biophys Biomol Struct* **28**: 319–365.

7

DISULFIDE BOND FORMATION IN PROKARYOTES AND EUKARYOTES

JAMES REGEIMBAL AND JAMES C.A. BARDWELL

DISULFIDES: THE BONDS THAT TIE

Almost 40 years ago, Anfinsen investigated how proteins fold, and he found that the information directing the folding of a protein into its correct three-dimensional structure is contained within its amino acid sequence. He treated ribonuclease, a small protein that contains four disulfide bonds, with a reductant to remove its disulfides and a denaturant to unfold it. He then removed the reductant and denaturant and let the ribonuclease refold in the presence of oxygen. Over a period of days a portion of this protein was able to refold to an active configuration. This classic experiment showed that the amino acid sequence of a protein contains sufficient information to direct its correct folding.

Although the information directing the folding of a protein is contained within the primary sequence, the native three-dimensional structure may depend on the correct formation of one or more disulfide bonds (Anfinsen, 1961, 1973; Gilbert, 1990). A disulfide bond is a covalent bond that forms between the thiol groups of two cysteine residues. Disulfide bonds are rarely found in intracellular proteins, and when they are present they often form transiently as part of the catalytic mechanism of an enzyme or as a means of regulating an enzyme's activity (Gilbert, 1990; Jakob et al., 1999). In contrast to intracellular proteins, disulfides are often found in secreted proteins (see Chapters 4 and 5). These disulfide bonds usually stabilize the

Protein Targeting, Transport & Translocation
ISBN 0-12-200731-X

fold of a protein, and are essential for proper structure. The disparity in the prevalence of disulfides in intracellular versus extracellular proteins is accounted for in part by the different redox potential of each environment. The formation of a disulfide bond is an oxidation reaction, and the reducing environment of the cell's cytosol inhibits the formation of disulfides. On the other hand, the more oxidizing environments of the periplasm of prokaryotes or the endoplasmic reticulum (ER) of eukaryotes, favors the maintenance of a disulfide bond once it is formed. For many years it was thought that this difference in redox potentials provides an adequate explanation for why disulfides form in the periplasm or the ER, and fail to form within the cytosol. More recently, it has become clear that there are specific catalytic systems in the periplasm and the ER that catalyze the formation of correct disulfides.

The following is a general scheme for the formation of a disulfide:

$$\text{R—SH} + \text{R}'\text{—SH} \underset{\text{Reduction}}{\overset{\text{Oxidation}}{\rightleftarrows}} \text{R—S—S—R}' + 2e^- + 2H^+ \qquad (1)$$

with e^- acceptor and e^- donor branches.

The forward direction of Equation (1) is an oxidation reaction with respect to the thiols, and the reverse direction is a reduction reaction with respect to the disulfide. As shown in Equation (1), the oxidation of thiols requires an appropriate electron acceptor and the reduction of a disulfide requires an electron donor.

The formation of a disulfide bond *in vivo* is a controlled and catalyzed process. The catalysis of disulfide bond formation occurs via a thiol–disulfide exchange reaction between a disulfide donor and a target protein. Within these reactions disulfides are neither created nor destroyed, they are simply transferred from one set of thiols to another. This transfer is depicted in Equation (2).

Disulfide donor (2)

In proteins that contain three or more cysteines, it is possible for a disulfide bond to form between the wrong pair of thiols. Incorrectly formed disulfides can be isomerized in a reaction that does not require the addition or elimination of electrons, the disulfides are simply rearranged. Disulfide rearrangement is depicted in Equation (3).

$$
\begin{matrix}
\text{R—S} & & \text{R—SH} \\
| \quad | & \text{Disulfide isomerase} & | \\
\text{R—S} & \underset{\longleftarrow}{\overset{\longrightarrow}{}} & \text{R—S} \\
| & & | \quad | \\
\text{R—SH} & & \text{R—S}
\end{matrix}
\tag{3}
$$

The chemical mechanism of thiol–disulfide exchange occurs via the nucleophilic attack by an incoming thiolate anion on one of the members of a disulfide, displacing the other member as a thiolate anion (Gilbert, 1990).

$$
\text{R}'\text{—S}^- + \underset{\overset{|}{R}}{\text{S}}\text{—S—R}'' \rightleftharpoons \left[\text{R}'\text{—S} \cdots \underset{\overset{|}{R}}{\text{S}} \cdots \text{S—R}'' \right]^{\ddagger} \rightleftharpoons \text{R}'\text{—S—}\underset{\overset{|}{R}}{\text{S}} + {}^-\text{S—R}''
\tag{4}
$$

Since the mechanism of disulfide exchange begins with the attack by a thiolate anion, the thiol must be deprotonated. In general, thiols with lower pK_a values will serve as better nucleophiles (will be more reactive) because they will be deprotonated at physiological pH. Thiol–disulfide exchange will also be enhanced if the leaving group is able to stabilize the negative charge of its thiolate anion. Though these factors will enhance disulfide exchange reactions, it is also important that the exchange is thermodynamically favorable. That is, the exchange reaction must occur 'down a redox gradient.' Electrons can only be donated from a molecule with a negative redox potential to a molecule with a more positive redox potential (Figure 7.1).

Summary: *Disulfides: the bonds that tie*

- A disulfide bond is a covalent bond that forms between the thiol groups of two cysteine residues within a single peptide, or across two different peptides.
- Disulfide bond formation is a redox reaction; thiols are oxidized to form a disulfide bond and when a disulfide is reduced, two thiols are recovered. Like all redox reactions, disulfide bond formation requires an electron donor and acceptor.
- Disulfides are rarely found in the cytosol, and when present they are usually part of an enzyme's catalytic mechanism or means of regulation. Extracellular proteins often have disulfides that stabilize the fold of the protein.

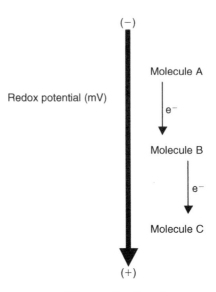

Figure 7.1 Electrons are passed from molecules of more negative redox potential to molecules of more positive redox potential.

- *In vivo* disulfide formation and isomerization involves specific protein catalysts that participate in thiol–disulfide exchange reactions.
- Disulfide exchange occurs down a redox gradient.

DISULFIDE BOND FORMATION *IN VIVO*

As stated before, many extracellular proteins require the formation of one or more disulfides for their correct structure and function. Disulfide bond formation and isomerization are catalyzed processes in both prokaryotes and eukaryotes, and are achieved via thiol–disulfide exchange reactions with specific disulfide donors and isomerases (Goldberger et al., 1963; Bardwell et al., 1991; Freedman et al., 1994). Over the past decade, much has been learned about disulfide bond formation in both prokaryotes and eukaryotes, and the specific pathways and proteins involved have been identified. We review what is currently known in both systems, beginning with the *Escherichia coli* system and then moving on to the eukaryotic system.

Disulfide bond formation in the *E. coli* periplasm

Disulfide bond formation in *E. coli* has been intensely studied over the past decade, and though there are still many questions to be answered, a great deal is known about how disulfides form *in vivo*. In *E. coli*, disulfide bond formation occurs in the periplasm, which is an extracytosolic compartment that is analogous to the ER in eukaryotes. In the periplasm, correct disulfide formation is achieved via the concerted effort of two distinct pathways. The oxidative pathway is responsible for *de novo* disulfide bond formation, while the isomerization pathway is responsible for isomerizing wrongly formed disulfides. To understand each pathway in detail, it is helpful first to describe the specific proteins in each pathway, and then to discuss how the proteins interact to achieve the overall activity of each pathway.

The oxidative pathway in E. coli
The oxidative pathway is responsible for *de novo* disulfide formation, and the molecule that acts as the immediate donor of disulfide bonds is DsbA. DsbA was the first protein in *E. coli* found to be involved in disulfide bond formation, and functions to donate disulfides to secreted proteins via thiol–disulfide exchange reactions (Bardwell et al., 1991). Prior to the identification of DsbA, it was widely believed that disulfide bond formation in prokaryotes is an 'uncatalyzed' process that occurs via spontaneous oxidation by oxygen or other small-molecule-oxidants in their growth media.

Strains lacking functional DsbA (*dsbA⁻* strains) show a severe deficiency in disulfide bond formation in periplasmic proteins. This is reflected in the loss of properly folded outer membrane protein A (OmpA), FlgI (flagellar P-ring protein), and alkaline phosphatase, all of which require one or more disulfide bonds for their correct folding. *dsbA⁻* strains also show increased sensitivity to the reductant dithiothreitol (DTT) and metal ions (Bardwell et al., 1991; Dailey and Berg, 1993; Missiakas et al., 1993; Rensing et al., 1997; Stafford et al., 1999). Because of the many phenotypes of a *dsbA⁻* mutant, DsbA is apparently the major oxidant responsible for forming disulfide bonds in periplasmic proteins (Bardwell et al., 1991). DsbA oxidizes proteins by donating a disulfide bond present at its active site to substrate proteins (Figure 7.2).

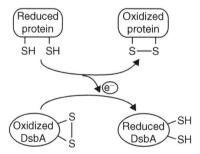

Figure 7.2   The oxidation of a target protein by oxidized DsbA. DsbA donates its active site disulfide to a substrate protein via an intermolecular thiol–disulfide exchange reaction. The substrate protein is released from the reaction in an oxidized state, and DsbA is released in a reduced state. In this way, DsbA is not a true enzyme since it is changed at the end of the reaction.

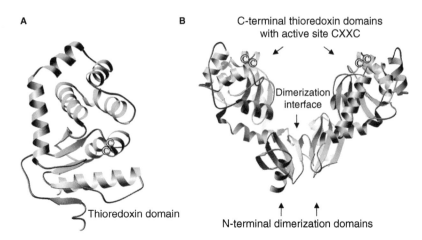

Figure 7.3 Crystal structures of DsbA (**A**) and DsbC (**B**). The 2.0 Å crystal structure of DsbA reveals that in addition to having a CXXC motif, DsbA also has a thioredoxin-like fold. Thus DsbA is a member of the thioredoxin superfamily. The 1.9 Å crystal structure of DsbC reveals that the overall structure of DsbC is a V-shaped homodimer. The N-terminal region of each peptide monomer forms a dimerization domain. The C-terminal domain of each monomer has an overall thioredoxin-like fold with a redox active CXXC motif. Thus DsbC is a member of the thioredoxin superfamily.

DsbA is a 21 kDa protein that has a redox active disulfide bond arranged in a CXXC (cysteine-X-X-cysteine) motif. This motif is characteristic of proteins involved in thiol–disulfide exchange reactions, and was first described in protein disulfide isomerase (PDI; see later section on 'Eukaryotic' disulfide bond formation) and the thiol-reductase thioredoxin. The crystal structure of DsbA has been solved to 2.0 Å, and reveals that DsbA has a thioredoxin-like structure (Martin et al., 1993) (Figure 7.3A). Thus DsbA is a member of the thioredoxin superfamily.

Despite having CXXC motifs and similar structures, DsbA and thioredoxin appear to have opposite functions. Thioredoxin acts as a cytosolic reductase whereas DsbA is the periplasmic protein-oxidant. In fact, DsbA is the most oxidizing protein known, with a calculated redox potential of $-120\,mV$, while the redox potential of thioredoxin is rather reducing at $-270\,mV$ (Wunderlich and Glockshuber, 1993; Zapun et al., 1993). DsbA's extreme oxidizing power is due to the fact that its active site disulfide makes its oxidized form unstable and reactive, while DsbA's reduced form is very stable (Zapun et al., 1993). Therefore, when DsbA donates its disulfide to a folding protein and becomes reduced, DsbA's fold is stabilized. This increase in stability favors the reduced form of DsbA, and allows the oxidized form to function as a powerful oxidant (disulfide donor).

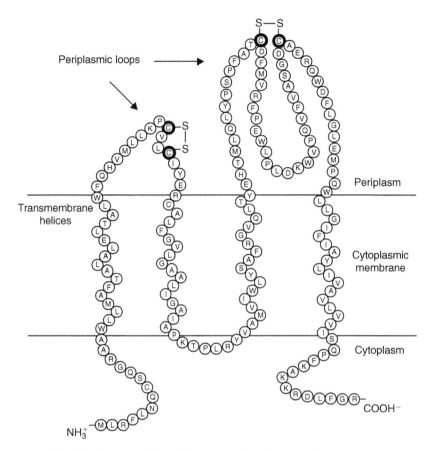

Figure 7.4   Topology of DsbB. DsbB is predicted to contain four transmembrane domains and two periplasmic loops. Each loop contains a pair of cysteines that are essential for the function of DsbB.

After oxidizing a target protein, DsbA is released from the reaction in a reduced state (Figure 7.2). In this way, DsbA is not a true enzyme, since it is changed at the end of the reaction. *In vivo*, however, DsbA acts as a catalyst because the membrane protein DsbB immediately reoxidizes reduced DsbA (Bardwell et al., 1993; Guilhot et al., 1995; Kishigami et al., 1995; Bader et al., 1999). DsbB deficient strains exhibit the same phenotypes as *dsbA⁻* strains, and they accumulate DsbA in its reduced form.

DsbB is a 20 kDa inner membrane protein that contains four transmembrane helices and two periplasmic loops (Jander et al., 1994) (Figure 7.4). Each periplasmic loop of DsbB contains a cysteine pair, with both pairs being essential for the activity of DsbB (Jander et al., 1994). The first pair of cysteines is arranged in a CXXC motif. Despite having a CXXC motif,

DsbB shows no other similarity with members of the thioredoxin super-family. The second pair of cysteines in DsbB are found in the second periplasmic loop, and appear to also form a redox active disulfide (Jander et al., 1994; Kobayashi et al., 1997; Kobayashi and Ito, 1999).

A mechanism for DsbB's oxidation of reduced DsbA has been proposed, and it involves a thiol–disulfide exchange cascade between DsbA and the two presumed disulfides of DsbB. It has been suggested that the second disulfide bond of DsbB directly reoxidizes reduced DsbA, leaving the thiols of the second disulfide in a reduced state (Guilhot et al., 1995; Kishigami et al., 1995; Kishigami and Ito, 1996). The second disulfide is then reoxidized by the CXXC disulfide of the first periplasmic domain, leaving the thiols of the CXXC in a reduced form (Kobayashi et al., 1997; Kobayashi and Ito, 1999). This mechanism is based on genetic evidence and still requires biochemical analysis.

Once DsbB oxidizes reduced DsbA, the question remaining is how DsbB is reoxidized. Because of DsbB's location in the membrane, it was originally speculated that DsbB might pass electrons to members of the electron transport chain (Bardwell, 1994). This was later proven to be the case (Bader et al., 1999; Kobayashi and Ito, 1999; Bader et al., 2000). Bader et al. (1999) successfully reconstituted the entire oxidative pathway *in vitro*, under both aerobic and anaerobic conditions, and showed directly that DsbB donates electrons to members of the respiratory chain (Bader et al., 1999) (Figure 7.5). Under aerobic conditions, DsbB transfers electrons to ubiquinone, which then donates the electrons to the terminal cytochrome oxidases, which ultimately reduce oxygen. Under anaerobic conditions, DsbB transfers electrons to menaquinone, which then donates the electrons to fumarate reductase or nitrate reductase. The same authors also showed that DsbB has at least one high-affinity binding site for quinones, and that DsbB donates electrons directly to quinone as part of its catalytic cycle (Bader et al., 2000). Therefore, DsbB generates disulfides by quinone reduction. This novel catalytic activity is apparently the major source of disulfides in prokaryotes.

By reconstituting the oxidative pathway, Bader et al. (1999) were able to show that disulfide bond formation is directly linked to the oxidative power of the electron transport system. This newly discovered connection between disulfide bond formation and electron transport enables *E. coli* to tap into the oxidative power of the electron transport chain, in order to ensure that proteins are properly folded in the periplasm.

Summary: *The oxidative pathway in* **E. coli**

- The oxidative pathway is responsible for forming disulfides *de novo* in proteins exported to the *E. coli* periplasm.

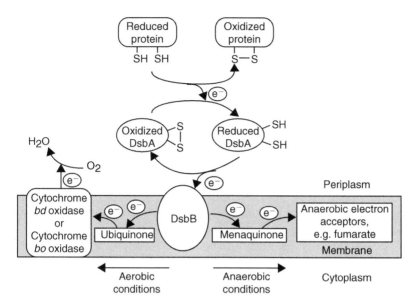

Figure 7.5   The complete oxidative pathway under aerobic and anaerobic conditions. When DsbA oxidizes a target protein it is released in a reduced state. For DsbA to function catalytically, it is then reoxidized by DsbB. Under aerobic conditions, DsbB then passes the electrons to ubiquinone. Ubiquinone passes the electrons down the remainder of the electron transport chain. Under anaerobic conditions, DsbB donates electrons to menaquinone. Menaquinone then passes the electrons to anaerobic electron acceptors. In this way, *E. coli* is able to ensure that disulfides are formed in the periplasm by linking their formation to the oxidative power of the electron transport chain.

- DsbA is a non-specific and very powerful protein-oxidant that is responsible for donating disulfides, via intermolecular thiol–disulfide exchange reactions, to the newly exported proteins. DsbA itself is not a true enzyme, since the initially oxidized protein is released from the reaction in a reduced state. But DsbA can function as a catalyst *in vivo* because it is immediately reoxidized by the inner membrane protein DsbB.
- After oxidizing DsbA, reduced DsbB donates electrons to members of the electron transport chain and becomes reoxidized. Therefore, disulfide bond formation is linked to the electron transport chain, and specifically, to quinone reduction.

Isomerase pathway (DsbC/DsbG-DsbD): disulfide isomerization
The second pathway involved in oxidative protein folding in the *E. coli* periplasm is the isomerization pathway. DsbA, being a relatively non-specific disulfide bond donor, can introduce non-native disulfides into a protein that contains three or more cysteines. To salvage proteins that contain

non-native disulfide bonds, *E. coli* employs the isomerization pathway to rearrange the disulfides and rescue proteins from folding traps.

The isomerization pathway consists of the proteins DsbC, DsbG and DsbD (Missiakas et al., 1994, 1995). DsbC and DsbG are both periplasmic proteins that are responsible for disulfide isomerization. It is unclear how these two proteins differ in their *in vivo* function and substrate recognition. For our purposes here, we will assume that DsbC and DsbG are essentially equivalent, and to simplify, we will discuss DsbC exclusively.

DsbC has been shown to be required for the proper folding of eukaryotic proteins with multiple disulfides that are expressed in *E. coli* (Rietsch et al., 1996). For instance, when mouse urokinase, an enzyme containing 12 disulfide bonds, is expressed in and targeted to the periplasm of a DsbC deficient strain, no properly folded urokinase can be detected. In wild-type *E. coli*, on the other hand, active mouse urokinase accumulates in the periplasm. In this way, the isomerase activity of DsbC is clearly necessary to achieve properly folded urokinase.

The crystal structure of DsbC has been solved to 1.9 Å (McCarthy et al., 2000), and shows that the overall structure of DsbC is a V-shaped homodimer (Figure 7.3B). The C-terminal region of each monomer has a thioredoxin-like fold with a redox active CXXC motif, placing DsbC in the thioredoxin superfamily. The N-terminal domain of each monomer is responsible for dimerization.

Like DsbA, DsbC has a very reactive CXXC, which is only slightly less oxidizing than the CXXC of DsbA. Despite this similarity with DsbA, DsbC has a completely different function than DsbA. DsbC rearranges disulfides, while DsbA donates disulfides. How is it possible for DsbA and DsbC, two proteins with similar structures and nearly identical redox potentials, to serve completely different *in vivo* functions? The different *in vivo* role of DsbC is a function of its redox state. In order for DsbC to act as a disulfide isomerase, DsbC has to be maintained in a *reduced* state. Recall that DsbB maintains DsbA in an *oxidized* state *in vivo*. When a protein contains a non-native disulfide bond, the reduced DsbC attacks one of the cysteines of the disulfide, forming a mixed disulfide between itself and the target protein (Figure 7.6). Once the non-native disulfide is reduced and the mixed disulfide between DsbC and the substrate protein forms, one of two things may happen: (1) a second cysteine in the substrate protein may attack to form the native disulfide bond, releasing DsbC in a reduced state. In this case, DsbC is acting as a true enzyme, since it is unchanged at the end of the reaction; (2) after DsbC forms a mixed disulfide with the substrate protein, the protein may be unable to form the native disulfide. Here, DsbC may form an intramolecular disulfide bond at its CXXC motif, releasing the substrate protein in a reduced state. In this case, DsbC would not be acting as a true enzyme, as it is changed at the end of the reaction, and actually is acting to reduce the non-native disulfide instead of isomerizing it. This

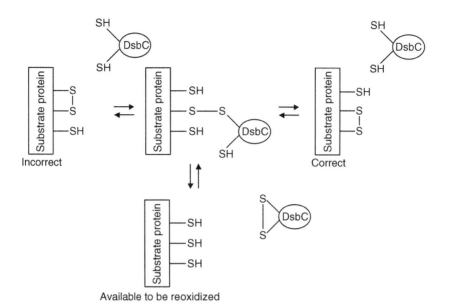

Available to be reoxidized

Figure 7.6 The isomerization of wrongly formed disulfides is achieved by DsbC. After reduced DsbC attacks an incorrect disulfide, forming an intermolecular disulfide between itself and the target protein, one of two things can occur: (1) The intermolecular disulfide between the target protein and DsbC is exchanged for the correct intramolecular disulfide in the target protein, releasing DsbC in a reduced state. (2) The intermolecular disulfide between the target protein and DsbC is exchanged for an intramolecular disulfide within DsbC itself, releasing the target protein in a reduced state and DsbC is in an oxidized state.

allows the protein to undergo another round of oxidation by DsbA. Although possible, the second mechanism is probably more rare since DsbC's redox potential is rather oxidizing. In either case, for DsbC to function properly, the protein must be maintained in a reduced state. This is achieved by the inner membrane protein DsbD.

DsbD is a 53 kDa inner membrane protein responsible for maintaining DsbC in a reduced state (Missiakas et al., 1995; Rietsch et al., 1996). DsbD deficient mutants accumulate DsbC in an oxidized state, and show isomerization defects similar to the ones observed in *dsbC⁻* mutants. DsbD consists of three domains, an N-terminal domain (α), a transmembrane domain (β) and the C-terminal thioredoxin-like domain (γ) (Figure 7.7). Genetic evidence suggests that DsbD delivers reducing equivalents to DsbC from the reducing power of the cytosolic thioredoxin system (Stewart et al., 1999; Chung et al., 2000) (Figure 7.7). In this model electrons move from cytosolic thioredoxin, via the inner membrane protein DsbD, to DsbC. In this way, DsbD is able to harness the reducing power of the cytosolic thioredoxin system, and move electrons through the membrane to maintain DsbC in a reduced state.

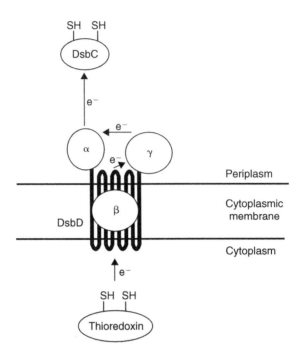

Figure 7.7   The predicted isomerization pathway. DsbD taps the reducing power of the cytosolic thioredoxin system, and delivers reducing equivalents (electrons) from the cytosol to the periplasm. By moving electrons through the membrane and donating them to DsbC, DsbD maintains DsbC in a reduced state.

Based on genetic evidence, a specific mechanism has been proposed for the movement of electrons through DsbD, which involves a complex thiol–disulfide exchange cascade (Katzen and Beckwith, 2000). In this mechanism, electrons are passed from thioredoxin to the β-domain of DsbD. Then through a thiol–disulfide cascade, the β-domain delivers electrons to the γ-domain, which then passes the electrons on to the α-domain. The α-domain is then responsible for delivering electrons to DsbC, thereby maintaining DsbC in a reduced state. Though the exact mechanism is still unclear, it is known that DsbD is able to tap the reducing power of the cytosol in order to maintain DsbC in a reduced state.

Summary: *Isomerase pathway (DsbC/DsbG-DsbD)*

- DsbA, being a non-specific and powerful oxidase, can introduce non-native disulfide bonds into proteins that contain three or more cysteines. To salvage these proteins from folding traps, *E. coli* employs the isomerization pathway.

- DsbC and DsbG are the isomerases responsible for shuffling the non-native disulfides to their native conformation. To function, DsbC and DsbG must be maintained in a reduced state. This is accomplished by DsbD.
- DsbD uses the reducing power of the cytosolic thioredoxin system, and delivers these reducing equivalents (electrons) to the periplasmic DsbC.

Oxidation and isomerization a possible futile cycle?
The activity of the oxidative pathway requires that DsbA is kept in an oxidized state *in vivo*, while the activity of the isomerization pathway requires that DsbC is kept in a reduced state *in vivo*. It is interesting that these two systems, which require exactly opposite redox states, can coexist in the periplasm. If there were any cross-talk between these two systems, there could be reduction of DsbA and oxidation of DsbC, resulting a futile cycle.

Bader et al. (2001) addressed the question of how *E. coli* can maintain two very similar pathways with opposite redox requirements in the same cellular compartment. The authors concluded that the two pathways are partitioned by the dimerization of DsbC. The authors found that mutations in the dimerization interface of DsbC, which prevented dimer formation, allowed the DsbC monomers to become oxidized by DsbB and partially complement a *dsbA⁻* mutation. Dimeric wild-type DsbC, on the other hand, is not a DsbB substrate (Bader et al., 2001). This indicates that the dimerization of DsbC protects it from DsbB oxidation, and serves as a means of separating these two pathways (Bader et al., 2001).

Eukaryotic disulfide bond formation

Most of the proteins that are secreted from a eukaryotic cell contain disulfide bonds. It is important that these bonds are formed and formed correctly. Thus the secretory compartments of a eukaryotic cell need both a source of oxidizing power and a method to ensure that the correct disulfides are formed. Recall that the original folding experiments performed by Anfinsen on ribonuclease were performed in the presence of oxygen. Oxygen present in the experiment was responsible for oxidizing the thiols of denatured ribonuclease, reforming the disulfides. Though oxygen could oxidize thiols to form a disulfide, Anfinsen correctly foresaw the need for a catalytic system *in vivo* to speed up the process of disulfide formation. He initiated a search for such a catalyst and succeeded in purifying a protein that came to be known as protein disulfide isomerase (PDI). This protein is a resident of the ER, the cellular compartment of eukaryotes where disulfides are formed. PDI is thought to have two roles, the oxidation of disulfides and the isomerization of incorrect disulfides (Noiva, 1999). The formation and rearrangement of disulfides is often the rate limiting step in protein folding *in vitro*, and since PDI can accelerate this process it is a true protein folding catalyst, the first one discovered.

The name for PDI emphasized its ability to isomerase incorrect disulfides. Indeed PDI, under the appropriate *in vitro* redox conditions, can be added to proteins whose disulfide bonds have formed incorrectly and PDI will speed up the acquisition of correct disulfides. Under other redox conditions, however, PDI will catalyze disulfide oxidation reactions (Noiva, 1999). How can one protein act both as an isomerase and as an oxidase? PDI, at its active sites, possesses pairs of cysteines, which are present in a CXXC motif. If these cysteines are in the oxidized form, PDI can act by donating its disulfides to secreted proteins, thus catalyzing their oxidation. In this way, PDI acts very similarly to the prokaryotic oxidase DsbA. If the CXXC motifs of PDI are present in a reduced form they can attack incorrectly formed disulfides. If PDI only transiently forms a mixed disulfide with the protein and then re-donates the disulfide to the protein, it has the potential of forming a different, possibly correct, disulfide in the protein (Woycechowsky and Raines, 2000). In this latter way PDI acts very much like the prokaryotic disulfide isomerase DsbC. As an isomerase PDI is acting as a true catalyst, it enters and leaves the reaction in the same form.

Summary: *Eukaryotic disulfide bond formation*

- Protein disulfide isomerase (PDI) catalyzes the formation of correct disulfides.
- PDI has both a disulfide oxidase activity that donates disulfides to proteins and a disulfide isomerase activity that rearranges incorrect disulfides.

The structure of protein disulfide isomerase
PDI is a protein of about 500 residues composed of five separate domains. These domains have been designated a, b, b', a' and c (Ferrari and Soling, 1999). Surprisingly, four of these domains, a, b, b' and a', all belong to the thioredoxin family. The domains a and a' are recognizably similar at a sequence level to thioredoxin and each contain a single redox active CXXC motif that can form a disulfide that links the adjacent cysteines. When the individual thioredoxin-like a and a' domains were expressed on their own, they were found to be capable of rapid disulfide exchange reactions. However, these individual domains lack the full isomerase activity exhibited by the complete PDI molecule (Darby and Creighton, 1995). It only became apparent that the b and b' domains are thioredoxin-like after the nuclear magnetic resonance (NMR) structure of b was solved. The b and b' domains do not contain thiol reactive CXXC motifs, but are important nonetheless in making PDI an effective isomerase, possibly because of a role in promoting peptide–peptide interactions. The domain with the greatest peptide binding activity, b', is the one that is most important for PDI's isomerase activity (Klappa et al., 1998).

Summary: *The structure of protein disulfide isomerase*

- The PDI protein consists of multiple domains; two of these contain active site CXXC motifs, the others probably help PDI to bind to partially unfolded proteins, and contribute to PDI's ability to isomerize incorrect disulfides.

What reoxidizes PDI?

Although for many years work on PDI emphasized its isomerase role, evidence from yeast has recently shown us that PDI is also important in acting as a carrier of oxidizing equivalents. PDI mutants accumulate secreted proteins in the reduced form, suggesting that PDI functions as an oxidant. Mixed disulfides can be detected between PDI and secreted proteins, and PDI is required for the oxidation of secreted proteins such as carboxypeptidase Y. If PDI acts as a source of oxidizing power, it will leave the disulfide exchange reaction in a reduced form, requiring then a factor that can reoxidize PDI. Otherwise the cell would waste a PDI molecule each time it needed to make a couple of disulfides. Two different genetic screens implicated a protein called Ero1 as the reoxidant of PDI (Frand and Kaiser, 1999; Pollard et al., 1998). Conditional mutants in the gene for this essential protein showed a decreased oxidative capacity. Secreted proteins that normally contained disulfide bonds accumulate in a reduced form in Ero1 mutants and these mutants are sensitive to the reductant DTT. PDI is also predominantly in a reduced form in this mutant, and mixed disulfides between PDI and Ero1 can be isolated. Based on this evidence a pathway has been proposed in which oxidizing equivalents flow from Ero1 to PDI and then on to the substrate proteins (Figure 7.8).

Ero1 is a membrane associated ER resident protein that contains flavin adenine dinucleotide (FAD) as a cofactor (Tu et al., 2000). Not only is oxidative folding *in vivo* very sensitive to FAD levels in the cell, but the oxidative folding system can be reconstituted using reduced protein, PDI, Ero1 and FAD. It has been suggested that FAD may be the source of oxidative power in eukaryotic systems, like quinones are the source in prokaryotic systems (Bader et al., 2000). Glutathione, a small molecule that was long thought to play a major role in oxidation in the ER, more probably acts as a reductant that removes incorrectly formed disulfides (Cuozzo and Kaiser, 1999).

Summary: *What oxidizes PDI?*

- Oxidizing equivalents are thought to flow from FAD to Ero1, to PDI and then on to substrate proteins.

Roles of PDI homologs

A wide variety of PDI homologs exist in the ER that appear to be functionally redundant with PDI. The best-characterized system is yeast, where

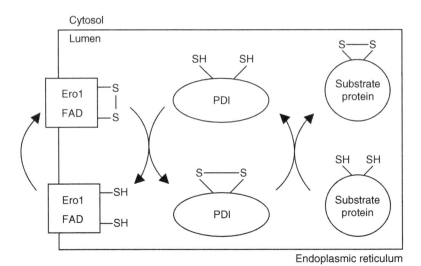

Figure 7.8   Proposed pathway for the reoxidation of PDI by Ero1. The membrane associated Ero1 uses the oxidizing power of FAD to reoxidize PDI, which allows PDI to function as a catalytic donor of disulfides.

five different thioredoxin-like proteins exist in the ER: PDI1, MPD1, MPD2, EUG1 and ESP1. PDI1, the classic protein disulfide isomerase, is an essential gene. All the others, when individually overexpressed, can at least partially compensate for the lethal growth defect associated with a deletion in the PDI1 gene (Tachibana and Stevens, 1992; Tachikawa et al., 1995, 1997; Wang and Chang, 1999). Their ability to restore viability to a *pdi*1-deleted strain when overexpressed, however, is also dependent on the presence of one or more of the other PDI homologs. These genetic results have been interpreted to mean that this family of thioredoxin-like proteins cooperate to catalyze both the formation and isomerization of disulfide bonds. Both the isomerase and oxidase functions appear to be essential (Norgaard et al., 2001). In the mammalian ER the number of putative PDIs is even higher than in yeast (Ferrari and Soling, 1999). The different PDI homologs differ in their domain organization. The stress-induced protein ERp72, for instance, has three redox active thioredoxin-like domains instead of the two found in 'normal' PDI (Dorner et al., 1990; Mazzarella et al., 1990). These mammalian PDI equivalents are capable of rescuing PDI1 null mutants in yeast, showing that they are essentially equivalent (Gunther et al., 1993; Laboissiere et al., 1995). The presence of several active site CXXCs within a single molecule may enhance the ability of the protein to isomerase disulfide bonds. Consistent with this the prokaryotic protein disulfide isomerase, DsbC, functions as a dimer of two thioredoxin like molecules, both of which contain a CXXC motif. This dimerization appears to be required for DsbC's isomerase activity (Sun and Wang, 2000). The DsbA protein, in contrast, has only one thioredoxin-like domain and functions only as an oxidase. It has essentially no isomerase activity (Zapun and Creighton, 1994).

OVERALL SUMMARY OF DISULFIDE BOND FORMATION

Disulfide bond formation and isomerization are catalyzed processes in both eukaryotes and prokaryotes. Disulfide bond formation occurs via a thiol–disulfide exchange reaction whereby a disulfide is transferred from the thioredoxin-like CXXC active site of disulfide donor to a substrate protein. The oxidizing equivalents necessary to form disulfide bonds in prokaryotes are generated by the reduction of quinones, while in eukaryotes the reduction of FAD may be involved. Disulfide isomerization requires that the incorrect disulfides be attacked using a reduced catalyst. This reaction is followed by the re-donation of the disulfide. This allows an alternate disulfide pairing. The disulfide isomerase needs to be reduced to be active. The ultimate source of these reducing equivalents in prokaryotes is NADPH via the thioredoxin system; in eukaryotes reduced glutathione may be involved. In prokaryotes there is a clear separation between the oxidative and isomerase pathways. In eukaryotes a large number of PDI homologs with both PDI and oxidase activities cooperate to catalyze proper disulfide bond formation.

REFERENCES

Anfinsen, C.B. (1973) Principles that govern the folding of protein chains. *Science* **181**: 223–230.

Anfinsen, C.B., Haber, E., Sela, M. and White, F.H. (1961) The kinetics of formation of native ribonuclease during oxidation of the reduced polypeptide chain. *Proc Natl Acad Sci USA* **47**: 1309–1314.

Bader, M., Muse, W., Ballou, D.P., Gassner, C. and Bardwell, J.C. (1999) Oxidative protein folding is driven by the electron transport system. *Cell* **98**: 217–227.

Bader, M.W., Xie, T., Yu, C.A. and Bardwell, J.C. (2000) Disulfide bonds are generated by quinone reduction. *J Biol Chem* **275**: 26082–26088.

Bader, M.W., Hiniker, A., Regeimbal, J. et al. (2001) Turning a disulfide isomerase into an oxidase: DsbC mutants that imitate DsbA. *EMBO J* **20**: 1555–1562.

Bardwell, J.C. (1994) Building bridges: disulphide bond formation in the cell. *Mol Microbiol* **14**: 199–205.

Bardwell, J.C., McGovern, K. and Beckwith, J. (1991) Identification of a protein required for disulfide bond formation *in vivo*. *Cell* **67**: 581–589.

Bardwell, J.C., Lee, J.O., Jander, G. et al. (1993) A pathway for disulfide bond formation *in vivo*. *Proc Natl Acad Sci USA* **90**: 1038–1042.

Chung, J., Chen, T. and Missiakas, D. (2000) Transfer of electrons across the cytoplasmic membrane by DsbD, a membrane protein involved in thiol–disulphide exchange and protein folding in the bacterial periplasm. *Mol Microbiol* **35**: 1099–1109.

Cuozzo, J.W. and Kaiser, C.A. (1999) Competition between glutathione and protein thiols for disulphide-bond formation. *Nat Cell Biol* **1**: 130–135.

Dailey, F. and Berg, H. (1993) Mutants in disulfide bond formation that disrupt flagellar assembly in *Escherichia coli*. *Proc Natl Acad Sci USA* **90**: 1043–1047.

Darby, N.J. and Creighton, T.E. (1995) Functional properties of the individual thioredoxin-like domains of protein disulfide isomerase. *Biochemistry* **34**: 11725–11735.

Dorner, A.J., Wasley, L.C., Raney, P. et al. (1990) The stress response in Chinese hamster ovary cells. Regulation of ERp72 and protein disulfide isomerase expression and secretion. *J Biol Chem* **265**: 22029–22034.

Ferrari, D.M. and Soling, H.D. (1999) The protein disulphide-isomerase family: unravelling a string of folds. *Biochem J* **339**: 1–10.

Frand, A.R. and Kaiser, C.A. (1999) Ero1p oxidizes protein disulfide isomerase in a pathway for disulfide bond formation in the endoplasmic reticulum. *Mol Cell* **4**: 469–477.

Freedman, R.B., Hirst, T.R. and Tuite, M.F. (1994) Protein disulphide isomerase: building bridges in protein folding. *Trends Biochem Sci* **19**: 331–336.

Gilbert, H.F. (1990) Molecular and cellular aspects of thiol-disulfide exchange. *Adv Enzymol Relat Areas Mol Biol* **63**: 69–172.

Goldberger, R.F., Epstein, C.J. and Anfinsen, C.B. (1963) Acceleration of reactivation of reduced bovine pancreatic ribonuclease by a microsomal system from rat liver. *J Biol Chem* **238**: 628–635.

Guilhot, C., Jander, G., Martin, N.L. and Beckwith, J. (1995) Evidence that the pathway of disulfide bond formation in *Escherichia coli* involves interactions between the cysteines of DsbB and DsbA. *Proc Natl Acad Sci USA* **92**: 9895–9899.

Gunther, R., Srinivasan, M., Haugejorden, S. et al. (1993) Functional replacement of the *Saccharomyces cerevisiae* Trg1/Pdi1 protein by members of the mammalian protein disulfide isomerase family. *J Biol Chem* **268**: 7728–7732.

Jakob, U., Muse, W., Eser, M. and Bardwell, J.C. (1999) Chaperone activity with a redox switch. *Cell* **96**: 341–352.

Jander, G., Martin, N.L. and Beckwith, J. (1994) Two cysteines in each periplasmic domain of the membrane protein DsbB are required for its function in protein disulfide bond formation. *EMBO J* **13**: 5121–5127.

Katzen, F. and Beckwith, J. (2000) Transmembrane electron transfer by the membrane protein DsbD occurs via a disulfide bond cascade. *Cell* **103**: 769–779.

Kishigami, S. and Ito, K. (1996) Roles of cysteine residues of DsbB in its activity to reoxidize DsbA, the protein disulphide bond catalyst of *Escherichia coli*. *Genes Cells* **1**: 201–208.

Kishigami, S., Akiyama, Y. and Ito, K. (1995) Redox states of DsbA in the periplasm of *Escherichia coli*. *FEBS Lett* **364**: 55–58.

Kishigami, S., Kanaya, E., Kikuchi, M. and Ito, K. (1995) DsbA–DsbB interaction through their active site cysteines. Evidence from an odd cysteine mutant of DsbA. *J Biol Chem* **270**: 17072–17074.

Klappa, P., Ruddock, L.W., Darby, N.J. and Freedman, R.B. (1998) The b' domain provides the principal peptide-binding site of protein disulfide isomerase but all domains contribute to binding of misfolded proteins. *EMBO J* **17**: 927–935.

Kobayashi, T. and Ito, K. (1999) Respiratory chain strongly oxidizes the CXXC motif of DsbB in the *Escherichia coli* disulfide bond formation pathway. *EMBO J* **18**: 1192–1198.

Kobayashi, T., Kishigami, S., Sone, M. et al. (1997) Respiratory chain is required to maintain oxidized states of the DsbA–DsbB disulfide bond formation system in aerobically growing *Escherichia coli* cells. *Proc Natl Acad Sci USA* **94**: 11857–11862.

Laboissiere, M.C., Chivers, P.T. and Raines, R.T. (1995) Production of rat protein disulfide isomerase in *Saccharomyces cerevisiae*. *Protein Expr Purif* **6**: 700–706.

Martin, J.L., Bardwell, J.C. and Kuriyan, J. (1993) Crystal structure of the DsbA protein required for disulphide bond formation *in vivo*. *Nature* **365**: 464–468.

Mazzarella, R.A., Srinivasan, M., Haugejorden, S.M. and Green, M. (1990) ERp72, an abundant luminal endoplasmic reticulum protein, contains three copies of the active site sequences of protein disulfide isomerase. *J Biol Chem* **265**: 1094–1101.

McCarthy, A.A., Haebel, P.W., Torronen, A. et al. (2000) Crystal structure of the protein disulfide bond isomerase, DsbC, from *Escherichia coli*. *Nat Struct Biol* **7**: 196–199.

Missiakas, D., Georgopoulos, C. and Raina, S. (1993) Identification and characterization of the *Escherichia coli* gene DsbB, whose product is involved in the formation of disulfide bonds *in vivo*. *Proc Natl Acad Sci USA* **90**: 7084–7088.

Missiakas, D., Georgopoulos, C. and Raina, S. (1994) The *Escherichia coli* dsbC (xprA) gene encodes a periplasmic protein involved in disulfide bond formation. *EMBO J* **13**: 2013–2020.

Missiakas, D., Schwager, F. and Raina, S. (1995) Identification and characterization of a new disulfide isomerase-like protein (DsbD) in *Escherichia coli*. *EMBO J* **14**: 3415–3424.

Noiva, R. (1999) Protein disulfide isomerase: the multifunctional redox chaperone of the endoplasmic reticulum. *Semin Cell Dev Biol* **10**: 481–493.

Norgaard, P., Westphal, V., Tachibana, C. et al. (2001) Functional differences in yeast protein disulfide isomerases. *J Cell Biol* **152**: 553–562.

Pollard, M.G., Travers, K.J. and Weissman, J.S. (1998) Ero1p: a novel and ubiquitous protein with an essential role in oxidative protein folding in the endoplasmic reticulum. *Mol Cell* **1**: 171–182.

Rensing, C., Mitra, B. and Rosen, B.P. (1997) Insertional inactivation of dsbA produces sensitivity to cadmium and zinc in *Escherichia coli*. *J Bacteriol* **179**: 2769–2771.

Rietsch, A., Belin, D., Martin, N. and Beckwith, J. (1996) An *in vivo* pathway for disulfide bond isomerization in *Escherichia coli*. *Proc Natl Acad Sci USA* **93**: 13048–13053.

Stafford, S.J., Humphreys, D.P. and Lund, P.A. (1999) Mutations in dsbA and dsbB, but not dsbC, lead to an enhanced sensitivity of *Escherichia coli* to Hg2+ and Cd2+. *FEMS Microbiol Lett* **174**: 179–184.

Stewart, E.J., Katzen, F. and Beckwith, J. (1999) Six conserved cysteines of the membrane protein DsbD are required for the transfer of electrons from the cytoplasm to the periplasm of *Escherichia coli*. *EMBO J* **18**: 5963–5971.

Sun, X.X. and Wang, C.C. (2000) The N-terminal sequence (residues 1-65) is essential for dimerization, activities and peptide binding of *Escherichia coli* DsbC. *J Biol Chem* **275**: 22743–22749.

Tachibana, C. and Stevens, T.H. (1992) The yeast EUG1 gene encodes an endoplasmic reticulum protein that is functionally related to protein disulfide isomerase. *Mol Cell Biol* **12**: 4601–4611.

Tachikawa, H., Takeuchi, Y., Funahashi, W. et al. (1995) Isolation and characterization of a yeast gene, MPD1, the overexpression of which suppresses inviability caused by protein disulfide isomerase depletion. *FEBS Lett* **369**: 212–216.

Tachikawa, H., Funahashi, W., Takeuchi, Y. et al. (1997) Overproduction of Mpd2p suppresses the lethality of protein disulfide isomerase depletion in a CXXC sequence dependent manner. *Biochem Biophys Res Commun* **239**: 710–714.

Tu, B.P., Ho-Schleyer, S.C., Travers, K.J. and Weissman, J.S. (2000) Biochemical basis of oxidative protein folding in the endoplasmic reticulum. *Science* **290**: 1571–1574.

Wang, Q. and Chang, A. (1999) Eps1, a novel PDI-related protein involved in ER quality control in yeast. *EMBO J* **18**: 5972–5982.

Woycechowsky, K.J. and Raines, R.T. (2000) Native disulfide bond formation in proteins. *Curr Opin Chem Biol* **4**: 533–539.

Wunderlich, M. and Glockshuber, R. (1993) Redox properties of protein disulfide isomerase (DsbA) from *Escherichia coli*. *Protein Sci* **2**: 717–726.

Zapun, A. and Creighton, T.E. (1994) Effects of DsbA on the disulfide folding of bovine pancreatic trypsin inhibitor and alpha-lactalbumin. *Biochemistry* **33**: 5202–5211.

Zapun, A., Bardwell, J.C. and Creighton, T.E. (1993) The reactive and destabilizing disulfide bond of DsbA, a protein required for protein disulfide bond formation *in vivo*. *Biochemistry* **32**: 5083–5092.

SUGGESTED READING

Fabianek, R.A., Hennecke, H. and Thony-Meyer, L. (2000) Periplasmic protein thiol:disulfide oxidoreductases of *Escherichia coli*. *FEMS Microbiol Rev* **24**: 303–316.

Frand, A.R., Cuozzo, J.W. and Kaiser, C.A. (2000) Pathways for protein disulphide bond formation. *Trends Cell Biol* **10**: 203–210.

Gilbert, H.F. (1990) Molecular and cellular aspects of thiol–disulfide exchange. *Adv Enzymol Relat Areas of Mol Biol* **63**: 69–172.

8

THE UNFOLDED PROTEIN RESPONSE

CARMELA SIDRAUSKI, JASON H. BRICKNER AND PETER WALTER

PREFACE

The endoplasmic reticulum (ER) is the gateway to the secretory pathway. Proteins which are destined either to be secreted from the cell or to reside in one of the compartments along the secretory pathway are translocated across the membrane of the ER. Proteins arrive in the ER lumen as linear unmodified polypeptides and must fold and undergo a series of post-translational modifications. The ER contains a group of specialized ER chaperones and modifying enzymes that assist in the folding and modification of newly translocated polypeptides.

The ER provides a unique protein folding environment that harbors high concentrations of protein chaperones and modifying enzymes. The major molecular chaperones implicated in folding are homologs of the large cytoplasmic heat shock proteins (Hsps) and are thought to function in an analogous fashion (Gething and Sambrook, 1992). Bip/GRP78, a member of the Hsp70 family of molecular chaperones, and GRP94, a member of the Hsp90 family, are thought to bind transiently to most nascent proteins as they are translocated into the ER lumen. In contrast to the cytosol, the ER is an oxidizing environment, allowing formation of disulfide bonds in secretory and membrane proteins, often coincident with protein folding. This reaction is catalyzed by protein disulfide isomerase. Similarly, asparagine-linked glycosyl chains are added by a glycosyl transferase complex. In many cases, carbohydrate addition and subsequent modifications help proteins achieve their correct conformation (Helenius, 1994). The ER also contains unique lectin-like proteins (calnexin and calreticulin) that recognize incompletely modified carbohydrate chains found on protein folding intermediates.

What happens if membrane and secretory proteins do not properly fold or assemble in the ER lumen? A quality control system exists that ensures that only folded proteins and fully assembled complexes continue their journey through the secretory pathway. This system protects cells from the potentially deleterious effects of delivery of partially misfolded polypeptides to their site of action. Protein folding intermediates and terminally misfolded proteins are retained in the ER lumen through interactions with chaperones.

Cells respond to the accumulation of misfolded proteins by transcriptionally upregulating ER-resident enzymes that catalyze the folding, assembly and modification of secretory proteins in the lumen. Thus, this 'unfolded protein response' (UPR) increases the folding capacity of the ER according to demand. The UPR signaling pathway senses the need for more chaperones and modifying enzymes in the ER and transmits the information across the ER membrane to the nucleus where transcription is activated. Thus, unlike the more commonly known signaling pathways that originate in the plasma membrane and sense changes outside the cell, the UPR senses and signals changes that occur in an organelle within the cell. Intracellular signaling pathways of this sort play a critical role in the maintenance of cellular homeostasis (reviewed by Nunnari and Walter, 1996).

OUTPUT OF THE UPR

The UPR remodels the secretory pathway to overcome ER stress

The UPR pathway was first described in mammalian cells where it was found to induce a set of ER resident proteins, called GRPs or glucose regulated proteins (Lee, 1987). This set of polypeptides is highly induced upon glucose starvation of mammalian cells. It was subsequently recognized that glucose deprivation leads to protein misfolding in the ER, probably because it impairs the glycosylation of secretory proteins. Among the glucose regulated proteins are BiP (GRP78) and GRP94, which are the major molecular chaperones involved in folding in the ER lumen. In addition, GRP170, an Hsp70-like protein, protein disulfide isomerase (PDI; which promotes correct disulfide-bond formation), and two PDI-related ER proteins (Erp72 and GRP58) were also found to be upregulated in mammalian cells (Mazarella et al., 1990; Chen et al., 1996). A similar set of genes was found to be induced by ER stress in *Saccharomyces cerevisiae*. They included BiP (encoded by KAR2), yeast PDI, Eug1 (a PDI-like protein), peptidylprolyl *cis-trans* isomerase Fkbp2, Lhs1p (an Hsp70-like protein) and Ero1p, a protein required for maintenance of the ER redox potential (Pollard et al., 1998).

The coordinate transcriptional induction of the UPR target genes suggests that they are regulated by a common transcription factor. Consistent with this idea, regulatory elements common to many of the UPR target genes have been identified in mammals and yeast. In mammalian cells, a unique sequence, ERSE (ER stress response element), is conserved in multiple promoters of the UPR target genes (Roy and Lee, 1999; Yoshida et al., 1998). Likewise, many of the UPR target genes identified in yeast shared a common regulatory element in their promoters, the unfolded protein response element (UPRE; Kohno et al., 1993; Mori et al., 1992). This 22 bp element was shown to be necessary and sufficient to activate transcription in response to the accumulation of unfolded proteins in the ER.

Genes regulated by a different upstream activating sequence, the UAS_{ino}, are also activated by the UPR (Cox et al., 1997). These genes include a number of enzymes involved in the biosynthesis of inositol and choline (Carman and Henry, 1989). Changes in the concentration of inositol, or in the levels of unfolded proteins in the ER, induce transcription of both UPRE- and UAS_{ino}-controlled genes. Because inositol and choline are precursors for phospholipid biosynthesis and because the ER is the major site of cellular lipid synthesis and membrane production, this may be a mechanism by which the cell is able to coordinately produce ER lumenal and membrane components. When unfolded proteins accumulate and lumenal chaperones are induced, the membranes to house the extra contents may also need to be increased. Conversely, when more lipid synthesis is needed and the ER membrane expands, the production of chaperones may need to be increased to maintain an appropriately high concentration of these folding enzymes.

The use of DNA microarrays to analyze transcriptional changes in response to unfolded proteins in the ER has recently been employed to identify additional targets of the yeast UPR (Travers et al., 2000). This technique simultaneously monitors changes in the mRNA levels of all yeast genes and has proven to be a very powerful tool in determining the total transcriptional response to a given condition or developmental stage. Of 6300 yeast genes, more than 350 are induced under conditions that activate the UPR.

The most striking revelation from this analysis is that the UPR regulates a battery of genes whose products are involved in many different aspects of ER and secretory pathway function (reviewed by Hampton, 2000). The genes of known function that are induced encode not only chaperones and protein modification enzymes but also a whole variety of factors implicated in protein translocation (and retro-translocation), lipid metabolism, glycosylation, ER-associated degradation (ERAD), ER to Golgi traffic, Golgi–ER retrieval and protein targeting to the vacuole and to the cell surface. The regulation of numerous classes of genes suggests that the cell responds

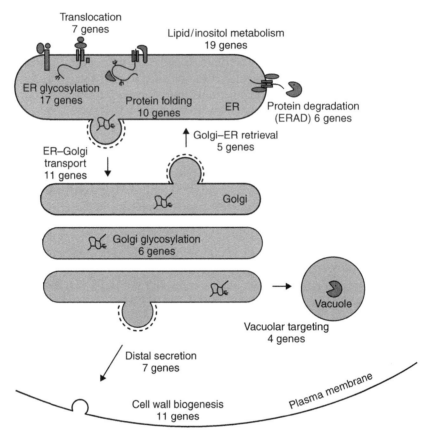

Figure 8.1 The UPR remodels the entire secretory pathway. The UPR increases transcription of genes encoding proteins involved in various aspects of the secretory pathway. The UPR targets are proteins involved in protein translocation, glycosylation, folding, ER-associated protein degradation (ERAD), lipid metabolism, ER–Golgi transport, Golgi–ER retrieval, protein targeting to the vacuole and the cell surface, and cell wall biogenesis.

to the accumulation of unfolded proteins in the ER by coordinately remodeling the entire secretory pathway in response to folding stress in the ER (Figure 8.1). When unfolded proteins accumulate in the ER, the UPR thus maximizes the likelihood that newly translocated proteins will fold and be properly modified (by upregulating chaperones and modifying enzymes) and enhances the capacity of the cell to translocate proteins out of the ER into the cytosol where they are degraded. By affecting the efficiency of folding, post-translational processing, export and degradation of secretory proteins, the load of unfolded proteins in the ER is diminished. In addition to promoting folding or degradation of unfolded proteins, the UPR also upregulates genes involved in both transport between the ER and the Golgi and vacuolar targeting. It may be that, when other mechanisms of disposal have become overwhelmed, the cell exploits anterograde vesicular transport out of the ER to target misfolded proteins to the vacuole for degradation (Hong et al., 1996).

The recent analysis of the numerous UPR target genes discovered using the DNA microarray technology has raised some interesting questions. First, many of the genes identified have no known function (Travers et al., 2000). By analogy to the known UPR targets, these new targets are excellent candidates for genes encoding factors with important functions in the secretory pathway. Future work can directly test this hypothesis. Second, unlike the earliest identified UPR target genes, the majority of the newly identified target genes do not contain the classical UPRE in their promoters. Further research is needed to determine what constitutes the minimal promoter elements that are responsible for regulation by the UPR and how they are distributed among the promoters of the various UPR target genes.

Whereas a comprehensive genomic analysis of the mammalian UPR has not been reported, the recent identification of diverse target genes suggests a similarly broad pattern of gene induction. Some of the newly discovered targets include the dolichol pathway (Doerrler and Lehrman, 1999), which is essential for protein glycosylation, and an ER calcium pump required for maintenance of calcium levels (Caspersen et al., 2000). Another known target is CHOP (GADD153), which is a pro-apoptotic transcription factor (Wang et al., 1998). As we will discuss later in this chapter, the UPR response in mammalian cells is broad and even more complex than it is in yeast cells.

ER-associated degradation (ERAD)

As mentioned in the last section, the coordinate production of ER chaperones and folding enzymes in response to ER stress is only one of the mechanisms the cell utilizes to get rid of the extra load of misfolded proteins in the ER. A second and complementary mechanism by which cells decrease the concentration of unfolded proteins in the ER is by upregulating their degradation. This process is called ER-associated degradation or ERAD (Brodsky and McCracken, 1999; Plemper and Wolf, 1999).

ERAD involves several steps: first, unfolded proteins are detected by ER-resident factors, which probably include the ER chaperones and processing enzymes that are targets of the UPR; second, unfolded proteins are targeted to the translocon for dislocation into the cytosol; and third, upon arrival in the cytosol, proteins are deglycosylated, ubiquitinated and degraded by the proteasome (reviewed by Brodsky and McCracken, 1999;

Plemper and Wolf, 1999). Several factors have been identified in screens for mutants deficient in ERAD and many of these proteins are targets of the UPR pathway. The ERAD deficient mutants are viable under normal growth conditions, but they are dependent on a functional UPR. Mutants defective for ERAD appear to be under constant folding stress, as indicated by the observation that the UPR is constitutively induced (Ng et al., 2000; Travers et al., 2000). Unlike wild-type strains, in strains with defects in ERAD, the UPR is essential for survival. This shows that ERAD is occurring all the time and that its loss results in an elevation of unfolded proteins substantial enough to be detected by the sensors of the UPR.

Interestingly, the UPR is also required for efficient degradation of unfolded proteins. Misfolded proteins are stabilized in UPR mutants (Casagrande et al., 2000; Friedlander et al., 2000; Travers et al., 2000). Furthermore, consistent with the transcriptional induction of ERAD genes by the UPR, activation of the UPR accelerates the rate of degradation of misfolded proteins, even in the absence of folding stress. This shows that the UPR confers tolerance to an increase in the amount of aberrant ER proteins not only by promoting their folding but also by enhancing their turnover. Therefore the UPR increases both the folding capacity of the ER and the degradation of terminally misfolded proteins through ERAD, mediating both possible fates of misfolded proteins in the ER, folding or degradation. These complementary mechanisms both prevent accumulation of unfolded proteins during normal growth and mitigate the toxicity of misfolded proteins if their accumulation cannot be prevented. The interplay of folding and degradation has not yet been examined carefully in mammalian cells but it seems likely that the tight relationship between the UPR and ERAD has been conserved.

Summary: *Output of the UPR*

- The UPR regulates a broad set of genes that affect several different processes in the cell. Despite its overwhelming breadth, the UPR results primarily in the remodeling of the secretory pathway (Figure 8.1).
- A subset of UPR-target genes encode proteins that function in the ER compartment. These proteins are involved in protein folding, glycosylation, translocation and ERAD and act in a concerted fashion to directly reduce the concentration of misfolded proteins.
- A second subset includes genes that encode factors involved in vesicular transport from the ER to the Golgi, vacuole and cell surface.
- Furthermore, the induction of phospholipid biosynthetic enzymes (inositol response) may generate new membranes, thereby increasing the volume of the ER to prepare the compartment to receive an increased load of ER-resident proteins.

THE UPR SIGNALING PATHWAY IN YEAST

The power of genetics identifies signaling components

Although the UPR was originally identified in mammalian cells, the signaling components were first identified in the yeast *S. cerevisiae*. The power of genetics in this simple model organism proved critical in the discovery of the UPR signal transducers. Genetic screens utilizing reporters driven by the UPR regulatory element (UPRE) led to the isolation of loss of function alleles of *IRE1* that block induction of these reporters (Cox et al., 1993; Mori et al., 1993). Ire1p is a transmembrane protein with an N-terminal ER-lumenal portion and a C-terminal cytosolic portion containing a kinase domain. The primary structure of Ire1p, which is similar to that of growth-factor receptor kinases, immediately suggested what its function might be in the UPR signaling cascade; the N-terminal lumenal domain might sense the accumulation of unfolded proteins, transmitting the signal across the ER membrane either to the cytoplasm or directly to the nucleus via the kinase domain. Thus, Ire1p was proposed to be the most upstream signaling component in the UPR pathway.

The second component of the UPR signaling pathway to be identified is Hac1p. *HAC1* was identified by three different genetic approaches (Cox and Walter, 1996; Mori et al., 1996; Nikawa et al., 1996). The deduced amino acid sequence of Hac1p also immediately suggested what its function might be in the unfolded protein response: Hac1p is a DNA-binding protein with homology to the leucine zipper family of transcription factors. This suggested that Hac1p might bind upstream of UPR target genes to affect their transcription. Consistent with this hypothesis, gel-shift analysis showed that Hac1p specifically binds to the UPRE sequence found in the promoters of the target genes of the pathway. This analysis identified Hac1p as the most downstream component of the UPR pathway in yeast.

The third UPR component identified was *RLG1*, which encodes tRNA ligase (Sidrauski et al., 1996). tRNA ligase is an essential protein that functions in splicing of tRNAs by ligating together the two tRNA halves generated by tRNA endonuclease. An allele of *RLG1*, *rlg1-100*, which was isolated in a synthetic lethal screen, is phenotypically indistinguishable from that of mutants in the other two UPR signaling molecules (Sidrauski et al., 1996). However, unlike Ire1p and Hac1p, the identification of Rlg1p as a component in the UPR pathway did not suggest an obvious role for this protein in the transduction of the UPR signal from the ER to the nucleus. The possibility that tRNA might be a component of the pathway was initially perplexing. However, the subsequent discovery that *HAC1* mRNA splicing was essential for transduction of the UPR signal (see below) led to the discovery of a novel role for tRNA ligase.

Surprising aspects of the UPR signaling pathway were already emerging through the identification and initial analysis of the mutants. Further research, using a combination of biochemistry and genetics, delineated more clearly how the different components were activated and how they transmitted the signal from the ER lumen to the nucleus.

Ire1p is a sensor of unfolded proteins

Based on its domain structure and sequence homology, Ire1p seemed well suited to transduce the unfolded protein signal across the ER membrane. The N-terminal lumenal domain would sense the accumulation of unfolded proteins, which, by analogy to other transmembrane kinases, would lead to activation of its cytoplasmic effector domain. Unfortunately, the amino terminal half of Ire1p shows no sequence homology to any other known family of proteins and thus provides no clue as to the ligand responsible for its induction. In contrast, the cytoplasmic/nuclear half contains a domain with similarity to serine/threonine protein kinases, followed by a carboxy-terminal tail that shows sequence similarity to RNase L (Bork and Sander, 1993). The functional significance of this homology to a ribonuclease remained a mystery until further analysis revealed that a RNA processing event is a key step in the UPR signaling pathway.

Activation of Ire1p resembles that of well-characterized growth factor kinase receptors. Immunoprecipitation experiments showed that Ire1p is phosphorylated *in vivo* and that its phosphorylation increases upon activation of the UPR (Shamu and Walter, 1996). Consistent with a regulatory role for this modification, point mutations in the catalytic site of the kinase domain of Ire1p, which abolish its kinase activity, block the UPR. These results indicate that Ire1p's activation leads to induction of its kinase activity, which in turn results in its autophosphorylation and activation of downstream events in the pathway. In addition, both genetic and biochemical data suggested that upon activation of the pathway, Ire1p oligomerizes leading to its *trans*-autophosphorylation by neighboring Ire1p molecules (Shamu and Walter, 1996; Welihinda and Kaufman, 1996).

How is the activation of Ire1p regulated? What is the identity of the ligand that binds to the lumenal domain of Ire1p? Recent experiments carried out in mammalian cells demonstrated that dimerization of Ire1p appears to be the sole requirement for its activation. Replacing the N-terminal lumenal domain of Ire1p with an unrelated leucine zipper dimerization motif yields a constitutively active chimeric protein (Liu et al., 2000). Two models for the regulation of Ire1p by unfolded proteins were proposed. The first is a positive regulatory model in which oligomerization is induced in response to binding of a ligand to the ER lumenal domain of Ire1p. The ligand could be simply unfolded proteins per se, a complex of an unfolded protein bound to an ER chaperone, or an as-yet-unidentified ligand generated by the increased activity of chaperones in the ER compartment. In this model, activation of Ire1p would resemble that of mammalian receptor kinases in the plasma membrane of the cell, which oligomerize upon ligand binding to the extracellular domain.

The second model involves a negative regulatory mode in which in the resting state, i.e., when there are few unfolded proteins in the ER, a chaperone (such as BiP) binds to Ire1p and prevents its dimerization. When the concentration of unfolded proteins increases, the chaperone would be titrated off Ire1p, allowing its oligomerization and activation. Several indirect observations support this second model. First, when an improperly folded mutant of the simian virus 5-hemagglutinin-neuraminidase glycoprotein that does not bind BiP accumulates in the ER, no induction of the UPR was detected. In contrast, when other misfolded mutants of the same protein that bind BiP accumulate in the ER, the UPR was induced (Ng et al., 1992). This suggested that the cell is monitoring the levels of free BiP in the ER. Accordingly, overexpression of BiP downregulates the UPR both in Chinese hamster ovary (CHO) cells and in yeast (Dorner et al., 1992; Kohno et al., 1993), and conversely, if BiP levels are artificially lowered by expressing a mutant BiP that is not retained in the ER, the UPR is activated (Beh and Rose, 1995; Sidrauski et al., 1996). Importantly, BiP was shown to form a complex with Ire1p in mammalian cells (Bertolotti et al., 2000). Accumulation of unfolded proteins promotes dissociation of BiP from the lumenal domain of Ire1p and the loss of BiP correlates with the formation of Ire1p homodimers. These results are consistent with the model in which BiP binds to Ire1p as a negative regulator of dimerization and is titrated off by the accumulation of unfolded proteins. However, BiP is present in the ER lumen at millimolar concentrations and thus it remains difficult to envision how small changes in the concentration of unfolded proteins would efficiently compete with Ire1p for BiP binding. It is possible that a modification in BiP, such as phosphorylation or ADP ribosylation, creates a specialized pool of BiP molecules that could be the primary target of the perturbation in the folding capacity of the ER. Thus, although BiP has been identified as a player, the mechanism by which BiP attenuates activation of Ire1p remains uncertain.

HAC1 mRNA is spliced upon ER stress

How does Ire1p activation lead to increased transcription of UPRE-containing genes? Is the transcriptional activity or the synthesis of Hac1p increased by the activation of Ire1p? By analogy to other signal transduction pathways that are mediated by kinases, the expectation was that the kinase cascade initiated by Ire1p would result in phosphorylation and activation of Hac1p. Unexpectedly, however, it was discovered that Ire1p regulates the *production* of Hac1p: Hac1p is synthesized only under

UPR-inducing conditions; in the absence of unfolded proteins, no Hac1p is detected (Cox and Walter, 1996; Kawahara et al., 1997). Because transcription of the *HAC1* gene does not depend upon UPR induction, the appearance of Hac1p must be due to a post-transcriptional regulatory event mediated by Ire1p.

Northern blot analysis of the *HAC1* mRNA prepared from cells under UPR-inducing and non-inducing conditions yielded a surprising result: *HAC1* mRNA prepared from unstressed cells migrates as a single 1.4 kb species that is converted to a smaller 1.2 kb species in ER-stressed cells. Sequencing revealed that, upon activation of the UPR pathway, a 252-nucleotide intron is removed from the 3′ end of the coding sequence of the *HAC1* mRNA. Splicing of this intron from *HAC1^u* mRNA (u for 'uninduced') results in the production of the smaller mRNA species called *HAC1^i* mRNA (i for 'induced'). Appearance of Hac1p strictly correlates with production of this spliced form of the mRNA. Ire1p is required for this splicing reaction: cells lacking *IRE1* fail to convert *HAC1^u* to *HAC1^i* mRNA upon UPR induction, and no Hac1p is produced. Moreover, constitutive expression of *HAC1^i* mRNA results in constitutive production of Hac1p and induction of the target genes, bypassing the requirement for Ire1p. Thus, splicing of *HAC1* mRNA is both required and sufficient to trigger induction of the UPR (Figure 8.2). This is the first example of a regulated mRNA processing step controlling the activity of a *bona fide* signal transduction pathway.

Ire1p and tRNA ligase comprise an unconventional splicing machinery

Two observations indicated that the splicing of *HAC1* mRNA was catalyzed by a non-conventional machinery. First, *HAC1* mRNA lacks the consensus sequences at the splice junctions that are common to all pre-mRNAs processed by the spliceosome. Second, mutational analysis demonstrated that a functional spliceosome, which is required for splicing of all other known pre-mRNAs, was not required for processing of *HAC1* mRNA (Sidrauski et al., 1996).

An important clue to the identity of the ribonuclease responsible for *HAC1* mRNA cleavage came from analysis of the sequence of Ire1p. As mentioned above, the cytoplasmic domain of Ire1p has similarity to mammalian nuclease, RNase L. RNase L is a soluble non-specific ribonuclease that is activated upon treatment of mammalian cells with interferon (Bork and Sander, 1993). Like Ire1p, RNase L contains a kinase domain, followed by a C-terminal extension that shows sequence homology to the 133-amino acid tail in Ire1p. Binding of its ligand to the N-terminus of RNase L allows its homodimerization and activation of the nuclease domain (Dong and Silverman, 1995; Zhou et al., 1993). This similarity, both in terms

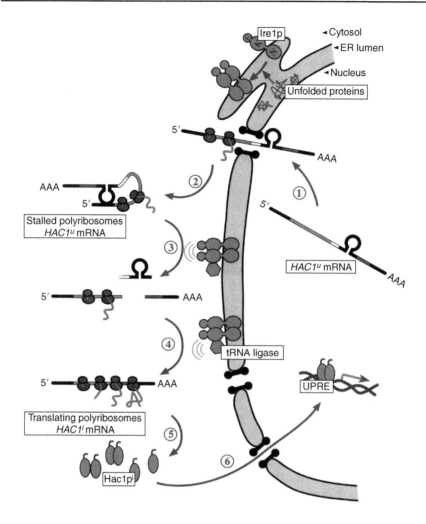

Figure 8.2 The unfolded protein response pathway in yeast. *HAC1ᵘ* mRNA is synthesized in the nucleus and is exported into the cytoplasm (step 1). Although *HAC1ᵘ* mRNA associates with polyribosomes, this mRNA is not translated (step 2). Base pairing between its intron and 5′ UTR blocks production of Hac1pᵘ. Accumulation of unfolded proteins in the ER lumen triggers the activation of Ire1p, a transmembrane serine-threonine kinase. When there are few unfolded proteins, the chaperone BiP binds to the N-terminal lumenal domain (N) of Ire1p preventing its oligomerization. When unfolded proteins accumulate, BiP is titrated off the Ire1p molecule, allowing its oligomerization and *trans*-autophosphorylation via the cytosolic kinase domain (K) and activation of the endonuclease activity in the C-terminal domain (E). The endonuclease cleaves *HAC1ᵘ* mRNA at both splice junctions (step 3) and the two exons are rejoined by tRNA ligase (step 4). After removal of the intron by the Ire1-mediated splicing reaction, *HAC1ⁱ* mRNA is efficiently translated into Hac1pⁱ (step 5), which then enters the nucleus and upregulates expression of genes containing the unfolded protein response element (UPRE; step 6). AAA, polyadenylation.

of sequence and mode of action, led to the hypothesis that Ire1p has both kinase and nuclease activities. Consistent with this hypothesis, deletion of the carboxy-terminal endonuclease domain of Ire1p impairs transmission of the unfolded protein response without affecting its kinase activity (Shamu and Walter, 1996).

By assaying the *in vitro* activity of a recombinant fusion protein containing the C-terminal half of the protein, which includes the kinase and tail domains, the nuclease activity of Ire1p was directly demonstrated (Sidrauski and Walter, 1997). This fusion protein phosphorylates itself and thus contains an active kinase domain. Moreover, it was shown to cleave an *in vitro* synthesized *HAC1* RNA substrate precisely at both its 5' and 3' splice junctions. The substrate requirements of this *in vitro* reaction are indistinguishable from the requirements of the *in vivo* splicing reaction: point mutations that change splice junction nucleotides in *HAC1* mRNA and abolished splicing *in vivo* also blocked cleavage *in vitro*.

While its closest relative, RNase L, is a non-specific endoribonuclease thought to help rid cells of invading viruses, the endoribonucleolytic activity of Ire1p is highly specific for *HAC1*^{u} mRNA. The splice junctions of *HAC1*^{u} mRNA share a common stem-loop structure that is required for cleavage: changes in sequence within the loop or disruption of the stem structure abolish cleavage. Moreover, this small and well-defined 7-nucleotide loop and stem can act as an Ire1p minisubstrate *in vitro* and, therefore, may define a unique 'signature motif' for Ire1p substrates (Gonzalez et al., 1999).

Ire1p is responsible for the cleavage step of the *HAC1* mRNA splicing reaction. How is splicing completed? As mentioned earlier, genetics provided the identity of the other half of this unique splicing machinery. In contrast to strains carrying mutations in spliceosomal components, *rlg1-100* strains exhibit a block in *HAC1* mRNA splicing (Sidrauski et al., 1996). Upon induction of the pathway, *HAC1*^{u} mRNA disappears in an *rlg1-100* mutant strain, but without the concomitant appearance of *HAC1*^{i} mRNA. In analogy to its role in tRNA splicing, it was proposed that tRNA ligase was involved in ligation of the two mRNA halves (Figure 8.2). Interestingly, the *rlg1-100* mutation does not affect ligation of tRNAs, which is essential for viability of the cell, but only impairs splicing of *HAC1* mRNA. The *rlg1-100* allele produces a protein with one amino acid substitution that does not map to any of the three distinct catalytic domains present in tRNA ligase. It remains to be determined how this mutation causes specific defects in *HAC1* mRNA processing while leaving tRNA processing unperturbed.

The *in vitro* system allowed confirmation of the role of tRNA ligase in processing of *HAC1* mRNA. Addition of purified tRNA ligase to the Ire1p *in vitro* cleavage reaction reconstituted splicing of the *HAC1* RNA substrate (Sidrauski and Walter, 1997). Thus, in contrast to the spliceosome, which utilizes a large number of proteins and small nuclear RNA molecules, processing of *HAC1* mRNA requires only two enzymes: Ire1p ribonuclease and tRNA ligase. *HAC1* mRNA splicing, therefore, resembles pre-tRNAs processing, which is catalyzed by the sequential action of tRNA endonuclease and tRNA ligase.

The role of the kinase domain of Ire1p in the transduction of the UPR signal is unclear. Ire1p kinase is activated upon induction of the UPR and this activity is required for transmission of the unfolded protein signal (Shamu and Walter, 1996). To date, however, the only known substrate of Ire1p's kinase activity is Ire1p itself. It is possible that the kinase activity directly stimulates Ire1p's endoribonucleolytic activity by phosphorylating the RNase domain or is required to stabilize an oligomeric conformation necessary for nuclease activity. Deciphering the precise role of the kinase activity of Ire1p remains an open and exciting question.

Translation of Hac1p is attenuated in unstressed cells

Splicing of *HAC1* mRNA is key to allowing production of Hac1p. How does removal of the intron result in appearance of Hac1p? The unspliced *HAC1*^{u} mRNA predicts a 230-amino acid protein, Hac1p^{u}. Splicing removes the nucleotide sequence encoding the last 10 C-terminal amino acids of Hac1p^{u} and replaces them with a new nucleotide sequence present in the second exon of *HAC1* mRNA, which encodes an alternative 18-amino acid tail. This 238-amino acid protein, Hac1p^{i}, encoded by the spliced mRNA, is the only form of Hac1p that can be detected in cells (Figure 8.2).

The lack of any detectable amounts of Hac1p^{u} in uninduced cells indicated that its production was somehow blocked. In principle, the intron in *HAC1*^{u} mRNA could block expression of Hac1p^{u} by preventing export of this mRNA from the nucleus to the cytoplasm. Alternatively, the intron might inhibit translation of *HAC1*^{u} mRNA. Finally Hac1p^{u} might be synthesized but it might be highly unstable and thus undetectable in unstressed cells. A series of experiments tested these possibilities. *In situ* hybridization showed that *HAC1*^{u} mRNA is primarily found in the cytosol, indicating that *HAC1*^{u} mRNA exits the nucleus, escaping controls that retain other pre-mRNAs in the nuclear compartment (Chapman and Walter, 1997). Consistent with this localization, *HAC1*^{u} mRNA co-fractionates with cytosolic polyribosomes. While these observations were consistent with the proposal that Hac1p^{u} might be translated and then rapidly degraded, both forms, Hac1p^{i} and Hac1p^{u}, were found to have identical half-lives (Chapman and Walter, 1997). The observation that *HAC1* mRNA was engaged by polyribosomes suggested that the intron regulates Hac1p protein production at the elongation step of translation. This would lead to the formation of a stable complex of *HAC1* mRNA bound by stalled ribosomes and is unlike other pathways which regulate translation by inhibiting the initiation step.

How does the *HAC1^u* mRNA intron cause the ribosomes engaged with the unspliced mRNA to stall? Analysis of the *HAC1* 5' untranslated region (UTR) and intron sequences revealed that they contain complementary sequences: the intron and the 5' UTR can interact by forming 16 Watson–Crick-type base pairs. In order to show the importance of this base pairing, the intron was mutated to disrupt this interaction. This modification led to efficient translation of Hac1p^u (Rüegsegger et al., 2001). Moreover, complementary mutations in the 5' UTR that restored base pairing prevented Hac1p^u synthesis. The *HAC1* 5' UTR and intron, therefore, act together through long-range base pairing to inhibit translation.

One important question is where within the cell does *HAC1* mRNA splicing take place. Ire1p could initiate cleavage of *HAC1* mRNA in either the cytoplasm or the nucleus, depending on whether it is localized to the ER or inner nuclear membrane, which are contiguous. Because tRNA ligase has been reported to localize to the nucleus where it interacts with the tightly membrane-associated tRNA endonuclease (Clark and Abelson, 1987), perhaps Ire1p signals directly from the ER lumen to the nucleus, where it functions with tRNA ligase to splice newly transcribed *HAC1* mRNA. Alternatively, a small fraction of the tRNA ligase molecules may reside in the cytoplasm and Ire1p may cleave *HAC1* mRNA that has been exported from the nucleus. It has been recently found that the pool of polysome-bound *HAC1^u* mRNA is a substrate of the splicing reaction, suggesting that processing of this mRNA takes place in the cytoplasm. Processing of this population of *HAC1^u* mRNA may allow a rapid burst in production of Hac1p^i upon conditions of ER stress, as the ribosomes may quickly resume translation after the intron is removed. Thus, the cell has a reservoir of substrate RNA that is preengaged with ribosomes, ready to be processed and immediately translated when a need for Hac1p^i arises.

Another recent finding is that removal of the *HAC1* intron not only allows expression of Hac1p but also leads to expression of a more potent transcriptional activator. It was shown that the C-terminal tail of Hac1p^i, when fused to an unrelated DNA-binding domain, serves as a potent transcriptional activation domain, whereas, fusion of the tail of Hac1p^u to the same DNA-binding domain results in a protein that is essentially inactive as a transcriptional activator, indicating that, in isolation, the tail of Hac1p^i, but not the tail of Hac1p^u, has the ability to promote transcription (Mori et al., 2000). Thus, the intron of *HAC1* mRNA plays two roles: it inhibits translation and it ensures that if translational attenuation is somehow circumvented, the protein produced will not be as efficient in eliciting a transcriptional response. It is tempting to speculate that, besides being a substrate of the splicing reaction, the ribosome-associated *HAC1^u* mRNA may give rise to Hac1p^u under different, as-yet-unknown conditions. Judged from its reduced transcriptional activity, Hac1p^u could potentially drive a different transcriptional program or have an entirely different role in the cell. While this remains an interesting possibility, to date there is no experimental evidence that Hac1p^u is ever produced.

Summary: *The UPR signaling pathway in yeast*

- Three components that act in a linear fashion are essential for transmission of the unfolded protein signal from the ER to the nucleus: Ire1p, tRNA ligase and Hac1p.
- The study of the yeast UPR revealed a surprising molecular pathway (Figure 8.2).
- Activation of the pathway results in regulated splicing of *HAC1* mRNA. This is the first example of a processing event acting as a key regulatory step in a signal transduction pathway.
- The regulated splicing of *HAC1* mRNA occurs by a non-conventional splicing mechanism that involves the action of a unique machinery constituted by a bifunctional kinase/ribonuclease, Ire1p, and tRNA ligase.
- Splicing results in removal of a translational attenuation signal present in the intron of *HAC1* mRNA allowing translation of Hac1p under conditions of ER stress.
- Based on the unusual nature of the reactions uncovered, it remained to be determined if this knowledge could be generalized to other eukaryotic cells.

THE UPR SIGNALING PATHWAY IN MAMMALS

In contrast to yeast, which contains only one ER membrane transducer of the unfolded protein signal, the mammalian UPR has several ER-proximal effectors: two homologs of yeast Ire1p, Ire1α and Ire1β, the transmembrane kinase PERK and the ER membrane-associated transcription factor ATF6. Under conditions of ER stress, these components appear to initiate parallel signaling pathways that ultimately result in a complex set of responses (Figure 8.3). The earliest response to ER stress is attenuation of global translation. Next, cells transcriptionally upregulate ER chaperones. Under conditions of continued stress, the cells arrest the cell cycle, and ultimately undergo apoptosis. In many cases, it is still unclear which of the proximal effectors is responsible for activation of a particular output. Furthermore, cross-talk between the different pathways suggests a complex circuitry of parallel, yet interconnected, signaling branches. Ultimately, there seem to be two possible outcomes upon activation of the mammalian UPR: adaptation to ER stress and survival, or activation of programmed cell death. Thus, the cell must integrate the information generated by these diverse outputs

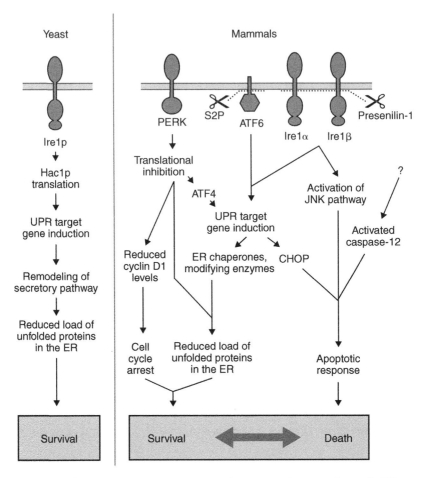

Figure 8.3 The unfolded protein response in yeast and mammalian cells. Whereas the yeast response to ER stress is a linear signal transduction pathway from Ire1p (left panel), the more complex response in higher eukaryotic cells is mediated by three ER kinases (Ire1α, Ire1β, and PERK) and a transmembrane transcription factor (ATF6), all of which are activated by accumulation of unfolded proteins in the ER. Under ER stress, cells attenuate overall translation, a response mediated by PERK. Subsequently, ER-resident chaperones and protein modifying enzymes are upregulated. The transcription factor ATF6 binds to the ERSE (ER stress response element) in the promoters of these genes, activating their transcription. The role of Ire1 in this response is still unclear. Another consequence of the activation of PERK is cell cycle arrest. Under continued ER stress, the apoptotic response is activated, due to upregulation of pro-apoptotic genes like CHOP, activation of the JNK cascade and proteolytic activation of caspase-12. A balance between these cytoprotective and cytotoxic pathways determines whether the cell will survive ER stress.

to determine its fate. In this chapter, we will introduce the different effectors and the pathways that are known to activate and then attempt to integrate the current knowledge of the complex relationships that exist between them.

Effectors of the UPR in mammalian cells

Ire1p homologs exist in higher eukaryotic cells
Metazoan homologs of yeast *IRE1* were first identified in *Drosophila melanogaster* and *Caenorhabditis elegans*. The sequences of these genes allowed the identification of two Ire1 homologs in mammals, referred to as Ire1α and Ire1β (Niwa et al., 1999; Tirasophon et al., 1998; Wang et al., 1998). Alignment of the predicted peptide sequence revealed that the kinase and C-terminal nuclease domains of Ire1 are both very similar to each other and highly conserved. In contrast, the N-terminal lumenal domain of Ire1 is more divergent, even between the two mammalian genes. Tissue expression studies revealed that, whereas Ire1α is ubiquitously expressed, in all cell types the expression of Ire1β is largely limited to the epithelia of the gut (Urano et al., 2000).

Several experiments indicate that both the unique endoribonucleolytic activity of Ire1 and the UPR-regulated splicing pathway are conserved in higher eukaryotic cells. First, purified recombinant Ire1α and Ire1β cytosolic domains are both active endoribonucleases capable of cleaving yeast *HAC1* mRNA *in vitro* at precisely the same positions and with the same structural requirements for substrate recognition as the yeast protein (Niwa et al., 1999). Second, as in yeast, overexpression of nuclease-dead forms of Ire1α inhibit, in a dominant negative manner, the induction of the BiP promoter upon ER stress (Tirasophon et al., 2000). Finally and most importantly, HeLa cells transiently expressing yeast *HAC1* mRNA can correctly process this mRNA in an ER-stress-dependent fashion. Taken together, these observations support the proposal that a non-conventional mRNA splicing event also mediates the mammalian UPR. Currently, however, no endogenous substrates of the nuclease activity of Ire1α or Ire1β have yet been identified. The search for such mammalian Ire1 substrate RNAs is an area of intensive research.

An additional level of regulation of the UPR was recently revealed by the observation that the mammalian Ire1 proteins undergo proteolysis in response to ER stress (Niwa et al., 1999). The proteolytic cleavage releases the cytoplasmic domain of Ire1p, which contains the kinase and nuclease domains, from the membrane. This effector domain then rapidly translocates into the nucleus of the cell. This was a surprising result, as processing of yeast Ire1p has never been observed. This observation may indicate that the function of the activated mammalian Ire1 proteins is more complex than that of yeast Ire1p. Both yeast and mammalian Ire1p could carry out mechanistically similar cytoplasmic splicing reactions, and the relocalization into the nucleus of mammalian Ire1p might reflect an additional role, such as the activation of the pro-apoptotic kinase cascades (see below).

The redistribution of Ire1 immunoreactivity from the cytosol to the nucleus upon activation of the pathway is apparent only when Ire1 is expressed at normal levels. It is not observable when Ire1 is overexpressed, suggesting that the capacity of the protease responsible for Ire1 cleavage is limiting. Interestingly, the nuclear localization of Ire1 requires presenilin-1 (PS1), a protein required for cleavage of both the developmentally regulated transcription factor Notch and the amyloid precursor protein (APP; reviewed by Selkoe, 1998). Cleavage of APP by PS1 generates $A\beta_{42}$, a major constituent of the amyloid plaques found in Alzheimer's disease patients. While the involvement of PS1 UPR induction remains controversial (Sato et al., 2000), it is potentially of great importance as it may suggest that the UPR may play some role in the pathogenesis of the Alzheimer's disease.

ATF6, a UPR transcription factor, is synthesized as an ER membrane protein and is proteolytically cleaved upon induction of the UPR
An as-yet-undiscovered mammalian Ire1-substrate mRNA may encode a transcription factor analogous to Hac1p in yeast. However, the mRNA encoding the only known mammalian UPR specific transcription factor to date, ATF6, does not undergo regulated splicing like that of yeast *HAC1*. ATF6 is a member of the ATF/CREB (cAMP regulatory element-binding protein) protein family and was isolated by screening for factors that bind to the ERSE (Haze et al., 1999), and like the mammalian Ire1 proteins, ATF6 is constitutively expressed as an ER membrane protein and undergoes proteolytic cleavage upon induction of the UPR (Haze et al., 1999; Yoshida et al., 2000). This cleavage liberates the cytosolic DNA binding/transcriptional activation domain, which is imported into the nucleus and induces transcription. Thus, rather than being regulated by splicing, the UPR transcriptional output in mammalian cells is controlled by a regulated proteolytic step.

The cleavage of ATF6 requires the site 1 (S1P) and site 2 (S2P) proteases (Ye et al., 2000). These proteins were originally identified as the proteases required for cleavage of the membrane-bound sterol-responsive element binding protein (SREBP) in response to cholesterol starvation. As is the case with ATF6, cleavage of SREBP leads to the release of the portion of the molecule that contains the DNA binding and transcriptional activation domains, which translocates to the nucleus inducing transcription of genes involved in cholesterol and fatty acid synthesis and uptake. SREBP cleavage is regulated by another protein, SREBP cleavage activating protein (SCAP), which, upon sterol starvation, escorts SREBP from the ER to the Golgi, where it undergoes proteolysis. ATF6 proteolysis, however, does not occur upon sterol deprivation, and is not dependent on the SCAP protein.

It is not known how cleavage of ATF6 is regulated by ER stress. It is possible that ATF6 has its own dedicated SCAP-like escort protein that transports it to the Golgi to be processed by S1P. By analogy to the SREBPs, ATF6 cleavage in response to the accumulation of unfolded proteins may be regulated at the level of exit from the ER.

PERK, another ER membrane signaling molecule, is responsible for translational attenuation

Upon induction of the UPR, mammalian cells repress global translation initiation rapidly (reviewed by Brostrom and Brostrom, 1998). Reduced protein synthesis rates thus result in a decrease in the load of newly synthesized secretory proteins, which are the substrates of the folding machinery in the ER. Translational repression provides an immediate and general response to ER stress that, in contrast to ER chaperone upregulation, does not depend on transcriptional induction. Translational inhibition is mediated by PERK (PKR-like ER kinase), which, like Ire1p, is a type I transmembrane protein kinase that is localized to the ER (Harding et al., 1999; Shi et al., 1998). Although the cytosolic kinase domain of PERK is only distantly related to that of Ire1p, the lumenal domain of PERK is similar to the sensor domain of Ire1p, suggesting that these proteins are coordinately regulated by the same ligand. In support of this model, recent experiments showed that the lumenal domains of these two proteins are functionally interchangeable in mediating the ER stress response. Furthermore, like Ire1, the lumenal domain of PERK was shown to form an ER stress-sensitive complex with the chaperone BiP (Bertolotti et al., 2000).

The cytoplasmic domain of PERK is most similar to the interferon-inducible RNA-dependent protein kinase PKR. This cytoplasmic protein kinase phosphorylates the α subunit of the eukaryotic translation initiation factor 2 (eIF2-α). Phosphorylated eIF2-α interferes with the formation of an active 43S translation initiation complex, inhibiting protein synthesis (reviewed by Clemens, 1996). Two other known eIF2-α kinases are the heme-regulated inhibitor (HRI) of erythroid cells and the general control of amino acid biosynthesis kinase, GCN2. PERK was shown to phosphorylate eIF2-α on the same serine residue as these other kinases *in vitro* (Harding et al., 1999). Furthermore, the kinase activity of PERK increases upon ER stress, as would be expected of the kinase responsible for transmitting the UPR translational attenuation signal. Recently, the role of PERK in UPR-induced overall translational inhibition was clearly established by showing that PERK$^{-/-}$ cells are unable to either phosphorylate eIF2-α or attenuate translation in response to accumulation of misfolded proteins in the ER (Harding et al., 2000b). PERK is highly expressed in the endocrine pancreas, an organ that is dedicated to protein secretion. This finding underscores the importance of the role of PERK in coupling ER folding to translation.

Physiological responses to unfolded proteins in the ER

Translational attenuation and the UPR

The ability of cells to attenuate translation in response to ER stress appears to be unique to higher eukaryotic cells. This strategy is employed by yeast cells only under conditions of amino acid starvation, in which Gcn2p protein kinase phosphorylates eIF2-α to inhibit overall translational initiation. Activation of the UPR in yeast does not result in a dramatic reduction in protein synthesis. Consistent with these observations, no PERK homolog is present in the yeast genome. In contrast, both the *Drosophila melanogaster* and the *Caenorhabditis elegans* genomes contain PERK homologs.

Another consequence of activation of the eIF2-α kinase Gcn2p in yeast is the transcriptional induction of a set of genes involved in amino acid biosynthesis and metabolism (reviewed by Hinnebusch, 1996). Paradoxically, the cell achieves activation of these genes by increasing the efficiency of translation of the mRNA encoding a transcriptional activator, Gcn4p. The translation of this message is normally inhibited by the translation of several short upstream open reading frames (uORFs) in its 5' UTR, thus preventing translation initiation at the *GCN4* open reading frame (Hinnebusch, 1997). Under starvation conditions in which translation initiation is inhibited, initiation at the AUG of the *GCN4* open reading frame is favored, leading to the synthesis of Gcn4p. This mechanism for transcriptional activation of gene expression upon amino acid starvation is conserved in mammalian cells. Translation of activating transcription factor 4 (ATF4) mRNA increases both upon amino acid starvation and under conditions of ER stress (Harding et al., 2000a). Like GCN4, the ATF4 mRNA contains a series of uORFs in its 5' UTR and eIF2-α phosphorylation is responsible for increased translation of this mRNA.

ATF4 may provide one of the connections between ER stress and the subsequent apoptotic response. ATF4 has been reported to activate CHOP (C/EBP homologous protein-10), a proapoptotic gene that is upregulated by ER stress. As expected, in PERK$^{-/-}$ cells, CHOP is not induced upon ER stress (Harding et al., 2000a). In contrast, BiP mRNA induction is unaffected, confirming that these mutant cells are able to mount a UPR. Interestingly, it was also noted that constitutive expression of ATF4 is not sufficient to bypass the requirement for PERK in CHOP induction, indicating that induction of CHOP requires a second ER stress-induced signal.

Induction of UPR target genes

As described earlier, the accumulation of unfolded proteins in the ER leads to the proteolytic liberation of ATF6 from the membrane, allowing it to translocate into the nucleus and activate transcription of UPR target genes. Two observations suggest that ATF6 proteolysis alone can account for a transcriptional response to ER stress. First, unregulated expression of the

DNA binding and transcriptional activation domain as a soluble protein leads to the constitutive induction of BiP and CHOP, indicating that production of the proteolytic fragment is sufficient for induction (Haze et al., 1999). Second, expression of cytosolic fragments of ATF6 lacking the activation domain blocks the induction of these genes in response to ER stress.

In contrast, the role of the mammalian Ire1α and Ire1β in the transcriptional induction of UPR target genes is not clear. Several observations support the notion that these proteins are important for induction of target genes. As is the case in yeast, overexpression of wild-type Ire1α or Ire1β leads to induction of BiP, whereas overexpression of kinase dead or nuclease dead forms of Ire1α blocks induction of BiP (Tirasophon et al., 1998, 2000). Thus, overexpression of Ire1α and Ire1β is sufficient to generate a UPR signal and this response depends upon both the kinase and nuclease activities of these proteins. Importantly, this effect is also observed when using a reporter gene under the control of an ATF6 binding site, suggesting that these proteins activate the ATF6 pathway (Wang et al., 2000). However, a recent observation strongly suggests that the Ire1 proteins are not essential for the UPR-dependent transcriptional response. Cell lines derived from IRE1α$^{-/-}$ and IRE1β$^{-/-}$ knockout and IRE1α$^{-/-}$ IRE1β$^{-/-}$ double knockout mouse embryos show no defect in the induction of BiP and CHOP in response to ER stress (F. Urano and D. Ron, unpublished observations). Therefore, although Ire1 proteins are able to induce the UPR, there must also be IRE1-independent pathways for regulation of gene expression by the UPR in mammals.

Cell cycle arrest during UPR induction
Further complexity in the mammalian system was revealed by the observation that activation of the UPR in tissue culture cells results in arrest of the cell cycle in the G1 phase (Brewer et al., 1999). The UPR-induced arrest is a consequence of inhibiting translation of cyclin D1. Cyclin D1-dependent kinase activity is rate-limiting for cell cycle progression and the accumulation of cyclin D1 is generally regulated at the level of transcription, protein degradation, and subcellular localization. In contrast, the UPR controls cell cycle progression by repressing translation of this cyclin, a phenomenon that is mediated by PERK (Brewer and Diehl, 2000; Niwa and Walter, 2000). Thus, repression of cyclin D1 synthesis is likely a consequence of the translational inhibition that takes place upon PERK induction.

What is the function of arresting the cell cycle during ER stress? The arrest in the G1 phase may prevent cells from progressing through the cell cycle before ER homeostasis is established (reviewed by Niwa and Walter, 2000). If adaptation to ER stress conditions through upregulation of ER chaperones is not possible, cells may then initiate a cell-death program. This notion may also explain the increased sensitivity of PERK$^{-/-}$ cells to treatments that cause ER stress. In addition to their inability to repress synthesis of secretory proteins, these cells do not have an expanded time window in which to cope with the increased load of unfolded proteins.

Apoptotic response to ER stress

Under conditions of continued ER stress, cells ultimately undergo apoptosis. This is mediated by at least three mechanisms. First, among the genes induced by the UPR is CHOP, a pro-apoptotic transcription factor. Second, activation of the UPR leads to the activation of the c-jun amino-terminal kinase pathway (JNKs) also referred to as 'stress activated protein kinases' or SAPKs (Urano et al., 2000). These pleiotropic signaling proteins are linked to many forms of stress and regulate gene expression leading to cell death. Finally, the ER-associated protease caspase-12, which belongs to a family of cysteine proteases that are critical mediators of cell death, is activated by ER stress (Nakagawa et al., 2000).

The Ire1 proteins play a key role in the apoptotic response by initiating two different pro-apoptotic pathways. First, overexpression of Ire1β induces transcription of the pro-apoptotic transcription factor CHOP and triggers programmed cell death in transfected cells (Wang et al., 1998). Second, while overexpression of Ire1α or Ire1β results in activation of JNK, activation of JNK is impaired in IRE1α$^{-/-}$ cells (Urano et al., 2000). In fact, the cytosolic domains of both Ire1α and Ire1β were shown to bind to the cytosolic adaptor protein TRAF2 (tumor necrosis factor receptor-associated factor 2), which is thought to recruit and activate proximal components of the JNK pathway. Interestingly, the kinase domain but not the nuclease domain of Ire1 is required for this interaction, suggesting that the kinase domain of Ire1p may have a role in promoting apoptosis which is independent of its putative effect on nuclease activity. Thus, in addition to their proposed role in the transcriptional induction of UPR target genes, the Ire1 proteins mediate the activation of SAPK pathways. Cells derived from IRE1α$^{-/-}$, IRE1β$^{-/-}$ or IRE1α$^{-/-}$ IRE1β$^{-/-}$ mice will allow the importance of Ire1 in induction of programmed cell death to be tested directly. If the Ire1 proteins are crucial in triggering this process, the mutant cells should have an impaired apoptotic response to prolonged ER stress.

One important (and often overlooked) complexity in understanding the UPR in metazoan organisms is that the various UPR components mentioned here are expressed differentially in different cell types. Thus, the response of cells to ER stress may differ depending on the tissue and/or cell studied and the balance between the protective effectors and death effectors may be unique to each cell type. The characteristic combination of UPR signaling branches that are activated in each cell type is likely to depend on the cell's particular needs and play a crucial role in determining its ability to survive ER stress.

Caspase-12 becomes proteolytically activated under conditions of ER stress. It is unclear how the activation of caspase-12 is regulated. Proteolytic activation of caspase-12 is independent of PERK (Harding et al., 2000b). Future work will address whether this proteolytic cleavage requires either activation of the Ire1 proteins or proteolysis of ATF6.

Summary: *The UPR signaling pathway in mammals*

- The UPR in yeast is a linear, and relatively well-understood, pathway. Activation of Ire1p leads to expression of Hac1p, which in turn leads to increased transcription of target genes. As we have just discussed, the mammalian UPR is far more complex and less well understood (Figure 8.3). ER stress activates several parallel signaling pathways mediated by at least three ER transmembrane kinases, Ire1α, Ire1β and PERK, as well as by ATF6, an ER membrane-anchored transcription factor.

- ER stress leads to repression of general protein synthesis, transcriptional upregulation of ER resident proteins involved in protein folding and processing and, eventually, to arrest of the cell cycle. Translational repression is an immediate response that limits further insult to the secretory pathway as the cell endeavors to eliminate misfolded proteins from the ER. Reduced levels of cyclin D1 prevent the cell from progressing through the cell cycle, expanding the window of time during which the cell can rid itself of the increase in misfolded polypeptides.

- Ultimately, a sustained UPR initiates a cytotoxic cascade. The pro-apoptotic transcription factor CHOP is a transcriptional target of the UPR. ER stress also activates the JNK pro-apoptotic signaling pathway through Ire1α and Ire1β. Finally, UPR induction leads to proteolytic activation of caspase-12, a critical mediator of ER stress induced apoptosis.

- Mammalian cells must integrate input from these pathways to negotiate two possible outcomes of ER stress: survival or death. It seems plausible that the timing of these various steps of the response to ER stress allows the cell to measure the severity of the insult, which may correlate with the length of the cell-cycle arrest, allowing it to either proceed with the cell cycle or activate the programmed cell-death pathway. If the measures taken to eliminate unfolded proteins fail, signaling continues and the apoptotic effectors of the UPR will eventually kill the cell.

CONCLUSIONS

Accumulation of unfolded proteins in the ER occurs under many different conditions. Recent studies suggest that cells are constantly preventing their accumulation by two complementary and co-regulated pathways that enhance folding and degradation. In addition, a number of diseases have been linked to the toxic effects of mutant proteins that accumulate in the

ER. Thus, the strategies that cells utilize in dealing with this kind of stress are of great importance to understanding how cells maintain their homeostasis during both normal growth and disease.

The characterization of the UPR pathway has revealed many surprises, both in its physiological complexities and in its novel mechanism of signal transduction. Work in yeast allowed rapid progress because genetic and biochemical avenues could be readily combined. The more recent discovery of the components of the mammalian UPR has revealed numerous additional complexities in this signaling pathway.

Despite these differences, many aspects of the UPR are conserved throughout evolution. Both in yeast and mammals, ER stress is sensed by transmembrane kinases that oligomerize and activate their cytosolic effector domains. Also, both the kinase and the site-specific endoribonuclease activity of Ire1p are conserved in the mammalian homologs. Transcription is regulated in both yeast and mammals by bZIP (basic leucine zipper) proteins of the ATF/CREB family. Finally, the scope of UPR transcriptional response is broad.

Superimposed on these similarities, significant differences have been uncovered in the mammalian UPR pathway (Figure 8.3). First, activation of the transcription factor ATF6 is controlled by regulated intramembrane proteolysis, rather than regulated splicing. Second, Ire1 appears to be proteolytically cleaved upon activation of the response. Third, in addition to transcriptional activation, the mammalian UPR downregulates overall translation and induces cell cycle arrest. Finally, the mammalian UPR leads to activation of programmed cell death pathways.

The existence of parallel UPR signaling pathways in mammalian cells makes study of the response challenging. Future work will define the relationships between activation of each of the upstream effectors and the outputs, as well as the potential cross-talk between the various signaling branches. It seems reasonable that if Ire1 and PERK are the only two sensors of the accumulation of unfolded proteins in the ER lumen, then they must somehow affect proteolysis of ATF6. Alternatively, an as-yet-unidentified protein may also detect changes in the folding capacity of the ER, leading to activation of ATF6. Studies of cell lines lacking the different UPR signaling components will be valuable both in understanding the role of these effectors and in determining their epistatic relationships.

Although a basic outline of the UPR pathway is now clear, there are a number of questions that remain unanswered in both the yeast and mammalian systems. We do not yet understand fully the most upstream UPR event, the activation of Ire1p and PERK by changes in the folding capacity of the ER and the role of chaperones like BiP in this event. Also, defining a role for Ire1p's kinase activity, the identity of the substrates of the mammalian Ire1 endoribonucleolytic activity, and determining of the physiological role of these proteins in higher eukaryotes are important fronts for future work. Finally, a most challenging question that remains to be addressed is how mammalian cells integrate the diverse outputs that are activated by ER stress and can lead to either adaptation or death.

ACKNOWLEDGMENTS

The authors thank members of the Walter laboratory for helpful comments on the manuscript. We also thank David Ron for sharing unpublished results. Work in the Walter laboratory is supported by grants from the National Institutes of Health. J.H.B. is supported by a Helen Hay Whitney postdoctoral fellowship. P.W. is an investigator of the Howard Hughes Medical Institute.

REFERENCES

Beh, C.T. and Rose, M.D. (1995) Two redundant systems maintain levels of resident proteins within the yeast endoplasmic reticulum. *Proc Natl Acad Sci USA* **92**: 9820–9823.

Bertolotti, A., Zhang, Y., Hendershot, L.M., Harding, H.P. and Ron, D. (2000) Dynamic interaction of BiP and ER stress transducers in the unfolded-protein response. *Nat Cell Biol* **2**: 326–332.

Bork, P. and Sander, C. (1993) A hybrid protein kinase-RNase in an interferon-induced pathway? *FEBS Lett* **334**: 149–152.

Brewer, J.W. and Diehl, J.A. (2000) PERK mediates cell-cycle exit during the mammalian unfolded protein response. *Proc Natl Acad Sci USA* **97**: 12625–12630.

Brewer, J.W., Hendershot, L.M., Sherr, C.J. and Diehl, J.A. (1999) Mammalian unfolded protein response inhibits cyclin D1 translation and cell-cycle progression. *Proc Natl Acad Sci USA* **96**: 8505–8510.

Brodsky, J.L. and McCracken, A.A. (1999) ER protein quality control and proteasome-mediated protein degradation. *Sem Cell Dev Biol* **10**: 507–513.

Brostrom, C.O. and Brostrom, M.A. (1998) Regulation of translational initiation during cellular responses to stress. *Prog Nucleic Acid Res Mol Biol* **58**: 79–125.

Carman, G.M. and Henry, S.A. (1989) Phospholipid biosynthesis in yeast. *Annu Rev Biochem* **58**: 635–669.

Casagrande, R., Stern, P., Diehn, M. et al. (2000) Degradation of proteins from the ER of *S. cerevisiae* requires an intact unfolded protein response pathway. *Mol Cell* **5**: 729–735.

Caspersen, C., Pedersen, P.S. and Treiman, M. (2000) The sarco/endoplasmic reticulum calcium-ATPase 2b is an endoplasmic reticulum stress-inducible protein. *J Biol Chem* **275**: 22363–22372.

Chapman, R.E. and Walter, P. (1997) Translational attenuation mediated by an mRNA intron. *Curr Biol* **7**: 850–859.

Chen, X., Easton, D., Oh, H. et al. (1996) The 170 kDa glucose regulated stress protein is a large HSP70-HSP110-like protein of the endoplasmic reticulum. *FEBS Lett* **380**: 68–72.

Clark, M.W. and Abelson, J. (1987) The subnuclear localization of tRNA ligase in yeast. *J Cell Biol* **105**: 1515–1526.

Clemens, M. (1996) Protein kinases that phosphorylate eIF2 and eIF2B, their role in eukaryotic cell translational control. In: Hershey, J., Mathews, M. and Sonenberg, N. (eds) *Translational Control*, pp. 199–244. Cold Spring Harbor, NY: Cold Spring Harbor Laboratory Press.

Cox, J.S. and Walter, P. (1996) A novel mechanism for regulating activity of a transcription factor that controls the unfolded protein response. *Cell* **87**: 391–404.

Cox, J.S., Shamu, C.E. and Walter, P. (1993) Transcriptional induction of genes encoding endoplasmic reticulum resident proteins requires a transmembrane protein kinase. *Cell* **73**: 1197–1206.

Cox, J.S., Chapman, R.E. and Walter, P. (1997) The unfolded protein response coordinates the production of ER protein and ER membrane. *Mol Biol Cell* **8**: 1805–1814.

Doerrler, W.T. and Lehrman, M.A. (1999) Regulation of the dolichol pathway in human fibroblasts by the endoplasmic reticulum unfolded protein response. *Proc Natl Acad Sci USA* **96**: 13050–13055.

Dong, B. and Silverman, R.H. (1995) 2-5A-dependent RNase molecules dimerize during activation by 2-5A*. *J Biol Chem* **270**: 4133–4137.

Dong, B. and Silverman, R.H. (1997) A bipartite model of 2–5A-dependent RNase L. *J Biol Chem* **272**: 22236–22242.

Dorner, A.J., Wasley, L.C. and Kaufman, R.J. (1992) Overexpression of GRP78 mitigates stress induction of glucose regulated proteins and blocks secretion of selective proteins in Chinese hamster ovary cells. *EMBO J* **11**: 1563–1571.

Friedlander, R., Jarosch, E., Urban, J., Volkwein, C. and Sommer, T. (2000) A regulatory link between ER-associated protein degradation and the unfolded-protein response. *Nat Cell Biol* **2**: 379–384.

Gething, M.-J. and Sambrook, J. (1992) Protein folding in the cell. *Nature* **355**: 33–45.

Gonzalez, T.N., Sidrauski, C., Dorfler, S. and Walter, P. (1999) Mechanism of non-spliceosomal mRNA splicing in the unfolded protein response pathway. *EMBO J* **18**: 3119–3132.

Hampton, R.Y. (2000) ER stress response: getting the UPR hand on misfolded proteins. *Curr Biol* **10**: R518–521.

Harding, H.P., Zhang, Y. and Ron, D. (1999) Protein translation and folding are coupled by an endoplasmic-reticulum-resident kinase. *Nature* **397**: 271–274.

Harding, H.P., Novoa, I.I., Zhang, Y. et al. (2000a) Regulated translation initiation controls stress-induced gene expression in mammalian cells. *Mol Cell* **6**: 1099–1108.

Harding, H.P., Zhang, Y., Bertolotti, A., Zeng, H. and Ron, D. (2000b) Perk is essential for translational regulation and cell survival during the unfolded protein response. *Mol Cell* **5**: 897–904.

Haze, K., Yoshida, H., Yanagi, H., Yura, T. and Mori, K. (1999) Mammalian transcription factor ATF6 is synthesized as a transmembrane protein and activated by proteolysis in response to endoplasmic reticulum stress. *Mol Biol Cell* **10**: 3787–3799.

Helenius, A. (1994) How N-linked oligosaccharides affect glycoprotein folding in the endoplasmic reticulum. *Mol Biol Cell* **5**: 253–265.

Hinnebusch, A.G. (1996) Translational control of GCN4: Gene-specific regulation by phosphorylation of eIF2. In: Hershey, J., Mathews, M. and Sonenberg, N. (eds)

Translational Control, pp. 199–244. Cold Spring Harbor, NY: Cold Spring Harbor Laboratory Press.

Hinnebusch, A.G. (1997) Translational regulation of yeast GCN4. A window on factors that control initiator-tRNA binding to the ribosome. *J Biol Chem* **272**: 21661–21664.

Hong, E., Davidson, A.R. and Kaiser, C. (1996) A pathway for targeting soluble misfolded proteins to the yeast vacuole. *J Cell Biol* **135**: 623–633.

Kawahara, T., Yanagi, H., Yura, T. and Mori, K. (1997) Endoplasmic reticulum stress-induced mRNA splicing permits synthesis of transcription factor Hac1p/Ern4p that activates the unfolded protein response. *Mol Biol Cell* **8**: 1845–1862.

Kohno, K., Normington, K., Sambrook, J., Gething, M.J. and Mori, K. (1993) The promoter region of the yeast KAR2 (BiP) gene contains a regulatory domain that responds to the presence of unfolded proteins in the endoplasmic reticulum. *Mol Cell Biol* **13**: 877–890.

Lee, A.S. (1987) Coordinated regulation of a set of genes by glucose and calcium ionophores in mammalian cells. *Trends Biochem Sci* **12**: 20–23.

Liu, C.Y., Schröder, M. and Kaufman, R.J. (2000) Ligand-independent dimerization activates the stress response kinases IRE1 and PERK in the lumen of the endoplasmic reticulum. *J Biol Chem* **275**: 24881–24885.

Mazarella, R.A., Srinivasan, M., Haugejordan, S.M. and Green, M. (1990) Erp72, an abundant lumenal endoplasmic reticulum protein, contains three copies of the active site sequences of protein disulfide isomerase. *J Biol Chem* **265**: 1094–1101.

Mori, K., Sant, A., Kohno, K. et al. (1992) A 22 bp cis-acting element is necessary and sufficient for the induction of the yeast KAR2 (BiP) gene by unfolded proteins. *EMBO J* **11**: 2583–2593.

Mori, K., Ma, W., Gething, M.-J. and Sambrook, J. (1993) A transmembrane protein with a cdc2+/CDC28-related kinase activity is required for signaling from the ER to the nucleus. *Cell* **74**: 743–756.

Mori, K., Kawahara, T., Yoshida, H., Yanagi, H. and Yura, T. (1996) Signalling from endoplasmic reticulum to nucleus: transcription factor with a basic-leucine zipper motif is required for the unfolded protein-response pathway. *Genes to Cells* **1**: 803–817.

Mori, K., Ogawa, N., Kawahara, T., Yanagi, H. and Yura, T. (2000) mRNA splicing-mediated C-terminal replacement of transcription factor Hac1p is required for efficient activation of the unfolded protein response. *Proc Natl Acad Sci USA* **97**: 4660–4665.

Nakagawa, T., Zhu, H., Morishima, N. et al. (2000) Caspase-12 mediates endoplasmic-reticulum-specific apoptosis and cytotoxicity by amyloid-beta. *Nature* **403**: 98–103.

Ng, D.T.W., Watowich, S.S. and Lamb, R.A. (1992) Analysis *in vivo* of GRP78-BiP/substrate interactions and their role in induction of the GRP78-BiP gene. *Mol Biol Cell* **3**: 143–155.

Ng, D.T., Spear, E.D. and Walter, P. (2000) The unfolded protein response regulates multiple aspects of secretory and membrane protein biogenesis and endoplasmic reticulum quality control. *J Cell Biol* **150**: 77–88.

Nikawa, J., Akiyoshi, M., Hirata, S. and Fukuda, T. (1996) *Saccharomyces cerevisiae IRE2/HAC1* is involved in *IRE1*-mediated KAR2 expression. *Nucl Acids Res* **24**: 4222–4226.

Niwa, M. and Walter, P. (2000) Pausing to decide. *Proc Natl Acad Sci USA* **97**: 12396–12397.

Niwa, M., Sidrauski, C., Kaufman, R.J. and Walter, P. (1999) A role for presenilin-1 in nuclear accumulation of Ire1 fragments and induction of the mammalian unfolded protein response. *Cell* **99**: 691–702.

Nunnari, J. and Walter, P. (1996) Regulation of organelle biogenesis. *Cell* **84**: 389–394.

Plemper, R.K. and Wolf, D.H. (1999) Retrograde protein translocation: ERADication of secretory proteins in health and disease. *Trends Biochem Sci* **24**: 266–270.

Pollard, M.G., Travers, K.J. and Weissman, J.S. (1998) Ero1p: a novel and ubiquitous protein with an essential role in oxidative protein folding in the endoplasmic reticulum. *Mol Cell* **1**: 171–182.

Roy, B. and Lee, A.S. (1999) The mammalian endoplasmic reticulum stress response element consists of an evolutionarily conserved tripartite structure and interacts with a novel stress-inducible complex. *Nucl Acids Res* **27**: 1437–1443.

Rüegsegger, U., Leber, J.H. and Walter, P. (2001) Block of *HAC1* mRNA translation by long-range base pairing is released by cytoplasmic splicing upon induction of the Unfolded Protein Response. *Cell* **107**: 103–114.

Sato, N., Urano, F., Leem, J.Y. et al. (2000) Upregulation of BiP and CHOP by the unfolded-protein response is independent of presenilin expression. *Nat Cell Biol* **2**: 863–870.

Selkoe, D.J. (1998) The cell biology of beta-amyloid precursor protein and presenilin in Alzheimer's disease. *Trends Cell Biol* **8**: 447–453.

Shamu, C.E. and Walter, P. (1996) Oligomerization and phosphorylation of the Ire1p kinase during intracellular signaling from the endoplasmic reticulum to the nucleus. *EMBO J* **15**: 3028–3039.

Shi, Y., Vattem, K.M., Sood, R. et al. (1998) Identification and characterization of pancreatic eukaryotic initiation factor 2 alpha-subunit kinase, PEK, involved in translational control. *Mol Cell Biol* **18**: 7499–7509.

Sidrauski, C. and Walter, P. (1997) The transmembrane kinase Ire1p is a site-specific endonuclease that initiates mRNA splicing in the unfolded protein response. *Cell* **90**: 1–20.

Sidrauski, C., Cox, J.S. and Walter, P. (1996) tRNA ligase is required for regulated mRNA splicing in the unfolded protein response. *Cell* **87**: 405–413.

Tirasophon, W., Welihinda, A.A. and Kaufman, R.J. (1998) A stress response pathway from the endoplasmic reticulum to the nucleus requires a novel bifunctional protein kinase/endoribonuclease (Ire1p) in mammalian cells. *Genes Dev* **12**: 1812–1824.

Tirasophon, W., Lee, K., Callaghan, B., Welihinda, A. and Kaufman, R.J. (2000) The endoribonuclease activity of mammalian IRE1 autoregulates its mRNA and is required for the unfolded protein response. *Genes Dev* **14**: 2725–2736.

Travers, K.J., Patil, C.K., Wodicka, L. et al. (2000) Functional and genomic analyses reveal an essential coordination between the unfolded protein response and ER-associated degradation. *Cell* **101**: 249–258.

Urano, F., Wang, X., Bertolotti, A. et al. (2000) Coupling of stress in the ER to activation of JNK protein kinases by transmembrane protein kinase IRE1. *Science* **287**: 664–666.

Wang, X.Z., Harding, H.P., Zhang, Y. et al. (1998) Cloning of mammalian Ire1 reveals diversity in the ER stress responses. *EMBO J* **17**: 5708–5717.

Wang, Y., Shen, J., Arenzana, N. et al. (2000) Activation of ATF6 and an ATF6 DNA binding site by the endoplasmic reticulum stress response. *J Biol Chem* **275**: 27013–27020.

Welihinda, A.A. and Kaufman, R.J. (1996) The unfolded protein response pathway in *Saccharomyces cerevisiae*. Oligomerization and *trans*-autophosphorylation of Ire1p (Ern1p) are required for kinase activation. *J Biol Chem* **271**: 18181–18187.

Ye, J., Rawson, R.B., Komuro, R. et al. (2000) ER stress induces cleavage of membrane-bound ATF6 by the same proteases that process SREBPs. *Mol Cell* **6**: 1355–1364.

Yoshida, H., Haze, K., Yanagi, H., Yura, T. and Mori, K. (1998) Identification of the *cis*-acting endoplasmic reticulum stress response element responsible for tran-scriptional induction of mammalian glucose-regulated proteins. Involvement of basic leucine zipper transcription factors. *J Biol Chem* **273**: 33741–33749.

Yoshida, H., Okada, T., Haze, K. et al. (2000) ATF6 activated by proteolysis binds in the presence of NF-Y (CBF) directly to the *cis*-acting element responsible for the mammalian unfolded protein response. *Mol Cell Biol* **20**: 6755–6767.

Zhou, A., Hassel, B.A. and Silverman, R.H. (1993) Expression cloning of 2-5A-dependent RNase: a uniquely regulated mediator of interferon action. *Cell* **72**: 753–765.

9

PROTEIN QUALITY CONTROL IN THE EXPORT PATHWAY: THE ENDOPLASMIC RETICULUM AND ITS CYTOPLASMIC PROTEASOME CONNECTION

ZLATKA KOSTOVA AND DIETER H. WOLF

INTRODUCTION: THE PLAYERS OF THE GAME

It is the unexpected, that creates the excitement in science! The endoplasmic reticulum (ER) is the organelle at which secretory and membrane proteins enter the central vacuolar system of the eukaryotic cell. This system comprises the ER itself, the Golgi apparatus, endosomes, lysosomes, intermediate transport compartments, and the plasma membrane (Rapoport et al., 1996). The ER is the organelle where newly synthesized proteins are folded to acquire their native structure. Proper folding enables them to exert their biological functions at their final destination: only correctly folded and assembled proteins are normally delivered to their site of action. This prerequisite necessitates the presence of a protein quality control system in the ER that can distinguish between properly and non-properly folded proteins. In 1999, the Nobel Prize for Physiology or Medicine was awarded to Günter Blobel for his seminal 'signal hypothesis' and its experimental proof of how proteins carrying address labels in the form of peptides reach their different destinations within the cell (Anderson and Walter, 1999). The numerous studies on protein entry into the ER had created a dogma that stated that polypeptides translocated into the ER via the Sec61 import channel were trapped in the secretory pathway unable to return into

Protein Targeting, Transport & Translocation
ISBN 0-12-200731-X

the cytoplasm (Blobel, 1995). The fall of this dogma came as a big surprise: a crucial step in the protein quality control process of the ER, the elimination of malfolded proteins, was discovered to be the retrotranslocation of such proteins through the Sec61 channel back into the cytoplasm where a degradation machinery is readily available to eliminate them.

Only a few years earlier, the discovery of this degradation machinery with its intricate structure and sophisticated degradation mechanism had been met with astonishment in the scientific community (Wolf, 2000). We now know that its function is vital for all eukaryotic cells from yeast to humans and that it plays an essential role in central regulatory events ranging from cell metabolism to cell cycle control and differentiation.

This powerful proteolytic machinery is a two-part structure. It consists of the *ubiquitin system*, which marks proteins for destruction, and the *26S proteasome*, which recognizes the marked proteins and degrades them (Hershko and Ciechanover, 1998; Hilt and Wolf, 1996, 2000). This machinery resides mainly in the cytoplasm but is also found within the nucleus of cells (Knecht and Rivett, 2000).

The molecule that targets selected proteins for destruction is *ubiquitin*. This 76 amino acid peptide is linked in an isopeptide bond to the protein's N-terminus or to the ε-amino group of internal lysine residues via its C-terminal glycine residue. The 'ubiquitination' process begins with an ubiquitin-activating enzyme (E1), which upon ATP consumption forms a thioester bond with the C-terminal glycine of ubiquitin. Subsequently, ubiquitin is transferred onto the active site cysteine residue of an ubiquitin-conjugating enzyme (Ubc, E2). In most cases, an ubiquitin ligase (E3) mediates the attachment of ubiquitin to the ε-amino groups of lysine residues of substrate proteins. This occurs either by direct transfer of ubiquitin from the ubiquitin-conjugating enzyme (E2) to the substrate or through the formation of an E3-ubiquitin intermediate (Hershko and Ciechanover, 1998). The last step is polyubiquitination: multiple ubiquitin units are added in tandem to a particular lysine residue, usually Lys48, of already bound ubiquitin.

The *26S proteasome* is the proteolytic machinery. It consists of a 20S cylindrical core unit, which contains six hydrolytic active sites within the inner core of the cylinder (Heinemeyer, 2000), and two 19S regulatory particles attached to it apically. The '19S cap' senses the polyubiquitin chain bound to the protein and the six ATPase subunits that are part of the 19S cap are believed to unfold the bound protein, uncover the channel of the 20S cylinder core and, after the release of the polyubiquitin chain, push the substrate through the 20S proteasome, where it is finally cleaved into peptides (Glickman et al., 2000).

Proteins destined for secretion or for residence within the compartments of the secretory pathway enter the ER via a channel known as the Sec61 translocon, composed of three integral membrane proteins (Sec61α, Sec61β, Sec61γ in mammalian cells; Sec61p, Sbh1p, Sss1p in yeast) (Corsi and

Schekman, 1996). Nascent polypeptides traverse the ER membrane in an unfolded state. During the subsequent folding process in the ER lumen, most proteins are modified by N-linked glycosylation and undergo disulfide bond formation. These reactions require the concerted action of a variety of ER resident enzymes and molecular chaperones. Properly modified and folded proteins are then packed into vesicles and transported to the Golgi apparatus from where they proceed toward their final destination. The large numbers of unfolded proteins that enter the ER require the presence of a sufficient and well-balanced chaperone equipment to protect them from aggregation and to keep them in a folding-competent state (Pryer et al., 1992). One of the most abundant proteins in the ER is BiP (Kar2p in yeast), a member of the Hsp70 family of chaperones, which transiently associates with secretory proteins and more permanently with misfolded or unassembled proteins (Rapoport et al., 1996). In addition to the classical chaperones, the major components of the ER folding and modification apparatus known to date are protein disulfide isomerases, peptidyl prolyl isomerases, lectin-like proteins such as calnexin and calreticulin, glycoprotein glucosyltransferases and carbohydrate hydrolases (Zapun et al., 1999) (Table 9.1).

The necessity for a proper protein folding apparatus in the ER is reflected in the existence of a sophisticated regulatory system, the *unfolded protein response* (UPR), capable of controlling the concentration of these auxiliary proteins (Chapman et al., 1998). An ER/nuclear envelope localized transmembrane kinase, Ire1p, senses the level of unfolded proteins through its ER lumenal domain. It is believed that this sensing is mediated by competition between the Ire1p lumenal domain and free unfolded proteins for binding to the major chaperone of the ER, Kar2p (BiP). When the pool of free Kar2p (BiP) is depleted due to binding to increasing amounts of unfolded proteins, Ire1p is activated by dimerization. This leads to a conformational change of the cytoplasmic domain of the protein which induces the non-canonical splicing of the HAC1 mRNA, giving rise to the synthesis of the transcription factor Hac1p, which, in turn, upregulates genes containing the UPR response element (Chapman et al., 1998).

ER QUALITY CONTROL: THE MECHANISM

'Nobody is perfect.' This characteristic of human beings can be traced back to single cells and molecular mechanisms. Inefficient folding, unbalanced subunit synthesis or mutations in secretory proteins are events which occur constantly and result in the failure of translocated polypeptides to assume their correct three-dimensional conformation (Bonifacino and Klausner, 1994). It is extremely important for the proper functioning of a cell that the ER avoid delivery of such misfolded, inactive proteins to their final location where their active counterparts should be present. Failure to do

Table 9.1   The major chaperones and enzymes participating in ER-quality control and protein degradation

Family	Protein	Description
Classical chaperones	*ER lumenal*	
	BiP (GRP78)/Kar2p	Hsp70 superfamily, DnaK homolog
	GRP94	Hsp90 family
	Sec63p	DnaJ homolog; interacts with BiP
	Cytosolic	
	Hsp70 and Hsp90	
	Ssa1p	Hsp70 superfamily
	Hsc70	
	Hdj2	DnaJ homolog
	CHIP	Cytosolic U-box protein; Co-chaperone of Hsp70 and Hsp90
Disulfide modifying proteins	PDI (ERp59)/Pdi1p	Protein disulfide isomerase
	Eug1p	Protein disulfide isomerase
	ERp57	Thiol oxidoreductase
	ERp72; PDIR	PDI-like proteins
	Ero1p	Oxidase of PDI; *De novo* disulfide bond formation
Peptidyl prolyl isomerases	FK506-binding proteins	
	Cyclophilins	
Lectin-like chaperones	CNX/Cne1p/Cnx1	Calnexin
	CRT	Calreticulin
	Mnl1p/Htm1p	Mannosidase-like protein
N-glycan modifying proteins	MNS1	α1,2-mannosidase
	GLS1	Glucosidase 1
	GLS2	Glucosidase 2
	UGGT (GT)	UDP-Glc:glycoprotein glucosyltransferase

For details and references refer to text and to Zapun et al. (1999).

so is undeniably expected to lead to impaired cell function. It is, therefore, imperative that the ER contain a system able to discriminate between properly modified, folded proteins and improperly modified, terminally misfolded proteins. This discrimination must be followed by retention of the malfolded proteins in the ER and subsequently by their degradation. The various recognition and targeting events as a whole have been referred to as the 'quality control' function of the ER (Hammond and Helenius, 1995). A number of studies have been carried out on individual mutated, malfolded or unassembled mammalian proteins (Bonifacino and Klausner, 1994; Hammond and Helenius, 1995). From these studies it became clear that quality control in the ER works by structural rather than functional criteria. For instance, distinct mutations in α1-antitrypsin, which leave the protein functionally intact, lead to ER retention and to degradation of the mutant molecule (Perlmutter, 1996). Another well-studied example is a mutated form of the cystic fibrosis transmembrane conductance regulator (CFTRΔF508): even though potentially active, the molecule is retained in the ER and is rapidly degraded (Jensen et al., 1995; Ward et al., 1995). When polypeptides are translocated into the ER, sorting between proteins that are naturally targeted to reside in the ER lumen and those that have to leave the organelle has to be achieved. The mechanism of 'quality control', then, has to distinguish between conformational variants of a particular protein.

The main aspects of protein quality control in the ER can be summarized as

1. Discovery of malfolded or unassembled proteins.
2. Retention of these proteins in the ER.
3. Delivery to the proteolytic system.
4. Protein degradation.

While our knowledge regarding the processes of (1) discovery of malfolded proteins and (2) their retention in the ER is very scarce, our understanding of (3) their delivery to the proteolytic system and (4) the mode of degradation has increased considerably in the last few years. As soon as a protein is translocated into the ER from the cytoplasm, it associates with ER chaperones and folding factors such as BiP (Kar2p in yeast), calnexin, protein disulfide isomerase (PDI), etc. These proteins not only act as folding factors, but also serve as retention anchors until the substrate protein attains its mature state. In addition to the components which are commonly needed for folding and retention in the ER, recently, a number of other, more substrate-specific factors have also been detected (Ellgaard et al., 1999). It is believed that the mechanism of discovery and retention of malfolded proteins is somehow determined through the interplay of the different folding factors and their respective substrates.

Masking and unmasking signals have been uncovered on certain mammalian membrane proteins. For example, charged residues within the transmembrane domain of the T-cell receptor were the first signals discovered that led to the degradation of unassembled subunits (Bonifacino et al., 1990). Two hydrophilic residues in the membrane spanning region of the μ heavy chain of cell surface IgM have been shown to induce ER retention in the absence of the κ light chain (Stevens et al., 1994). In case of the human high-affinity IgE receptor, the cytoplasmic domain of the α chain contains a dilysine retention signal, which keeps the unassembled subunit in the ER: this signal becomes nonfunctional following assembly with the γ chain, allowing export of the fully assembled receptor. As a result, through such recognition mechanisms, the ER retention machinery is capable of performing the necessary quality control and of discriminating between properly assembled and unassembled receptors (Letourneur et al., 1995). Quality control of the heterooligomeric ATP sensitive K^+ channel in

Drosophila melanogaster is thought to take place via a similar masking–unmasking mechanism: each subunit of this channel contains a cytosolic Arg-Lys-Arg (RKR) sequence, which leads to retention or retrieval when unmasked (Zerangue et al., 1999). Interestingly, a similar motif (RXR) present in the cytoplasmic domain of CFTR is responsible for blocking the transport of misfolded CFTRΔF508 to the plasma membrane (Ellgaard et al., 1999; Gilbert et al., 1998). Recently, it has been determined that CHIP, a newly identified cytosolic U-box protein, functions together with Hsc70 to sense the folded state of CFTR and targets the aberrant forms for degradation (Meacham et al., 2001). In the case of membrane spanning proteins the issue of discovery of a malfolded protein is complicated by the fact that recognition of the misfolding must take place on either the lumenal or cytoplasmic face of the ER membrane, depending on the topological location of the misfolding.

A chaperone-mediated retention mechanism in the ER lumen is especially well characterized for glycoproteins. In mammalian cells, two homologous lectins, calnexin and calreticulin, bind to nearly all soluble and membrane glycoproteins imported into this compartment. They specifically associate with glycoproteins that carry monoglucosylated trimming intermediates of their N-linked glycans (Ellgaard et al., 1999; Hammond et al., 1994; Helenius et al., 1997). Initially, preassembled N-glycans are co-translationally linked to the polypeptide chain as a 14 unit, branched oligosaccharide structure. During folding, two of the outermost three glucose residues of the N-linked oligosaccharide are rapidly trimmed by glucosidases I and II. Exposure of the innermost glucose leads to the binding of the monoglucosylated glycoprotein to calnexin or calreticulin. This complex, then, associates with a thiol oxidoreductase, Erp57p, and forms transient mixed-disulfides with the enzyme. The remaining glucose residue is eventually trimmed by glucosidase II and the complex dissociates releasing a protein with the carbohydrate structure $Man_9GlcNAc_2$. If the released protein is still not properly folded, it is recognized by a UDP-glucose, glycoprotein glucosyltransferase (GT), and becomes re-glucosylated. This leads to a new round of calnexin/calreticulin binding, which prevents the glycoprotein's escape from the ER. If this cycle persists due to the inability of the protein to attain its final folded conformation, mannosidase I cleaves the α1,2-linked mannose of the middle branch, generating $Man_8GlcNAc_2$ which is re-glucosylated by GT. The resulting carbohydrate structure, $Glc_1Man_8GlcNAc_2$, once again binds to calnexin/calreticulin. However, since $Glc_1Man_8GlcNAc_2$ is a poor substrate of glucosidase II, the glycoprotein is no longer rapidly released from the lectin and is subsequently delivered to the degradation machinery (Ellgaard et al., 1999; Liu et al., 1999). Recently, it has been reported that the degradation of a *cog* thyroglobulin mutant and of unassembled immunoglobulin μ and J chains is also dependent on cleavage of the terminal mannose from the central branch by mannosidase I. However, in these cases, calnexin and calreticulin do not seem to play a major role in timing the degradation process (Fagioli and Sitia, 2001; Tokunaga et al., 2000). These different findings indicate the existence of more than one pathway that coordinates the degradation of misfolded glycoproteins.

In recent years, studies on a model eukaryote, the yeast *Saccharomyces cerevisiae*, have set the pace in the field. Research on four mutated proteins, the peptide pheromone α-factor, the lysosomal carboxypeptidase yscY (CPY*), the translocon component Sec61p, and the ABC transporter (Pdr5*), has paved the way for our current understanding of how these malfolded proteins are eliminated from the ER (Brodsky and McCracken, 1999; Plemper and Wolf, 1999; Sommer and Wolf, 1997). Mutated (Gly255Arg) carboxypeptidase yscY (CPY*) is completely imported into the ER lumen and fully N-glycosylated (Plemper et al., 1999b); however, further export to the vacuole is impaired. Clearance of the ER-retained mutant protein requires the action of glucosidases I and II and α1,2-mannosidase (Jakob et al., 1998; Knop et al., 1996b). The carbohydrate structure $Man_8GlcNac_2$, with a mannose missing from the middle branch, was specifically identified as a necessary signal for the degradation of CPY* (Jakob et al., 1998). Removal of this particular mannose most likely generates a signal that is recognized by a lectin-like receptor, which targets the protein for degradation. As for higher eukaryotic cells, mannose trimming is a rather slow process and may determine the time frame within which the glycoprotein has to complete folding. If proper folding is not achieved, the protein is degraded (Jakob et al., 1998). Calnexin, however, is not involved in targeting CPY* for degradation (Knop et al., 1996b). Furthermore, *S. cerevisiae* does not possess a UDP-glucose: glycoprotein glucosyltransferase (Helenius and Aebi, 2001). Very recently a new α-mannosidase-like protein (Mnl1p/Htm1p) has been described in *S. cerevisiae*, which is devoid of α-mannosidase activity but is necessary for CPY* degradation. This protein may function as a lectin recognizing the oligosaccharide structure $Man_8GlcNac_2$ and targeting CPY* for degradation (Nakatsukasa et al., 2001; Jakob et al., 2001). A homologous lectin (EDEM) has also been found in mice (Hosokawa et al., 2001). Nevertheless, glycosylated proteins are not the only ones targeted for ER degradation. Studies with two unglycosylated yeast proteins, the mutated pro-α-factor (McCracken and Brodsky, 1996) and the mutant Sec61 translocon protein Sec61–2p (Biederer et al., 1996), revealed that they too undergo rapid ER degradation. Interestingly, despite the absence of any N-linked oligosaccharides, elimination of the mutated pro-α-factor depends on the presence of calnexin (McCracken and Brodsky, 1996).

These results suggest that a great deal of plasticity can be expected for ER-associated degradation (ERAD) targeting mechanisms. Elimination of CPY* also requires the presence of the Hsp70 chaperone Kar2p (BiP) (Plemper et al., 1997). The requirement for Kar2p (BiP) has also been observed for the ER-associated degradation of mutated pro-α-factor (Brodsky et al., 1999) together with protein disulfide isomerase (PDI) (Gillece et al., 1999). In contrast to its function in the elimination of soluble, ER-lumenal proteins, Kar2p (BiP) does not seem to be necessary for the degradation of misfolded membrane proteins. Pdr5*, a mutant form of the polytopic plasma membrane ABC transporter Pdr5, is retained in the ER membrane and its elimination proceeds even in the presence of a non-functional Kar2p (BiP) (Plemper et al., 1998). In addition, the concentration of Ca^{2+} ions in the ER has proved to be critical for the elimination of soluble CPY* but not for the integral membrane protein Pdr5*; in fact, deletion of the ER/Golgi Ca^{2+} pump Pmr1p severely affected degradation of CPY* but not Pdr5* (Dürr et al., 1998; R.K. Plemper and D.H. Wolf, unpublished).

Summary: *ER quality control: the mechanism*

- Protein quality control in the ER works by structural criteria.
- Discovery of malfolded proteins occurs via masking and unmasking signals in proteins.
- ER localized enzymes, chaperones and lectins are responsible for recognition and retention of malfolded proteins in the ER.
- Enzyme catalyzed carbohydrate modification might work as the timer for glycoprotein degradation.

THE ERAD MACHINERY

The mechanism for the elimination of malfolded or unassembled proteins from the ER had remained a mystery for a long time. It was believed that ER-localized proteinases and peptidases were responsible for the elimination of these proteins (Bonifacino and Klausner, 1994). However, since proteins enter the ER in an unfolded state and the folding process in the ER takes time, the existence of an unspecific and aggressive proteolytic apparatus in the ER was hard to imagine.

A first clue to the nature of the proteolytic system involved in the elimination of ER membrane proteins came from studies on an ER-membrane-located ubiquitin-conjugation enzyme in yeast (Ubc6p). Ubc6p, residing in the ER membrane and with a cytosolic active site, when mutated, restored the protein translocation competence of *sec61-2* mutant cells at the restrictive temperature. This finding led to the speculation that the protein translocation defect of the temperature sensitive *sec61-2* mutants is linked to ubiquitin-mediated proteolysis (Sommer and Jentsch, 1993). This hypothesis turned out to be true: it was demonstrated that the Sec61-2 mutant protein was degraded at the restrictive temperature in a ubiquitin-dependent fashion. The ubiquitin-conjugating enzymes Ubc6p and Ubc7p were responsible for synthesis of the polyubiquitin chain necessary for proteasomal degradation (Biederer et al., 1996). The involvement of the cytoplasmically located proteasome in the degradation of malfolded ER membrane proteins was determined upon investigation of the fate of a mutated ATP binding cassette (ABC) transporter in human cells, the ΔF508 cystic fibrosis transmembrane conductance regulator (CFTR), the molecular cause of cystic fibrosis. Although partially active, the ΔF508 CFTR protein is not delivered to the plasma membrane; instead, it is retained in the ER membrane and it is rapidly degraded via the ubiquitin–proteasome pathway (Jensen et al., 1995; Ward et al., 1995).

These findings indicated that elimination of membrane proteins, which had undergone ER-quality control, occurred via the cytoplasmic ubiquitin–proteasome system. This seemed to be quite a reasonable and natural solution because membrane proteins expose large domains into the cytoplasm, which can be easily attacked by the proteolytic system residing in this cellular compartment. However, a number of questions concerning the mechanism of removal of membrane proteins still remained. Cytoplasmic domains could be shaved off by the proteasome but how are transmembrane and ER-lumenal domains removed? Likewise, how are malfolded or unassembled soluble, ER lumenal proteins eliminated? Studies on CPY* and pro-α-factor created an answer: remarkably, the elimination of these two proteins was also dependent on the proteasome (Hiller et al., 1996; Werner et al., 1996). While degradation of mutated pro-α-factor was independent of ubiquitination (Werner et al., 1996), elimination of CPY* required the action of ubiquitin-conjugation (E2) enzymes: ER membrane bound Ubc6p and, more prominently, Ubc7p were necessary for its degradation (Hiller et al., 1996). ER membrane recruitment and activation of the soluble, cytoplasmic Ubc7p by the ER membrane anchored Cue1 protein turned out to be a prerequisite for efficient CPY* degradation (Biederer et al., 1997).

Recently, another E2, Ubc1p, has also been found to be partially involved in ubiquitin-conjugation onto CPY* (Friedländer et al., 2000). The missing ubiquitin-protein ligase (E3) of the degradation process was identified as the RING-H2 finger motif containing ER membrane protein Der3p/Hrd1p (Bays et al., 2001; Bordallo et al., 1998; Bordallo and Wolf, 1999; Deak and Wolf, 2001; Hampton et al., 1996). As all the components of the proteolytic machinery functioned in the cytoplasm of the cell, how this proteolytic machinery and the proteins destined for degradation came together became a crucial question. As import of proteins into the ER was believed to be a one-way process (Blobel, 1995), the possibility that this process could be reversed came as a surprise. The discovery that signal sequence cleaved, ER-glycosylated CPY* could be found on the cytoplasmic face of the ER membrane in an ubiquitinated state, and was degraded by the proteasome, clearly indicated that retrograde transport of CPY* to the cytoplasm must have occurred (Hiller et al., 1996). In addition, the finding that elimination of mutated, signal sequence cleaved pro-α-factor as well as a mutated human α1-proteinase inhibitor expressed in yeast depended on the cytoplasmic proteasome strongly supported the idea of a retrograde protein transport mechanism to the cytoplasm (Werner et al., 1996). This then raised the question of how these malfolded proteins were retrograde transported across the ER membrane into the cytoplasm. The import channel, the Sec61 translocon, was an obvious candidate for the passage of malfolded proteins across the ER membrane. Genetic studies in yeast, utilizing *sec61* mutants, gave the answer: under conditions where import of CPY* into the ER lumen was completely normal, degradation of CPY* was considerably slowed down in *sec61-2* mutants (Plemper et al., 1997). *In vitro* studies with two other alleles of *SEC61* (*sec61-32* and *sec61-41*) uncovered a considerably delayed degradation of mutated pro-α-factor (Pilon et al., 1997).

These studies directly implicated the translocon in the retrograde transport process. The more recent isolation of *SEC61* mutations specifically defective in the retrograde transport of mutated pro-α-factor (Zhou and Schekman, 1999) and CPY* (Wilkinson et al., 2000) corroborates the previous findings which link Sec61p to the retrotranslocation event. As expected from a defect in dislocation of malfolded proteins from the ER (Knop et al., 1996a), the *sec61* mutants tested also exhibited an induced unfolded protein response (Wilkinson et al., 2000; Zhou and Schekman, 1999). The assessment of the Sec61p translocon as the export channel for soluble ER proteins was soon extended to polytopic membrane proteins. Degradation of a mutated ABC-transporter, Pdr5*, which cannot reach the plasma membrane but is retained in the ER, was also dependent on the presence of an intact Sec61 channel (Plemper et al., 1998). The observation that the degradation of Pdr5* occurred in a highly processive manner, without formation of any intermediate cleavage products, gave an indication for the mechanism by which polytopic membrane proteins could be degraded. It is very likely that the transmembrane domains diffuse laterally into the Sec61 translocon, from which they become extracted into the cytoplasm where they undergo hydrolysis by the 26S proteasome (Plemper et al., 1998; Plemper and Wolf, 1999). It is known that CPY* leaves the translocon completely after import into the ER (Plemper et al., 1999b), which implies that retrograde transport back to the cytoplasm requires a new targeting mechanism. Interaction with the ER lumenal chaperone Kar2p and its partner Sec63p (Plemper et al., 1997), as well as binding to PDI (Gillece et al., 1999), are possibly part of this process. For glycoproteins like CPY*, the lectin-like ER membrane protein Mnl1p/Htm1p (Nakatsukasa et al., 2001; Jakob et al., 2001) may also be part of the targeting chain. Additionally, two recently identified ER membrane proteins are required for

ERAD: Der1p (Knop et al., 1996a) and Hrd3p (Hampton et al., 1996). While Der1p is only needed for the degradation of soluble, ER lumenal proteins (Knop et al., 1996a; Plemper et al., 1998), Hrd3p is required for the elimination of soluble and membrane proteins alike (Hampton et al., 1996; Plemper et al., 1998). Hrd3p interacts with the ubiquitin-protein ligase Der3p/Hrd1p and it is believed to play a role in the signaling events between the ER lumen and the cytoplasmic ubiquitin-proteasome machinery (Deak and Wolf, 2001; Gardner et al., 2000; Plemper et al., 1999a). The establishment of a genetic interaction between Hrd3p and the Sec61 channel may be indicative of the existence of specific retrotranslocation channels, different from import channels (Plemper et al., 1999a). In this way, the two-way protein traffic across the ER membrane may be achieved by two distinct subsets of translocons.

Recent studies in yeast point to the fact that ER degradation and secretion are interconnected processes. Mutants defective in genes necessary for protein transport between the ER and the Golgi compartments (*ERV29, SEC12, SEC18, SEC23, UFE1*) block degradation of ER lumenal CPY* but not membrane spanning substrates (Caldwell et al., 2001; C. Taxis, F. Vogel and D.H. Wolf, unpublished). This observation poses the following questions: (i) are soluble ERAD substrates eliminated in some unidentified post-ER vesicle? (ii) is ER to Golgi traffic necessary for an efficient ERAD process? Electron microscopy studies have not given any evidence for the existence of a new post-ER vesicle type, which could be involved in the degradation of CPY*. Consequently, this implies that functional ER to Golgi traffic is necessary for an efficient CPY* degradation (C. Taxis, F. Vogel and D.H. Wolf, unpublished).

ER degradation in mammalian cells seems to follow the same principles as in yeast. An important finding was the ubiquitin–proteasome dependent degradation of the mutated ΔF508 CFTR protein retained in the ER membrane (Jensen et al., 1995; Ward et al., 1995). Polyubiquitination was found to precede membrane extraction of ΔF508 CFTR (Xiong et al., 1999). Immunoprecipitation studies discovered complex formation between CFTR and the Sec61 complex. The amount of Sec61–CFTR complex increased after inhibition of the proteasome, giving strong indication for Sec61 translocon-mediated dislocation of CFTR (Bebök et al., 1998). Another crucial discovery concerned the misuse of the ER degradation machinery by cytomegalovirus to escape immune detection (Wiertz et al., 1996a, 1996b). In this case, two proteins encoded by the human cytomegalovirus, US2 and US11, force the major histocompatibility (MHC) class I heavy chain to leave the ER after complete import and glycosylation. MHC chains become deglycosylated and degraded via the proteasome. Breakdown intermediates of MHC class I polypeptides were found to co-immunoprecipitate with the translocon subunit Sec61β. Another well-studied example of ERAD in mammalian cells is the elimination of a truncated variant of the rough ER-specific type I transmembrane glycoprotein ribophorin I (RI332) (de Virgilio et al., 1998). This mutant protein is released into the cytosol and degraded by the proteasome in an ubiquitin-dependent fashion. The process was shown to rely on an active ubiquitin-activating enzyme (E1) and binding of the protein to the Sec61 complex was also determined.

Recently, the murine homologs of the yeast ubiquitin-conjugating enzymes Ubc6p and Ubc7p have been described (Tiwari and Weissman, 2001). The authors communicate that MmUbc7p indeed plays a crucial role in the degradation of unassembled T-cell receptor subunits (Table 9.2 lists a number of studied ERAD substrates; Figure 9.1 summarizes the translocation event and the steps leading to the degradation of CPY*).

Deglycosylation: a necessary step prior to degradation?

Interestingly, glycosylated proteins destined for degradation via the ubiquitin–proteasome system leave the ER in a glycosylated form (Hiller et al., 1996; Wiertz et al., 1996b). There are many examples where de-N-glycosylated intermediates of the overall degradation process can be detected, indicating that these proteins are deglycosylated in the cytoplasm prior to degradation by the proteasome (Bebök et al., 1998; de Virgilio et al., 1998; Halaban et al., 1997; Hughes et al., 1997; Huppa and Ploegh, 1997; Johnston et al., 1998; Mosse et al., 1998; Wiertz et al., 1996a; Yang et al., 1998; Yu et al., 1997). A peptide N-glycanase is believed to be responsible for this process. In yeast the *PNG1* gene has been identified recently. It encodes a soluble protein, which exhibits N-glycanase activity. The enzyme is highly conserved between eukaryotes from yeast to humans. Most of the protein is found in the nucleus, some of it in the cytosol. *PNG1* is not an essential gene, but its product is required for efficient degradation of misfolded CPY* (Suzuki et al., 2000). It was found that Png1p binds to Rad23p, a protein that interacts with the proteasome via its ubiquitin-like (Ubl) domain. Clearly, Rad23p seems to escort Png1p to the proteasome. It is hypothesized that the association of the 26S proteasome and Png1p produce a complex in which de-N-glycosylation and proteolysis of the unwound protein substrates could be accomplished in an efficient manner (Suzuki et al., 2001).

Summary: *The ERAD machinery*

- Malfolded proteins are retrogradely transported from the ER back to the cytosol.
- The Sec61 translocon allows retrograde transport of malfolded proteins back into the cytoplasm.
- Glycoproteins are de-N-glycosylated in the cytosol prior to degradation.
- Polyubiquitination targets malfolded proteins for degradation via the cytosolic proteasome.

Table 9.2 Examples of ERAD substrates

Protein	Description
ERAD sustrates studied in *S. cerevisiae*	
pro-α factor	Peptide pheromone
A1PiZ	Human α_1-proteinase inhibitor
CPY*	Carboxypeptidase Y
Fur4p	Uracil permease
Hmg2p	HMG-CoA reductase
Pdr5*	ABC (ATP binding cassette) transporter
Sec61p	Translocon component
Ste6p	a-Factor transporter (ABC transporter)
Vph1p	V-ATPase subunit
Medically relevant ERAD substrates	
α1-antitrypsin (α1-ATZ or A1PiZ)	Lung emphysema; liver disease
Apolipoprotein B100 (apoB100)	Abetalipoproteinemia
Aquaporin2 (AQP2)	Nephrogenic diabetes insipidus
Cystic fibrosis transmembrane regulator (CFTR)	Cystic fibrosis
Hydroxymethylglutaryl-CoA reductase (HMG-R)	Cholesterolemia; heart disease
Golgi P-type ATPase (Wilson protein)	Wilson disease
Insulin receptor	Diabetes mellitus
Low-density lipoprotein receptor	Familial hypercholesterolemia
Lysosomal α-galactosidase A	Fabry disease
Prion protein (PrP^{Sc})	Neurodegenerative disorders (Creutzfeldt–Jakob disease)
Thyroglobulin (Tg)	Congenital hypothyroid goiter
Tyrosinase	Malignant melanoma
Von Willebrand factor	Von Willebrand's disease
Receptors and immunoglobulins	
T-cell receptor α chain	Degraded when unassembled
Human δ-opioid receptor	Degraded due to folding difficulties
Human high-affinity IgE receptor	Untrimmed glycans activate ER quality control
Cell surface IgM μ-chain	Degraded in the absence of light chain assembly
Opportunistic uses of ERAD	
Viruses escape immuno-detection via ERAD	
MHC class I heavy chain	HCMV US2 and US11 mediated degradation
CD4 receptor	HIV-1 Vpu induced degradation
Bacterial and plant toxins attach the host cell via ERAD	
Cholera toxin A1 chain (CTA1)	
Pertussis toxin	
Ricin A chain (RTA)	
Shiga toxin	
Other substrates	
Ribophorin I (RI332)	Subunit of the oligosaccharyl-transferase complex
ATP-sensitive K$^+$ channel	From *Drosophila melanogaster*

For details and references refer to text.

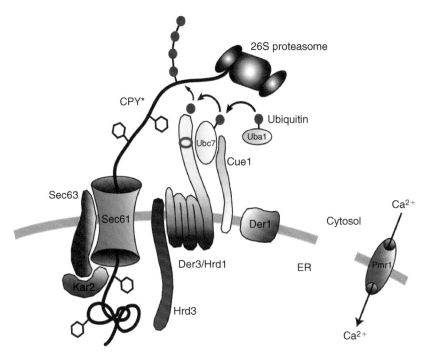

Figure 9.1   Protein quality control of the ER. The 'classical' ERAD pathway. Proteins which fail to fold properly (CPY*) are scanned and retrotranslocated through the Sec61 translocon, polyubiquitinated at the ER membrane via the ubiquitin-activating enzyme Uba1, the ubiquitin-conjugating enzyme Ubc7, and the ubiquitin-protein ligase Der3/Hrd1 complexed to Hrd3. Polyubiquitination is followed by rapid degradation of the protein via the 26S proteasome (modified after Deak and Wolf, 2001; cover page *J Biol Chem 276*, issue 14, 6 April 2001, by permission of the American Society for Biochemistry and Molecular Biology).

- The ubiquitination machinery resides at the cytosolic face of the ER membrane.
- Proteasome degradation is highly processive.

ENZYME REGULATION VIA THE ER QUALITY CONTROL PATHWAY

It is interesting to note that the ER quality control machinery is also used by the cell to regulate the level of some ER enzymes. One of the best-studied examples is the downregulation of 3-hydroxy-3-methylglutaryl CoA reductase (HMG-R), a key regulatory enzyme in sterol synthesis. In yeast, two

HMG-R enzymes exist, both localized at the ER membrane. HMG-R2 is subject to rapid degradation, while the isozyme HMG-R1 is a relatively stable protein. Farnesylpyrophosphate is the primary signal for HMG-R2 degradation in yeast (Hampton and Bhakta, 1997) as well as for the degradation of the mammalian homolog HMG-R (Meigs and Simoni, 1997). Degradation occurs via the ubiquitin–proteasome system (Hampton and Bhakta, 1997; Hampton et al., 1996; Ravid et al., 2000). The ubiquitin-conjugating enzyme Ubc7p, the ubiquitin-protein ligase Der3p/Hrd1p, and Hrd3p, components crucial for the degradation of misfolded soluble and transmembrane proteins, have also been identified as the necessary elements for the degradation of yeast HMG-R2 (Bays et al., 2001; Gardner et al., 2000; Hampton and Bhakta, 1997; Hampton et al., 1996). Entry of HMG-R2 into the ER quality control pathway is regulated by means of signals from the mevalonate pathway. These signals cause allosteric changes that lead to the conversion of the enzyme from a stable protein into a quality control substrate (Cronin et al., 2000).

Summary: *Enzyme regulation via the ER quality control pathway*

- 3-Hydroxy-3-methylglutaryl CoA reductase is regulated via the ER quality control pathway.

ALTERNATIVE QUALITY CONTROL PATHWAYS

It appears that the majority of the ER-associated degradation substrates are eliminated from the ER via retrograde transport through the Sec61 translocon, are ubiquitinated by ER membrane-associated ubiquitin-conjugating enzymes (E2s) and a specific ubiquitin-protein ligase (E3), and are finally degraded by the cytoplasmic proteasome. This is generally accepted as the classical ER-associated degradation pathway. Interestingly, Ubc6p, the ubiquitin-conjugating enzyme embedded in the ER membrane and involved in some degradation events in yeast (Biederer et al., 1996; Hiller et al., 1996; Sommer and Jentsch, 1993), is itself a naturally short-lived protein (Walter et al., 2001). Its rapid, proteasomal degradation is independent of the Sec61 translocon. Moreover, although its turnover depends on Ubc7p and Cue1p, which recruits this E2 to the ER membrane, degradation of Ubc6p is independent of the ubiquitin-protein ligase Der3p/Hrd1p that is usually involved in Ubc7p dependent degradation events. Der3p/Hrd1p's interaction and signaling partner Hrd3p does not take part in this degradation event either. Recent experiments uncovered the involvement of a new ubiquitin-protein ligase of the ER, Doa10p, in the degradation event of Ubc6p (Swanson et al., 2001). In contrast to the Sec61p-Der3p/Hrd1p-Hrd3p dependent elimination of membrane proteins, for which membrane extraction and degradation cannot be separated, an accumulating degradation intermediate, consisting of a membrane fragment of Ubc6p, could be detected in proteasomal mutants. This may indicate that the proteasome is directly involved in the membrane extraction process (Walter et al., 2001). Ubc6p belongs to the class of tail-anchored proteins (Kutay et al., 1993), which differ from all other membrane proteins so far investigated in their mode of membrane insertion. Their membrane anchors insert post-translationally, independent of Sec61p (Kutay et al., 1995). The C-terminal transmembrane anchor of Ubc6p is thought to protrude into the ER lumen only by its last lysine residue (Sommer and Jentsch, 1993). It is this feature that may have generated a modified, alternate ER degradation pathway. It will be interesting to see if this alternate degradation pathway holds true for other members of this class of proteins.

The vacuolar ATPase (V-ATPase) of the yeast *S. cerevisiae* is a hetero-oligomeric complex which assembles on the ER membrane. In cells lacking Vma22p, assembly of the complex fails and the 100 kDa membrane subunit Vph1p is rapidly degraded. Degradation is dependent on ubiquitination and the proteasome. However, neither the 'classical' ER ubiquitin-conjugating enzymes Ubc6p and Ubc7p, nor the 'classical' ER ubiquitin-protein ligase Der3p/Hrd1p and its partner Hrd3p, are required. Degradation of Vph1p is slowed down in mutants simultaneously deleted in a combination of the ubiquitin-conjugating enzymes Ubc1p and Ubc4p or Ubc2p and Ubc4p. The involvement of the translocon component Sec61p in Vph1p dislocation has not been tested (Hill and Cooper, 2000). Similar results, namely the lack of an absolute requirement for Ubc7p and Der3p/Hrd1p for degradation of Vph1p, have also been reported by Wilhovsky et al. (2000). They also tested the fusion protein Deg1-HMG-R1, containing the Deg1 degradation domain of the MATα2 repressor fused to the stable HMG-R1 isozyme. This fusion protein underwent Ubc7p-dependent degradation. However, the ubiquitin-protein ligase Der3p/Hrd1p and Hrd3p were rather ineffective in the degradation process. As the newly discovered ubiquitin-protein ligase Doa10p is required for Deg1 mediated ubiquitination (Swanson et al., 2001) it is this E3 which might trigger the elimination of Deg1-HMG-R1. The same features were found for the degradation of a mutated uracil permease (Wilhovsky et al., 2000).

A completely different route of elimination was found for a fusion protein containing a mutated form of the N-terminal domain of the λ-repressor protein and the secreted protein invertase. This fusion protein, expressed in *S. cerevisiae*, was not targeted for proteasomal degradation in the cytosol but, instead, it was delivered to the vacuole by receptor-mediated transport. Specifically, the Golgi receptor Vps10p, responsible for targeting carboxypeptidase yscY to the vacuole, was found to transport the hybrid protein to the vacuole where it was degraded (Hong et al., 1996). Evidently, the scanning system of the ER is unable either to discover this protein as a misfolded protein or to deliver it to the cytosolic ubiquitin–proteasome system. Future studies will be necessary to further clarify the cellular strategies employed to recognize and to destroy malfolded ER proteins.

Summary: *Alternative quality control pathways*

- A few examples of ERAD substrate have been found that do not follow the 'classical' ERAD pathway: an ER-membrane tail-anchored protein does not need the Sec61 translocon for dislocation, nor does it use the ER-membrane located ubiquitin ligase Der3p/Hrd1p-Hrd3p complex for degradation. The E3 Doa10p is used instead.
- ERAD for another substrate does not use the 'classical' ER-associated ubiquitin-conjugating enzymes (Ubc6p-Ubc7p), nor the ER membrane located ubiquitin-ligase Der3p/Hrd1p-Hrd3p complex.
- A third substrate does not undergo cytoplasmic proteasomal degradation but is transported to the vacuole (lysosome) instead.

PROTEIN QUALITY CONTROL AND DISEASE: OPPORTUNISTIC USES OF ERAD

Toxins reach the host cell's cytosol via ERAD

Some proteins, like the bacterial cholera toxin, Shiga toxin, pertussis toxin and the plant toxin ricin, are highly toxic to mammalian cells. They are commonly referred to as AB toxins, because of their structural organization. In general, the A moiety has enzymatic activity and modifies a cellular target upon entry into the host cytosol, causing drastic changes on cellular physiology or even cell death. These toxins are all part of a class that reaches the host cytosol via retrotranslocation through the Golgi apparatus and eventually through the ER (Falnes and Sandvig, 2000). An increasing body of research suggests that these toxins take advantage of the translocation machinery in the ER membrane to reach the cytosol. The best-studied examples are ricin and cholera toxin. Glycosylated ricin molecules were recovered from the cytoplasmic fraction, which indicated that they had passed through the ER. In addition, a fraction of the glycosylated ricin was co-immunoprecipitated with anti-Sec61 antibodies (Wesche et al., 1999). Furthermore, when the ricin A chain was expressed and imported into the ER of yeast, mutations in proteasomal subunits or in Sec61p inhibited its export and caused a decrease in its degradation (Simpson et al., 1999). Using a cell-free system, it has been demonstrated that the A1 subunit of cholera toxin is exported from microsomes in an ATP dependent manner (Schmitz et al., 2000). Co-immunoprecipitation studies found association of cholera toxin A1 subunit with Sec61p during export. In addition, the blocking of the Sec61 translocon by nascent polypeptides arrested during import strongly inhibited export of cholera toxin A1 chain (Schmitz et al., 2000). A very recent report directly implicates PDI in the unfolding of cholera toxin A1 chain in a new way. PDI was shown to act as a redox-driven chaperone, which binds the A1 chain when in the reduced state and releases it when oxidized. It is hypothesized that PDI delivers the cholera toxin A1 chain to the Sec61 channel for export either by itself or via a downstream chaperone (Tsai et al., 2001). Taken together, these data indicate that ricin and cholera toxin (mis)use the ER quality control machinery for their translocation into the cytosol where they carry out their detrimental activities.

Viral strategies to escape the host defense

The cellular immune response against invading viruses is primarily based on the elimination of virus-infected cells by cytotoxic $CD8^+$ T-lymphocytes. MHC class I molecules present antigenic peptide fragments of viral proteins on the surface of infected cells. $CD8^+$ cells recognize the MHC class I antigen complexes through specific receptors and trigger the death of infected cells. Viral peptides are loaded on MHC class I molecules in the ER, and these antigen-presenting complexes are then transported through the secretory pathway to the cell surface. The pressure exerted by the immune system on viruses has led them to acquire, in the course of their evolution, a set of strategies with which they can elude cytolytic T-cells. Human cytomegalovirus (HCMV), for example, has evolved a rescue mechanism which prevents cell surface presentation of its antigenic peptides, by destroying the host's MHC class I heavy chains. HCMV encodes two ER-targeted transmembrane proteins, US2 and US11, which interact with MHC class I chains and trigger their dislocation and extraction from the ER membrane. Dislocated MHC class I molecules are then degraded by the proteasome in a ubiquitination-dependent manner (Shamu et al., 1999; Wiertz et al., 1996a, 1996b).

HIV, on the other hand, has developed a strategy to reduce the level of its receptor CD4 on the surface of infected cells. The HIV-1 encoded Vpu protein triggers the formation of a ternary complex consisting of CD4, Vpu and h-βTrCP at the ER membrane (Margottin et al., 1998). h-βTrCP is a WD repeat protein containing an F-box motif, which binds and recruits Skp1, acting as a protein ubiquitin ligase. As a consequence nascent CD4 molecules are unable to leave the ER and are rapidly degraded by the proteasome (Margottin et al., 1998). There is also evidence that Vpu may initiate the degradation of MHC class I molecules (Kerkau et al., 1997). However, in all these cases, the details of the mechanism employed and the additional components necessary for degradation must still be elucidated.

Cystic fibrosis

Cystic fibrosis is one of the most prevalent genetic disorders, affecting approximately one in two thousand live births among populations of Caucasian or northern European descent. It manifests itself as a series of severe bronchopulmonary disorders and pancreatic insufficiency, caused by a defective chloride (Cl^-/HCO_3^-) channel, the CFTR. The maturation of this polytopic membrane protein in the ER is very inefficient and only about 25% of the wild-type protein actually reaches the plasma membrane of epithelial cells. Mutant forms of CFTR never leave the ER and are degraded. The majority of patients suffering from cystic fibrosis carry the CFTRΔF508 allele. This mutant form of the protein is unable to fold properly, which affects its trafficking through the secretory pathway: CFTRΔF508 is completely retained in the ER membrane and rapidly eliminated by the ubiquitin–proteasome system. Just like most proteasome substrates, this CFTR mutant is polyubiquitinated (Bebök et al., 1998; Jensen et al., 1995; Ward et al., 1995) and like Pdr5* of yeast it is extracted from the ER membrane via the Sec61 channel (Bebök et al., 1998; Xiong et al., 1999). Interestingly, CFTRΔF508 is a potentially functional protein; however, its ER retention and subsequent degradation hinder its role as a Cl^-/HCO_3^- channel.

Lung emphysema

The Z mutant allele of human α1-antitrypsin (α1-AT), an abundant serum glycoprotein, causes a number of severe phenotypes: individuals homozygous for this allele suffer from lung emphysema, sometimes combined with chronic liver disease and hepatocellular carcinoma. The mutant α1-AT protein does not fold properly, is retained in the ER and it is most probably degraded by the proteasome (Qu et al., 1996). When expressed in yeast, mutant α1-AT is, in fact, subject to proteasomal degradation after escape from the ER (Werner et al., 1996).

Malignant melanoma

Melanin biosynthesis takes place in a post-Golgi compartment known as the melanosome. Loss of tyrosinase, a key enzyme in melanin biosynthesis, leads to formation of malignant melanomas. Analysis of different mice and human tyrosinase mutants revealed that these variants were retained in the ER, were core glycosylated, had prolonged association with calnexin/calreticulin, then were ubiquitinated and rapidly degraded by the proteasome (Halaban et al., 1997).

Neurodegenerative diseases

Wilson disease is caused by an inherited disorder in copper metabolism marked by neuronal degeneration and hepatic cirrhosis. The protein which, when mutated, is responsible for Wilson disease, the Wilson protein, is a copper transporting P-type ATPase localized in the *trans*-Golgi network. Mutated and, therefore, misfolded versions of this protein seem to be rapidly recognized by the ER quality control machinery and degraded (Payne et al., 1998).

It has been suggested that aberrant regulation of protein biogenesis and ER membrane insertion could result in prion diseases such as bovine spongiform encephalopathy (BSE), scrapie or Creutzfeldt–Jakob disease (CJD). An inefficient ER-proof-reading is believed to contribute to the development of these neurodegenerative disorders. In all cases, a highly conserved 209 amino acid glycoprotein, the prion protein (PrP^c), is the agent responsible for pathogenesis. The normal cellular function of PrP^c is still unclear. Expression and accumulation of an abnormal isoform of PrP (Prp^{Sc}) in the brain is thought to be responsible for such diseases (Prusiner, 1997). It has been proposed that the Prp^{Sc} form is capable of converting the cellular form (PrP^c) into PrP^{Sc}, leading to further aggregation. PrP^c can be synthesized in different topological forms: either as a soluble protein completely residing in the ER lumen or as an integral membrane protein (Yost et al., 1990). Biochemical evidence suggests that the transmembrane ^{Ctm}PrP variant is potentially pathogenic. Under normal conditions, this isoform is presumed to be rapidly degraded by ERAD. However, under certain, not yet well-defined circumstances, ^{Ctm}PrP can escape destruction and is delivered to post-ER compartments where it accumulates and causes disease (Hegde et al., 1998). To date, 23 pathogenic PrP mutations have been reported, associated with three phenotypes: CJD, fatal familial insomnia, and Gerstmann–Sträussler–Scheinker disease (GSS). The cause of GSS is an amber mutation at position 145 (Y145 stop), resulting in a truncated Prp^{145} variant which is imported into the ER, but it is then degraded by the proteasome indicating the action of the ER quality control system. The Prp^{145} variant contains an uncleaved signal peptide, which predisposes it for aggregation once degradation by the proteasome becomes inefficient (Zanusso et al., 1999). This may explain the later onset of GSS development in patients carrying the Prp^{145} form. Another PrP variant leading to GSS is PrP^{217} (Q217R). The majority of PrP^{217} escape ER quality control and accumulate in post-Golgi compartments in an aggregated form (Jin et al., 2000). However, a fraction of PrP^{217} proteins retain the C-terminal glycosylphosphatidyl inositol (GPI) anchor signal peptide due to some defects in GPI anchor addition. This variant, known as PrP^{32}, does not exit the ER but interacts with BiP and it is subsequently degraded by the proteasome. A recent development in the field is the finding that juvenile Parkinsonism is a result of ER-stress caused by the accumulation of an insoluble form of the putative G protein-coupled transmembrane protein, the Pael receptor. This receptor is a substrate of the RING-finger ubiquitin-protein ligase (E3) Parkin, which interacts with the human orthologs of Ubc6 and Ubc7, residing in the ER membrane. In juvenile Parkinson patients the

Parkin gene is mutated leading to the accumulation of unfolded Pael receptor with subsequent loss of dopaminergic neurons (Imai et al., 2001 and
references therein). (For a more comprehensive list of ER quality control
related diseases, see Helenius, 2001.)

Summary: *Protein quality control and disease*

- Toxins (mis-)use ERAD components for transport to the cytoplasm.
- Viruses evade immune detection using ERAD to destroy components of
  the immune system.
- Some diseases (e.g. cystic fibrosis) develop because of a 'hypersensitive'
  ER quality control system.
- Prion diseases develop on the basis of escape from the ER quality control.

PROTEIN QUALITY CONTROL IS ESSENTIAL

As is apparent from the prion diseases, the folding state of a protein can be
decisive between health and disease. It is easy to imagine, then, that the
fine-tuning of the rate of synthesis of a given protein and its rate of folding
and degradation becomes essential for a healthy metabolism. If the rate of
synthesis exceeds the combined rates of folding and degradation, a fraction
of the proteins will be unable to progress through the secretory pathway
and their accumulation will eventually disturb cellular functions. The appearance of such problems is exacerbated when proteins carry mutations which
interfere with protein folding or when the proteins themselves induce unfolding. Environmental stress, such as heat, metal ions, and oxidation, may also
interfere with protein folding.

In conclusion, since the overall ER quality control mechanism is not simply a locally restricted phenomenon but is coupled as a whole to protein
transport through the ER membrane via the Sec61 import channel, two
steps are absolutely critical for the elimination of misfolded or unassembled
proteins. The first is the extraction of the degradation substrate from the ER
membrane or lumen, and the second is the degradation by the cytosolic
ubiquitin–proteasome system. While inefficiency in the extraction process
will lead to accumulation of malfolded proteins in the ER, reduced degradation with normal dislocation will cause their deposition and aggregation
in the cytosol. Both events will undoubtedly result in severe cellular phenotypes and, in the case of higher eukaryotes, in a variety of serious diseases.

Deletion of genes specifically involved in ERAD is not lethal in yeast
(Plemper and Wolf, 1999). It has been previously reported that accumulation
of aberrant proteins in the ER leads to upregulation of several ER chaperones by the unfolded protein response (Chapman et al., 1998). An increase in
the level of misfolded proteins activates the kinase/nuclease Ire1p, located

in the ER/nuclear membrane, which subsequently triggers the formation of the transcription factor Hac1. Hac1 induces the expression of target UPR genes (Kawahara et al., 1997; Sidrauski and Walter, 1997). The first indication that ERAD and UPR are intertwined came from experiments showing that specific expression of a single misfolded protein, CPY*, in a yeast ER-degradation defective *DER1* deletion strain, leads to a measurable induction of the UPR (Knop et al., 1996a). Significantly, mutations in ER degradation and UPR cause synthetic phenotypes: yeast strains defective in the ERAD genes (*UBC7, DER3/ HRD1, HRD3*) and simultaneously deficient in *IRE1* or *HAC1* exhibit a significant growth phenotype. Induction of ER stress by high temperature, tunicamycin or reducing agents, which results in an increase in misfolded proteins, also leads to cell death (Friedländer et al., 2000; Travers et al., 2000). ERAD and UPR are closely coordinated: efficient ER degradation requires an intact UPR and UPR induction enhances the potential of ER degradation. In *S. cerevisiae*, upregulation of the transcription of 381 genes under UPR-inducing stress conditions has been identified by DNA microarray technology; 173 of these genes are of unknown function. All of the gene products described so far that are specifically required for folding and ERAD are among the known genes (Travers et al., 2000). It is of utmost importance to unravel the function in ER quality control of the yet unknown genes to finally reach a more complete understanding of this intricate vital process.

Summary: *Protein quality control is essential*

- ERAD and UPR are intertwined.
- A malfuctioning ER quality control leads to disease and cell death.

HISTORICAL NOTES

Historical Note 1

Until the late 1970s, intracellular proteolytic events were thought to be exclusively associated with the lysosomal (vacuolar) compartment of the cell. This compartment was considered to be the 'gut' of the cell, digesting cellular waste. There was no indication that a proteolytic system responsible for protein regulation or degradation of misfolded proteins could reside in any other compartment of the eukaryotic cell. In 1978, Ciechanover et al. discovered a small protein, later identified as ubiquitin, as a component of an ATP-dependent proteolytic system present in reticulocytes. Two years later, in 1980, Hershko et al. showed by *in vitro* experiments that ubiquitin was covalently linked to protein substrates via an ATP-requiring reaction, suggesting that proteins were marked for degradation through their conjugation to ubiquitin. Before 1985, the E1–E2–E3 enzymology of ubiquitin-conjugation had been deciphered by Hershko, Ciechanover and co-workers only by *in vitro* experiments (Hershko, 1996). In 1984, Finley et al. provided the first evidence for the requirement of ubiquitin for protein degradation in living cells. The first degradation signal of ubiquitin-linked protein degradation, known as the N-end rule, was described in 1986 (Bachmair et al., 1986). The first physiological function of the ubiquitin system was identified when the *S. cerevisiae* proteins Rad6 and Cdc34 – key components of DNA repair and cell cycle control – were found to be ubiquitin conjugating enzymes (Jentsch et al., 1987; Goebl et al., 1988). In 1989, Chau et al. discovered that a polyubiquitin chain was a prerequisite for protein degradation. However, the link between ubiquitin-targeted protein degradation and the degradation machinery itself remained obscure for a long time. In 1980 and 1981, Wilk and Orlowsky isolated and characterized an enzyme, which they called the 'multicatalytic protease complex'. This enzyme complex of 700 kDa consisted of multiple subunits and contained three major proteolytic activities capable of cleaving after hydrophobic neutral, acidic, and basic residues (Wilk and Orlowski, 1980; Orlowski and Wilk, 1981). In 1986, Hough et al. identified an ATP-dependent proteinase from reticulocyte lysates, which degraded ubiquitinated lysozyme *in vitro*, and in 1987, the same researchers purified two high molecular mass proteases that sedimented at 20S and 26S. They identified that the 20S protease was identical to the multicatalytic protease described by Wilk and Orlowski, and the 26S complex the protease able to degrade ubiquitinated lysozyme *in vitro* (Hough et al., 1987). In 1988, Arrigo et al. named this multicatalytic protease complex the 'proteasome'. Electron optic and genetic studies revealed that the proteolytically active 'proteasome' was the core of the 26S protease and was shaped as a cylinder composed of four stacked rings, with the subunit composition $\alpha7\beta7\beta7\alpha7$ (Zwickl et al., 1992; Pühler et al., 1992; Schauer et al., 1993). Consequently, this central proteolytic core particle was called the 20S proteasome, while 26S proteasome referred to the entire structure consisting of the 20S proteasome to which large particles were attached on each end.

Studies conducted in *S. cerevisiae* were crucial in expanding our understanding of the proteasome in many ways. The existence of proteasomes in yeast, then still called the multicatalytic protease complex, was discovered via mutants deficient in vacuolar proteases (Achstetter et al., 1984). X-ray analysis and genetic dissection of the yeast proteasome revealed both the exact structure and the catalytic mechanism of the eukaryotic 20S proteasome, identifying the three active site β-subunits that conferred different specificities to the enzyme (Chen and Hochstrasser, 1996; Heinemeyer et al., 1997; Groll et al., 1997). The physiological function of the proteasome was also described for the first time in *S. cerevisiae*. In 1991, Heinemeyer et al. following the isolation of mutants defective in different subunits of the 20S proteasome, identified this protease as an essential cellular component involved in the degradation of ubiquitinated proteins *in vivo* and as the machinery responsible for coping with malfolded proteins that accumulate due to stresses induced by temperature or amino acid analogues (Heinemeyer et al., 1991).

Historical Note 2

Although the existence of a second proteolytic system, different than the lysosome and capable of selective degradation of abnormal proteins, was suspected as early as the 1970s (Schimke, 1970), a function for the ER – the site for synthesis, folding, and assembly of secretory proteins – as a compartment associated with protein degradation was not suggested until the 1980s.

An early indication for ER degradation came from within the ER itself. Already in the 1970s it was established that, even though most ER proteins are relatively stable and have long half-lives, the ER is a highly dynamic environment. Most enzymatic activities are regulated through cycles of synthesis and degradation. One typical and well-studied example was HMG-CoA reductase (HMG-R), which catalyzes the conversion of HMG-CoA to mevalonate, a key step in sterol synthesis. Early studies indicated that HMG-R underwent regulated degradation controlled by the metabolic status of the cell and that proteolysis proceeded by a non-lysosomal pathway (Edwards and Gould, 1972; Faust et al., 1982; Orci et al., 1984). In 1992, Meigs and Simoni showed that degradation of HMG-R was not affected by an ER-to-Golgi transport block.

As studies on the fate of ER-resident HMG-R continued, some secretory proteins, such as the parathyroid hormone and immunoglobulins, were found to be degraded to some extent shortly after their synthesis. In 1981, Sidman showed that mutant forms of IgM were specifically degraded intracellularly by a lysosome-independent process, while normal molecules produced by the same B-cell hybridomas were efficiently secreted. Another interesting finding was that, in non-secreting cells, both the soluble and transmembrane forms of IgMs were degraded without acquiring Golgi modifications of their carbohydrate structures, suggesting a pre-Golgi site for degradation (Dulis et al., 1982). In 1987, Sitia et al. described the localization of such proteins to the ER cisternae. Studies on T-cell receptor subunits were crucial in implicating the ER in the degradation process and in further emphasizing the selectivity of the system. It was determined that the unassembled subunits TCR-α, TCR-β and CD3-δ exhibited a very short half-life and were rapidly degraded, while the unassembled TCR-ϵ and CD3-ζ were stable (Lippincott-Schwartz et al., 1988; Bonifacino et al., 1989; Klausner and Sitia, 1990). In 1991, Stafford and Bonifacino presented evidence that the ER was the site of degradation.

Parallel studies on the differential fate of the two subunits of asialogly-coprotein (Amara et al., 1989) and of the two forms of apolipoprotein B (Sato et al., 1990; Furukawa et al., 1992), the HIV-1 induced degradation of the CD4 receptor (Willey et al., 1992a, 1992b), and degradation of transport-impaired mutants of α1-antitrypsin (Le et al., 1990), among other examples, further strengthened the notion of a selective proteolytic process associated with the ER.

It was initially thought that degradation of regulated or unassembled ER proteins occurred through ER-resident proteinases (Bonifacino and Klausner, 1994). The finding in yeast, in 1993, that the conditionally lethal growth phenotype of cells defective in a subunit of the translocon at restrictive temperature was suppressed by deletion of a gene coding for an ER-membrane-bound ubiquitin–conjugating enzyme (UBC6), pointed to the fact that the ubiquitin system was involved in the degradation of ER membrane proteins (Sommer and Jentsch, 1993). Two years later, the cause of cystic fibrosis was determined to be the ubiquitin–proteasome dependent proteolysis of mutant forms of the CFTR, which were retained in the ER (Jensen et al., 1995; Ward et al., 1995). These findings substantiated the involvement of the ubiquitin–proteasome system in the degradation of ER membrane proteins but did not explain the mechanism of the process. A first hint that dislocation of proteins from the ER membrane was a prerequisite for degradation came from studies on cytomegalovirus protein induced degradation of the MHC I heavy chain (Wiertz et al., 1996). Finally, the breakthrough for the understanding of the ER degradation process came from genetic and biochemical studies in yeast: ER-lumenal proteins, when mutated, were shown to be retrogradely transported to the cytosol where they were degraded via the ubiquitin–proteasome system (Hiller et al., 1996; Werner et al., 1996). Co-immunoprecipitation and genetic studies (Wiertz et al., 1996a; Plemper et al., 1997; Pilon et al., 1997) implicated the Sec61 translocon as the retrograde transport channel.

ACKNOWLEDGMENTS

The authors thank E. Tosta and P. Deak for help with the preparation of the manuscript. The work of the authors was supported by the Deutsche Forschungsgemeinschaft, Bonn, the German Federal Ministry of Education and Research within the framework of the German–Israeli Project Cooperation (DIP) and the Fonds der Chemischen Industrie, Frankfurt.

REFERENCES

Achstetter, T., Ehmann, C., Osaki, A. and Wolf, D.H. (1984) Proteolysis in eukaryotic cells. Proteinase yscE, a new yeast peptidase. *J Biol Chem* **259**: 13344–13348.

Anderson, D. and Walter, P. (1999) Blobel's nobel: a vision validated. *Cell* **99**: 557–558.

Amara, J.F., Lederkremer, G. and Lodish, H.F. (1989) Intracellular degradation of unassembled asialoglycoprotein receptor subunits: a pre-Golgi, nonlysosomal endoproteolytic cleavage. *J Cell Biol* **109**: 3315–3324.

Arrigo, A.P., Tanaka, K., Goldberg, A.L. and Welch, W.J. (1988) Identity of the 19S 'prosome' particle with the large multifunctional protease complex of mammalian cells (the proteasome). *Nature* **331**(6152): 192–194.

Bachmair, A., Finley, D. and Varshavsky, A. (1986) *In vivo* half-life of a protein is a function of its amino-terminal residue. *Science* **234**: 179–186.

Bays, N.W., Gardner, R.G., Seelig, L.P., Joazeiro, C.A. and Hampton, R.Y. (2001) Hrd1p/Der3p is a membrane-anchored ubiquitin ligase required for ER-associated degradation. *Nat Cell Biol* **3**: 24–29.

Bebök, Z., Mazzochi, C., King, S.A., Hong, J.S. and Sorscher, E.J. (1998) The mechanism underlying cystic fibrosis transmembrane conductance regulator transport from the endoplasmic reticulum to the proteasome includes Sec61beta and a cytosolic, deglycosylated intermediary. *J Biol Chem* **273**: 29873–29878.

Biederer, T., Volkwein, C. and Sommer, T. (1996) Degradation of subunits of the Sec61p complex, an integral component of the ER membrane, by the ubiquitin-proteasome pathway. *EMBO J* **15**: 2069–2076.

Biederer, T., Volkwein, C. and Sommer, T. (1997) Role of Cue1p in ubiquitination and degradation at the ER surface. *Science* **278**: 1806–1809.

Blobel, G. (1995) Unidirectional and bidirectional protein traffic across membranes. *Cold Spring Harb Symp Quant Biol* **60**: 1–10.

Bonifacino, J.S. and Klausner, R.D. (1994) Degradation of proteins retained in the endoplasmic reticulum. In: Ciechanover, A.J. and Schwartz, A.L. (eds) *Modern Cell Biology*, Vol. 15. *Cellular Proteolytic Systems*, pp. 137–160. New York: John Wiley and Sons.

Bonifacino, J.S., Suzuki, C.K., Lippincott-Schwartz, J., Weissman, A.M. and Klausner, R.D. (1989) Pre-Golgi degradation of newly synthesized T-cell antigen receptor chains: intrinsic sensitivity and the role of subunit assembly. *J Cell Biol* **109**: 73–83.

Bonifacino, J.S., Cosson, P. and Klausner, R.D. (1990) Colocalized transmembrane determinants for ER degradation and subunit assembly explain the intracellular fate of TCR chains. *Cell* **63**: 503–513.

Bordallo, J. and Wolf, D.H. (1999) A RING-H2 finger motif is essential for the function of Der3/Hrd1 in endoplasmic reticulum associated protein degradation in the yeast *Saccharomyces cerevisiae*. *FEBS Lett* **448**: 244–248.

Bordallo, J., Plemper, R.K., Finger, A. and Wolf, D.H. (1998) Der3p/Hrd1p is required for endoplasmic reticulum-associated degradation of misfolded lumenal and integral membrane proteins. *Mol Biol Cell* **9**: 209–222.

Brodsky, J.L. and McCracken, A.A. (1999) ER protein quality control and proteasome-mediated protein degradation. *Semin Cell Dev Biol* **10**: 507–513.

Brodsky, J.L., Werner, E.D., Dubas, M.E. et al. (1999) The requirement for molecular chaperones during endoplasmic reticulum-associated protein degradation demonstrates that protein export and import are mechanistically distinct. *J Biol Chem* **274**: 3453–3460.

Caldwell, S.R., Hill, K.J. and Cooper, A.A. (2001) Degradation of ER quality control substrate requires transport between the ER and Golgi. *J Biol Chem* **276**: 23296–23303.

Chapman, R., Sidrauski, C. and Walter, P. (1998) Intracellular signaling from the endoplasmic reticulum to the nucleus. *Annu Rev Cell Dev Biol* **14**: 459–485.

Chau, V., Tobias, J.W., Bachmair, A. et al. (1989) A multiubiquitin chain is confined to specific lysine in a targeted short-lived protein. *Science* **243**(4898): 1576–1583.

Chen, P. and Hochstrasser, M. (1996) Autocatalytic subunit processing couples active site formation in the 20S proteasome to completion of assembly. *Cell* **86**: 961–972.

Corsi, A.K. and Schekman, R. (1996) Mechanism of polypeptide translocation into the endoplasmic reticulum. *J Biol Chem* **271**: 30299–30302.

Cronin, S.R., Khoury, A., Ferry, D.K. and Hampton, R.Y. (2000) Regulation of HMG-CoA reductase degradation requires the P-type ATPase Cod1p/Spf1p. *J Cell Biol* **148**: 915–924.

de Virgilio, M., Weninger, H. and Ivessa, N.E. (1998) Ubiquitination is required for the retro-translocation of a short-lived luminal endoplasmic reticulum glycoprotein to the cytosol for degradation by the proteasome. *J Biol Chem* **273**: 9734–9743.

Deak, P.M. and Wolf, D.H. (2001) Membrane topology and function of Der3/Hrd1p as a ubiquitin-protein ligase (E3) involved in endoplasmic reticulum degradation. *J Biol Chem* **276**: 10663–10669.

Dulis, B.H., Kloppel, T.M., Grey, H.M. and Kubo, R.T. (1982) Regulation of catabolism of IgM heavy chains in a B lymphoma cell line. *J Biol Chem* **257**: 4369–4374.

Dürr, G., Strayle, J., Plemper, R. et al. (1998) The medial-Golgi ion pump Pmr1 supplies the yeast secretory pathway with Ca^{2+} and Mn^{2+} required for glycosylation, sorting, and endoplasmic reticulum-associated protein degradation. *Mol Biol Cell* **9**: 1149–1162.

Edwards, P.A. and Gould, R.G. (1972) Turnover rate of hepatic 3-hydroxy-3-methylglutaryl coenzyme A reductase as determined by use of cycloheximide. *J Biol Chem* **247**: 1520–1524.

Ellgaard, L., Molinari, M. and Helenius, A. (1999) Setting the standards: quality control in the secretory pathway. *Science* **286**: 1882–1888.

Fagioli, C. and Sitia, R. (2001) Glycoprotein quality control in the endoplasmic reticulum. Mannose trimming by endoplasmic reticulum mannosidase I times the proteasomal degradation of unassembled immunoglobulin subunits. *J Biol Chem* **276**: 12885–12892.

Falnes, P.O. and Sandvig, K. (2000) Penetration of protein toxins into cells. *Curr Opin Cell Biol* **12**: 407–413.

Faust, J.R., Luskey, K.L., Chin, D.J., Goldstein, J.L. and Brown, M.S. (1982) Regulation of synthesis and degradation of 3-hydroxy-3-methylglutaryl- coenzyme A reductase by low density lipoprotein and 25-hydroxycholesterol in UT-1 cells. *Proc Natl Acad Sci USA* **79**: 5205–5209.

Finley, D., Ciechanover, A. and Varshavsky, A. (1984) Thermolability of ubiquitin-activating enzyme from the mammalian cell cycle mutant ts85. *Cell* **37**(1): 43–55.

Friedländer, R., Jarosch, E., Urban, J., Volkwein, C. and Sommer, T. (2000) A regulatory link between ER-associated protein degradation and the unfolded-protein response. *Nat Cell Biol* **2**: 379–384.

Furukawa, S., Sakata, N., Ginsberg, H.N. and Dixon, J.L. (1992) Studies of the sites of intracellular degradation of apolipoprotein B in Hep G2 cells. *J Biol Chem* **267**: 22630–22638.

Gardner, R.G., Swarbrick, G.M., Bays, N.W. et al. (2000) Endoplasmic reticulum degradation requires lumen to cytosol signaling. Transmembrane control of Hrd1p by Hrd3p. *J Cell Biol* **151**: 69–82.

Gilbert, A., Jadot, M., Leontieva, E., Wattiaux-De Coninck, S. and Wattiaux, R. (1998) Delta F508 CFTR localizes in the endoplasmic reticulum–Golgi intermediate compartment in cystic fibrosis cells. *Exp Cell Res* **242**: 144–152.

Gillece, P., Luz, J.M., Lennarz, W.J., de La Cruz, F.J. and Romisch, K. (1999) Export of a cysteine-free misfolded secretory protein from the endoplasmic reticulum for degradation requires interaction with protein disulfide isomerase. *J Cell Biol* **147**: 1443–1456.

Glickman, M.H., Rubin, D.M., Larsen, C.N., Schmidt, M. and Finley, D. (2000) The regulatory particle of the yeast proteasome. In: Hilt, W. and Wolf, D.H. (eds) *Proteasomes: The World of Regulatory Proteolysis*, pp. 71–90. Austin, TX: Landes Bioscience, Georgetown/Eurekah.com.

Goebl, M.G., Yochem, J., Jentsch, S. et al. (1988) The yeast cell cycle gene CDC34 encodes a ubiquitin-conjugating enzyme. *Science* **241**: 1331–1335.

Groll, M., Ditzel, L., Lowe, J. et al. (1997) Structure of 20S proteasome from yeast at 2.4 A resolution. *Nature* **386**: 463–471.

Halaban, R., Cheng, E., Zhang, Y. et al. (1997) Aberrant retention of tyrosinase in the endoplasmic reticulum mediates accelerated degradation of the enzyme and contributes to the dedifferentiated phenotype of amelanotic melanoma cells. *Proc Natl Acad Sci USA* **94**: 6210–6215.

Hammond, C. and Helenius, A. (1995) Quality control in the secretory pathway. *Curr Opin Cell Biol* **7**: 523–529.

Hammond, C., Braakman, I. and Helenius, A. (1994) Role of N-linked oligosaccharide recognition, glucose trimming, and calnexin in glycoprotein folding and quality control. *Proc Natl Acad Sci USA* **91**: 913–917.

Hampton, R.Y. and Bhakta, H. (1997) Ubiquitin-mediated regulation of 3-hydroxy-3-methylglutaryl-CoA reductase. *Proc Natl Acad Sci USA* **94**: 12944–12948.

Hampton, R.Y., Gardner, R.G. and Rine, J. (1996) Role of 26S proteasome and HRD genes in the degradation of 3-hydroxy-3-methylglutaryl-CoA reductase, an integral endoplasmic reticulum membrane protein. *Mol Biol Cell* **7**: 2029–2044.

Hegde, R.S., Mastrianni, J.A., Scott, M.R. et al. (1998) A transmembrane form of the prion protein in neurodegenerative disease. *Science* **279**: 827–834.

Heinemeyer, W. (2000) Active sites and assembly of the 20S proteasome. In: Hilt, W. and Wolf, D.H. (eds) *Proteasomes: The World of Regulatory Proteolysis*, pp. 48–70. Austin, TX: Landes Bioscience, Georgetown/Eurekah.com.

Heinemeyer, W., Fischer, M., Krimmer, T., Stachon, U. and Wolf, D.H. (1997) The active sites of the eukaryotic 20S proteasome and their involvement in subunit precursor processing. *J Biol Chem* **272**: 25200–25209.

Heinemeyer, W., Kleinschmidt, J.A., Saidowsky, J., Escher, C. and Wolf, D.H. (1991) Proteinase yscE, the yeast proteasome/multicatalytic-multifunctional proteinase: mutants unravel its function in stress induced proteolysis and uncover its necessity for cell survival. *EMBO J* **10**: 555–562.

Helenius, A. (2001) Quality control in the secretory assembly line. *Phil Trans R Soc Lond B* **356**: 147–150.

Helenius, A. and Aebi, M. (2001) Intracellular functions of N-linked glycans. *Science* **291**: 2364–2369.

Helenius, A., Trombetta, E.S., Hebert, D.N. and Simons, J.F. (1997) Calnexin, calreticulin and the folding of glycoproteins. *Trends Cell Biol* **7**: 193–200.

Hershko, A. (1996) Lessons from the discovery of the ubiquitin system. *Trends Biochem Sci* **21**: 445–449.

Hershko, A. and Ciechanover, A. (1998) The ubiquitin system. *Annu Rev Biochem* **67**: 425–479.

Hill, K. and Cooper, A.A. (2000) Degradation of unassembled Vph1p reveals novel aspects of the yeast ER quality control system. *EMBO J* **19**: 550–561.

Hiller, M.M., Finger, A., Schweiger, M. and Wolf, D.H. (1996) ER degradation of a misfolded luminal protein by the cytosolic ubiquitin-proteasome pathway. *Science* **273**: 1725–1728.

Hilt, W. and Wolf, D.H. (1996) Proteasomes: destruction as a programme. *Trends Biochem Sci* **21**: 96–102.

Hilt, W. and Wolf, D.H. (eds) (2000) *Proteasomes: The World of Regulatory Proteolysis.* Austin, TX: Landes Bioscience, Georgetown/Eurekah.com.

Hong, E., Davidson, A.R. and Kaiser, C.A. (1996) A pathway for targeting soluble misfolded proteins to the yeast vacuole. *J Cell Biol* **135**: 623–633.

Hosokawa, N., Wada, I., Hasegawa, K. et al. (2001) A novel ER α-mannosidase-like protein accelerates ER-associated degradation. *EMBO Rep* **2**: 415–422.

Hughes, E.A., Hammond, C. and Cresswell, P. (1997) Misfolded major histocompatibility complex class I heavy chains are translocated into the cytoplasm and degraded by the proteasome. *Proc Natl Acad Sci USA* **94**: 1896–1901.

Hough, R., Pratt, G. and Rechsteiner, M. (1987) Purification of two high molecular weight proteases from rabbit reticulocyte lysate. *J Biol Chem* **262**(17): 8303–8313.

Huppa, J.B. and Ploegh, H.L. (1997) The alpha chain of the T cell antigen receptor is degraded in the cytosol. *Immunity* **7**: 113–122.

Imai, Y., Soda, M., Inoue, H. et al. (2001) An unfolded putative transmembrane polypeptide, which can lead to endoplasmic reticulum stress is a substrate of Parkin. *Cell* **105**: 891–902.

Jakob, C.A., Burda, P., Roth, J. and Aebi, M. (1998) Degradation of misfolded endoplasmic reticulum glycoproteins in *Saccharomyces cerevisiae* is determined by a specific oligosaccharide structure. *J Cell Biol* **142**: 1223–1233.

Jakob, C.A., Bodiner, D., Spirig, U. et al. (2001) Htm1p, a mannosidase-like protein, is involved in glycoprotein degradation in yeast. *EMBO Rep* **2**: 423–430.

Jensen, T.J., Loo, M.A., Pind, S. et al. (1995) Multiple proteolytic systems, including the proteasome, contribute to CFTR processing. *Cell* **83**: 129–135.

Jentsch, S., McGrath, J.P. and Varshavsky, A. (1987) The yeast DNA repair gene RAD6 encodes a ubiquitin-conjugating enzyme. *Nature* **329**: 131–134.

Jin, T., Gu, Y., Zanusso, G. et al. (2000) The chaperone protein BiP binds to a mutant prion protein and mediates its degradation by the proteasome. *J Biol Chem* **275**: 38699–38704.

Johnston, J.A., Ward, C.L. and Kopito, R.R. (1998) Aggresomes: a cellular response to misfolded proteins. *J Cell Biol* **143**: 1883–1898.

Kawahara, T., Yanagi, H., Yura, T. and Mori, K. (1997) Endoplasmic reticulum stress-induced mRNA splicing permits synthesis of transcription factor Hac1p/Ern4p that activates the unfolded protein response. *Mol Biol Cell* **8**: 1845–1862.

Kerkau, T., Bacik, I., Bennink, J.R. et al. (1997) The human immunodeficiency virus type 1 (HIV-1) Vpu protein interferes with an early step in the biosynthesis of major histocompatibility complex (MHC) class I molecules. *J Exp Med* **185**: 1295–1305.

Klausner, R.D. and Sitia, R. (1990) Protein degradation in the endoplasmic reticulum. *Cell* **62**: 611–614.

Knecht, E. and Rivett, A.J. (2000) Intracellular localization of proteasomes. In: Hilt, W. and Wolf, D.H. (eds) *Proteasomes: The World of Regulatory Proteolysis*, pp. 176–185. Austin, TX: Landes Bioscience, Georgetown/Eurekah.com.

Knop, M., Finger, A., Braun, T., Hellmuth, K. and Wolf, D.H. (1996a) Der1, a novel protein specifically required for endoplasmic reticulum degradation in yeast. *EMBO J* **15**: 753–763.

Knop, M., Hauser, N. and Wolf, D.H. (1996b) N-Glycosylation affects endoplasmic reticulum degradation of a mutated derivative of carboxypeptidase yscY in yeast. *Yeast* **12**: 1229–1238.

Kutay, U., Hartmann, E. and Rapoport, T.A. (1993) A class of membrane proteins with a C-terminal anchor. *Trends Cell Biol* **3**: 72–75.

Kutay, U., Ahnert-Hilger, G., Hartmann, E., Wiedenmann, B. and Rapoport, T.A. (1995) Transport route for synaptobrevin via a novel pathway of insertion into the endoplasmic reticulum membrane. *EMBO J* **14**: 217–223.

Le, A., Graham, K.S. and Sigers, R.N. (1990) Intracellular degradation of the transport-impaired human PiZ α1-antitrypsin variant. Biochemical mapping of the degradative event among compartments of the secretory pathway. *J Biol Chem* **265**: 14001–14007.

Letourneur, F., Hennecke, S., Demolliere, C. and Cosson, P. (1995) Steric masking of a dilysine endoplasmic reticulum retention motif during assembly of the human high affinity receptor for immunoglobulin E. *J Cell Biol* **129**: 971–978.

Lippincott-Schwartz, J., Bonifacino, J.S., Yuan, L.C. and Klausner, R.D. (1988) Degradation from the endoplasmic reticulum: disposing of newly synthesized proteins. *Cell* **54**: 209–220.

Liu, Y., Choudhury, P., Cabral, C.M. and Sifers, R.N. (1999) Oligosaccharide modification in the early secretory pathway directs the selection of a misfolded glycoprotein for degradation by the proteasome. *J Biol Chem* **274**: 5861–5867.

Margottin, F., Bour, S.P., Durand, H. et al. (1998) A novel human WD protein, h-beta TrCp, that interacts with HIV-1 Vpu connects CD4 to the ER degradation pathway through an F-box motif. *Mol Cell* **1**: 565–574.

McCracken, A.A. and Brodsky, J.L. (1996) Assembly of ER-associated protein degradation *in vitro*: dependence on cytosol, calnexin, and ATP. *J Cell Biol* **132**: 291–298.

Meacham, G.C., Patterson, C., Zhang, W., Younger, J.M. and Cyr, D.M. (2001) The Hsc70 co-chaperone CHIP targets immature CFTR for proteasomal degradation. *Nat Cell Biol* **3**: 100–105.

Meigs, T.E. and Simoni, R.D. (1997) Farnesol as a regulator of HMG-CoA reductase degradation: characterization and role of farnesyl pyrophosphatase. *Arch Biochem Biophys* **345**: 1–9.

Meigs, T.E. and Simoni, R.D. (1992) Regulated degradation of 3-hydroxy-3-methylglutaryl-coenzyme A reductase in permeabilized cells. *J Biol Chem* **267**(19): 13547–13552.

Mosse, C.A., Meadows, L., Luckey, C.J. et al. (1998) The class I antigen-processing pathway for the membrane protein tyrosinase involves translation in the endoplasmic reticulum and processing in the cytosol. *J Exp Med* **187**: 37–48.

Nakatsukasa, K., Nishikawa, S., Hosokawa, N., Nagata, K. and Endo, T. (2001) Mnl1p, an alpha-mannosidase-like protein in yeast *Saccharomyces cerevisiae*, is required for endoplasmic reticulum-associated degradation of glycoproteins. *J Biol Chem* **276**: 8635–8638.

Orci, L., Brown, M.S., Goldstein, J.L., Garcia-Segura, L.M. and Anderson, R.G. (1984) Increase in membrane cholesterol: a possible trigger for degradation of HMG CoA reductase and crystalloid endoplasmic reticulum in UT-1 cells. *Cell* **36**: 835–845.

Orlowski, M. and Wilk, S. (1981) A multicatalytic protease complex from pituitary that forms enkephalin and enkephalin containing peptides. *Biochem Biophys Res Commun* **101**: 814–822.

Payne, A.S., Kelly, E.J. and Gitlin, J.D. (1998) Functional expression of the Wilson disease protein reveals mislocalization and impaired copper-dependent trafficking of the common H1069Q mutation. *Proc Natl Acad Sci USA* **95**: 10854–10859.

Perlmutter, D.H. (1996) Alpha-1-antitrypsin deficiency: biochemistry and clinical manifestations. *Ann Med* **28**: 385–394.

Pilon, M., Schekman, R. and Romisch, K. (1997) Sec61p mediates export of a misfolded secretory protein from the endoplasmic reticulum to the cytosol for degradation. *EMBO J* **16**: 4540–4548.

Plemper, R.K. and Wolf, D.H. (1999) Retrograde protein translocation: ERADication of secretory proteins in health and disease. *Trends Biochem Sci* **24**: 266–270.

Plemper, R.K., Bohmler, S., Bordallo, J., Sommer, T. and Wolf, D.H. (1997) Mutant analysis links the translocon and BiP to retrograde protein transport for ER degradation. *Nature* **388**: 891–895.

Plemper, R.K., Egner, R., Kuchler, K. and Wolf, D.H. (1998) Endoplasmic reticulum degradation of a mutated ATP-binding cassette transporter Pdr5 proceeds in a concerted action of Sec61 and the proteasome. *J Biol Chem* **273**: 32848–32856.

Plemper, R.K., Bordallo, J., Deak, P.M. et al. (1999a) Genetic interactions of Hrd3p and Der3p/Hrd1p with Sec61p suggest a retro-translocation complex mediating protein transport for ER degradation. *J Cell Sci* **112**: 4123–4134.

Plemper, R.K., Deak, P.M., Otto, R.T. and Wolf, D.H. (1999b) Re-entering the translocon from the lumenal side of the endoplasmic reticulum. Studies on mutated carboxypeptidase yscY species. *FEBS Lett* **443**: 241–245.

Prusiner, S.B. (1997) Prion diseases and the BSE crisis. *Science* **278**: 245–251.

Pryer, N.K., Wuestehube, L.J. and Schekman, R. (1992) Vesicle-mediated protein sorting. *Annu Rev Biochem* **61**: 471–516.

Pühler, G., Weinkauf, S., Bachmann, L. et al. (1992) Subunit stoichiometry and three-dimensional arrangement in proteasomes from *Thermoplasma acidophilum*. *EMBO J* **11**: 1607–1616.

Qu, D., Teckman, J.H., Omura, S. and Perlmutter, D.H. (1996) Degradation of a mutant secretory protein, alpha1-antitrypsin Z, in the endoplasmic reticulum requires proteasome activity. *J Biol Chem* **271**: 22791–22795.

Rapoport, T.A., Jungnickel, B. and Kutay, U. (1996) Protein transport across the eukaryotic endoplasmic reticulum and bacterial inner membranes. *Annu Rev Biochem* **65**: 271–303.

Ravid, T., Doolman, R., Avner, R., Harats, D. and Roitelman, J. (2000) The ubiquitin-proteasome pathway mediates the regulated degradation of mammalian 3-hydroxy-3-methylglutaryl-coenzyme A reductase. *J Biol Chem* **275**: 35840–35847.

Sato, R., Imanaka, T., Takatsuki, A. and Takano, T. (1990) Degradation of newly synthesized apolipoprotein B-100 in a pre-Golgi compartment. *J Biol Chem* **265**: 11880–11884.

Schauer, T.M., Nesper, M., Kehl, M. et al. (1993) Proteasomes from *Dictyostelium discoideum*: characterization of structure and function. *J Struct Biol* **111**: 135–147.

Schimke, R.T. (1970) Regulation of protein degradation in mammalian tissues. In: Munro, H.N. (ed.), *Mammalian Protein Metabolism* Vol. 4, pp. 177–228. New York: Academic Press.

Schmitz, A., Herrgen, H., Winkeler, A. and Herzog, V. (2000) Cholera toxin is exported from microsomes by the Sec61p complex. *J Cell Biol* **148**: 1203–1212.

Shamu, C.E., Story, C.M., Rapoport, T.A. and Ploegh, H.L. (1999) The pathway of US11-dependent degradation of MHC class I heavy chains involves a ubiquitin-conjugated intermediate. *J Cell Biol* **147**: 45–58.

Sidman, C. (1981) B lymphocyte differentiation and the control of IgM mu chain expression. *Cell* **23**(2): 379–389.

Sidrauski, C. and Walter, P. (1997) The transmembrane kinase Ire1p is a site-specific endonuclease that initiates mRNA splicing in the unfolded protein response. *Cell* **90**: 1031–1039.

Simpson, J.C., Roberts, L.M., Romisch, K., Davey, J., Wolf, D.H. and Lord, J.M. (1999) Ricin A chain utilises the endoplasmic reticulum-associated protein degradation pathway to enter the cytosol of yeast. *FEBS Lett* **459**: 80–84.

Sitia, R., Neuberger, M.S. and Milstein, C. (1987) Regulation of membrane IgM expression in secretory B cells: translational and post-translational events. *EMBO J* **6**(13): 3969–3977.

Sommer, T. and Jentsch, S. (1993) A protein translocation defect linked to ubiquitin conjugation at the endoplasmic reticulum. *Nature* **365**: 176–179.

Sommer, T. and Wolf, D.H. (1997) Endoplasmic reticulum degradation: reverse protein flow of no return. *FASEB J* **11**: 1227–1233.

Stafford, F.J. and Bonifacino, J.S. (1991) A permeabilized cell system identifies the endoplasmic reticulum as a site of protein degradation. *J Cell Biol* **115**(5): 1225–1236.

Stevens, T.L., Blum, J.H., Foy, S.P., Matsuuchi, L. and DeFranco, A.L. (1994) A mutation of the mu transmembrane that disrupts endoplasmic reticulum retention. Effects on association with accessory proteins and signal transduction. *J Immunol* **152**: 4397–4406.

Suzuki, T., Park, H., Hollingsworth, N.M., Sternglanz, R. and Lennarz, W.J. (2000) PNG1, a yeast gene encoding a highly conserved peptide:N-glycanase. *J Cell Biol* **149**: 1039–1052.

Suzuki, T., Park, H., Kwofie, M.A. and Lennarz, W.J. (2001) Rad23 provides a link between the Png1 deglycosylating enzyme and the 26S proteasome in yeast. *J Biol Chem* **276**: 21601–21607, published online on March 20.

Swanson, R., Locher, M. and Hochstrasser, M. (2001) A conserved ubiquitin ligase of the nuclear envelope endoplasmic reticulum that functions in both ER-associated and matα2 repressor degradation. *Genes Dev* **15**: 2660–2674.

Tiwari, S. and Weissman, A.M. (2001) Endoplasmic reticulum (ER)-associated degradation of T cell receptor subunits. Involvement of ER-associated ubiquitin-conjugating enzymes (E2s). *J Biol Chem* **276**: 16193–16200.

Tokunaga, F., Brostrom, C., Koide, T. and Arvan, P. (2000) Endoplasmic reticulum (ER)-associated degradation of misfolded N-linked glycoproteins is suppressed upon inhibition of ER mannosidase I. *J Biol Chem* **275**: 40757–40764.

Travers, K.J., Patil, C.K., Wodicka, L. et al. (2000) Functional and genomic analyses reveal an essential coordination between the unfolded protein response and ER-associated degradation. *Cell* **101**: 249–258.

Tsai, B., Rodighiero, C., Lencer, W.I. and Rapoport, T.A. (2001) Protein disulfide isomerase acts as a redox-dependent chaperone to unfold cholera toxin. *Cell* **104**: 937–948.

Walter, J., Urban, J., Volkwein, C. and Sommer, T. (2001) Sec61p independent degradation of the tail-anchored membrane protein Ubc6p. *EMBO J* **20**: 3124–3131.

Ward, C.L., Omura, S. and Kopito, R.R. (1995) Degradation of CFTR by the ubiquitin-proteasome pathway. *Cell* **83**: 121–127.

Werner, E.D., Brodsky, J.L. and McCracken, A.A. (1996) Proteasome-dependent endoplasmic reticulum-associated protein degradation: an unconventional route to a familiar fate. *Proc Natl Acad Sci USA* **93**: 13797–13801.

Wesche, J., Rapak, A. and Olsnes, S. (1999) Dependence of ricin toxicity on translocation of the toxin A-chain from the endoplasmic reticulum to the cytosol. *J Biol Chem* **274**: 34443–34449.

Wiertz, E.J., Jones, T.R., Sun, L. et al. (1996a) The human cytomegalovirus US11 gene product dislocates MHC class I heavy chains from the endoplasmic reticulum to the cytosol. *Cell* **84**: 769–779.

Wiertz, E.J., Tortorella, D., Bogyo, M. et al. (1996b) Sec61-mediated transfer of a membrane protein from the endoplasmic reticulum to the proteasome for destruction. *Nature* **384**: 432–438.

Wilhovsky, S., Gardner, R. and Hampton, R. (2000) HRD gene dependence of endoplasmic reticulum-associated degradation. *Mol Biol Cell* **11**: 1697–1708.

Wilk, S. and Orlowski, M. (1980) Cation-sensitive neutral endopeptidase: isolation and specificity of the bovine pituitary enzyme. *J Neurochem* **35**: 1172–1182.

Wilkinson, B.M., Tyson, J.R., Reid, P.J. and Stirling, C.J. (2000) Distinct domains within yeast Sec61p involved in post-translational translocation and protein dislocation. *J Biol Chem* **275**: 521–529.

Willey, R.L., Maldarelli, F., Martin, M.A. and Strebel, K. (1992a) Human immunodeficiency virus type 1 Vpu protein induces rapid degradation of CD4. *J Virol* **66**: 7193–7200.

Willey, R.L., Maldarelli, F., Martin, M.A. and Strebel, K. (1992b) Human immunodeficiency virus type 1 Vpu protein regulates the formation of intracellular gp160-CD4 complexes. *J Virol* **66**: 226–234.

Wolf, D.H. (2000) Proteasomes: A historical retrospective. In: Hilt, W. and Wolf, D.H. (eds) *Proteasomes: The World of Regulatory Proteolysis*, pp. 1–7. Austin, TX: Landes Bioscience, Georgetown/Eurekah.com.

Xiong, X., Chong, E. and Skach, W.R. (1999) Evidence that endoplasmic reticulum (ER)-associated degradation of cystic fibrosis transmembrane conductance regulator is linked to retrograde translocation from the ER membrane. *J Biol Chem* **274**: 2616–2624.

Yang, M., Omura, S., Bonifacino, J.S. and Weissman, A.M. (1998) Novel aspects of degradation of T cell receptor subunits from the endoplasmic reticulum (ER) in T cells: importance of oligosaccharide processing, ubiquitination, and proteasome-dependent removal from ER membranes. *J Exp Med* **187**: 835–846.

Yost, C.S., Lopez, C.D., Prusiner, S.B., Myers, R.M. and Lingappa, V.R. (1990) Non-hydrophobic extracytoplasmic determinant of stop transfer in the prion protein. *Nature* 343: 669–672.

Yu, H., Kaung, G., Kobayashi, S. and Kopito, R.R. (1997) Cytosolic degradation of T-cell receptor alpha chains by the proteasome. *J Biol Chem* 272: 20800–20804.

Zanusso, G., Petersen, R.B., Jin, T. et al. (1999) Proteasomal degradation and N-terminal protease resistance of the codon 145 mutant prion protein. *J Biol Chem* 274: 23396–23404.

Zapun, A., Jakob, C.A., Thomas, D.Y. and Bergeron, J.J. (1999) Protein folding in a specialized compartment: the endoplasmic reticulum. *Structure* 7: R173–182.

Zerangue, N., Schwappach, B., Jan, Y.N. and Jan, L.Y. (1999) A new ER trafficking signal regulates the subunit stoichiometry of plasma membrane K(ATP) channels. *Neuron* 22: 537–548.

Zhou, M. and Schekman, R. (1999) The engagement of Sec61p in the ER dislocation process. *Mol Cell* 4: 925–934.

Zwickl, P., Grziwa, A., Puhler, G. et al. (1992) Primary structure of the *Thermoplasma* proteasome and its implications for the structure, function, and evolution of the multicatalytic proteinase. *Biochemistry* 31: 964–972.

SUGGESTED READING

Blobel, G. (1995) Unidirectional and bidirectional protein traffic across membranes. *Cold Spring Harb Symp Quant Biol* 60: 1–10.

Bonifacino, J.S., Cosson, P. and Klausner, R.D. (1990) Colocalized transmembrane determinants for ER degradation and subunit assembly explain the intracellular fate of TCR chains. *Cell* 63: 503–513.

Deak, P.M. and Wolf, D.H. (2001) Membrane topology and function of Der3/Hrd1p as a ubiquitin-protein ligase (E3) involved in endoplasmic reticulum degradation. *J Biol Chem* 276: 10663–10669.

Hammond, C. and Helenius, A. (1995) Quality control in the secretory pathway. *Curr Opin Cell Biol* 7: 523–529.

Hiller, M.M., Finger, A., Schweiger, M. and Wolf, D.H. (1996) ER degradation of a misfolded luminal protein by the cytosolic ubiquitin-proteasome pathway. *Science* 273: 1725–1728.

Plemper, R.K., Bohmler, S., Bordallo, J., Sommer, T. and Wolf, D.H. (1997) Mutant analysis links the translocon and BiP to retrograde protein transport for ER degradation. *Nature* 388: 891–895.

Schmitz, A., Herrgen, H., Winkeler, A. and Herzog, V. (2000) Cholera toxin is exported from microsomes by the Sec61p complex. *J Cell Biol* 148: 1203–1212.

Sommer, T. and Jentsch, S. (1993) A protein translocation defect linked to ubiquitin conjugation at the endoplasmic reticulum. *Nature* 365: 176–179.

Ward, C.L., Omura, S. and Kopito, R.R. (1995) Degradation of CFTR by the ubiquitin-proteasome pathway. *Cell* 83: 121–127.

Wiertz, E.J., Tortorella, D., Bogyo, M., Yu, J., Mothes, W., Jones, T.R., Rapoport, T.A. and Ploegh, H.L. (1996) Sec61-mediated transfer of a membrane protein from the endoplasmic reticulum to the proteasome for destruction. *Nature* 384: 432–438.

10

TRANSLOCATION OF PROTEINS INTO MITOCHONDRIA

THORSTEN PRINZ, NIKOLAUS PFANNER AND
KAYE N. TRUSCOTT

SUMMARY

Most mitochondrial proteins are synthesized in the cytosol and subsequently imported into the organelle. Protein translocation machineries in the outer and inner mitochondrial membranes are responsible for the specific recognition and import of preproteins. The preprotein translocase of the outer membrane (TOM) consists of several receptors and the general import pore that allows all types of preproteins to cross the outer membrane. The inner membrane contains two distinct translocases. The presequence translocase (TIM23 complex) and matrix Hsp70 are required for preprotein translocation into the matrix. The carrier translocase (TIM22 complex) mediates the insertion of hydrophobic proteins into the inner membrane.

PREFACE

Eukaryotic cells are compartmentalized into numerous organelles such as endoplasmic reticulum, chloroplasts, peroxisomes and mitochondria. Since most cellular proteins are nuclear-encoded and synthesized in the eukaryotic cytosol, mechanisms for protein targeting and import into these organelles are essential. This chapter will provide an overview of the principles of protein import into mitochondria. Mitochondria are self-replicating,

Protein Targeting, Transport & Translocation
ISBN 0-12-200731-X

semiautonomous organelles that contain two functionally distinct membranes, the outer membrane which defines the boundary of the organelle and the structurally complex inner membrane which is the major site of cellular energy transduction. These membranes enclose two soluble compartments, the intermembrane space and the mitochondrial matrix. The mitochondrial matrix contains a complete genetic apparatus and translation machinery. However, the mitochondrial DNA encodes only a few proteins, mainly of the respiratory chain complexes of the mitochondrial inner membrane.

More than 98% of mitochondrial proteins are encoded in the nucleus, synthesized on cytosolic ribosomes and imported post-translationally into this organelle. The nuclear-encoded mitochondrial proteins can be roughly divided into two groups with respect to the nature of their targeting information. These preproteins contain either an N-terminal signal sequence or internal targeting information (Schatz and Dobberstein, 1996; Neupert, 1997; Pfanner et al., 1997). With the help of cytosolic chaperones, precursor proteins are guided to the mitochondrial surface where they engage the action of specialized translocation components. Mitochondria contain three distinct protein import systems, one in the outer membrane and two in the inner membrane (Figure 10.1). A multi-subunit protein complex of the mitochondrial outer membrane, the so-called *t*ranslocase of the *o*uter *m*embrane (TOM), forms the central entry gate for virtually all mitochondrial preproteins. More recently, two multi-subunit protein complexes of the mitochondrial inner membrane have been discovered, termed the *t*ranslocases of the *i*nner *m*embrane (TIM). Protein import components of mitochondria are designated Tom or Tim, followed by their apparent molecular mass in kilodaltons (kDa), e.g. Tom40 or Tim23. Most translocation components are conserved from lower to higher eukaryotes, i.e. from yeast to human (Mori and Terada, 1998; Voos et al., 1999). Due to the ease of genetic manipulation and the relatively simple culture conditions, the yeast *Saccharomyces cerevisiae* is most commonly used as the model organism for mitochondrial protein import studies. Many studies have also been performed with the filamentous fungus *Neurospora crassa*. Table 10.1 gives an overview on the important discoveries of mitochondrial protein import within the past 25 years. The details will be discussed in the following sections.

MITOCHONDRIAL TARGETING SEQUENCES

What guides proteins to mitochondrial compartments? Two major classes of preproteins can be distinguished, preproteins with presequences and preproteins with internal targeting signals.

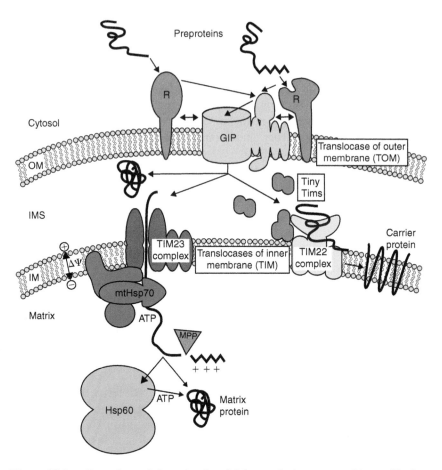

Figure 10.1   Overview of the mitochondrial protein import machinery. Nuclear-encoded mitochondrial precursor proteins bind to receptors (R) on the surface of the mitochondria. From these receptors they are transferred to the general import pore (GIP), which provides the channel for translocation across the outer membrane (OM). Proteins destined for the mitochondrial matrix are subsequently translocated through a channel in the inner membrane (IM) that is made up by the TIM23 complex. Upon arrival of precursor proteins in the matrix, Hsp70 binds them, and the presequence is cleaved off by the mitochondrial processing peptidase (MPP) (Hawlitschek et al., 1988). Subsequently, some proteins interact with Hsp60 to reach their native folding state (Ostermann et al., 1989). Alternatively, proteins destined for the IM interact with the TIM22 complex and subsequently move laterally into the membrane. From the GIP these precursor proteins are shuttled through the intermembrane space (IMS) by 'tiny' Tim proteins to the TIM22 complex from where they reach their final destination in the IM. The translocation of proteins into the matrix or insertion into the IM requires a membrane potential ($\Delta\Psi$) across the IM.

Table 10.1   Important discoveries of protein translocation into mitochondria

Year	Discovery	References
1976/77	Post-translational import of preproteins into mitochondria of the fungus *Neurospora crassa*	Hallermayer et al., 1977
1979	Import and proteolytic processing of preproteins by yeast mitochondria	Maccecchini et al., 1979
1979	Identification of preproteins without presequences	Zimmerman et al., 1979
1984/85	A presequence is sufficient as mitochondrial targeting signal	Hurt et al., 1984; Horwich et al., 1985
1985	Import of preproteins via translocation contact sites	Schleyer and Neupert, 1985
1986	Folded proteins are not imported into mitochondria	Eilers and Schatz, 1986
1986	Protein import requires both a membrane potential and ATP	Pfanner and Neupert, 1986
1987	Import pathway of hydrophobic proteins into mitochondria	Pfanner et al., 1987
1988	Identification of the mitochondrial processing peptidase MPP	Hawlitschek et al., 1988
1989	Identification of the first import receptor, Tom20 (previously named MOM19)	Söllner et al., 1989
1989/90	First import component essential for cell viability, Tom40 (previously named Isp42)	Baker et al., 1990
1990	Matrix Hsp70 is essential for translocation and unfolding of preproteins	Kang et al., 1990
1991	Preproteins in transit are exposed to the intermembrane space	Hwang et al., 1991; Rassow and Pfanner, 1991
1992	First inner membrane component of import machinery, Tim44 (previously named Mpi1, Mim44 or Isp45)	Maarse et al., 1992; Scherer et al., 1992
1994	Matrix Hsp70 and Tim44 cooperate as import motor	Kronidou et al., 1994; Rassow et al., 1994; Schneider et al., 1994
1996/97	Identification of the carrier translocase, Tim22 and Tim54	Sirrenberg et al., 1996; Kerscher et al., 1997
1998	First intermembrane space components of the import machinery, Tim10 (Mrs11) and Tim12 (Mrs5)	Koehler et al., 1998a; Sirrenberg et al., 1998
1998	Reconstitution of the protein import channel of the outer membrane	Hill et al., 1998; Künkele et al., 1998
1999	Active unfolding of preproteins by mitochondria	Huang et al., 1999; Voisine et al., 1999
2000	High resolution structure of receptor (Tom20 binding groove) with amphipathic presequences	Abe et al., 2000
2001	Reconstitution of protein import channels of the inner membrane	Truscott et al., 2001

Hsp, heat shock protein; TIM, translocase of inner membrane; TOM, translocase of outer membrane.

Many mitochondrial precursor proteins contain N-terminal targeting signals. Experiments with chimeric constructs or with gene deletions have shown that, in general, sufficient information for both import and sorting is present within the cleavable N-terminus of these preproteins (presequences) (Hurt et al., 1984; Horwich et al., 1985). For example, the cytosolic enzyme dihydrofolate reductase fused to the first 22 amino acid residues (presequence) of cytochrome c oxidase subunit IV was imported into the mitochondrial matrix. Presequences are ~10–80 amino acid residues long and in most cases are proteolytically removed by the heterodimeric mito-chondrial processing peptidase (MPP) once they are imported into the matrix (Hawlitschek et al., 1988). In addition to positively charged amino acid residues, these presequences contain a high proportion of hydrophobic and hydroxylated amino acid residues. However, amino acid sequence conservation could not be found amongst presequences but rather a com-mon structural motif that is a positively charged amphipathic α-helix (von Heijne, 1986). Indeed, chemically synthesized peptides corresponding to mitochondrial presequences have been shown to form an α-helix in the presence of lipid or detergent (Roise et al., 1986).

Some proteins destined for the intermembrane space such as cytochrome b_2 contain a bipartite presequence at the N-terminus (Hartl et al., 1987). In addition to the matrix-targeting signal these proteins harbor a second signal peptide consisting of a few basic residues and a stretch of hydrophobic residues that serves to sort the preprotein to the intermembrane space. Two sorting models for intermembrane space proteins have been proposed (Glick et al., 1992). According to the 'conservative sorting' model, the precursor protein is first imported into the matrix, then cleaved by MPP, re-exported as an intermediate form into the inner membrane and finally cleaved to the mature protein by the inner membrane peptidase (Imp) (Hartl et al., 1987). Alternatively, the 'stop-transfer' model suggests that the preprotein is arrested in the inner membrane by the hydrophobic sorting peptide, followed by lateral movement into the lipid bilayer without com-plete entry of the preprotein into the matrix (Glick et al., 1992). Conse-quently, only the presequence protrudes into matrix and is cleaved off by MPP. The current evidence indicates that preproteins with a bipartite pre-sequence are sorted by a stop-transfer-like mechanism (Glick et al., 1992; Gärtner et al., 1995).

Numerous mitochondrial preproteins, however, do not carry cleavable presequences, but contain internal targeting signals in the mature protein part. Proteins of this type are for example the carrier proteins of the inner membrane, some intermembrane space proteins and the outer membrane proteins (Schatz and Dobberstein, 1996; Neupert, 1997; Pfanner et al., 1997). The exact nature of internal targeting signals is not known; there is no obvious consensus sequence. Interestingly, some of these proteins, like the abundant carrier proteins, not only contain one targeting signal, but also possess multiple targeting signals distributed over the entire preprotein (Endres et al., 1999; Wiedemann et al., 2001).

The mitochondrial preproteins are usually imported in a post-translational manner, i.e. after their complete synthesis on cytosolic polysomes. The preproteins interact with cytosolic factors such as the molecular chaperone Hsp70 or the mitochondrial import stimulating factor (MSF) of mammalian cells in order to prevent misfolding and aggregation. The cytosolic transport complexes keep preproteins in an import-competent conformation and deliver them to the mitochondrial outer membrane receptors Tom20 or Tom70 (Komiya et al., 1997). The release of preproteins from the cytosolic chaperones can require the hydrolysis of ATP (Komiya et al., 1997). Although the cytosolic factors show some preference for subsets of preproteins, they function only as chaperones and not as specific targeting factors. The signaling information for the specific targeting of preproteins to mitochondria is only contained in the targeting sequences of the preproteins and is decoded by membrane-bound receptors of the mitochondria.

Summary: *Mitochondrial targeting sequences*

- More than 98% of mitochondrial proteins are encoded by nuclear genes and synthesized as precursors on cytosolic ribosomes. The precursor proteins are usually imported into mitochondria in a post-translational manner.
- Many mitochondrial precursor proteins carry cleavable N-terminal targeting signals. These presequences form positively charged amphipathic α-helices and direct the targeting of the proteins to receptors on the mitochondrial surface and the subsequent translocation across both mitochondrial membranes into the matrix. The presequences are cleaved off by the mitochondrial processing peptidase.
- A second class of precursor proteins comprises proteins with targeting signals within the mature protein part. In particular, hydrophobic membrane proteins, e.g. inner membrane carrier proteins, are synthesized as non-cleavable precursor proteins with multiple internal signals.

TRANSLOCATION MACHINERY OF THE OUTER MEMBRANE

Since porin, the major pore-forming protein of the mitochondrial outer membrane, is not permeable to proteins but only to molecules of up to 5–10 kDa, mitochondrial preproteins must be selectively translocated through this membrane by the TOM complex. So far seven major proteins belonging to this multi-subunit complex have been identified in yeast (Figure 10.2 and Table 10.2).

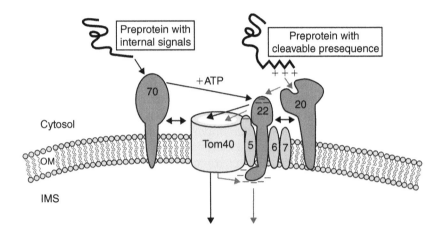

Figure 10.2   Schematic representation of receptor-mediated targeting and translocation of preproteins ('binding chain' hypothesis) by the translocase of the outer membrane (TOM). Preproteins with a cleavable positively charged presequence bind initially to the binding groove of Tom20 facilitated by hydrophobic interactions. From there they are transferred via Tom22 and Tom5 (ionic interactions) to the translocation channel formed by Tom40. After passage through the outer membrane (OM), these preproteins interact with the intermembrane space (IMS) domain of Tom22. Alternatively, preproteins with internal targeting signals bind first to Tom70 and are then transferred in an ATP-dependent manner to Tom22. Subsequently, these preproteins insert via Tom5 into the Tom40 translocation pore.

Import receptors

From cytosolic chaperones preproteins are transferred to one or more of the receptor proteins Tom20, Tom22, or Tom70. Tom70 and Tom20 are anchored to the outer membrane via N-terminal transmembrane domains and expose large soluble C-terminal domains to the cytosol (Söllner et al., 1989; Brix et al., 1997), whereas Tom22 is a tail anchored protein, i.e the N-terminus of Tom22 is exposed to the cytosol while it is anchored to the membrane near its C-terminus with a small functional domain protruding into the intermembrane space (Kiebler et al., 1993). The isolated cytosolic domains of these receptors bind mitochondrial preproteins but with different specificity (Brix et al., 1997). (i) Tom20 predominantly binds preproteins with N-terminal presequence, but is also capable of binding some preproteins with internal targeting sequences. The binding of preproteins to Tom20 is enhanced by the addition of salt, pointing towards an interaction of a hydrophobic nature (Brix et al., 1997). Abe et al. (2000) determined the high-resolution structure of a part of Tom20, including the binding groove, in complex with a presequence peptide. This exciting study showed that the presequence indeed formed an amphipathic α-helix and contacted the receptor via the hydrophobic side chains of several amino acid residues.

Table 10.2 Components of the TOM machinery

Component	Deletion phenotype (yeast)	Localization	Proposed function	References
Tom70	Slow growth	N-terminal membrane anchor, OM	Receptor for preproteins with internal signals	Steger et al., 1990; Brix et al., 2000
Tom40	Lethal	Integral (multiple β-strands), OM	GIP	Baker et al., 1990; Hill et al., 1998
Tom22	Strong growth reduction	Single membrane anchor, OM	Receptor and organizer of GIP complex	Kiebler et al., 1993; Bolliger et al., 1995; van Wilpe et al., 1999
Tom20	Slow growth	N-terminal membrane anchor, OM	Receptor for proteins with presequence	Söllner et al., 1989; Schneider et al., 1991; Ramage et al., 1993
Tom7	Slight growth reduction	Integral, OM	Dissociation of TOM complex	Hönlinger et al., 1996
Tom6	Slight growth reduction	Integral, OM	Assembly of Tom22 with Tom40	Dekker et al., 1998
Tom5	Slight growth reduction	C-terminal membrane anchor, OM	Transfer of proteins from receptor to GIP	Dietmeier et al., 1997

When the genes for two different Tom proteins are deleted in the yeast *Saccharomyces cerevisiae*, strong synthetic growth defects are observed. In many cases, double deletions are lethal for the cells.

GIP, general import pore; OM, outer membrane; TomX, subunit of translocase of outer membrane.

(ii) After recognition by Tom20, presequence-containing preproteins are transferred to Tom22, probably via a direct contact between the cytosolic domains of the receptors (van Wilpe et al., 1999). Tom22 interacts with the positively charged surface of presequences via electrostatic forces (Bolliger et al., 1995; Brix et al., 1997) and directs the preproteins into the general import pore (GIP) of the outer membrane (Kiebler et al., 1993; van Wilpe et al., 1999). Tom22 functions as the central receptor and is able to compensate partially for the loss of the two other receptors, Tom20 and Tom70 (Lithgow et al., 1994). (iii) Tom70 mainly interacts with proteins containing internal targeting information, in particular the preproteins of hydrophobic membrane proteins such as the metabolite carriers of the inner membrane (Brix et al., 1997, 2000). In the absence of preproteins Tom70 forms a dimer. The presence of a preprotein with multiple signals apparently induces an oligomerization of Tom70 dimers. In the case of a carrier preprotein, three Tom70 dimers bind to one and the same preprotein (Wiedemann et al., 2001). We propose that Tom70 not only functions as receptor, but is likely to fulfill a chaperone-like function by binding to multiple sites of the hydrophobic preproteins (Brix et al., 2000; Wiedemann et al., 2001). In this way Tom70 guides the transfer of preproteins from the cytosolic transport complexes to the general import pore and prevents their aggregation.

Tom70 contains seven tetratrico-peptide repeat (TPR) motifs that are degenerate repeats with a length of 34 amino acid residues, forming helix-turn-helix motifs (Brix et al., 2000). TPR motifs are present in many different proteins, including molecular chaperones. They form scaffolds by stacking to each other in an ordered manner and are involved in protein–protein interactions (Scheufler et al., 2000). The exact role of the TPR motifs in Tom70 is unknown; they may function in the interaction of Tom70 with preproteins or other Tom proteins. Interestingly, Tom20 contains only a single TPR motif that participates in forming the binding site for presequences (Abe et al., 2000).

General import pore (GIP) complex

The stable core of the outer membrane translocase if formed by a complex of ~400 kDa, termed the GIP complex. It contains the central receptor Tom22 and four additional subunits, Tom40, Tom7, Tom6 and Tom5 (Dekker et al., 1998; Ahting et al., 1999; Meisinger et al., 2001). The two initial receptors Tom20 and Tom70 are more loosely associated with this 400 kDa complex. The TOM complex was isolated from the fungus *Neurospora crassa* and analyzed by electron microscopy, revealing that one complex probably contains two to three pores of ~2 nm diameter (Künkele et al., 1998; Ahting et al., 1999).

Which protein is responsible for the formation of the translocation pore in the outer membrane? Tom40 is the only mitochondrial outer membrane protein that is strictly essential for cell viability under all growth conditions (Baker et al., 1990) (Table 10.2). It forms a β-barrel protein, traversing the outer membrane multiple times with β-strands. When purified Tom40 was reconstituted into liposomes, it formed a cation-selective pore with an effective diameter of ~2.0–2.2 nm and provided a specific binding site for presequences (Hill et al., 1998). This pore size is sufficient for the translocation of a polypeptide chain in α-helical conformation, even a preprotein in a loop formation (i.e. two α-helices) can be transported through the Tom40 pore.

Tom5 is linked to the outer membrane by a single membrane anchor and its negatively charged N-terminal segment is exposed to the cytosol (Dietmeier et al., 1997). Tom5 is directly associated with Tom40 and plays an important role during import of all types of preproteins. Tom5 is thought to mediate the transfer of preproteins from the receptors to the GIP and assist the insertion of preproteins into the translocation pore. Tom5 may have receptor-like properties. Indeed, some preproteins destined for the intermembrane space do not need any of the classical receptors for import, but directly interact with Tom5 and are then translocated through the Tom40 pore (Kurz et al., 1999).

The other two small Tom proteins, Tom6 and Tom7, are also integral membrane proteins, but do not come into direct contact with preproteins. Tom6 and Tom7 modulate the assembly and dissociation of the TOM machinery. Tom6 promotes association of the receptors with the GIP, in particular the association of Tom22 with Tom40 (Dekker et al., 1998; van Wilpe et al., 1999). In the absence of Tom22, the TOM complex is dissociated into 100 kDa subcomplexes containing a dimer of Tom40 with a single pore. A deletion of the gene for Tom6 leads to a similar dissociation of the TOM complex into 100 kDa units. Tom7 functions in part in an antagonistic manner to Tom6 since it favors a dissociation of the TOM complex (Hönlinger et al., 1996; Model et al., 2001). Tom7 is required for the sorting of proteins into the outer membrane; e.g. the import of the major outer membrane protein porin requires Tom7 (Hönlinger et al., 1996; Krimmer et al., 2001). It is thought that Tom7 is involved in the opening of the TOM complex, allowing the lateral release of preproteins into the outer membrane.

Binding chain hypothesis

Several Tom proteins, in particular Tom22 and Tom5, contain clusters of negatively charged amino acid residues in their cytosolic domains. Furthermore, Tom22 and Tim23 (of the inner membrane translocase) expose domains with a net negative charge to the intermembrane space. This property of import components led to the 'acid chain' hypothesis, which proposes that the translocation of preproteins occurs via the sequential interaction of positively charged presequences with negatively charged clusters of import components positioned along the import pathway

(Bolliger et al., 1995; Dietmeier et al., 1997; Komiya et al., 1998). A presequence-containing preprotein sequentially interacts with Tom20, Tom22 and Tom5. Following translocation through the import pore formed by Tom40 the presequence binds to the intermembrane space domain of Tom22 (Bolliger et al., 1995; Moczko et al., 1997). Tim23 probably acts as a *cis* binding site for the preprotein at the inner membrane (Bauer et al., 1996; Komiya et al., 1998).

Recent studies showed that the preproteins interact with Tom proteins not only via electrostatic forces, but also via other types of non-covalent interactions such as hydrophobic interactions and hydrogen bonds, e.g. in the case of Tom20, Tom40 and further Tom proteins (Brix et al., 1997; Abe et al., 2000; Yano et al., 2000; Meisinger et al., 2001). Therefore the 'acid chain' hypothesis has been extended to the 'binding chain' hypothesis, i.e. that numerous preprotein binding sites exist along the import route, including each type of non-covalent interactions (Meisinger et al., 2001). The affinity of an individual binding site for preproteins is relatively low (Brix et al., 1997; Komiya et al., 1998). However, the chain of several binding sites together provides a high specificity of the mitochondrial import system. The binding chain thus serves as a guiding system to direct preproteins across the mitochondrial outer membrane and to the inner membrane (Figure 10.2).

Summary: *Translocation machinery of the outer membrane*

- Presequence-containing preproteins are recognized by two import receptors of the outer mitochondrial membrane, Tom20 and Tom22.
- Precursor proteins with internal targeting signals preferentially interact with the receptor Tom70.
- The β-barrel protein Tom40 forms the channel (general import pore) across the outer membrane for virtually all types of precursor proteins.
- The Tom proteins are assembled in a large dynamic protein complex of the outer membrane (TOM complex). Small Tom proteins assist in the transfer of precursor proteins from receptors to the import channel (Tom5) or modulate the assembly and dissociation of the TOM complex (Tom6 and Tom7).

TRANSLOCATION MACHINERIES OF THE INNER MEMBRANE

Early studies using mitochondria with a ruptured outer membrane demonstrated the existence of a protein translocation system in the inner membrane that acted independently of the TOM translocase (Hwang et al., 1989). In total eleven Tim proteins and two matrix chaperones have been identified that are involved in translocation of preproteins into or across the inner membrane (Table 10.3).

It is evident that two distinct TIM complexes exist in the inner membrane; the TIM23 complex (presequence translocase) is required for the import of presequence-containing preproteins, while the TIM22 complex (carrier translocase) is required for the import and insertion of integral inner membrane proteins.

The presequence translocase and energetics of translocation into the matrix

The 90 kDa TIM23 core complex consists of the integral membrane proteins Tim23 and Tim17 (Dekker et al., 1997) (Figure 10.3). The membrane-spanning domain of Tim23 is homologous to Tim17. Each protein is predicted to span the inner membrane four times in an α-helical conformation (Dekker et al., 1993; Emtage and Jensen, 1993; Kübrich et al., 1994). Despite their homology and topological similarity neither protein can substitute for the loss of the other since each is essential for cell viability.

The N-terminal domain of Tim23 contains a leucine-zipper motif that is potentially involved in its dimerization (Bauer et al., 1996). Experimental evidence indicates that upon contact with a preprotein the Tim23 dimer dissociates, probably leading to the opening of the inner membrane channel (Bauer et al., 1996). When purified Tim23 is reconstituted in liposomes it forms a cation-selective channel with specificity for mitochondrial presequences. The inner diameter of the Tim23 pore is ~1.3 nm, i.e. only one α-helix of a polypeptide chain will be able to pass through the pore, indicating that preproteins translocated across the inner mitochondrial membrane must be substantially unfolded, at least to the level of individual α-helices (Truscott et al., 2001). Because of its homology to Tim23, it is expected that Tim17 may also form a channel. However, the specific role of Tim17 in the presequence translocase is not known.

Protein import across and into the inner membrane strictly depends on the membrane potential, $\Delta\Psi$. The net negative charge on the matrix side of the inner membrane creates an electrophoretic force on the positively charged presequences and thereby contributes to their translocation across the inner membrane (Martin et al., 1991). In addition, $\Delta\Psi$ seems to activate and open the channel formed by Tim23 (Bauer et al., 1996; Truscott et al., 2001). The membrane potential is essential only during the initial steps of import through the TIM channel and not for the translocation of the mature portion of the preprotein (Schleyer and Neupert, 1985; Martin et al., 1991).

The translocation of the major portion of a preprotein into the matrix requires the hydrolysis of ATP to power the matrix heat shock protein Hsp70 and its cooperation with the inner membrane protein Tim44. When the preprotein emerges from the TIM23 translocation channel it interacts

Table 10.3 The TIM machineries and interacting components

Component	Deletion phenotype (yeast)	Localization	Proposed function	References
Presequence translocase (TIM23 complex)				
Tim44	Lethal	IM associated from matrix side	Membrane anchor for mtHsp70	Maarse et al., 1992; Scherer et al., 1992; Kronidou et al., 1994; Rassow et al., 1994; Schneider et al., 1994
Tim23	Lethal	Integral, IM	Preprotein translocation through IM	Dekker et al., 1993; Emtage and Jensen, 1993
Tim17	Lethal	Integral, IM	Preprotein translocation through IM	Kübrich et al., 1994; Maarse et al., 1994; Ryan et al., 1994
mtHsp70	Lethal	Matrix	Translocation motor	Kang et al., 1990; Voisine et al., 1999
Mge1	Lethal	Matrix	Nucleotide exchange factor for mtHsp70	Bolliger et al., 1994; Schneider et al., 1996

Carrier translocase (TIM22 complex)

Tim54	Lethal	Integral, IM	Insertion of polytopic proteins into IM	Kerscher et al., 1997
Tim22	Lethal	Integral, IM	Insertion of polytopic proteins into IM	Sirrenberg et al., 1996
Tim18	Slight growth reduction	Integral, IM	Insertion of polytopic proteins into IM	Kerscher et al., 2000; Koehler et al., 2000
Tim13	Normal growth	Soluble, IMS	Import of the precursor of Tim23; complex with Tim8	Koehler et al., 1999; Kurz et al., 1999; Davis et al., 2000; Paschen et al., 2000
Tim12	Lethal	Peripheral IM protein (intermembrane space side)	IM insertion of carrier proteins	Koehler et al., 1998a; Sirrenberg et al., 1998
Tim10	Lethal	Soluble, IMS	Shuttle of carrier proteins from TOM to TIM; complex with Tim9	Koehler et al., 1998a; Sirrenberg et al., 1998
Tim9	Lethal	Soluble, IMS	Shuttle of carrier proteins from TOM to TIM; complex with Tim10	Koehler et al., 1998b; Adam et al., 1999
Tim8	Normal growth	Soluble, IMS	Import of the precursor of Tim23; complex with Tim13	Koehler et al., 1999; Kurz et al., 1999; Davis et al., 2000; Paschen et al., 2000

Hsp, heat shock protein; IM, inner membrane; IMS, intermembrane space; TimX, subunit of translocase of inner membrane.

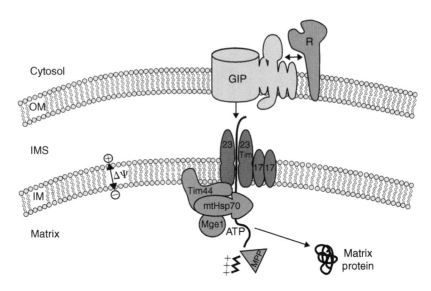

Figure 10.3 Schematic representation of the molecular organization of the TIM23 complex (presequence translocase) in the mitochondrial inner membrane (IM). After passage through the outer membrane (OM) the preprotein inserts into an IM channel formed by Tim23. Tim23 is associated with the integral membrane protein Tim17 forming a translocation complex. Protein import across the IM is dependent on the membrane potential ($\Delta\Psi$). On the matrix side of the IM the preprotein interacts with the import motor, which is composed of mitochondrial Hsp70 (mtHsp70) and membrane-bound Tim44. In an ATP-dependent manner the preprotein is driven into the matrix by mtHsp70. The protein Mge1 serves as a nucleotide exchange factor for mtHsp70. Finally, the mitochondrial processing peptidase (MPP) cleaves off the positively charged presequence and the mature protein is folded in the matrix.

with Tim44 on the matrix side of the inner membrane (Scherer et al., 1992; Blom et al., 1993). Tim44 is a peripheral membrane protein although there is some evidence that its C-terminus is exposed to the intermembrane space (Maarse et al., 1992). Tim44 acts together with the matrix chaperone mtHsp70 and its co-chaperone, the nucleotide exchange factor Mge1, as an import motor for matrix proteins. MtHsp70 forms a 1:1 complex with Tim44 that is dissociated by binding of ATP (Kronidou et al., 1994; Rassow et al., 1994; Schneider et al., 1994, 1996; Moro et al., 1999; Voisine et al., 1999).

Currently, two controversial models are discussed that describe the role of Tim44 and mtHsp70 in protein import. The 'Brownian ratchet or trapping' model proposes that the TOM and TIM23 complexes form passive diffusion channels through which the preprotein slides back and forth by Brownian motion. On the matrix side of the inner membrane the preprotein is bound by mtHsp70, preventing retrograde movement into the channel, and the preprotein is trapped by mtHsp70 in the matrix (Schneider et al., 1994, 1996; Moro et al., 1999). Thus, mtHsp70 is proposed to fulfill a more passive role in protein import (Gaume et al., 1998). In contrast, the 'pulling' model predicts that Tim44-bound mtHsp70 binds the incoming preprotein and then subsequent ATP-binding to mtHsp70 induces a conformational change, resulting in an active pulling force on the preprotein (Horst et al., 1996; Matouschek et al., 1997). The idea that mtHsp70 actively unfolds preproteins gained experimental support through the use of a model protein that in solution starts to unfold in its middle portion. When fused to a mitochondrial presequence, however, the model protein unfolded with the N-terminal portion first, demonstrating that the mitochondrial import system changed the unfolding pathway of the protein and thus actively catalyzed its unfolding (Huang et al., 1999). Indeed, the kinetics of unfolding of proteins by the mitochondrial Hsp70 system were significantly faster than their unfolding in solution, indicating an active unfolding function of mitochondria (Matouschek et al., 1997; Lim et al., 2001). By the use of yeast mutants in mtHsp70 it could be shown that actually both mechanisms, pulling and trapping, cooperate in the import of preproteins. Loosely folded preproteins can be imported by a trapping mechanism, while preproteins with tightly folded domains additionally require a pulling force generated by conformational changes of mtHsp70 (Voos et al., 1996; Voisine et al., 1999).

Although both the TOM and TIM translocases can act independently of each other for the import of proteins *in vitro* (Hwang et al., 1989, 1991; Rassow and Pfanner, 1991), a preprotein in transit physically connects both translocases in so-called translocation contact sites (Schleyer and Neupert, 1985). In the presence of a preprotein that has been accumulated in a two-membrane-spanning fashion, the 90 kDa TIM23 complex can associate with the 400 kDa TOM complex forming a large supercomplex (Dekker et al., 1997). It is not known if transient contacts between TOM and TIM complexes also exist in the absence of preproteins, e.g. via hypothetical adaptor proteins of the intermembrane space. Very recently, evidence was presented indicating that the N-terminus of the inner membrane protein Tim23 is integrated into the outer membrane, thereby linking both mitochondrial membranes (Donzeau et al., 2000). However, no connection of Tim23 to the TOM machinery was detected and thus the role of this surprising topology of Tim23 for import of preproteins is unclear.

The carrier translocase

As soon as members of the mitochondrial carrier protein family, e.g. the ADP/ATP carrier (AAC) or the phosphate carrier, and other polytopic inner membrane proteins enter the intermembrane space they are shuttled by one of two soluble 70 kDa protein complexes to the TIM22 complex of the inner membrane (Figure 10.4). These intermembrane space complexes are formed by the so-called 'tiny Tim' proteins and contain either Tim9 and

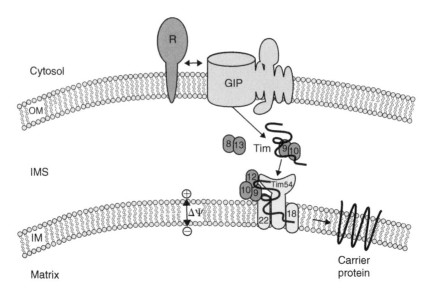

Figure 10.4   Schematic representation of the molecular organization of the TIM22 complex (carrier translocase) in the mitochondrial inner membrane (IM). After passage through the outer membrane (OM) the hydrophobic carrier preprotein is shuttled by a complex of the 'tiny' Tim proteins Tim9 and Tim10 through the intermembrane space (IMS) to the TIM22 complex. This complex consists of the membrane proteins Tim54, Tim22 and Tim18 as well as the peripherally attached Tim12. The TIM22 complex mediates the membrane insertion of preproteins in a membrane potential ($\Delta\Psi$)-dependent step. Subsequently, the carrier preprotein moves laterally into the IM and assembles into its active form.

Tim10 or Tim8 and Tim13 (Koehler et al., 1998a, 1998b, 1999; Sirrenberg et al., 1998; Adam et al., 1999). All tiny Tim proteins contain zinc-finger-like motifs that seem to promote protein–protein interactions. Each complex is thought to function as a specific chaperone that mediates the transport of a subset of inner membrane proteins across the intermembrane space (Leuenberger et al., 1999). Tim9 and Tim10 are both essential for cell viability, while cells lacking Tim8 and Tim13 are still viable (Table 10.3) (Koehler et al., 1998a, 1998b; Adam et al., 1999). The crucial function of the Tim9–Tim10 complex for the import of inner membrane proteins was deduced from the observation that the functional loss of Tim10 impairs translocation of carrier proteins across the outer membrane (Koehler et al., 1998a; Sirrenberg et al., 1998). Tim8 and Tim13 play a more special role, in particular for import of the precursor of Tim23 (Koehler et al., 1999; Davis et al., 2000; Paschen et al., 2000). Interestingly, yeast Tim8 is the homolog of the human DDP1 protein (Koehler et al., 1999). A mutation in DDP1 causes the deafness dystonia syndrome in humans. Thus, a defect in mitochondrial protein import may be responsible for this disease.

The 300 kDa TIM22 complex of the inner membrane consists of the integral membrane proteins Tim22, Tim54 and Tim18, and the peripheral membrane protein Tim12 (Sirrenberg et al., 1996, 1998; Koehler et al., 1998a, 2000; Kerscher et al., 2000). Tim12 is homologous to the tiny Tim proteins of the intermembrane space. Tim12 probably functions as the docking point for the 70 kDa intermembrane space complexes. Indeed, small amounts of Tim9 and Tim10 are found associated with Tim12 at the inner membrane (Adam et al., 1999). A mutation in Tim12 blocks insertion of carrier proteins into the inner membrane (Koehler et al., 1998a). Tim22 is homologous to Tim23 and Tim17 of the TIM23 complex (Sirrenberg et al., 1996) and probably forms a channel for the insertion of preproteins. Tim54 spans the inner membrane 1–2 times and exposes its C-terminus to the intermembrane space (Kerscher et al., 1997). Both Tim54 and Tim22 are required for the insertion of carrier proteins into the inner membrane and like Tim23 are essential for cell viability. The non-essential Tim18 may act in the assembly and stabilization of the TIM22 complex rather than being directly involved in the insertion of proteins into the inner membrane (Koehler et al., 2000).

The mechanism of the action of the carrier translocase is still open. We know that the inner membrane potential is required for insertion of preproteins into the translocase (Pfanner et al., 1987; Endres et al., 1999; Wiedemann et al., 2001). The import channel of the carrier translocase may then open laterally to release the preproteins into the lipid phase of the inner membrane.

Summary: *Translocation machineries of the inner membrane*

- The inner membrane contains two distinct translocases for the import of precursor proteins.
- Presequence-containing preproteins are transported through the presequence translocase (TIM23 complex), consisting of the integral membrane proteins Tim23 and Tim17, the peripheral membrane protein Tim44, and the attached molecular chaperone matrix Hsp70.
- Transport via the presequence translocase requires two energy sources. The membrane potential (negative on the inside) drives translocation of the positively charged presequences and activates the channel protein Tim23. Matrix Hsp70 functions as an ATP-dependent import motor in cooperation with Tim44 and the co-chaperone Mge1 (a nucleotide exchange factor).
- Hydrophobic precursor proteins with multiple internal targeting signals, e.g. carrier proteins, are directed by small Tim proteins (Tim9–Tim10) through the aqueous intermembrane space and inserted into the inner membrane by the TIM22 complex (carrier translocase), consisting of Tim54, Tim22, Tim18 and Tim12.
- Transport via the TIM22 complex requires the presence of a membrane potential, but no addition of ATP.

BIOGENESIS OF THE MITOCHONDRIAL
PREPROTEIN TRANSLOCASES

All Tom and Tim proteins are themselves encoded by nuclear genes and are synthesized as precursors on cytosolic polysomes. Only a few of these precursors contain classical presequence-type targeting signals; most precursors of Tom or Tim proteins contain internal targeting signals at distinct locations of the mature proteins. It has been found that none of these import components is able to direct its own precursor to mitochondria. For example, the precursor of the receptor Tom20 does not use the mature receptor Tom20, but other Tom proteins for its specific targeting to mitochondria (Schneider et al., 1991). It can therefore be excluded that the mistargeting of a few receptor molecules (e.g. Tom20 to another cell organelle) would subsequently lead to the mistargeting of many more receptor molecules and other mitochondrial preproteins to the wrong organelle.

An interesting example is the precursor of Tom40, which requires nearly all other Tom proteins for its proper import and assembly into the TOM complex (Model et al., 2001). Two receptors, Tom20 and Tom22, are required to target the precursor of Tom40 to mitochondria. Subsequently two intermediate complexes, surprisingly one of them located on the intermembrane space side, are formed during assembly of Tom40 into the 400 kDa TOM complex. On its assembly pathway, the precursor of Tom40 successively associates with Tom5, Tom6 and Tom22. Remarkably, newly synthesized Tom40 can assemble into preexisting TOM complexes (Rapaport and Neupert, 1999; Model et al., 2001). A continuous exchange between the mature TOM complex and the late assembly intermediate, promoted by Tom7 and Tom6, seems to represent the key mechanism in the assembly of new subunits into an active complex. Taken together, this study suggests a dynamic behavior of the TOM complex that is modulated by the small regulatory Tom proteins 6 and 7.

The biogenesis of the Tim proteins of the carrier pathway also revealed several surprises (Kurz et al., 1999). First, the precursors of the tiny Tim proteins of the intermembrane space do not require any of the classical receptors on the mitochondrial surface. The precursors simply interact with Tom5 and are then translocated through Tom40 to their functional destination. Second, the precursors of Tim22 and Tim54 combine distinct portions of the two major import pathways (presequence pathway and carrier pathway) to new import routes. On the one hand, the precursor of Tim22 is recognized by the receptor Tom20-like preproteins with a presequence, but upon passage through the GIP follows the import route of carrier proteins via the TIM22 complex. On the other hand, the precursor of Tim54 is recognized by Tom70-like carrier proteins and is translocated through the GIP, but then uses the presequence pathway via the TIM23 complex. Thus, a crossing-over between the presequence pathway and the carrier pathway can occur at the level of the GIP (Kurz et al., 1999).

Summary: *Biogenesis of the mitochondrial preprotein translocases*

- All Tom and Tim proteins are encoded by nuclear genes and must be imported into mitochondria.
- The precursors of most Tom and Tim proteins do not carry classical presequence-type targeting signals, but contain various internal targeting signals. These precursors use numerous variations of the main import pathways and can assemble into preexisting translocase complexes.

PERSPECTIVES

Many open questions remain in the field of mitochondrial protein import. What is the exact information contained in targeting signals? How are the intramitochondrial sorting signals decoded and which components are responsible for directing preproteins to the correct mitochondrial subcompartment? Recent evidence shows that the mitochondrial inner membrane contains at least one more protein translocase. The Oxa1 complex is involved in the export of proteins from the matrix into the inner membrane (He and Fox, 1997; Hell et al., 1998). The exact role of Oxa1 and putative additional translocation components in sorting of mitochondrially encoded and nuclear encoded proteins will be the subject of future research.

A pressing issue will be the elucidation of the structural organization of the translocases, of receptors and channels. We would like to know which energy sources drive the translocation of preproteins across the outer membrane and how the import motor of mtHsp70 and Tim44 functions at a molecular level. How can the inner membrane translocation channels be kept tight for ions during the transport of polypeptide chains? Finally, the mechanisms of cooperation and regulation of the translocases of both membranes will pose important questions for the future.

REFERENCES

Abe, Y., Shodai, T., Muto, T. et al. (2000) Structural basis of presequence recognition by the mitochondrial protein import receptor Tom20. *Cell* **100**: 551–560.

Adam, A., Endres, M., Sirrenberg, C. et al. (1999) Tim9, a new component of the TIM22·54 translocase in mitochondria. *EMBO J* **18**: 313–319.

Ahting, U., Thun, C., Hegerl, R. et al. (1999) The TOM core complex: the general protein import pore of the outer membrane of mitochondria. *J Cell Biol* **147**: 959–968.

Baker, K.P., Schaniel, A., Vestweber, D. and Schatz G. (1990) A yeast mitochondrial outer membrane protein essential for protein import and cell viability. *Nature* **348**: 605–609.

Bauer, M.F., Sirrenberg, C., Neupert, W. and Brunner M. (1996) Role of Tim23 as voltage sensor and presequence receptor in protein import into mitochondria. *Cell* **87**: 33–41.

Blom, J., Kübrich, M., Rassow, J. et al. (1993) The essential yeast protein MIM44 (encoded by *MPI1*) is involved in an early step of preprotein translocation across the mitochondrial inner membrane. *Mol Cell Biol* **13**: 7364–7371.

Bolliger, L., Deloche, O., Glick, B.S. et al. (1994) A mitochondrial homolog of bacterial GrpE interacts with mitochondrial hsp70 and is essential for viability. *EMBO J* **13**: 1998–2006.

Bolliger, L., Junne, T., Schatz, G. and Lithgow T. (1995) Acidic receptor domains on both sides of the outer membrane mediate translocation of precursor proteins into yeast mitochondria. *EMBO J* **14**: 6318–6326.

Brix, J., Dietmeier, K. and Pfanner N. (1997) Differential recognition of preproteins by the purified cytosolic domains of the mitochondrial import receptors Tom20, Tom22, and Tom70. *J Biol Chem* **272**: 20730–20735.

Brix, J., Ziegler, G.A., Dietmeier, K. et al. (2000) The mitochondrial import receptor Tom70: identification of a 25 kDa core domain with a specific binding site for preproteins. *J Mol Biol* **303**: 479–488.

Davis, A.J., Sepuri, N.B., Holder, J., Johnson, A.E. and Jensen, R.E. (2000) Two intermembrane space TIM complexes interact with different domains of Tim23p during its import into mitochondria. *J Cell Biol* **150**: 1271–1282.

Dekker, P.J., Keil, P., Rassow, J. et al. (1993) Identification of MIM23, a putative component of the protein import machinery of the mitochondrial inner membrane. *FEBS Lett* **330**: 66–70.

Dekker, P.J., Martin, F., Maarse, A.C. et al. (1997) The Tim core complex defines the number of mitochondrial translocation contact sites and can hold arrested preproteins in the absence of matrix Hsp70-Tim44. *EMBO J* **16**: 5408–5419.

Dekker, P.J., Ryan, M.T., Brix, J. et al. (1998) Preprotein translocase of the outer mitochondrial membrane: molecular dissection and assembly of the general import pore complex. *Mol Cell Biol* **18**: 6515–6524.

Dietmeier, K., Hönlinger, A., Bömer, U. et al. (1997) Tom5 functionally links mitochondrial preprotein receptors to the general import pore. *Nature* **388**: 195–200.

Donzeau, M., Kaldi, K., Adam, A. et al. (2000) Tim23 links the inner and outer mitochondrial membranes. *Cell* **101**: 401–412.

Eilers, M. and Schatz G. (1986) Binding of a specific ligand inhibits import of a purified precursor protein into mitochondria. *Nature* **322**: 228–232.

Emtage, J.L. and Jensen, R.E. (1993) *MAS6* encodes an essential inner membrane component of the yeast mitochondrial protein import pathway. *J Cell Biol* **122**: 1003–1012.

Endres, M., Neupert, W. and Brunner M. (1999) Transport of the ADP/ATP carrier of mitochondria from the TOM complex to the TIM22·54 complex. *EMBO J* **18**: 3214–3221.

Gärtner, F., Bömer, U., Guiard, B. and Pfanner N. (1995) The sorting signal of cytochrome b_2 promotes early divergence from the general mitochondrial import pathway and restricts the unfoldase activity of matrix Hsp70. *EMBO J* **14**: 6043–6057.

Gaume, B., Klaus, C., Ungermann, C. et al. (1998) Unfolding of preproteins upon import into mitochondria. *EMBO J* **17**: 6497–6507.

Glick, B.S., Brandt, A., Cunningham, K. et al. (1992) Cytochromes c_1 and b_2 are sorted to the intermembrane space of yeast mitochondria by a stop-transfer mechanism. *Cell* **69**: 809–822.

Hallermayer, G., Zimmermann, R. and Neupert, W. (1977) Kinetic studies on the transport of cytoplasmically synthesized proteins into the mitochondria in intact cells of *Neurospora crassa*. *Eur J Biochem* **81**: 523–532.

Hartl, F.U., Ostermann, J., Guiard, B. and Neupert, W. (1987) Successive translocation into and out of the mitochondrial matrix: targeting of proteins to the intermembrane space by a bipartite signal peptide. *Cell* **51**: 1027–1037.

Hawlitschek, G., Schneider, H., Schmidt, B. et al. (1988) Mitochondrial protein import: identification of processing peptidase and of PEP, a processing enhancing protein. *Cell* **53**: 795–806.

He, S. and Fox, T.D. (1997) Membrane translocation of mitochondrially coded Cox2p: distinct requirements for export of N and C termini and dependence on the conserved protein Oxa1p. *Mol Biol Cell* **8**: 1449–1460.

Hell, K., Herrmann, J.M., Pratje, E., Neupert, W. and Stuart, R. A. (1998) Oxa1p, an essential component of the N-tail protein export machinery in mitochondria. *Proc Natl Acad Sci USA* **95**: 2250–2255.

Hill, K., Model, K., Ryan, M.T. et al. (1998) Tom40 forms the hydrophilic channel of the mitochondrial import pore for preproteins. *Nature* **395**: 516–521.

Hönlinger, A., Bömer, U., Alconada, A. et al. (1996) Tom7 modulates the dynamics of the mitochondrial outer membrane translocase and plays a pathway-related role in protein import. *EMBO J* **15**: 2125–2137.

Horst, M., Oppliger, W., Feifel, B., Schatz, G. and Glick, B.S. (1996) The mitochondrial protein import motor: dissociation of mitochondrial hsp70 from its membrane anchor requires ATP binding rather than ATP hydrolysis. *Protein Sci* **5**: 759–767.

Horwich, A.L., Kalousek, F., Mellman, I. and Rosenberg, L.E. (1985) A leader peptide is sufficient to direct mitochondrial import of a chimeric protein. *EMBO J* **4**: 1129–1135.

Huang, S., Ratliff, K.S., Schwartz, M.P., Spenner, J.M. and Matouschek, A. (1999) Mitochondria unfold precursor proteins by unraveling them from their N-termini. *Nat Struct Biol* **6**: 1132–1138.

Hurt, E.C., Pesold-Hurt, B. and Schatz, G. (1984) The cleavable prepiece of an imported mitochondrial protein is sufficient to direct cytosolic dihydrofolate reductase into the mitochondrial matrix. *FEBS Lett* **178**: 306–310.

Hwang, S., Jascur, T., Vestweber, D., Pon, L. and Schatz, G. (1989) Disrupted yeast mitochondria can import precursor proteins directly through their inner membrane. *J Cell Biol* **109**: 487–493.

Hwang, S.T., Wachter, C. and Schatz, G. (1991) Protein import into the yeast mitochondrial matrix. A new translocation intermediate between the two mitochondrial membranes. *J Biol Chem* **266**: 21083–21089.

Kang, P.J., Ostermann, J., Shilling, J. et al. (1990) Requirement for hsp70 in the mitochondrial matrix for translocation and folding of precursor proteins. *Nature* **348**: 137–143.

Kerscher, O., Holder, J., Srinivasan, M., Leung, R.S. and Jensen, R.E. (1997) The Tim54p–Tim22p complex mediates insertion of proteins into the mitochondrial inner membrane. *J Cell Biol* **139**: 1663–1675.

Kerscher, O., Sepuri, N.B. and Jensen, R.E. (2000) Tim18p is a new component of the Tim54p–Tim22p translocon in the mitochondrial inner membrane. *Mol Biol Cell* **11**: 103–116.

Kiebler, M., Keil, P., Schneider, H. et al. (1993) The mitochondrial receptor complex: a central role of MOM22 in mediating preprotein transfer from receptors to the general insertion pore. *Cell* **74**: 483–492.

Koehler, C.M., Jarosch, E., Tokatlidis, K. et al. (1998a). Import of mitochondrial carriers mediated by essential proteins of the intermembrane space. *Science* **279**: 369–373.

Koehler, C.M., Merchant, S., Oppliger, W. et al. (1998b). Tim9p, an essential partner subunit of Tim10p for the import of mitochondrial carrier proteins. *EMBO J* **17**: 6477–6486.

Koehler, C.M., Leuenberger, D., Merchant, S. et al. (1999) Human deafness dystonia syndrome is a mitochondrial disease. *Proc Natl Acad Sci USA* **96**: 2141–2146.

Koehler, C.M., Murphy, M.P., Bally, N.A. et al. (2000) Tim18p, a new subunit of the TIM22 complex that mediates insertion of imported proteins into the yeast mitochondrial inner membrane. *Mol Cell Biol* **20**: 1187–1193.

Komiya, T., Rospert, S., Schatz, G. and Mihara, K. (1997) Binding of mitochondrial precursor proteins to the cytoplasmic domains of the import receptors Tom70 and Tom20 is determined by cytoplasmic chaperones. *EMBO J* **16**: 4267–4275.

Komiya, T., Rospert, S., Koehler, C. et al. (1998) Interaction of mitochondrial targeting signals with acidic receptor domains along the protein import pathway: evidence for the 'acid chain' hypothesis. *EMBO J* **17**: 3886–3898.

Krimmer, T., Rapaport, D., Ryan, M.T. et al. (2001) Biogenesis of porin of the outer mitochondrial membrane involves an import pathway via receptors and the general import pore of the TOM complex. *J Cell Biol* **152**: 289–300.

Kronidou, N.G., Oppliger, W., Bolliger, L. et al. (1994) Dynamic interaction between Isp45 and mitochondrial hsp70 in the protein import system of the yeast mitochondrial inner membrane. *Proc Natl Acad Sci USA* **91**: 12818–12822.

Kübrich, M., Keil, P., Rassow, J. et al. (1994) The polytopic mitochondrial inner membrane proteins MIM17 and MIM23 operate at the same preprotein import site. *FEBS Lett* **349**: 222–228.

Künkele, K.P., Heins, S., Dembowski, M. et al. (1998) The preprotein translocation channel of the outer membrane of mitochondria. *Cell* **93**: 1009–1019.

Kurz, M., Martin, H., Rassow, J., Pfanner, N. and Ryan, M.T. (1999) Biogenesis of Tim proteins of the mitochondrial carrier import pathway: differential targeting mechanisms and crossing over with the main import pathway. *Mol Biol Cell* **10**: 2461–2474.

Leuenberger, D., Bally, N.A., Schatz, G. and Koehler, C.M. (1999) Different import pathways through the mitochondrial intermembrane space for inner membrane proteins. *EMBO J* **18**: 4816–4822.

Lim, J.H., Martin, F., Guiard, B., Pfanner, N. and Voos, W. (2001) The mitochondrial Hsp70-dependent import system actively unfolds preproteins and shortens the lag phase of translocation. *EMBO J* **20**: 941–950.

Lithgow, T., Junne, T., Suda, K., Gratzer, S. and Schatz, G. (1994) The mitochondrial outer membrane protein Mas22p is essential for protein import and viability of yeast. *Proc Natl Acad Sci USA* **91**: 11973–11977.

Maarse, A.C., Blom, J., Grivell, L.A. and Meijer, M. (1992) *MPI1*, an essential gene encoding a mitochondrial membrane protein, is possibly involved in protein import into yeast mitochondria. *EMBO J* **11**: 3619–3628.

Maarse, A.C., Blom, J., Keil, P., Pfanner, N. and Meijer, M. (1994) Identification of the essential yeast protein MIM17, an integral mitochondrial inner membrane protein involved in protein import. *FEBS Lett* **349**: 215–221.

Maccecchini, M.L., Rudin, Y., Blobel, G. and Schatz, G. (1979) Import of proteins into mitochondria: precursor forms of the extramitochondrially made F_1-ATPase subunits in yeast. *Proc Natl Acad Sci USA* **76**: 343–347.

Martin, J., Mahlke, K. and Pfanner, N. (1991) Role of an energized inner membrane in mitochondrial protein import: $\Delta\Psi$ drives the movement of presequences. *J Biol Chem* **266**: 18051–18057.

Matouschek, A., Azem, A., Ratliff, K. et al. (1997) Active unfolding of precursor proteins during mitochondrial protein import. *EMBO J* **16**: 6727–6736.

Meisinger, C., Ryan, M.T., Hill, K. et al. (2001) Protein import channel of the outer mitochondrial membrane: a highly stable Tom40-Tom22 core structure differentially interacts with preproteins, small Tom proteins, and import receptors. *Mol Cell Biol* **21**: 2337–2348.

Moczko, M., Bömer, U., Kübrich, M. et al. (1997) The intermembrane space domain of mitochondrial Tom22 functions as a *trans* binding site for preproteins with N-terminal targeting sequences. *Mol Cell Biol* **17**: 6574–6584.

Model, K., Meisinger, C., Prinz, T. et al. (2001) Multistep assembly of the protein import channel of the mitochondrial outer membrane. *Nat Struct Biol* **8**: 361–370.

Mori, M. and Terada, K. (1998) Mitochondrial protein import in animals. *Biochim Biophys Acta* **1403**: 12–27.

Moro, F., Sirrenberg, C., Schneider, H.C., Neupert, W. and Brunner, M. (1999) The TIM17·23 preprotein translocase of mitochondria: composition and function in protein transport into the matrix. *EMBO J* **18**: 3667–3675.

Neupert, W. (1997) Protein import into mitochondria. *Annu Rev Biochem* **66**: 863–917.

Ostermann, J., Horwich, A.L., Neupert, W. and Hartl, F.U. (1989) Protein folding in mitochondria requires complex formation with hsp60 and ATP hydrolysis. *Nature* **341**: 125–130.

Paschen, S.A., Rothbauer, U., Kaldi, K. et al. (2000) The role of the TIM8-13 complex in the import of Tim23 into mitochondria. *EMBO J* **19**: 6392–6400.

Pfanner, N. and Neupert, W. (1986) Transport of F_1-ATPase subunit β into mitochondria depends on both a membrane potential and nucleoside triphosphates. *FEBS Lett* **209**: 152–156.

Pfanner, N., Tropschug, M. and Neupert, W. (1987) Mitochondrial protein import: nucleoside triphosphates are involved in conferring import-competence to precursors. *Cell* **49**: 815–823.

Pfanner, N., Craig, E.A. and Hönlinger, A. (1997) Mitochondrial preprotein translocase. *Annu Rev Cell Dev Biol* 1997; **13**: 25–51.

Ramage, L., Junne, T., Hahne, K., Lithgow, T. and Schatz, G. (1993) Functional cooperation of mitochondrial protein import receptors in yeast. *EMBO J* **12**: 4115–4123.

Rapaport, D. and Neupert, W. (1999) Biogenesis of Tom40, core component of the TOM complex of mitochondria. *J Cell Biol* **146**: 321–331.

Rassow, J. and Pfanner, N. (1991) Mitochondrial preproteins en route from the outer membrane to the inner membrane are exposed to the intermembrane space. *FEBS Lett* **293**: 85–88.

Rassow, J., Maarse, A.C., Krainer, E. et al. (1994) Mitochondrial protein import: biochemical and genetic evidence for interaction of matrix hsp70 and the inner membrane protein MIM44. *J Cell Biol* **127**: 1547–1556.

Roise, D., Horvath, S.J., Tomich, J.M., Richards, J.H. and Schatz, G. (1986) A chemically synthesized pre-sequence of an imported mitochondrial protein can form an amphiphilic helix and perturb natural and artificial phospholipid bilayers. *EMBO J* **5**: 1327–1334.

Ryan, K.R., Menold, M.M., Garrett, S. and Jensen, R.E. (1994) SMS1, a high-copy suppressor of the yeast mas6 mutant, encodes an essential inner membrane protein required for mitochondrial protein import. *Mol Biol Cell* **5**: 529–538.

Schatz, G. and Dobberstein, B. (1996) Common principles of protein translocation across membranes. *Science* **271**: 1519–1526.

Scherer, P.E., Manning-Krieg, U.C., Jeno, P., Schatz, G. and Horst, M. (1992) Identification of a 45-kDa protein at the protein import site of the yeast mitochondrial inner membrane. *Proc Natl Acad Sci USA* **89**: 11930–11934.

Scheufler, C., Brinker, A., Bourenkov, G. et al. (2000) Structure of TPR domain-peptide complexes: critical elements in the assembly of the Hsp70-Hsp90 multichaperone machine. *Cell* **101**: 199–210.

Schleyer, M. and Neupert, W. (1985) Transport of proteins into mitochondria: translocational intermediates spanning contact sites between outer and inner membranes. *Cell* **43**: 339–350.

Schneider, H., Söllner, T., Dietmeier, K. et al. (1991) Targeting of the master receptor MOM19 to mitochondria. *Science* **254**: 1659–1662.

Schneider, H.C., Berthold, J., Bauer, M.F. et al. (1994) Mitochondrial Hsp70/MIM44 complex facilitates protein import. *Nature* **371**: 768–774.

Schneider, H.C., Westermann, B., Neupert, W. and Brunner, M. (1996) The nucleotide exchange factor MGE exerts a key function in the ATP-dependent cycle of mt-Hsp70–Tim44 interaction driving mitochondrial protein import. *EMBO J* **15**: 5796–5803.

Sirrenberg, C., Bauer, M.F., Guiard, B., Neupert, W. and Brunner, M. (1996) Import of carrier proteins into the mitochondrial inner membrane mediated by Tim22. *Nature* **384**: 582–585.

Sirrenberg, C., Endres, M., Fölsch, H. et al. (1998) Carrier protein import into mitochondria mediated by the intermembrane proteins Tim10/Mrs11 and Tim12/Mrs5. *Nature* **391**: 912–915.

Söllner, T., Griffiths, G., Pfaller, R., Pfanner, N. and Neupert, W. (1989) MOM19, an import receptor for mitochondrial precursor proteins. *Cell* **59**: 1061–1070.

Steger, H.F., Söllner, T., Kiebler, M. et al. (1990) Import of ADP/ATP carrier into mitochondria: two receptors act in parallel. *J Cell Biol* **111**: 2353–2363.

Truscott, K.N., Kovermann, P., Geissler, A. et al. (2001) A presequence- and voltage-sensitive channel of the mitochondrial preprotein translocase formed by Tim23. *Nat Struct Biol* **8**: 1074–1082.

van Wilpe, S., Ryan, M.T., Hill, K. et al. (1999) Tom22 is a multifunctional organizer of the mitochondrial preprotein translocase. *Nature* **401**: 485–489.

Voisine, C., Craig, E.A., Zufall, N. et al. (1999) The protein import motor of mitochondria: unfolding and trapping of preproteins are distinct and separable functions of matrix Hsp70. *Cell* **97**: 565–574.

von Heijne, G. (1986) Mitochondrial targeting sequences may form amphiphilic helices. *EMBO J* **5**: 1335–1342.

Voos, W., von Ahsen, O., Müller, H. et al. (1996) Differential requirement for the mitochondrial Hsp70–Tim44 complex in unfolding and translocation of preproteins. *EMBO J* **15**: 2668–2677.

Voos, W., Martin, H., Krimmer, T. and Pfanner, N. (1999) Mechanisms of protein translocation into mitochondria. *Biochim Biophys Acta* **1422**: 235–254.

Wiedemann, N., Pfanner, N. and Ryan, M.T. (2001) The three modules of ADP/ATP carrier cooperate in receptor recruitment and translocation into mitochondria. *EMBO J* **20**: 951–960.

Yano, M., Hoogenraad, N., Terada, K. and Mori, M. (2000) Identification and functional analysis of human Tom22 for protein import into mitochondria. *Mol Cell Biol* **20**: 7205–7213.

Zimmerman, R., Paluch, U., Sprinzl, M. and Neupert, W. (1979) Cell-free synthesis of the mitochondrial ADP/ATP carrier protein of *Neurospora crassa*. *Eur J Biochem* **99**: 247–252.

SUGGESTED READING

Bauer, M.F., Hofmann, S., Neupert, W. and Brunner, M. (2000) Protein translocation into mitochondria: the role of TIM complexes. *Trends Cell Biol* **10**: 25–31.

Gabriel, K., Buchanan, S.K. and Lithgow, T. (2001) The alpha and the beta: protein translocation across mitochondrial and plastid outer membranes. *Trends Biochem Sci* **26**: 36–40

Koehler, C.M., Merchant, S. and Schatz, G. (1999) How membrane proteins travel across the mitochondrial intermembrane space. *Trends Biochem Sci* **24**: 428–432.

Matouschek, A., Pfanner, N. and Voos, W. (2000) Protein unfolding by mitochondria: the Hsp70 import motor. *EMBO Reports* **1**: 404–410.

Pfanner, N. and Geissler, A. (2001) Versatility of the mitochondrial protein import machinery. *Nat Rev* **2**: 339–349.

Schatz, G. and Dobberstein, B. (1996) Common principles of protein translocation across membranes. *Science* **271**: 1519–1526.

Voos, W., Martin, H., Krimmer, T. and Pfanner, N. (1999) Mechanisms of protein translocation into mitochondria. *Biochim Biophys Acta* **1422**: 235–254.

11

THE IMPORT AND SORTING OF PROTEIN INTO CHLOROPLASTS

JÜRGEN SOLL, COLIN ROBINSON AND LISA HEINS

INTRODUCTION

The characteristic organelles of plant cells are plastids. Depending on the tissue and developmental stage, plastids have different functions such as the synthesis of carotenoids in chromoplasts, which mainly reside in petals and fruits. In roots amyloplasts contain large starch grains as a metabolic reservoir. In dark-grown tissues other than roots etioplasts are formed, whose characteristic feature is a crystal-like structure, the prolamellar body. Etioplasts convert to chloroplasts upon illumination, and photosynthesis becomes their basic function, namely assimilation of carbon dioxide and the generation of oxygen. All plastids evolve from undifferentiated, semi-autonomous proplastids. They contain their own genome but it comprises only a subset of the genetic information necessary for the biogenesis of diverse plastids. During development from a free-living cyanobacterial ancestor into highly specialized organelles, most of the endosymbiont's genes have been transferred to the host nucleus (Martin et al., 1998). Current estimations reveal that the chloroplast proteome of *Arabidopsis thaliana* comprises up to 2500 proteins, only 87 of which are encoded by the chloroplast genome (Abdallah et al., 2000). Therefore, the biogenesis of plastids requires the coordinated transport and assembly of nuclear and chloroplast encoded proteins. The difficult venture becomes even more challenging due to the complex structure of chloroplasts: six subcompartments can be distinguished. Despite considerable differences in structure, chloroplasts like all other plastids are bounded by an envelope consisting

Protein Targeting, Transport & Translocation
ISBN 0-12-200731-X

of the outer and the inner membrane and the intermembrane space. Photosynthesis is carried out at the thylakoid membranes, a dissipating internal membrane system, which separates the thylakoid lumen and the stroma.

CHECKPOINT OF CHLOROPLAST BIOGENESIS: PROTEIN TRANSLOCATION AT THE ENVELOPE MEMBRANES

Located at the interface between plastids and the surrounding cytosol, the envelope membranes hold a key position in the coordination of cell differentiation and plastid development as well as in the regulation of chloroplast biogenesis. A proteinaceous import apparatus, at both the outer (Toc complex) and inner (Tic complex) envelope membrane, has been developed to reimport from the cytosol proteins encoded by the transferred genes. Most proteins destined for the chloroplasts are made in the cytosol with a cleavable, N-terminal presequence. Together with cytosolic factors they target the preprotein first to the organellar surface and then into the stroma, apparently using a general import pathway for transport across the envelope membranes (Figure 11.1). The cleavage of the presequence in the stroma completes the translocation of preproteins, and further routing to their final location within the plastid is achieved by additional targeting signals.

Traffic control in the cytosol

The cytosol should not be imagined as an open space. Instead proteins destined for chloroplasts are like people trying to reach work during rush hour. Since the biogenesis of several organelles is dependent on nuclear-encoded proteins, specific routing information is required that helps chloroplast proteins to escape confusion. Meanwhile the proteins have to maintain a loose-folded state to ensure import-competent conditions. Based on experiments using radiolabeled preproteins for import into isolated chloroplasts the primary structure of the targeting signal has been investigated (Figure 11.2). Specific cytosolic factors and molecular chaperones associated with the presequences are identified by using immobilized polypeptides that are incubated with cytosolic extracts.

Multiple functions of the presequence
The principal information for specific and sufficient targeting of nuclear-encoded chloroplast proteins is localized in their amino acid sequence: most of the known chloroplast proteins contain cleavable, N-terminal

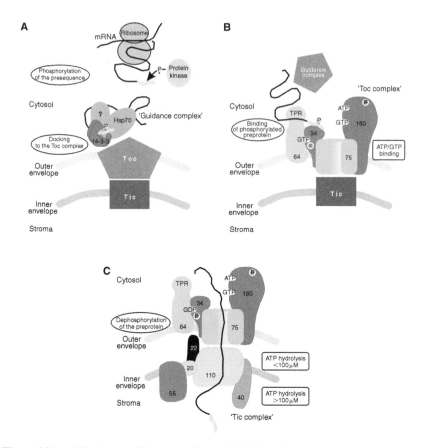

Figure 11.1 The 'general import pathway' of chloroplasts.

presequences. The extreme N-terminal part of presequences is rich in uncharged amino acids; the middle part contains predominantly positively charged and hydroxylated residues such as serine and threonine. Finally, the C-terminal part is enriched in arginines (for review, see Bruce, 2000).

Since presequences lack a strict consensus sequence, a specific modification might help to increase the fidelity of the chloroplast signal. *In vitro* serine and threonine function as a target of a cytosolic protein kinase, which exclusively phosphorylates chloroplast but not mitochondrial presequences (Waegemann and Soll, 1996). The phosphorylation stimulates not only the binding of the presequence to a cytosolic, hetero-oligomeric complex (May and Soll, 2000), but also the interaction with the isolated import receptor Toc34 (Sveshnikova et al., 2000). However, the phosphorylated preprotein is prevented from passing through the import apparatus (Waegemann and Soll, 1996) and consequently, a not yet established phosphatase activity,

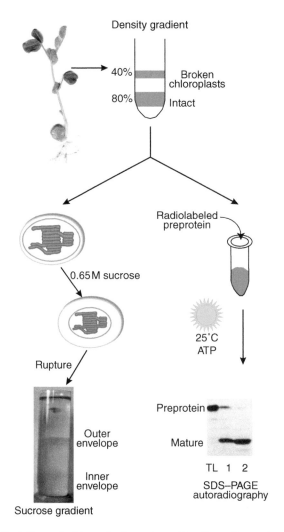

Figure 11.2 Isolation of chloroplasts and purification of membrane fractions. Chloroplasts are isolated from pea leaves of 10- to 12-day-old plants. After separation of cell debris by filtration and centrifugation intact chloroplasts are purified on silica sol gradients. A standard import assay contains chloroplasts equivalent to 20 µg chlorophyll and *in vitro* translated, radiolabeled precursor. The mixture is incubated with ATP in the light at 25°C for roughly 10 min. Then re-isolated chloroplasts are treated with protease to remove those preproteins that bound to the surface but did not enter. Once inside the chloroplasts, the preprotein is processed to its mature size and protected from protease treatment. Finally, chloroplasts are solubilized with detergent and the proteins are subjected to sodium dodecyl sulfate polyacrylamide gel electrophoresis (SDS–PAGE) and autoradiography (TL: translation product; lane 1: import; lane 2: protease treated import). Fractions of outer and inner envelope membranes are recovered on sucrose gradients after osmotic shrinking of intact chloroplast followed by rupturing with a dounce homogenizor.

presumably localized at the outer envelope membrane, becomes necessary. Systematic deletions of internal parts of the ferredoxin presequence provide evidence that at the chloroplast surface in addition to Toc34 another receptor, Toc160, contributes to the initial recognition of the N-terminal part of the presequence (Rensink et al., 2000). Accordingly, both Toc160 and Toc34 may cooperate in recognition of the presequence although a sequence of molecular steps has not yet been established. Another mechanism supporting the specific interaction of presequences with the chloroplast surface might be represented by a conformational change in a lipid environment. In an aqueous solution presequences are largely unstructured (Pilon et al., 1992), but placed in a membrane-mimicking solvent such as trifluorethanol the presequences of ferredoxin (Fd) and of the Rubisco activase adopt helix-coil structures (Lancelin et al., 1994; Krimm et al., 1999). In a lipid environment preFd can form a helix at the N- and C-terminus, respectively (Wienk et al., 1999). Studies using artificial membranes consisting of chloroplast-specific lipids have demonstrated that the N-terminus of preFd inserts into monogalactosyl-diacylglycerol containing surfaces while the C-terminus is involved in the insertion into a negatively charged lipid surface (Pilon et al., 1995; Pinnaduwage and Bruce, 1996). Furthermore, chloroplasts isolated from an *Arabidopsis* mutant deficient in digalactosyl-diacylglycerol lose the ability to import stroma-targeting preproteins (Chen and Li, 1998). Taken together these results support the idea that the interaction of a preprotein with the chloroplast surface may be facilitated by chloroplast-specific galactolipids, the negatively charged sulfoquinovosyl-diglycerol and the anionic phospholipid phosphatidylglycerol.

The preprotein and molecular chaperones
Molecular chaperones such as members of the Hsp70 family have been shown to maintain preproteins in an unfolded, import-competent conformation after synthesis at cytosolic ribosomes. Using preLHCP (light harvesting chlorophyll *a/b*-binding protein) purified after expression in *Escherichia coli* its efficient import into chloroplasts is dependent on the presence of soluble proteins from leaf extract. The purified, cytosolic Hsp70 could partially substitute for the leaf extract, thus leading to the conclusion that at least one or more unknown components are required to ensure efficient translocation of preLHCP (Waegemann et al., 1990). Although Hsp70 stimulates the import of the thylakoid membrane-spanning LHCP, the import of other preproteins that are soluble stromal proteins such as Fd and the small subunit of Rubisco (SSU) is not stimulated by Hsp70.

Interestingly, several findings suggest that a strong unfolding activity at the chloroplast surface exists (for review, see Heins et al., 1998). The preSSU becomes more and more protease sensitive from binding to the organellar surface until it reaches a final protease-resistant location inside the chloroplasts (Waegemann and Soll, 1991). Further evidence for the unfolded state of the preprotein is derived from the physical properties of the purified translocation pore at the outer envelope membrane of chloroplasts, Toc75. Its small diameter probably prevents the passage of folded preproteins (Hinnah et al., 1997). However, a preprotein comprising preSSU and the bovine-pancreatic-trypsin inhibitor, which remains tightly folded during translocation, is translocated at the envelope membranes (Clark and Theg, 1997). Because the bovine-pancreatic-trypsin inhibitor domain extends to about 18 Å in diameter and 30 Å in length, it may be concluded that *in situ* the translocation pore could acquire different conformations to allow the passage of globular proteins. In addition to maintaining an unfolded protein structure, Hsp70 seems to interact specifically with the presequence of chloroplast preproteins (Ivey III et al., 2000). Statistical analysis of the CHLPEP database (von Heijne et al., 1991) reveals that more than 75% of the presequences are predicted to contain an Hsp70 binding site at the N-terminus and C-terminus. This observation leads to the speculative suggestion that Hsp70 localized at the cytosolic leaflet of the outer envelope membrane and in the intermembrane space (Marshall et al., 1990; Waegemann and Soll, 1991; Kourtz and Ko, 1997) could function as a chloroplast translocation molecular motor pulling the presequence like Hsp70 in the matrix of mitochondria (Ivey III et al., 2000). Concerning binding of Hsp70 to presequences, chloroplast and mitochondrial import of preproteins seems similar, although a function of Hsp70 as a translocation motor has not yet been established in chloroplasts.

Guided tour to the chloroplast surface
Nevertheless, the interaction of a preprotein with molecular chaperones does probably not describe an organelle-specific mechanism. Besides the putative Hsp70 binding sites the presequence contains a preprotein kinase phosphorylation motif that shows strong similarity to a motif specifically recognized by a member of the 14-3-3 class of molecular chaperones. Translated in wheat germ only, phosphorylated chloroplast preproteins associate with a high molecular 'guidance complex' containing 14-3-3 proteins, Hsp70 and maybe further components (Figure 11.1A), whereas after translation of a chloroplast preprotein in animal reticulocyte lysate neither phosphorylation nor binding of a 14-3-3 protein is observed (Waegemann and Soll, 1996; May and Soll, 2000). In contrast, a mitochondrial presequence, which is not capable of being phosphorylated (Waegemann and Soll, 1996), does not co-purify with the plant guidance complex of wheat germ (May and Soll, 2000). The removal of the preprotein kinase phosphorylation motif/14-3-3-binding motif prevents the formation of the guidance complex, but import of the preprotein into isolated chloroplasts is still possible (May and Soll, 2000; Waegemann and Soll, 1996). Certainly, the presence of the guidance complex stimulates the import of chloroplast preproteins up to ten-fold (May and Soll, 2000).

These complex findings raise a number of questions concerning the physiological significance of this chloroplast-specific phenomenon: Does this mechanism maintain a more import-competent state of chloroplast preproteins ensuring efficient targeting of preproteins *in vivo*? Or does this mechanism mediate between nuclear information and chloroplast biogenesis?

Preprotein translocation at the chloroplast envelope membranes

Hetero-oligomeric protein complexes at both the outer and the inner envelope membrane facilitate the joint translocation of preproteins containing a cleavable presequence. The composition and functions of the translocon at the outer envelope membrane and at the inner envelope membrane of chloroplasts were approached by different experimental techniques. Using preproteins linked to a radiolabeled crosslinking reagent and isolated chloroplasts for an import experiment the identification of putative components of the Toc or Tic complex next to the preprotein was achieved. After the preprotein has associated, the transfer of the radioactive label onto a target protein is induced allowing its identification by amino acid sequencing. A second approach focuses on the purification of an intact Toc or Tic complex after solubilization of either isolated envelope membranes or chloroplasts (Figure 11.2) without or together with preproteins which then can be arrested at the envelope membranes. Next, the protein complexes are isolated by native electrophoresis, sucrose density gradients and size exclusion chromatography, or by affinity chromatography using a tagged preprotein.

However useful these experiments may be, their limiting factors should be taken into consideration. Usually, crosslinking reagents consist of a chemically reactive group at either end linked by a spacer that is variable in length. Because of this spacer, proteins that are in close proximity may be crosslinked although a functional relation does not exist. Likewise, the purification of the import apparatus is probably hampered by unspecific interactions, because *in vitro* the composition of Toc and Tic complex varies depending on the experimental conditions such as the detergent used or ionic strength of the isolation buffer. In addition, the growth conditions of the plants are not the same in all laboratories contributing to our knowledge of protein translocation across the chloroplast envelope membranes. As a result, the following model may reflect different stages of plant development and does not claim to be a final one.

Energetic way of looking at protein translocation at the envelope

The initial interaction of the preprotein with the chloroplast surface is argued to be first mediated by a specific preprotein–lipid interaction and then by a preprotein–receptor interaction. While the first step of recognition is independent on energy and reversible, the hydrolysis of low concentrations of

ATP ($10–100\,\mu M$) results in tight binding of the preprotein to the import apparatus, but the preprotein remains still protease accessible (Waegemann and Soll, 1991). Thus the preprotein does not enter the inside of the chloroplast, but is already physically linked to components of the Toc and the Tic complex (Caliebe et al., 1997; Nielsen et al., 1997). This result indicates that the preprotein engages both the Toc and the Tic complex simultaneously, though the presence of a preprotein is not a prerequisite for the association of distinct components of the Toc and Tic complex (Nielsen et al., 1997). So far, no evidence exists that any of the Toc or Tic components drives the preprotein translocation by ATP hydrolysis. Instead Hsp70 at the cytosolic leaflet and in the intermembrane space has been suggested to have the ATP-hydrolyzing activity observed (Waegemann and Soll, 1991; Kourtz and Ko, 1997). Finally, ATP hydrolysis ($>0.1\,mM$) in the stroma is required to accomplish preprotein translocation. In contrast to mitochondria a membrane potential is not involved (Theg et al., 1989). A member of another class of molecular chaperones, Hsp100, which associates stably with the Tic complex, is thought to assist in pulling the preproteins across the inner envelope membrane (Nielsen et al., 1997).

As a unique feature chloroplast preprotein translocation requires GTP (Young et al., 1999). Toc160 and Toc34 have GTP-binding sites (Hirsch et al., 1994; Kessler et al., 1994) and the GTP-bound form of Toc34 has been demonstrated to stimulate the interaction with the preprotein. Furthermore, GTP- or ATP-dependent protein kinase activity occurs phosphorylating Toc34 (Sveshnikova et al., 2000). In particular, since the preprotein and Toc34 interact dependent on a GTP-, GDP-bound form of Toc34 (Sveshnikova et al., 2000), a role in regulation of preprotein–receptor interaction has been attributed to GTP.

The Toc complex
In the past, components of the chloroplast protein import apparatus have been referred to as OEP (*o*uter *e*nvelope *p*rotein) or IAP (*i*mport *a*ssociated *p*rotein). Now, the acronym Toc stands for *t*ranslocon at the *o*uter membrane of *c*hloroplasts, and the numbers designate the calculated molecular mass of the given protein in kilodaltons (Schnell et al., 1997). The Toc complex comprises four components, Toc34, Toc64, Toc75 and Toc160 formerly known as Toc86, that together constitute a prominent portion of the outer envelope proteins (for review, see Vothknecht and Soll, 2000) (Figure 11.1B).

Toc64 is the most recently identified component of the Toc complex (Sohrt and Soll, 2000), but it might be first at the interaction of the Toc complex with the cytosolic guidance complex containing a preprotein. Preliminary results support the notion that the C-terminal portion of Toc64 might be involved in docking the guidance complex, because the C-terminus contains three repeats of a tetratricopeptide-repeat motif, which protrudes out of the outer envelope membrane into the cytosol. Further examples of the participation of the tetratricopeptide-repeat motif in protein–protein interaction involved especially in protein translocation are the mitochondrial import receptor Tom70 and the peroxisomal protein Pex5 (for review, see Vothknecht and Soll, 2000).

Next, the preprotein is probably presented to the receptors at the chloroplast surface, Toc34 and Toc160, which contain structurally related GTP-binding domains at the N-terminus exposed to the cytosol. Both proteins are highly susceptible to proteolytic activity (Hirsch et al., 1994; Kessler et al., 1994; Seedorf et al., 1995). Toc160 was originally identified as Toc86, representing a C-terminal, proteolytic fragment of 86 kDa that is likely to occur during the standard chloroplast isolation procedure. Although the presence of intact Toc160 significantly stimulates the translocation of radiolabeled preprotein into chloroplasts, its 86 kDa fragment seems to facilitate preprotein translocation *in vitro* as well (Bölter et al., 1998a). The notion that the 86 kDa polypeptide could function as a preprotein receptor derives from several observations. As shown by label-transfer crosslinking experiments, preSSU interacts with the 86 kDa polypeptide in an initial, energy-independent step (Perry and Keegstra, 1994). Furthermore, Fab fragments of antibodies against the 86 kDa polypeptide applied to isolated chloroplast prior to import prevent binding of a preprotein to the chloroplast surface (Hirsch et al., 1994). *In vivo*, an *Arabidopsis* mutant deficient in Toc160 is unable to grow on soil and is highly affected especially in import of proteins involved in photosynthesis (Bauer et al., 2000).

Now the precise role of the acidic, N-terminal portion of Toc160 has to be reinvestigated, especially because two further proteins, designated Toc132 and Toc120, have been identified that show significant homology to the C-terminal portion of Toc160 (Bauer et al., 2000). The reasons for the existence of the 120 and the 132 kDa protein remains puzzling. Although the molecular mechanism is not yet clear, both Toc160 and Toc34 mediate the initial binding of the preprotein to the Toc complex and the proof-reading of this event. Isolated Toc34 without the transmembrane domain was used to investigate the molecular interaction with a preprotein (Sveshnikova et al., 2000). As shown by co-immunoprecipitation with antibodies against Toc34 under these conditions the phosphorylated preprotein binds to isolated Toc34 with a five- to ten-fold higher affinity than the non-phosphorylated. Using outer envelope fractions Toc34 is specifically phosphorylated by an ATP- or GTP-dependent protein kinase and as a result GTP-binding of Toc34 is prevented. Then, the interaction of the GDP-bound form of Toc34 with phosphorylated preprotein significantly decreases, indicating that the phosphorylated preprotein may be released by a GTP-GDP exchange at Toc34. In *Arabidopsis* a second GTP-binding protein, Toc33, with high sequence similarity to Toc34 has been identified (Jarvis et al., 1998). The disruption of this gene is not lethal, but protein import is strongly affected, especially in early developmental stages of the plant, a time when Toc33 is highly expressed and Toc34 is present at lower levels.

The existence of Toc160 and two similar proteins as well as that of Toc34 and Toc33 raises the question whether chloroplast biogenesis is maintained by different protein import apparatus that adapt to the developmental or physiological needs of the plant. Multiple forms of Toc34 may form the initial stage for regulation of protein import particularly due to its close proximity to Toc75, as demonstrated by the formation of a covalent disulfide bridge *in vitro* (Seedorf et al., 1995). Several lines of evidence corroborate the idea that Toc75 forms the central preprotein translocation pore. Proteolytic treatment of either chloroplasts or outer membrane vesicles does not cause degradation of Toc75, indicating that the protein is deeply embedded in the membrane (Tranel et al., 1995). In the presence of low ATP concentrations (<0.1 mM), at a later stage of import, Toc75 and a radiolabeled preprotein form the major crosslinked product (Perry and Keegstra, 1994). Electrophysiological experiments using purified Toc75 expressed in *Escherichia coli* provided the best evidence supporting its function as a translocation pore (Hinnah et al., 1997). Reconstitution in liposomes results in the formation of a water-filled channel, adopting a β-barrel conformation similar to that of bacterial pore proteins. Furthermore, Toc75 mainly recognizes the preprotein but not the mature form; thus Toc75 may contain its own preprotein binding site (Perry and Keegstra, 1994; Hinnah et al., 1997).

Even if Toc75 shows structural similarity to transport components of the mitochondria or endoplasmic reticulum, like most of the chloroplast, membranous translocation components it does not share significant amino acid sequence similarity. However, in the complete genome sequence of the cyanobacterium *Synechocystis* PCC6803 (Kaneko et al., 1996) is an open reading frame (slr1227) encoding an amino acid sequence that shows about 55% similarity to Toc75 of plants (Heins et al., 1998). Likewise the plant protein synToc75 reconstitutes to liposomes as a pore, which consists of β-sheets and reacts specifically with a chloroplast presequence (Bölter et al., 1998b). Interestingly, both proteins are localized at the outer membrane, and its disruption in *Synechocystis* appears to be lethal (Reumann et al., 1999). Because chloroplasts have originated from a free-living cyanobacterial ancestor, the obvious conclusion is that syn75 may be part of a yet unknown transport system. The chloroplast import apparatus developed later, fitting together components such as Toc160 and Toc34 newly evolved by the eukaryotic host and cyanobacterial components.

The Tic complex
The Toc and Tic components jointly facilitate preprotein translocation (Lübeck et al., 1996; Nielsen et al., 1997). While considerable steps towards understanding the function and interaction of the Toc components have been made, the number, the arrangement and the function of the Tic

components is less clear. So far, the Tic complex probably comprises Tic110, Tic55, Tic40, Tic22 and Tic20 (for review, see Keegstra and Cline, 1999; Vothknecht and Soll, 2000) (Figure 11.1C).

Due to its physical proximity to components of the Toc complex as demonstrated by co-immunoprecipitation, Tic110 was the first component of the Tic complex to be identified (Kessler and Blobel, 1996; Lübeck et al., 1996). Different hypothetical functions of Tic110 are suggested with the support of experiments mapping the topology by selective protease treatment of either isolated inner envelope vesicles or mainly intact chloroplasts (Lübeck et al., 1996; Jackson et al., 1998). The protease trypsin is able to penetrate the outer envelope membrane, and then only proteins of the inner envelope membrane protruding into the intermembrane space are accessible to the protease; meanwhile the inner envelope membrane remains intact. Finally, membrane fractions are obtained after osmotic lysis in the presence of protease inhibitors. This experiment has been carried out in different laboratories with contradictory results: (i) Since the large C-terminal portion is protease-sensitive, the main portion of Tic110 protruding into the intermembrane space may be involved in the interaction of the Toc and Tic complex during translocation of the preprotein. This idea sustained support by the co-immunoprecipitation of Toc75 and Tic110 with an antibody against the preprotein trapped to the import apparatus by crosslinking (Lübeck et al., 1996). (ii) Under only slightly altered conditions chosen by Jackson and co-workers (1998) Tic110 is highly resistant to protease treatment. Therefore, a function of Tic110 mainly in recruiting molecular chaperones such as cpn60 or Hsp100 in the stroma is favored. Crosslinking experiments revealed earlier that these chaperones interact with Tic110 at a later stage of preprotein translocation (Kessler and Blobel, 1996; Nielsen et al., 1997).

Using blue-native gel electrophoresis Tic110 co-migrates with Tic55, an unusual component that contains an iron-sulfur center (Caliebe et al., 1997). In addition, antibodies against Tic55 co-immunoprecipitate Tic110. Further support that Tic55 is indeed a functional component of the Tic complex is provided by the finding that Tic55 co-purifies together with an affinity-tagged preprotein and Toc and Tic components after locking the preprotein to the translocon by energy limitation. The idea that Tic55 may act as a redox-sensor arises from the susceptibility of preprotein translocation to diethylpyrocarbonate, which has been demonstrated to modify the iron-sulfur center of Rieske-type proteins. As a result the preprotein is halted at the inner envelope membrane.

Furthermore, Tic110 is co-immunoprecipitated by antibodies against Tic40, a component showing similarity to molecular chaperones at the C-terminal portion (Stahl et al., 1999). There has been confusion concerning the molecular structure and localization of this component. Since in intact chloroplasts a crosslink product of a 44 kDa protein, localized at both envelope membranes, and a preprotein was obtained, it was named first

Cim/Com44 (Ko et al. 1995). Next, it was renamed Toc36, although supporting experimental results have not been provided (Schnell et al., 1997). Finally, a pea clone has been identified, which encodes a significant N-terminal extension (Stahl et al., 1999) compared to the initial clone of *Brassica napus* (Ko et al., 1995). The N-terminus of the pea protein was determined by amino acid sequencing, and its localization at the inner envelope membrane was accessed by import of its radiolabeled preprotein, leading to the present name Tic40 (Stahl et al., 1999). Complementation of a SecA deficient *E. coli* mutant, which is defective in protein secretion (Pang et al., 1997), and the amino acid sequence provide a clue that Tic40 may act as membrane-localized molecular chaperone. However, the real function of Tic40 during preprotein remains elusive.

Two small components of the Tic complex, Tic22 and Tic20, have been identified by label transfer crosslink experiments, indicating that both proteins are in close proximity to the preprotein during translocation. Both preproteins are unusual in that their presequences contain a considerable amount of negatively charged amino acids (Kouranov et al., 1998). Tic20 is predicted to be an integral membrane protein and, therefore, the authors speculate that Tic20 may represent the so far unknown translocation pore, while in the intermembrane space Tic22 takes over the preprotein from the Toc complex. However, these assumptions lack experimental evidence. In conclusion research is quite far away from understanding the molecular function of preprotein translocation at the Tic complex.

Protein insertion into the outer and inner envelope membrane

Since the general import pathway mainly imports proteins with an N-terminal presequence, a different pathway has to exist for chloroplast destined protein without a cleavable presequence, namely most outer envelope proteins. Outer envelope proteins identified so far comprise proteins (i) containing a single, α-helical membrane anchor, (ii) spanning the membrane with only β-sheets, (iii) spanning the membrane with both, α-helical and β-sheets (for a review, see Soll and Tien, 1998).

The finding that in contrast to the general import pathway, the insertion of almost all outer envelope proteins is independent of ATP or protease-sensitive membrane components leads to the assumption that insertion of these proteins occurs spontaneously or that unknown protease-resistant components are involved. Nevertheless, the molecular mechanism forming the basis of the insertion remains puzzling: (i) which factors determine the orientation of the N- and C-terminus, (ii) how are the proteins assembled prior to or during their insertion into the membrane? The best-studied mechanism is the insertion of Toc34. Probably due to the content of hydrophobic amino acids the C-terminal portion of Toc34 functions as an α-helical membrane anchor (Seedorf et al., 1995). However, the distribution of charged amino acids adjacent to the membrane anchor influences the orientation of Toc34 as shown by *in vitro* import of mutants into chloroplasts and mapping the resulting topology by protease treatment (May and Soll, 1998). Furthermore, *in vitro* GTP stimulates the efficiency of Toc34 insertion into the membrane (Chen and Schnell, 1997; Tsai et al., 1999). This stimulating effect may be due to the nucleotide-binding induced conformational change of Toc34 as a prerequisite of protein assembly (Chen and Schnell, 1997), or other unknown, GTP-dependent proteins may be involved in insertion (Tsai et al., 1999).

While these results open a door to the molecular insertion-mechanism of outer envelope proteins spanning the membrane once, the means of insertion of proteins with multiple membrane spans is less clear (for review, see Soll and Tien, 1998). Toc75 is an exception since it contains an unusual N-terminal bipartite presequence whose N-terminal portion employs the general import pathway. After cleavage of the N-terminal part in the stroma, the C-terminal portion directs Toc75 back to the outer envelope membrane by an unknown mechanism (Tranel and Keegstra, 1996). Proteins of the inner envelope membrane contain two different classes of targeting information. An N-terminal cleavable presequence achieves the targeting to the organellar surface and entrance into the stroma, while the membrane targeting information is contained within the mature protein. Evidence exists that this signal is localized at hydrophobic, probably membrane-spanning domains, but its character has not yet been established (Brink et al., 1995; Lübeck et al., 1997). Two pathways are conceivable. The hydrophobic signal may function as a stop-transfer signal that induces the release of the pre-protein from the import apparatus directly to membrane insertion. On the other hand, complete translocation of the preprotein to the stroma may be a prerequisite for insertion to the inner envelope membrane. Using a tight timetable during import of radiolabeled preprotein into isolated chloroplasts, a fusion protein consisting of the N-terminal portion of Tic110 and the mature SSU was mainly found in the stroma at early time points, supporting the latter hypothesis (Lübeck et al., 1997). So far, stromal factors or a protein complex at the inner envelope membrane that assist insertion have not been described.

INTRAORGANELLAR TARGETING OF IMPORTED THYLAKOID PROTEINS

A high proportion of imported chloroplast proteins undergo further targeting steps to reach the thylakoid network, which carries out the critical functions of light capture, photosynthetic electron transport and photophosphorylation. These processes are mediated by highly abundant membrane-bound complexes, notably photosystems I and II (PSI, PSII), the cytochrome b_6–f complex and the ATP synthase. However, there is now clear evidence that the soluble phase enclosed by the thylakoid membrane (the thylakoid lumen) also houses a significant number of proteins, certainly in excess of 30 and possibly over 100, although not all are directly engaged in photosynthesis (Kieselbach et al., 1998). The targeting of these proteins has attracted a considerable amount of interest and a combination of *in vitro* reconstitution experiments and genetic approaches have given us a good idea of at least some of the important pathways.

All of the available evidence indicates that a common basic pathway operates for the initial targeting of thylakoid proteins across the envelope membranes (see above) and they appear so far to resemble imported stromal proteins in key respects (Cline et al., 1993). Accordingly, all known cytosolically synthesized thylakoid proteins are synthesized with amino-terminal 'envelope transit' signals that mediate binding to the Toc machinery and subsequent translocation into the stroma (reviewed in Dalbey and Robinson, 1999). However, at least four distinct pathways have been identified for the subsequent targeting of these proteins into and across the thylakoid membrane, and these targeting pathways and translocation mechanisms will be discussed in the remainder of this chapter.

TWO DISTINCT PATHWAYS FOR THE TARGETING OF THYLAKOID LUMEN PROTEINS

The biogenesis of lumenal proteins has been of particularly intense interest because these proteins must cross all three chloroplast membranes to reach their site of function – an unusually complex targeting pathway. Initial studies into this area in the late 1980s suggested a common basic import pathway, but the story has become far more complex and interesting in the last few years.

All known imported lumenal proteins are synthesized with bipartite presequences which have been shown in several cases to comprise two targeting signals in tandem. The amino-terminal section is a typical 'envelope transit' signal which specifies translocation into the stroma, and several reports have confirmed that these signals are structurally and functionally equivalent to the presequences of imported stromal proteins (e.g. Hageman et al., 1990). The second signals, however, are very different to the basic, hydrophilic signals that specify chloroplast import. Instead, they can all be divided into three distinct domains: a charged amino-terminal (N-) domain, hydrophobic (H-) core domain and more polar carboxy-terminal (C-) domain ending with an Ala-X-Ala consensus motif (reviewed in Dalbey and Robinson, 1999; Robinson and Bohuis, 2001). These signals, which closely resemble bacterial 'signal' peptides (see below) are exposed once the envelope transit signals are removed in the stroma, and several early studies showed that they contain essential information specifying translocation across the thylakoid membrane (e.g. Hageman et al., 1990).

Early studies in this area showed that, consistent with the presence of dual targeting signals, the targeting of thylakoid proteins can be divided into two basic steps. The precursor proteins are first transported into the stroma, where the amino-terminal signal is usually removed (although not always, as shown for PsaN; Nielsen et al., 1994). This generates a transient stromal intermediate form that can often be visualized using certain inhibitors (see below). Unexpectedly, however, lumenal proteins are targeted by two completely different pathways despite possessing apparently similar targeting signals that all resemble bacterial signal peptides. A subset of proteins are transported by a Sec-related system which clearly resembles prokaryotic Sec pathways. These pathways act in bacteria to export newly synthesized proteins into the periplasm (reviewed in Dalbey and Robinson, 1999), and substrates for this pathway bear amino-terminal signal peptides with the 3-domain structure described above. The export process involves the operation of soluble cytoplasmic chaperone proteins such as SecB, which prevent the substrates from folding prior to export, the peripheral SecA ATPase, and a membrane-bound SecYEG translocon through which SecA threads the substrate using the energy from ATP hydrolysis. A number of lumenal proteins are transported across the chloroplast thylakoid membrane by a similar mechanism involving stromal SecA together with thylakoid-localized SecY and SecE homologs (Yuan et al., 1994; Laidler et al., 1995; Schuenemann et al., 1999). There appears, however, to be no SecB homolog in chloroplasts and this is perhaps unsurprising because *secB* genes appear to be absent in the cyanobacteria studied to date.

Properties of the Sec-independent pathway for lumenal proteins

The existence of a chloroplastic Sec pathway makes evolutionary sense because this organelle is widely accepted to have evolved from a cyanobacterial-type prokaryote, and Sec components are present in the thylakoids as well as the plasma membrane in cyanobacteria (interestingly, sequencing of the *Synechocystis* genome reveals only a single set of *sec* genes, raising the question of how lumenal and periplasmic proteins are correctly sorted). However, the last few years have witnessed the characterization of a second targeting pathway, initially in chloroplasts but latterly in many bacteria.

In vitro studies in chloroplasts showed quite clearly that a large number of lumenal proteins were transported by a mechanism that did not require any stromal proteins or nucleoside triphosphates, but which depended entirely on the thylakoidal proton gradient (Mould and Robinson, 1991; Cline et al.,

1992). Most importantly, the choice of pathway is largely, if not completely, dictated by the type of presequence present, and Sec-dependent mature proteins can be quantitatively re-routed onto the ΔpH-dependent pathway by attaching an appropriate targeting signal (Robinson et al., 1994; Henry et al., 1994). This finding was initially surprising because all known lumenal-targeting signals are basically similar to well-defined bacterial signal peptides, but later studies showed subtle differences that are critical in determining pathway specificity. First, signals for the ΔpH-dependent system bear an invariant and essential twin-arginine motif just prior to the H-domain (Chaddock et al., 1995) and secondly, targeting by this pathway requires the presence of a highly hydrophobic residue two or three positions into the H-domain (Brink et al., 1998).

The first component of the ΔpH-dependent system was identified following the isolation of a maize mutant, termed *hcf106* (for high chlorophyll fluorescence; Voelker and Barkan, 1995). Further studies and sequencing of the gene (Settles et al., 1997) showed that it encoded a small, single-span thylakoid membrane protein unrelated to any known protein. However, homologous genes are present in the majority of sequenced bacterial genomes, at that time designated as unassigned reading frames. The *Escherichia coli* genome contains three genes homologous to *hcf106*; these genes have been termed *tat* genes (for twin-arginine translocase) and the organization of the *E. coli tat* genes is shown in Figure 11.3. Disruption of the *hcf106* homologs, *tatA*, *tatE* and *tatB*, has shown that these genes do indeed encode components of a Sec-independent export system, although interestingly, these Hcf106 homologs play two distinct roles in the export process. TatA and TatE play similar roles and Tat-dependent export can occur using either of these subunits. However, TatA appears to be far more abundant and its disruption has accordingly more severe consequences (Sargent et al., 1998, 1999). TatB is also essential for Tat-dependent export (Weiner et al., 1998) and the next gene in the operon, *tatC*, likewise encodes an essential element of the system (Bogsch et al., 1998). The fourth gene, *tatD*, does not appear to encode an important element of the Tat system and its position in the operon is something of a mystery (Wexler et al., 2000).

All of the available data suggest that a basically similar Tat system operates in thylakoids; Hcf106 is believed to be homologous to TatB, an apparent TatA homolog has been cloned (Walker et al., 1999) and a cDNA clone encoding a chloroplastic TatC homolog has also been cloned. Antibodies to thylakoidal Tat components effectively inhibit Tat-dependent import into thylakoids, indicating their involvement in this pathway (Mori et al., 1999). A TatABC complex has recently been purified from *E. coli* (Bolhuis et al., 2000, 2001) and at present there is no evidence for the existence of further subunits, possibly indicating that the Tat system complex is a remarkably simple system in terms of subunit composition. However, the *E. coli*

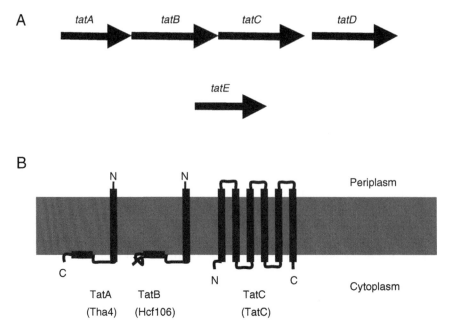

Figure 11.3 Genes and polypeptides involved in the Tat pathway. A, gene organization in *Escherichia coli*. The known genes are mostly organized in a *tatABCD* operon as shown. The *tatABC* genes encode important (TatA) or essential (TatB and TatC) membrane-bound components whereas the *tatD* gene encodes a cytoplasmic protein whose activity is not important for Tat-dependent export. The monocistronic *tatE* gene encodes a TatA homolog but this gene is expressed at low levels and the function of the encoded protein is largely dispensable. B, Predicted protein structure. The predicted basic structures of the *E. coli* Tat proteins are illustrated and the names of the corresponding plant thylakoid homologs are also shown (Tha4, Hcf106, TatC). Note that Tha4 and Hcf106 have yet to be experimentally confirmed as TatA/TatB homologs. TatA and B are predicted to contain a short periplasmic amino-terminal domain and a single membrane-spanning region followed by a short amphipathic region and cytoplasmic domain. TatC is predicted to contain six transmembrane helices with short interconnecting loop regions.

complex has not yet been shown to be active, and thus the possibility of further subunits can not yet be excluded.

Summary

- Lumenal proteins are transported across the thylakoid membrane by two completely distinct systems.
- The choice of pathway is dictated by the type of presequence present.
- The thylakoid membrane contains both a Sec-type system and a novel Tat system that is also present in many bacteria.

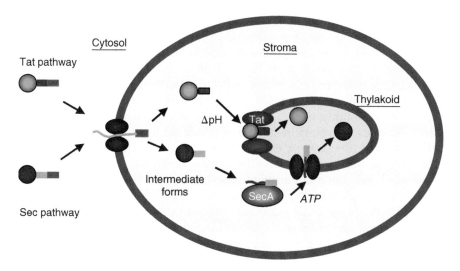

Figure 11.4 Pathways for the targeting of thylakoid lumen proteins in chloroplasts. The diagram illustrates the differing characteristics of the Tat- and Sec-dependent pathways. Proteins targeted by both pathways are synthesized with bipartite presequences containing envelope transit signals (red boxes) and thylakoid-targeting signals in tandem. The thylakoid-targeting signals of 23K and plastocyanin (ΔpH- and Sec-dependent substrates, respectively), are depicted by blue or green boxes. The envelope transit peptides of both precursors are recognized by a protein transport system in the envelope membranes which facilitates translocation into the stroma. The envelope transit signals are usually removed in the stroma and the intermediate forms are then targeted by separate pathways. Tat-dependent proteins (such as 23K) are thought to refold in the stroma before being transported in a folded form by a ΔpH-driven Tat system; other lumenal proteins, such as 33 K and plastocyanin, are transported by the Sec route involving stromal SecA, ATP hydrolysis and a membrane-bound Sec translocon. After translocation, substrates on both pathways are processed to the mature forms by the thylakoidal processing peptidase.

The Tat complex transports folded proteins

The properties, abilities and mechanism of the Tat system are all very different to those of Sec systems. Whereas Sec systems 'thread' their substrates through the SecYEG complex in a largely unfolded state (reviewed in Chapter 4) there is now overwhelming evidence that the Tat complex is able to transport large globular proteins in a fully folded state. This was shown directly for the thylakoidal system in two different studies. In one (Clark and Theg, 1997) it was shown that bovine pancreatic trypsin inhibitor could be transported by the Tat pathway using a Tat-type targeting signal, even when the protein was internally crosslinked to prevent unfolding. In the second study, dihydrofolate reductase was found to be efficiently transported across the thylakoid membrane by the Tat system after binding a folate analog in the active site (Hynds et al., 1998). Figure 11.4 illustrates and contrasts the basic features of the Tat- and Sec-dependent targeting pathways in chloroplasts.

Studies on bacterial Tat systems provide further compelling evidence for the transport of folded proteins. In *E. coli*, the primary substrates for the Tat export pathway are periplasmic proteins that bind one of a range of redox cofactors including FeS and molybdopterin centers (Berks, 1996; Santini et al., 1998). Notably, these cofactors are only inserted in the cytoplasm (using complex enzymatic machinery), which means that the proteins are obliged to fold to a large extent in this compartment. These proteins must, therefore, be transported across the plasma membrane in a folded form, providing good, albeit circumstantial, evidence that bacterial Tat systems are able to transport large globular proteins in a folded form.

Not all Tat substrates bind cofactors and this certainly applies to the thylakoid substrates, very few of which bind any form of cofactor. Hence, it is probably the case that the second role of the Tat system is simply to transport those substrates that fold too rapidly and/or tightly for the Sec system to handle efficiently.

Summary

- The Tat system has the unique ability to transport folded proteins across the thylakoid membrane or the plasma membrane in bacteria.
- It operates to transport proteins that either must fold prior to translocation (e.g. to bind cofactors) or which apparently fold too rapidly or tightly for the Sec system to handle.

THE SRP-DEPENDENT PATHWAY FOR THE TARGETING OF THYLAKOID MEMBRANE PROTEINS

The targeting of thylakoid membrane proteins has also been studied with some enthusiasm and two further pathways have been identified for these hydrophobic proteins following their import into the plastid (reviewed in Dalbey and Robinson, 1999). The first of these requires signal recognition particle (SRP) which, in bacteria, has been shown to be required for the insertion of a range of plasma membrane proteins (see Chapter 4). SRP in *E. coli* is a complex of a 48 kDa subunit and 4.5S RNA molecule which binds preferentially to highly hydrophobic regions such as those found in newly synthesized membrane proteins. In at least some cases, this factor helps to insert membrane proteins together with another factor, FtsY, and the membrane-bound SecYEG complex (see Chapter 4 for detailed review). Both SRP and FtsY hydrolyze GTP during their active cycle.

The chloroplastic SRP is somewhat different in structural terms, in that it lacks RNA and is instead composed of a 54 kDa subunit (homologous to the *E. coli* 48 kDa subunit) and a novel 43 kDa subunit (Li et al., 1995; Schuenemann et al., 1998; DeLille et al., 2000). This SRP is required for the insertion of the major light-harvesting chlorophyll-binding protein, Lhcb1, and other studies have shown the involvement of a chloroplastic FtsY homolog in this process (Tu et al., 1999; Kogata et al., 1999).

Oxa1-related proteins have been recently shown to play a central role in both the thylakoidal and bacterial SRP-dependent targeting pathways. Oxa1p is a mitochondrial inner membrane protein that is important for the insertion of a range of inner membrane proteins, particularly those that insert from the matrix side (Herrmann et al., 1997; Hell et al., 2001). Thylakoids contain a homologous protein, Albino3 (Alb3), that is essential for the insertion of the SRP-dependent Lhcb1 protein (Moore et al., 2000) and the related YidC protein has recently been shown to be essential for the efficient insertion of a range of inner membrane proteins in *E. coli* (Samuelson et al., 2000). Interestingly, the YidC protein was also shown to be required for the insertion of several SRP-*independent* proteins, such as M13 procoat, demonstrating that it plays a wide-ranging role in membrane protein biogenesis. This finding suggests that YidC may exist as two distinct pools, one associated with the Sec apparatus and involved in the more complex insertion pathways and a separate pool in which YidC acts as a translocon in its own right. Figure 11.5 depicts a model for the insertion of proteins by the SRP-dependent pathway but it should be emphasized that the precise events taking place at the thylakoid membrane are still unclear. As yet, there is no evidence that the thylakoidal SecYEG system is involved in the SRP pathway, and antibodies to SecY were indeed found to block the import of lumenal Sec substrates but not the SRP-dependent Lhcb1 protein (Mori et al., 1999).

SEVERAL THYLAKOID PROTEINS ARE INSERTED BY AN SRP-INDEPENDENT, POSSIBLY 'SPONTANEOUS' PATHWAY

In vitro assays have been developed to study a large number of thylakoid membrane proteins and these studies have produced a surprising result: most of the proteins studied do not require any of the known targeting apparatus (reviewed in Dalbey and Robinson, 1999). Several of these proteins form a rather novel group because they are synthesized with bipartite presequences that very much resemble those of imported lumenal proteins; these substrates include subunit II of the CF_0 assembly and subunits W, X and Y of photosystem II. These proteins insert efficiently into thylakoids in

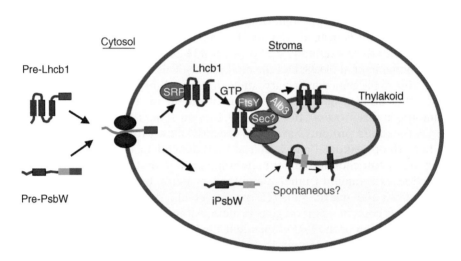

Figure 11.5 Models for the 'assisted' and 'spontaneous' targeting pathways for thylakoid membrane proteins. The SRP pathway: Lhcb1 is a chlorophyll-binding protein that spans the membrane three times (membrane-spanning regions depicted as gray boxes). It is synthesized in the cytosol as a precursor containing a single envelope transit peptide. This directs the protein across the envelope membranes in an unfolded form, after which it interacts with stromal signal recognition particle, SRP. SRP then delivers the protein to translocation apparatus in the thylakoid membrane (possibly the SecYEG translocon, but this is not certain) and, together with FtsY, facilitates insertion by a GTP-dependent process. A further factor, Alb3, is required for full insertion. Spontaneous pathway: PsbW is a single-span protein synthesized in the cytosol with a bipartite presequence comprising an envelope transit peptide followed by a hydrophobic signal-type peptide. Translocation across the envelope probably takes place by the standard route after which the transit peptide is removed and the protein then inserts into the thylakoid membrane in a loop conformation. The thylakoidal processing peptidase cleaves on the lumenal face of the membrane, generating the mature PsbW protein.

the absence of SRP, SecA, FtsY or nucleoside triphosphates (NTPs), and proteolysis of thylakoids destroys the membrane-bound Sec and Tat translocons but has no effect on the insertion of these proteins (Michl et al., 1994; Kim et al., 1998; Robinson et al., 1996). After insertion, cleavage by thylakoidal processing peptidase releases the mature protein. In the absence of any identifiable targeting factor or energy requirements, it has been suggested that these proteins insert spontaneously into the thylakoid membrane, in which case the role of the signal peptide (the second domain in the bipartite presequence) may be to provide a second hydrophobic region to assist insertion as a loop intermediate (Thompson et al., 1999). This targeting pathway is illustrated in Figure 11.5.

The CF_0II-type presequences are highly unusual because only one other membrane protein (M13 procoat in *E. coli*) has been found to be synthesized with a signal-type peptide yet inserted without the involvement of Sec- or Tat-type translocons (reviewed by Dalbey and Robinson, 1999). However, M13 procoat insertion does depend on YidC and is not therefore spontaneous in the true sense, whereas recent studies reveal that the above thylakoid proteins do not depend on the homologous Alb3 protein (Woolhead, Thompson, Moore, Henry and Robinson, manuscript submitted). Hence, these proteins may insert directly into the bilayer without the input of any other protein, although this remains to be experimentally confirmed or refuted.

Other proteins, lacking cleavable signal peptides, are also inserted by SRP/Sec-independent pathways, and perhaps surprisingly these proteins include relatives of the well-studied Lhcb1 protein. An early light-induced protein (Elip) and the PsbS are both members of the large chlorophyll-binding protein family but these proteins are able to insert by SRP/Sec/NTP-independent mechanisms in chloroplasts (Kim et al., 1999). Thus, of those thylakoid proteins studied to date, only Lhcb1 has been shown to require SRP and membrane-bound translocation machinery, although the related Lhcb5 protein was found to require stromal extract and NTPs in the study by Kim et al. (1999) and is thus probably likewise inserted by the SRP pathway.

Summary

- A subset of thylakoid membrane proteins are inserted using an SRP/Fts Y-dependent pathway which, like the Sec/Tat systems, was inherited from the cyanobacterial-type progenitor of chloroplasts.
- Other membrane proteins are inserted using a much simpler, apparently spontaneous mechanism.

HISTORICAL NOTE

The first thylakoid lumen proteins were cloned in the late 1980s and it became clear that they possessed a 'signal'-type peptide as well as a chloroplast import signal. This in itself suggested that a Sec-type system may operate in the thylakoid; the presence of another, separate pathway was not envisaged because bacteria were thought to export proteins using only the Sec pathway. In the mid-1990s the first chloroplast Sec proteins were cloned but biochemical studies at the same time firmly showed two different translocation mechanisms. The situation was clarified in 1997 when the first component of the Sec-independent system was cloned: the presence of homologs in many bacteria suggested for the first time that bacteria too may contain a second translocation system.

REFERENCES

Abdallah, F., Salamini, F. and Leister, D. (2000) A prediction of the size and evolutionary origin of the proteome of chloroplasts of *Arabidopsis. Trends Plant Sci* **5**: 141–142.

Bauer, J., Chen, K., Hiltbunner, A. et al. (2000) The major protein import receptor of plastids is essential for chloroplast biogenesis. *Nature* **403**: 203–207.

Berks, B.C. (1996) A common export pathway for proteins binding complex redox cofactors? *Mol Microbiol* **22**: 393–404.

Bogsch, E.G., Sargent, F, Stanley, N.R. et al. (1998) An essential component of a novel bacterial protein export system with homologues in plastids and mitochondria. *J Biol Chem* **273**: 18003–18006.

Bolhuis, A., Bogsch, E.G. and Robinson, C. (2000) Subunit interactions in the twin-arginine translocase complex of *Escherichia coli. FEBS Lett* **472**: 88–92.

Bolhuis, A., Mathers, J.E., Thomas, J.D. et al. (2001) TatB and TatC form a structural and functional unit of the twin-arginine translocase from *Escherichia coli. J Biol Chem* **276**: 20213–20219.

Bölter, B., May, T. and Soll, J. (1998a) A protein import receptor in pea chloroplasts, Toc86, is only a proteolytical fragment of a larger polypeptide. *FEBS Lett* **441**: 59–62.

Bölter, B., Soll, J., Schulz, A. et al. (1998b) Origin of a chloroplast protein importer. *Proc Natl Acad Sci USA* **95**: 15831–15836.

Brink, S., Fischer, K., Klösgen, R.B. et al. (1995) Sorting of nuclear encoded chloroplast membrane proteins to the envelope and the thylakoid membrane. *J Biol Chem* **270**: 20808–20815.

Brink, S., Bogsch, E.G., Edwards, W.R. et al. (1998) Targeting of thylakoid proteins by the ΔpH-driven twin-arginine translocation pathway requires a specific signal in the hydrophobic domain in conjunction with the twin-arginine motif. *FEBS Lett* **434**: 425–430.

Bruce, B.D. (2000) Chloroplast transit peptides: structure, function and evolution. *Trends Cell Biol* **10**: 440–446.

Caliebe, A., Grimm, R., Kaiser, G. et al. (1997) The chloroplastic protein import machinery contains a Rieske-type iron-sulfur cluster and a mononuclear iron-binding protein. *EMBO J* **16**: 7342–7350.

Chaddock, A.M., Mant, A., Karnauchov, I. et al. (1995) A new type of signal peptide: central role of a twin-arginine motif in transfer signals for the ΔpH-dependent thylakoidal protein translocase. *EMBO J* **14**: 2715–2722

Chen, K. and Schnell, D.J. (1997) Insertion of the 34-kDa chloroplast protein import component, IAP34, into the chloroplast outer membrane is dependent on its intrinsic GTP-binding capacity. *J Biol Chem* **272**: 6614–6620.

Chen, L.J. and Li, H.M. (1998) A mutant deficient in the plastid lipid DGD is defective in protein import into chloroplasts. *Plant J* **16**: 33–39.

Clark, S.A. and Theg, S.M. (1997) A folded protein can be transported across the chloroplast envelope and thylakoid membranes. *Mol Biol Cell* **8**: 923–934.

Cline, K., Ettinger, W.F. and Theg, S.M. (1992) Protein-specific energy requirements for protein transport across or into thylakoid membranes. Two lumenal proteins are transported in the absence of ATP. *J Biol Chem* **267**: 2688–2696.

Cline, K., Henry, R., Li, C. and Yuan, J. (1993). Multiple pathways for protein transport into or across the thylakoid membrane. *EMBO J* **12**: 4105–4114.

Dalbey, R.E. and Robinson, C. (1999) Protein translocation into and across the bacterial plasma membrane and the plant thylakoid membrane. *Trends Biochem Sci* **24**: 17–22.

DeLille, J., Peterson, E.C., Johnson, T. et al. (2000) A novel precursor recognition element facilitates posttranslational binding to the signal recognition particle in chloroplasts. *Proc Natl Acad Sci USA* **97**: 1926–1931.

Hageman, J., Baecke, C., Ebskamp, M. et al. (1990) Protein import and sorting inside the chloroplast are independent processes. *Plant Cell* **2**: 479–494.

Heins, L., Collinson, I. and Soll, J. (1998) The protein translocation apparatus of chloroplast envelopes. *Trends Plant Sci* **3**: 56–61.

Hell, K., Neupert, W. and Stuart, R.A. (2001) Oxa1p acts as a general membrane insertion machinery for proteins encoded by mitochondrial DNA. *EMBO J* **20**: 1281–1288.

Henry, R., Kapazoglou, A., McCaffery, M. and Cline, K. (1994) Differences between lumen targeting domains of chloroplast transit peptides determine pathway specificity for thylakoid transport. *J Biol Chem* **269**: 10189–10192.

Herrmann, J.M., Neupert, W. and Stuart, R.A. (1997) Insertion into the mitochondrial inner membrane of a polytopic protein, the nuclear-encoded Oxa1p. *EMBO J* **16**: 2217–2226.

Hinnah, S.C., Hill, K., Wagner, R. et al. (1997) Reconstitution of a chloroplast protein import channel. *EMBO J* **16**: 7351–7360.

Hirsch, S., Muckel, E., Heemeyer, F. et al. (1994) A receptor component of the chloroplast protein translocation machinery. *Science* **266**: 1989–1992.

Hynds, P.J., Robinson, D. and Robinson, C. (1998) The Sec-independent twin-arginine translocation system can transport both tightly folded and malfolded proteins across the thylakoid membrane. *J Biol Chem* **273**: 34868–34874.

Ivey III, R.A., Subramanian, C. and Bruce, B.D. (2000) Identification of a Hsp70 recognition domain within the Rubisco small subunit transit peptide. *Plant Physiol* **122**: 1289–1299.

Jackson, D.T., Froehlich, J.E. and Keegstra, K. (1998) The hydrophilic domain of Tic110, an inner envelope membrane component of the chloroplastic protein translocation apparatus, faces the stromal compartment. *J Biol Chem* **273**: 16583–16588.

Jarvis, P., Chen, L.J., Li, H.M. et al. (1998) An *Arabidopsis* mutant defective in the plastid general protein import apparatus. *Science* **282**: 100–103.

Kaneko, T., Sato, S., Kotani, H. et al. (1996) Sequence analysis of the genome of the unicellular cyanobacterium *Synechocystis* sp. strain PCC6803. II. Sequence determination of the entire genome and assignment of potential protein-coding regions. *DNA Res* **3**: 109–136.

Keegstra, K. and Cline, K. (1999) Protein import and routing systems of chloroplasts. *Plant Cell* **11**: 557–570.

Kessler, F. and Blobel, G. (1996) Interaction of the protein import and folding machineries of the chloroplast. *Proc Natl Acad Sci USA* **93**: 7684–7689.

Kessler, F., Blobel, G., Patel, H.A. et al. (1994) Identification of two GTP-binding proteins in the chloroplast protein import machinery. *Science* **266**: 1035–1039.

Kieselbach, T., Hagman, Å., Andersson, B. and Schröder, W.P. (1998) The thylakoid lumen of chloroplasts: isolation and characterisation. *J Biol Chem* **273**: 6710–6716.

Kim, S.J., Robinson, C. and Mant, A. (1998) Sec/SRP-independent insertion of two thylakoid membrane proteins bearing cleavable signal peptides. *FEBS Lett* **424**: 105–108.

Kim, S.J., Jansson, S., Hoffman, N.E. et al. (1999) Distinct 'assisted' and 'spontaneous' mechanisms for the insertion of polytopic chlorophyll-binding proteins into the thylakoid membrane. *J Biol Chem* **274**: 4715–4721.

Ko, K., Budd, D., Wu, C. et al. (1995) Isolation and characterization of a cDNA clone encoding a member of the Com/Cim44 envelope components of the chloroplast protein import apparatus. *J Biol Chem* **270**: 28601–28608.

Kogata, N., Nishio, K., Hirohashi, T. et al. (1999) Involvement of a chloroplast homologue of the signal recognition particle receptor protein, FtsY, in protein targeting to thylakoids. *FEBS Lett* **329**: 329–333.

Kouranov, A., Chen, X., Fuks, B. et al. (1998) Tic20 and Tic22 are new components of the protein import apparatus at the chloroplast inner envelope membrane. *J Cell Biol* **143**: 991–1002.

Kourtz, L. and Ko, K. (1997) The early stage of chloroplast protein import involves Com70. *J Biol Chem* **272**: 32264–32271.

Krimm, I., Gans, P., Hernandez, J.F. et al. (1999) A coil-helix instead of a helix-coil motif can be induced in a chloroplast transit peptide from *Chlamydomonas reinhardtii*. *Eur J Biochem* **265**: 171–180.

Laidler, V., Chaddock, A.M., Knott, T.G. et al. (1995) A SecY homolog in *Arabidopsis thaliana*. Sequence of a full-length cDNA clone and import of the precursor protein into chloroplasts. *J Biol Chem* **270**: 17664–17667.

Lancelin, J.M., Bally, I., Arlaud, G.J. et al. (1994) NMR structures of ferredoxin chloroplastic transit peptide from *Chlamydomonas reinhardtii* promoted by trifluoroethanol in aqueous solution. *FEBS Lett* **343**: 261–266.

Li, X., Henry, R., Yuan, J. et al. (1995) A chloroplast homologue of the signal recognition particle subunit SRP54 is involved in the post-translational integration of a protein into thylakoid membranes. *Proc Natl Acad Sci USA* **92**: 3789–3793.

Lübeck, J., Soll, J., Akita, M et al. (1996) Topology of IEP110, a component of the chloroplastic protein import machinery present in the inner envelope membrane. *EMBO J* **15**: 4230–4238.

Lübeck, J., Heins, L. and Soll, J. (1997) A nuclear-coded chloroplastic inner envelope protein uses a soluble sorting intermediate upon import into the organelle. *J Cell Biol* **137**: 1279–1286.

Marshall, J.S., DeRocher, A.E., Keegstra K et al. (1990) Identification of heat shock protein hsp70 homologues in chloroplasts. *Proc Natl Acad Sci USA* **87**: 374–378.

Martin, W., Stoebe, B., Goremykin, V. et al. (1998) Gene transfer to the nucleus and the evolution of chloroplasts. *Nature* **393**: 162–165.

May, T. and Soll, J. (1998) Positive charges determine the topology and functionality of the transmembrane domain in the chloroplastic outer envelope protein Toc34. *J Cell Biol* **141**: 895–904.

May, T. and Soll, J. (2000) 14-3-3 proteins form a guidance complex with chloroplast precursor proteins in plants. *Plant Cell* **12**: 53–63.

Michl, D., Robinson, C., Shackleton, J.B. et al. (1994) Targeting of proteins to the thylakoids by bipartite presequences: CF_0II is imported by a novel, third pathway. *EMBO J* **13**: 1310–1317.

Moore, M., Harrison, M.S., Peterson, E.C. and Henry, R. (2000) Chloroplast Oxa1p homolog albino3 is required for post-translational integration of the light harvesting chlorophyll-binding protein into thylakoid membranes. *J Biol Chem* **275**: 1529–1532.

Mori, H., Summer, E.J., Ma, X. and Cline, K. (1999) Component specificity for the thylakoidal Sec and delta pH-dependent protein transport pathways. *J Cell Biol* **146**: 45–55.

Mould, R.M. and Robinson, C. (1991) A proton gradient is required for the transport of two lumenal oxygen-evolving proteins across the thylakoid membrane. *J Biol Chem* **266**: 12189–12193.

Nielsen, V.S., Mant, A., Knoetzel, J. et al. (1994) Import of barley photosystem I subunit N into the thylakoid lumen is mediated by a bipartite presequence lacking an intermediate processing site; role of the delta pH in translocation across the thylakoid membrane. *J Biol Chem* **269**: 3762–3766.

Nielsen, E., Akita, M., Davila-Aponte, J. et al. (1997) Stable association of chloroplastic precursors with the protein translocation complexes that contain proteins from both envelope membranes and a stromal hsp100 molecular chaperone. *EMBO J* **16**: 935–946.

Pang, P., Meathrel, K. and Ko, K. (1997) A component of the chloroplast protein import apparatus functions in bacteria. *J Biol Chem* **272**: 25623–25627.

Perry, S.E. and Keegstra, K. (1994) Envelope membrane proteins that interact with chloroplast precursor proteins. *Plant Cell* **6**: 93–105.

Pilon, M., Rietveld, A.G., Weisbeek, P.J. et al. (1992) Secondary structure and folding of a functional chloroplast precursor protein. *J Biol Chem* **267**: 19907–19913.

Pilon, M., Wienk, H., Sips, W. et al. (1995) Functional domains of the ferredoxin transit sequence involved in chloroplast import. *J Biol Chem* **270**: 3882–3893.

Pinnaduwage, P. and Bruce, B.D. (1996) *In vitro* interaction between a chloroplast transit peptide and chloroplast outer envelope lipids is sequence-specific and lipid-class dependent. *J Biol Chem* **271**: 32907–32915.

Rensink, W.A., Schnell, D.J. and Weisbeek, P.J. (2000) The transit sequence of ferredoxin contains different domains for translocation across the outer and inner membrane of the chloroplast envelope. *J Biol Chem* **275**: 10265–10271.

Reumann, S., Davila-Aponte, J. and Keegstra, K. (1999) The evolutionary origin of the protein-translocation channel of chloroplastic envelope membranes: identification of a cyanobacterial homolog. *Proc Natl Acad Sci USA* **96**: 784–789.

Robinson, C., Cai, D., Hulford, A. et al. (1994) The presequence of a chimeric construct dictates which of two mechanisms are utilized for translocation across the thylakoid membrane: evidence for the existence of two distinct translocation systems. *EMBO J* **13**: 279–285.

Robinson, D., Karnauchov, I., Herrmann, R.G. et al. (1996) Protease-sensitive thylakoidal import machinery for the Sec-, ΔpH- and signal recognition particle-dependent protein targeting pathways, but not for CF_0II integration. *Plant J* **10**: 149–155.

Robinson, C. and Bolhuis, A. (2001) Protein targeting by the twin-arginine translocation pathway. *Nature Reviews Molec Cell Biol* **2**: 350–355.

Samuelson, J.C., Chen, M., Jiang, F. et al. (2000) YidC mediates membrane protein insertion in bacteria. *Nature* **406**: 637–641.

Santini, C.L., Ize, B., Chanal, A. et al. (1998) A novel Sec-independent periplasmic protein translocation pathway in *Escherichia coli*. *EMBO J* **17**: 101–112.

Sargent, F., Bogsch, E.G., Stanley, N.R. et al. (1998) Overlapping functions of components of a bacterial Sec-independent export pathway. *EMBO J* **17**: 3640–3650.

Sargent, F., Stanley, N.R., Berks, B.C. and Palmer, T. (1999) Sec-independent protein translocation in *Escherichia coli*. A distinct and pivotal role for the TatB protein. *J Biol Chem* **274**: 36073–36082.

Schnell, D.J., Blobel, G., Keegstra, K. et al. (1997) A consensus nomenclature for the protein-import components of the chloroplast envelope. *Trends Cell Biol* **7**: 303–304.

Schuenemann, D., Gupta, S., Persello-Cartieaux, F. et al. (1998) A novel signal recognition particle targets light-harvesting proteins to the thylakoid membranes. *Proc Natl Acad Sci USA* **95**: 10312–10316.

Schuenemann, D., Amin, P., Hartmann, E. and Hoffman, N.E. (1999) Chloroplast SecY is complexed to SecE and involved in the translocation of the 33-kDa, but not the 23-kDa subunit of the oxygen-evolving complex. *J Biol Chem* **274**: 12177–12182.

Seedorf, M., Waegemann, K. and Soll, J. (1995) A constituent of the chloroplast import complex represents a new type of GTP-binding protein. *Plant J* **7**: 401–411.

Settles, M.A., Yonetani, A., Baron, A. et al. (1997) Sec-independent protein translocation by the maize Hcf106 protein. *Science* **278**: 1467–1470.

Sohrt, K. and Soll, J. (2000) Toc64, a new component of the protein translocon of chloroplasts. *J Cell Biol* **148**: 1213–1221.

Soll, J. and Tien, R. (1998) Protein translocation into and across the chloroplastic envelope membranes. *Plant Mol Biol* **38**: 191–207.

Stahl, T., Glockmann, C., Soll, J. et al. (1999) Tic40, a new 'old' subunit of the chloroplast protein import translocon. *J Biol Chem* **274**: 37467–37472.

Sveshnikova, N., Soll, J. and Schleiff, E. (2000) Toc34 is a preprotein receptor regulated by GTP and phosphorylation. *Proc Natl Acad Sci USA* **97**: 4973–4978.

Theg, S.M., Baeuerle, C., Olsen, L.J. et al. (1989) Internal ATP is the only energy requirement for the translocation of precursor proteins across chloroplastic membranes. *J Biol Chem* **264**: 6730–6736.

Thompson, S.J., Robinson, C and Mant, A. (1999) Dual signal peptides mediate the Sec/SRP-independent insertion of a thylakoid membrane polyprotein, PsbY. *J Biol Chem* **274**: 4059–4066.

Tranel, P.J. and Keegstra, K. (1996) A novel, bipartite transit peptide targets OEP75 to the outer membrane of the chloroplastic envelope. *Plant Cell* **8**: 2093–2104.

Tranel, P.J., Froehlich, J., Goyal, A. et al. (1995) A component of the chloroplastic protein import apparatus is targeted to the outer envelope membrane via a novel pathway. *EMBO J* **14**: 2436–2446.

Tsai, L.Y., Yu, S.L. and Li, H.M. (1999) Insertion of atToc34 into chloroplastic outer membrane is assisted by at least two proteinaceous components in the import system. *J Biol Chem* **274**: 18735–18740.

Tu, C.J., Schuenemann, D. and Hoffman, N.E. (1999) Chloroplast FtsY, chloroplast signal recognition particle, and GTP are required to reconstitute the soluble phase of light-harvesting chlorophyll protein transport into thylakoid membranes. *J Biol Chem* **274**: 27219–27224.

Voelker, R. and Barkan, A. (1995) Two nuclear mutations disrupt distinct pathways for targeting proteins to the chloroplast thylakoid. *EMBO J* **14**: 3905–3914.

von Heijne, G., Hirai, T., Klösgen, R.B. et al. (1991) CHLPEP: a database of chloroplast transit peptides. *Plant Mol Biol Rep* **9**: 104–126.

Vothknecht, U. and Soll, J. (2000) Protein import: the hitchhikers guide into chloroplasts. *Biol Chem* **381**: 887–897.

Waegemann, K. and Soll, J. (1991) Characterization of the protein import apparatus in isolated outer envelopes of chloroplasts. *Plant J* **1**: 149–158.

Waegemann, K. and Soll, J. (1996) Phosphorylation of the transit sequence of chloroplast precursor proteins. *J Biol Chem* **271**: 6545–6554.

Waegemann, K., Paulsen, H. and Soll, J. (1990) Translocation of proteins into isolated chloroplasts requires cytosolic factors to obtain import competence. *FEBS Lett* **261**: 89–92.

Walker, M.B., Roy, L.M., Coleman, E. et al. (1999) The maize tha4 gene functions in sec-independent protein transport in chloroplasts and is related to hcf106, tatA, and tatB. *J Cell Biol* **147**: 267–276.

Weiner, J.H., Bilous, P.T., Shaw, G.M. et al. (1998) A novel and ubiquitous system for membrane targeting and secretion of cofactor-containing proteins. *Cell* **93**: 93–101.

Wexler, M., Sargent, F., Jack, R.L. et al. (2000) TatD is a cytoplasmic protein with DNase activity. No requirement for TatD family proteins in Sec-independent protein export. *J Biol Chem* **275**: 16717–16722.

Wienk, H.J.L., Czisch, M. and De Kruijff, B. (1999) The structural flexibility of the preferredoxin transit peptide. *FEBS Lett* **453**: 318–326.

Young, M.E., Keegstra, K. and Froehlich, J.E. (1999) GTP promotes the formation of early-import intermediates but is not required during the translocation step of protein import into chloroplasts. *Plant Physiol* **121**: 237–244.

Yuan, J., Henry, R., McCaffery, M. and Cline, K. (1994) SecA homolog in protein transport within chloroplasts: evidence for endosymbiont-derived sorting. *Science* **266**: 796–798.

12

IMPORT OF PROTEINS INTO PEROXISOMES

SURESH SUBRAMANI, VINCENT DAMMAI,
PARTHA HAZRA, IVET SURIAPRANATA AND
SOOJIN LEE

DISCOVERY AND FUNCTIONS OF PEROXISOMES

What are peroxisomes and why are they studied?

The peroxisomes were the last of the major subcellular organelles to be discovered. They were described initially as 'microbodies' or small, single-membrane-bound structures surrounding a granular matrix in the cytoplasm of mouse proximal kidney tubules (Rhodin, 1954). The organelle was renamed the 'peroxisome' in recognition of the fact that this compartment houses enzymes involved in both the generation and degradation of (hydrogen) peroxide (De Duve and Baudhuin, 1966).

Peroxisomes are ubiquitous in all eukaryotic cells and range in cross-sectional diameter between 0.1 and 1.0 μM. Although it is now recognized that this organelle is essential for survival of humans (Fujiki, 2000; Gould and Valle, 2000), and that it plays a central role in many aspects of the metabolism of lipids (Wanders and Tager, 1998), the cataloging of the enzymatic roles of peroxisomes is not yet complete. The organelle is of particular interest for many reasons:

1. It is among the least complex of the subcellular organelles, a point of special significance in this post-genomic era in which the goal of identification and characterization of all the components of a subcellular organelle seems attainable.

Protein Targeting, Transport & Translocation
ISBN 0-12-200731-X

2. Its contribution to cellular metabolism, particularly of lipids, is of interest to biochemists (Wanders and Tager, 1998).
3. An understanding of its assembly and turnover is central to the problem of organelle homeostasis and its regulation by nutritional cues.
4. Its role in at least 17 human disorders, most of which are fatal, has sparked the interest of biomedical scientists (Fujiki, 2000; Gould and Valle, 2000).
5. Its evolutionary origin in relation to other subcellular compartments remains a mystery.

Among the features of peroxisomes that make them tractable as an experimental system for study are their ease of identification, purification, inducibility and dispensability at the single-cell level under the correct growth conditions.

Metabolic functions of peroxisomes

What do peroxisomes do that makes them essential for human survival? An obvious clue is their role in many aspects of lipid metabolism. These activities include (i) the β-oxidation of fatty acids, (ii) fatty acid α-oxidation, (iii) synthesis of cholesterol and other isoprenoids, (iv) ether–phospholipid synthesis and (v) biosynthesis of polyunsaturated fatty acids (Wanders and Tager, 1998). Table 12.1 lists the major metabolic functions of peroxisomes, and evolutionarily similar organelles called glyoxysomes (in plants, fungi) and glycosomes (in Kinetoplastida). The metabolic functions of peroxisomes can vary with the cell type and the nutritional milieu that these cells are in (Wanders and Tager, 1998). This responsiveness of cells to their growth environment is reflected by the induction of peroxisome biogenesis under certain circumstances (Kunau et al., 1993), and the autophagic degradation

Table 12.1   Major metabolic functions of peroxisomes

	Peroxisomes (mammals)	Glyoxysomes (plants)	Glyoxysomes or peroxisomes (yeasts, fungi)	Glycosomes (Kinetoplastida)
Peroxide metabolism	+	+	+	−
Ether phospholipid synthesis	+	−	−	+
β-oxidation of fatty acids	+	+	+	+
Glyoxylate cycle	−	+	+	−
Photorespiration	−	+	−	−
Glycolysis	−	−	−	+

From Wiemer and Subramani (1994). Reprinted with permission from Academic Press.

of excess peroxisomes by lysosomes or vacuoles (Dunn, 1994; Tuttle and Dunn, 1995). This ease of manipulation of peroxisome biogenesis and degradation makes the organelle very interesting. One can study not only these processes per se, but also the pathways that integrate extracellular signals and translate them into intracellular responses.

Summary: *Discovery and functions of peroxisomes*

- Peroxisomes are the simplest subcellular organelles found in virtually all eukaryotic cells. They are related to the glyoxysomes of plants and fungi, and the glycosomes of Kinetoplastida, in terms of their metabolic functions and mechanisms of biogenesis.
- Peroxisomes house enzymes involved in the production and degradation of hydrogen peroxide, as well as a variety of lipid metabolic pathways.
- Peroxisomes are intimately involved in many human diseases and are excellent models for the study of organelle biogenesis and turnover.

OVERVIEW OF PEROXISOME BIOGENESIS

Unlike the nucleus, mitochondrion and chloroplast, peroxisomes have no DNA. Therefore, all of their polypeptide components are encoded by nuclear genes, synthesized on cytoplasmic polyribosomes and then transported post-translationally to peroxisomes. Proteins destined for the peroxisome matrix or membrane possess distinct targeting signals that engage signal sequence receptors to drive their transport to their final subcellular destination. In addition to proteins, the peroxisomal membrane also contains lipids, which are believed to be derived from the endoplasmic reticulum, but the mechanism by which this process occurs has not been defined. Consequently, the sections below are devoted only to the import of proteins, and not lipids, to the peroxisomal matrix and membrane.

Import signals on peroxisomal matrix and membrane proteins

The first of these peroxisome targeting signals (PTSs) was discovered in 1987 in firefly luciferase, and named PTS1 (Gould et al., 1987, 1989). It is a conserved, C-terminal tripeptide, SKL or its variants, that is completely necessary and sufficient for protein targeting into the peroxisome matrix or lumen (Figure 12.1). It is the most abundant signal involved in protein transport to the peroxisome matrix. Subsequently, using a protein lacking a PTS1 sequence, another nonapeptide, named PTS2, was also found to target proteins to the peroxisome matrix (Osumi et al., 1991; Swinkels et al., 1991). This sequence, located near the NH_2 terminus or at internal locations in

Matrix proteins

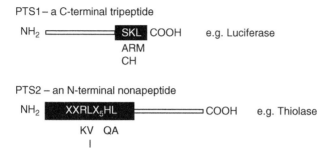

Membrane proteins

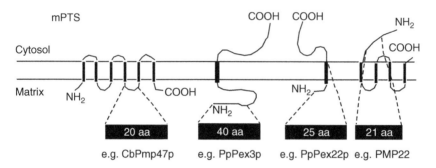

Figure 12.1   Targeting signals used by peroxisomal matrix and membrane proteins. The letters in the black boxes denote the consensus amino acid sequences in the one letter code, or the length of the PTS. Some, but not all, of the common amino acid variants are also shown below these sequences. The mPTSs may lie on the lumenal or cytosolic side of the peroxisomal membrane and may or may not include adjacent transmembrane domains shown as vertical black rectangles.

proteins, is used by a smaller set of peroxisomal matrix proteins, and is conserved in evolution, as is the PTS1 sequence. Together, PTS1 and PTS2 account for the transport of most proteins into the peroxisome matrix.

Peroxisomal membrane proteins (PMPs) embedded in the organelle membrane use membrane PTSs (mPTSs) for their targeting. These have been characterized in several proteins but have little in common except for a basic region (Baerends et al., 2000; Dyer et al., 1996; Subramani et al., 2000). They may or may not include an adjacent transmembrane domain, and can reside on either the lumenal or cytosolic sides of the peroxisomal membrane (Figure 12.1).

Import of proteins into the peroxisomal matrix

Recognition of the PTSs by receptors in the cytosol

Early experiments on the peroxisomal targeting of reporter and endo-genous proteins in the late 1980s showed that similar PTSs were used from yeast to humans (Gould et al., 1989). This led to the idea that the signals, and perhaps the machinery, involved in the sorting of peroxisomal proteins would be conserved in evolution. Consequently, in the early 1990s several groups around the world began using yeasts, and later Chinese hamster ovary (CHO) cells, to obtain mutations in *PEX* genes involved in peroxi-some biogenesis (Subramani, 1993). The genetic strategies (Subramani, 1993) defined many complementation groups, *PEX* genes and proteins (collectively called peroxins) involved in peroxisome assembly. As this work progressed, it became obvious that cells from a variety of human patients suffering from peroxisome biogenesis disorders also fell into multiple complementation groups like the CHO cell mutants (Slawecki et al., 1995).

These studies have led to the discovery of at least 23 *PEX* genes (Table 12.2). The analysis of the phenotypes of some of these mutants led directly to the cloning of the genes encoding the PTS1 and PTS2 receptors. For example, the *pex5* mutant was selectively impaired in only the import of PTS1-containing proteins and not in the import of PTS2- or mPTS-containing polypeptides. Not surprisingly it has a mutation in the PTS1 receptor encoded by the *PEX5* gene (McCollum et al., 1993). In an analo-gous fashion, the *PEX7* gene was shown to encode the PTS2 receptor (Marzioch et al., 1994; Zhang and Lazarow, 1995). Biochemical receptor-ligand studies and/or yeast two-hybrid experiments showed that the PTS1 and PTS2 sequences interacted with Pex5p and Pex7p, respectively (Rehling et al., 1996; Terlecky et al., 1995). The two receptors have structural motifs, seven tetratricopeptide repeats in Pex5p and WD40 repeats in Pex7p, both of which are involved in protein (poly)peptide interactions (Figure 12.2).

Docking of the PTS receptor/cargo complex at the peroxisomal membrane

The PTS1 and PTS2 receptors, Pex5p and Pex7p respectively, have been studied in yeasts, mammals, plants and parasites. In most cases, they are primarily cytosolic and only partially peroxisome-associated. The receptors can recognize PTS-containing proteins in the cytosol, but this cargo must then be delivered to the peroxisome (Subramani et al., 2000). The PTS receptors bound to cargo interact with a docking complex comprising Pex3p, Pex13p, Pex14p and Pex17p (Albertini et al., 1997; Elgersma et al., 1996; Erdmann and Blobel, 1996; Gould et al., 1996; Huhse et al., 1998; Johnson et al., 2001; Snyder et al., 1999b). The receptors interact with Pex13p and Pex14p, but not directly with Pex17p which associates with Pex14p (Huhse et al., 1998) (Figure 12.3). Thus the PTS1 and PTS2 import

Table 12.2 *PEX* genes and their characteristics

PEX gene	Peroxin characteristics
PEX1	AAA ATPase
PEX2	C_3HC_4 zinc-biding PMP, part of putative translocation subcomplex
PEX3	PMP required for membrane biogenesis, part of docking subcomplex
PEX4	Peroxisome-associated ubiquitin-conjugating enzyme
PEX5	PTS1 receptor, shuttles from the cytosol to or into peroxisomes
PEX6	AAA ATPase
PEX7	PTS2 receptor
PEX8	Lumenal protein containing a PTS2 and/or PTS1
PEX9	PMP of unknown function
PEX10	C_3HC_4 zinc-binding PMP, part of putative translocation subcomplex
PEX11	Involved in peroxisome proliferation but is an MCFA transporter
PEX12	C_3HC_4 zinc-binding PMP, binds the PTS1 receptor, part of putative translocation subcomplex
PEX13	SH3-containing PMP, binds the PTS1 receptor, part of docking subcomplex
PEX14	Membrane docking protein for Pex5p and Pex7p, part of docking subcomplex
PEX15	Peroxisomal integral membrane protein
PEX16	Intraperoxisomal peripheral membrane protein
PEX17	Found in PTS-receptor docking complex with Pex14p
PEX18	Involved with Pex21p in Pex7p-mediated targeting
PEX19	Predominantly cytosolic, interacts with many PMPs on peroxisome
PEX20	Required for thiolase oligomerization and import
PEX21	Involved with Pex18p in Pex7p-mediated targeting
PEX22	Anchors Pex4p to the peroxisomal membrane
PEX23	Involved in matrix protein import

This list is compiled from peroxins found in different organisms (adapted from Subramani et al., 2000 with permission from the *Annual Review of Biochemistry*, Volume 69, © 2000, by Annual Reviews, www.AnnualReviews.org). Although most are conserved, not every one of these has been found in all organisms.

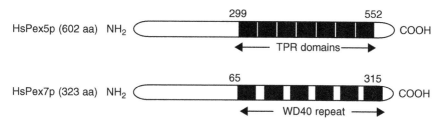

Figure 12.2 Schematic structures of the PTS1 receptor, Pex5p, and the PTS2 receptor, Pex7p, from humans. The TPR domains bind the PTS1 peptide, and the WD40 repeats are involved in protein–protein interactions.

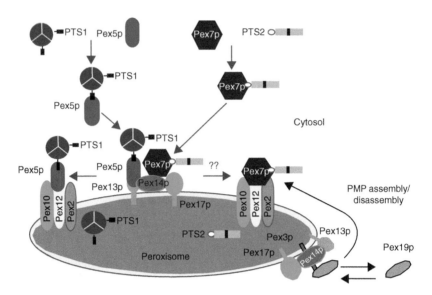

Figure 12.3   Model for import of peroxisomal matrix proteins. PTS1- and PTS2-containing proteins are recognized in the cytosol by their specific receptors, Pex5p and Pex7p, respectively. The receptor–cargo complexes then migrate to the peroxisomal membrane where they interact with the docking subcomplex consisting of Pex3p, Pex13p, Pex14p and Pex17p. In yeasts, Pex5p interacts with both Pex13p and Pex14p, whereas Pex7p interacts with Pex14p. The Pex5p–cargo complex then interacts with components of the putative translocation subcomplex, comprising the zinc-binding proteins Pex2p, Pex10p and Pex12p. Since PTS1 and PTS2 proteins probably use a common translocation pathway, it is presumed (denoted by ??), but not proven, that the Pex7p–cargo complex also interacts with the putative translocation subcomplex. The cargo then enters the peroxisome matrix. Pex19p, a predominantly cytosolic protein farnesylated in several species, binds to Pex3p and to several PMPs (see text) and is postulated to aid the assembly/disassembly of PMP subcomplexes. The diagram also shows a unique feature of peroxisomal protein import – the lack of the requirement for protein unfolding during membrane translocation. Cargo consisting of mixed oligomers of subunits with and without PTSs can enter the peroxisomal matrix.

pathways are unique in the cytosol but converge at the docking complex on the peroxisomal membrane.

Although the cargo dependence of the interaction of Pex5p and Pex7p has not been analyzed in detail, it has been reported that PTS1/Pex5p complexes have a higher affinity for Pex14p than does Pex5p alone. In contrast, PTS1/Pex5p complexes have a lower affinity for Pex13p than does Pex5p alone, leading to the idea that cargo-bound PTS receptors dock first with Pex14p, and then with Pex13p perhaps after cargo release (Urquhart et al., 2000). In the absence of Pex14p, no Pex5p docking occurs on the peroxisomal membrane (Otera et al., 2000).

PTS receptor/peroxin interactions downstream of the docking complex
Recent experiments have shown that two zinc-binding, ring-finger proteins, Pex10p and Pex12p, are in a complex and interact with Pex5p downstream of the docking complex. In their absence, Pex5p still interacts with the docking complex (Chang et al., 1999b; Okumoto et al., 2000). It has been postulated that these zinc-binding proteins might serve as the translocation machinery that facilitates cargo transit past the peroxisomal membrane, but additional evidence is needed to confirm this point (Chang et al., 1999b) (Figure 12.3). We have found that in *Pichia pastoris*, a third member of this family, Pex2p, is also part of a complex with Pex10p and Pex12p (Hazra, Suriapranata and Subramani, unpublished data). Some of the proteins in this complex may play a role in PTS receptor cycling to the cytosol. For example, mutations in Pex2p cause accumulation of some Pex5p inside the peroxisomes (Otera et al., 2000).

Shuttling of the PTS1 receptor to the peroxisome
Although most studies support predominantly cytosolic, and only partially peroxisome-associated, locations for Pex5p and Pex7p, there are reports of their intraperoxisomal localization (Szilard et al., 1995; van der Klei et al., 1998; Zhang and Lazarow, 1996). The differing subcellular locations of the PTS receptors have led to several models for their action during matrix protein import. For example, the intraperoxisomal Pex5p and Pex7p led to the suggestion that these receptors act from within the organelle to pull cargo into the matrix (Zhang and Lazarow, 1996). The analogy used here is the mechanism by which intramitochondrial Hsp70 functions as an ATP-dependent chaperone during mitochondrial matrix protein import (see Chapter 10). However, there is no experimental evidence for such a role for Pex5p or Pex7p.

The alternative model is one in which the PTS receptors recognize cargo in the cytosol and shuttle them to the peroxisome (Rachubinski and Subramani, 1995). Several published lines of evidence support a cargo-shuttling role for Pex5p. First, Pex5p is known to bind cargo in the cytosol, and interact with peroxins of the docking (Pex13p and Pex14p) and putative translocation subcomplexes (Pex10p and Pex12p) (Subramani et al., 2000). Second, loss of proteins of the putative translocation subcomplex leads to accumulation of Pex5p at docking sites on the peroxisome surface (Chang et al., 1999b; Dodt and Gould, 1996). Third, conditions that inhibit protein translocation into the peroxisome matrix (low temperature or ATP depletion) cause accumulation of Pex5p on the peroxisome membrane (Dodt and Gould, 1996). Pex5p returns to its predominantly cytosolic location when conditions favoring matrix protein translocation are restored (Dodt and Gould, 1996). Finally, in certain human patient cell lines, Pex5p was found to be intraperoxisomal (Chang et al., 1999b; Otera et al., 2000). Except for the last finding, these studies while favoring a cargo-shuttling role for Pex5p, do not address whether Pex5p delivers its PTS1 cargo to the peroxisomal membrane and returns to the cytosol (simple-shuttle model), or whether it enters the peroxisome matrix with its cargo, releases the cargo in the matrix and returns to the cytosol (extended shuttle model) (Rachubinski and Subramani, 1995).

Protein translocation across the peroxisomal membrane
One of the distinguishing features of peroxisomal protein import is the fact that fully folded (Walton et al., 1995) and oligomerized proteins (Glover et al., 1994; Hausler et al., 1996; Lee et al., 1997; McNew and Goodman, 1994), as well as 9 nm gold particles coupled to a PTS1 peptide (Walton et al., 1995) can traverse the peroxisomal membrane. A beautiful demonstration of this point is that protein subunits lacking a PTS can enter the peroxisomes 'piggy back' in association with other subunits that do possess a PTS. This suggests that the size of the translocation pore must be large enough to accommodate oligomeric proteins. Such a pore has never been seen on the peroxisomal membrane, but a plausible explanation might be that the pore may be transiently assembled only when cargo/receptor complexes dock on the peroxisomal membrane.

Compatibility of the extended shuttle model for Pex5p with different locations for this receptor
The varying subcellular locations reported for both Pex5p and Pex7p in various organisms represent a paradox which is easily clarified by the extended shuttle model for Pex5p function. The steady-state levels of Pex5p in the cytosol or the peroxisome matrix are controlled by two rate constants – one for its entry into peroxisomes (k1), and the other for its exit (k2). It is easy to see that if k1 is much greater than k2, most of the Pex5p would be intra-peroxisomal, while if k2 is much greater than k1, the receptor would be primarily cytosolic. More importantly, irrespective of its steady-state localization, the recycling receptor would still be able to shuttle cargo from the cytosol to the peroxisome. It is even conceivable that environmental conditions, or the nutritional milieu, may influence the steady-state locations of the receptors in the cytosol versus the peroxisomal compartments (Rachubinski and Subramani, 1995).

Link between the PTS1 and PTS2 pathways in mammals
In yeasts, mutations in the *PEX5* and *PEX7* genes affect only the PTS1 and PTS2 import pathways, respectively, and these proteins exist as single iso-forms. However, in mammalian cells, there are splice variants of Pex5p, resulting in the synthesis of short (Pex5pS) and long (Pex5pL) isoforms, which differ by an internal insertion of 37 amino acids. These isoforms can form both homo- and heteromeric dimers. Both isoforms, when expressed separately, support import of PTS1 proteins, but only Pex5pL restores import of PTS2

proteins. The basis of the involvement of mammalian Pex5pL in the PTS2 import pathway seems to be that only Pex5pL, and not Pex5pS, interacts with Pex7p, and is necessary for the transfer of the cargo/Pex7p complex from the cytosol to the peroxisomal membrane. In yeasts, Pex5p and Pex7p bind Pex14p on the peroxisomal membrane independently, rather than being interdependent, as in mammalian cells. A role for Pex5pL in PTS2 import in mammalian cells is supported by genetic evidence wherein a *pex5*-deficient CHO cell line, ZPG231, was defective only in the import of PTS2, and not PTS1 proteins. It had an S214F mutation in *PEX5* that specifically disrupted the interaction between Pex5pL and Pex7p (Fujiki, 2000).

In vitro *systems for the analysis of peroxisomal matrix protein import*
Unlike the transport of proteins into many other subcellular compartments, there is no natural biochemical hallmark of peroxisomal matrix protein import, such as glycosylation or proteolytic removal of a signal sequence. This initially hampered the development of *in vitro* systems, but several of these are now available based on the following strategies (Subramani et al., 2000):

- Import of radiolabeled substrates into purified peroxisomes and protease resistance of the imported material.
- Microinjection of substrates into mammalian cells and monitoring of import by indirect immunofluorescence using antibodies to the import substrate.
- Import of substrates into semi-permeabilized mammalian cells followed by detection of import using indirect immunofluorescence with antibodies to the import substrate.
- A quantitative ELISA-based assay for the import of biotinylated import substrates into the peroxisomal matrix.

These studies show that the import of both PTS1- and PTS2-containing substrates is time-, temperature-, signal-, PTS receptor-, ATP- and cytosol-dependent and the requirement for some membrane peroxins, such as Pex14p, and Hsp40 and Hsp70 is also evident (Legakis and Terlecky, 2001; Terlecky et al., 2001).

Other proteins involved in matrix protein import and assembly
Heat shock proteins of the DnaK (Hsp70) and DnaJ (djp1p) are needed in the cytosol for the import of PTS1-containing proteins (Hettema et al., 1998; Preisig-Muller et al., 1994; Walton et al., 1994), but in view of the fact that protein unfolding is not a prerequisite for matrix protein import, the exact role of the chaperones is unclear. In the glyoxysomes of watermelons, there is an intraperoxisomal Hsp70 that is targeted to the organelle by a PTS2 sequence, but the function of this protein in import or protein assembly is unknown (Wimmer et al., 1997).

Two other cytosolic peroxins, Pex18p and Pex21p, are involved in *Saccharomyces cerevisiae* in the import of PTS2-containing proteins (Purdue et al., 1998). These are in a complex with Pex7p in the cytosol. Another protein, Pex20p, is needed in the cytosol for the dimerization and import of thiolase, a PTS2-containing protein (Titorenko et al., 1998). This protein interacts directly with Pex8p, an intraperoxisomal peroxin. The targeting of Pex8p is not dependent on the presence of Pex20p. In the absence of Pex8p, thiolase and Pex20p associate with the peroxisomes, and are protected from the action of proteases, leading to the suggestion that Pex20p may accompany thiolase into the peroxisome. Pex8p may play a role in the dissociation of Pex20p from thiolase or in the recycling of Pex20p to the cytosol (Smith and Rachubinski, 2001). It is worth noting that Pex8p also interacts with Pex5p, which is related to Pex20p via their N-terminal regions (Rehling et al., 2000; Titorenko et al., 1998).

Summary: *Import of proteins into the peroxisomal matrix*

- Proteins involved in peroxisomal protein import and biogenesis are called peroxins and are encoded by *PEX* genes. This nomenclature supersedes old names for these genes such as *PAS*, *PER*, *PAY*, etc.
- Protein targeting to peroxisomes is mediated by PTSs. The PTS1 and PTS2 are involved in protein targeting to the organelle matrix, while the mPTS targets proteins to the peroxisome membrane.
- The PTS1 and PTS2 sequences are recognized in the cytosol by the PTS receptors, Pex5p and Pex7p, respectively.
- The PTS receptor/cargo complex interacts on the peroxisomal membrane with the docking subcomplex comprising the peroxins Pex3p, Pex13p, Pex14p and Pex17p.
- The PTS receptor/cargo complex (shown for Pex5p) is transferred to a putative translocation subcomplex, consisting of the three ring-finger, zinc-binding proteins, Pex2p, Pex10p and Pex12p.
- Unlike the transport of proteins into many other organelles, folded and oligomerized proteins are transported across the peroxisomal membrane. The PTSs do not need to be part of the primary amino acid sequence of the protein, and even non-proteinaceous gold particles can be transported into the peroxisome matrix when coupled to a PTS1 peptide.
- In addition to the proteins of the docking and translocation subcomplexes, several other peroxins and chaperones are involved in the import and/or assembly of peroxisomal matrix proteins.

Import of peroxisomal membrane proteins

Relative to our knowledge of the import of peroxisomal matrix proteins, we know very little about the mechanism of insertion of proteins into the peroxisomal membrane. The targeting of most PMPs is believed to be post-translational and directly from the cytosol to the peroxisome (Lazarow and Fujiki, 1985). There is good evidence for these points in the case of three mammalian proteins, Pex2p, PMP22 and PMP70 (Diestelkotter and Just, 1993; Imanaka et al., 1996). The targeting of all integral PMPs is mediated by mPTSs. The mPTS receptor has not been identified, partly because there is no simple phenotypic selection for mutants deficient only in the mPTS import pathway. Because many PMPs are necessary also for the PTS1 and PTS2 import pathways, any mutant deficient in the mPTS receptor would be impaired in all peroxisomal import pathways. The lack of a consensus mPTS sequence in PMPs has also made biochemical approaches to finding the mPTS receptor unsuccessful. There has been a suggestion that the pre-dominantly cytosolic, farnesylated protein Pex19p binds to many PMPs, and may be the mPTS receptor (Sacksteder et al., 2000). However, there is also evidence that this is not the case (Snyder et al., 2000) because several PMPs are inserted into the membranes of peroxisome intermediates in the absence of Pex19p, and the Pex19p-binding sites on several PMPs do not overlap with the mPTS regions. Additionally Pex19p appears to act after PMP insertion into the peroxisomal membrane to assemble/disassemble PMP complexes. There is evidence for the formation of dynamic subcomplexes in the peroxisomal membrane, but the mechanisms by which these are regulated is unclear (Subramani et al., 2000).

For the few PMPs whose targeting has been analyzed *in vitro*, the process is independent of ATP, but dependent on cytosolic factors (Just and Diestelkotter, 1996; Pause et al., 1997). How the PMPs synthesized in the cytosol are kept from aggregating via their transmembrane domains has been addressed only partially by the finding that in mammalian systems, the cytosolic TCP1 (T-complex protein 1) ring complex (a chaperonin) is associated with PMP22 and other factors that could, in principle, include the mPTS receptor (Pause et al., 1997).

Peroxins implicated in PMP import and/or assembly
Four peroxins, Pex3p, Pex16p, Pex17p and Pex19p, are thought to play roles in PMP import and/or assembly (Subramani et al., 2000). Pex3p is a PMP involved in the earliest stages of peroxisome membrane biogenesis (Wiemer et al., 1996). In its absence, no peroxisome biogenesis intermediates have been detected and many other PMPs are unstable (Hettema et al., 2000). One function for Pex3p might be that it is the protein on the peroxisomal membrane with which Pex19p docks (Snyder et al., 1999a). As described above, this protein is believed to play a role in the assembly/disassembly of PMP complexes, after their insertion in the peroxisomal membrane (Snyder et al., 2000). Pex16p has been shown, in mammalian cells, to be involved in the early stages of peroxisomal membrane assembly, because in its absence, no detectable biogenesis intermediates have been found (South and Gould,

1999). However, the absence of detectable intermediates should be interpreted with caution because it is a negative result subject to the methods and reagents used to identify the intermediates. Additionally, although there are reports of no peroxisome intermediates in *pex16* and *pex19* mutants in certain cell types, these intermediates have been detected in other model organisms (Eitzen et al., 1997; Snyder et al., 1999a). Pex17p, a component of the docking subcomplex, plays a role in the efficient insertion of Pex3p into the peroxisomal membrane, but how it does this is not known (Snyder et al., 1999b).

Summary: *Import of peroxisomal membrane proteins*

- Peroxisomal membrane proteins use mPTSs for their targeting to the organelle membrane.
- Most PMPs are believed to be made in the cytosol and imported post-translationally from the cytosol to the peroxisome.
- The mPTS receptor and the translocation pores have not been defined unequivocally.
- PMP import is energy independent.
- Several peroxins (Pex3p, Pex16p, Pex17p, Pex19p) are involved in the import and/or assembly of PMP subcomplexes in the peroxisomal membrane, but their precise functions are unclear.
- Molecules (e.g. TCP1 ring complex) performing chaperone-like functions have been implicated in PMP import.
- Subcomplexes involving PMPs are formed dynamically in the peroxisomal membrane.

Unique features of peroxisomal protein import

In comparison with the transport of proteins across other organelle membranes, there are many features that set peroxisomal protein import apart.

1. Unlike the transport of proteins across membranes of the mitochondrion, endoplasmic reticulum or chloroplast where unfolded proteins cross the membrane, in the case of peroxisomes fully folded and oligomerized proteins can be imported (Walton et al., 1995). In fact, proteins lacking a true PTS can gain 'piggy-back' entry into the peroxisome matrix, in association with a homomeric or heteromeric protein containing a PTS (Glover et al., 1994; Hausler et al., 1996; Lee et al., 1997; McNew and Goodman, 1994).

2. In contrast to the endoplasmic reticulum, mitochondrion and chloroplast, the pathways for the import of matrix and membrane proteins are distinct for peroxisomes.

3. Heat shock proteins, such as Hsp40 and Hsp70, are needed for peroxisomal matrix protein import but these chaperones are not involved in keeping the cargo protein in the unfolded state (Hettema et al., 1998; Preisig-Muller et al., 1994; Walton et al., 1994).
4. None of the peroxins (with the exception of the AAA-family members Pex1p and Pex6p) are related to proteins involved in polypeptide transport into other organelles (Faber et al., 1998).

PEROXISOME INHERITANCE AND BIOGENESIS INTERMEDIATES

Peroxisome inheritance to daughter cells

In mammalian cells, peroxisome distribution in the cell and its movement are dependent on microtubules. The majority of the peroxisomes are associated with microtubules, except at the time of mitosis (Rapp et al., 1996; Schrader et al., 1996; Wiemer et al., 1997). Using the jellyfish green fluorescent protein appended to a PTS1 sequence (GFP-SKL), the organelle can be monitored in living cells undergoing mitosis. Peroxisomes are distributed to daughter cells in a 'stochastic' manner, meaning that with multiple, randomly distributed organelles there is a high probability that each daughter cell will get some peroxisomes during cell division. Upon completion of cytokinesis, the process of constitutive peroxisome division is believed to restore the required number of peroxisomes to each cell.

Metabolic control of peroxisome proliferation

The number, size and proliferation of peroxisomes can vary in response to nutritional cues and the enzymatic content of the organelle (Chang et al., 1999a; Sakai et al., 1998). In yeast, as well as in mammalian cells, the peroxin Pex11p was found to cause peroxisome proliferation upon overexpression (Erdmann and Blobel, 1995; Marshall et al., 1995, 1996; Sakai et al., 1995; Schrader et al., 1998). In the absence of this protein, peroxisome division, and consequently proliferation, was impaired causing the accumulation of giant peroxisomes. However, this protein was shown recently to be involved in the β-oxidation of medium chain fatty acids (MCFA). In *S. cerevisiae* specifically, it is required for the transport of MCFA across the peroxisomal membrane prior to its activation by the acyl-CoA synthetase, Faa2p (van Roermund et al., 2000). Therefore, a peroxisomal membrane transporter can affect peroxisome proliferation, although a second function for Pex11p in this process has not been ruled out. It has been proposed that the MCFA oxidation pathway regulates the level of a signaling molecule that modulates peroxisome number. This idea is supported by the report that the *Candida boidinii* peroxisomal membrane protein, PMP47, which is required for oxidation of MCFA in yeast peroxisomes, probably as an ATP carrier, is also necessary for normal peroxisome proliferation (Nakagawa et al., 2000).

In mouse knock-outs of the peroxisomal acyl-CoA oxidase gene, there is excess peroxisome proliferation (Chang et al., 1999a; Fan et al., 1998). Here it is believed that this enzyme, when active, keeps the level of fatty-acyl-CoA and other PPARα ligands in check (Fan et al., 1998). The signaling and feedback mechanisms by which the metabolic activities of a subcellular organelle such as the peroxisome can modulate organelle proliferation are fascinating topics that deserve further study.

Peroxisomes can proliferate constitutively during normal growth of cells, and they can also be induced in response to metabolic need. For example, in yeasts, the β-oxidation of fatty acids occurs in the peroxisomes, so the organelle can be induced to proliferate upon growth on oleate. In *S. cerevisiae*, the induction is mediated by the oleate-responsive, *trans*-acting factors Oaf1 and Pip2(Oaf2) (Karpichev and Small, 1998; Small et al., 1997). Similarly, peroxisome biogenesis is induced in yeast cells that are respiration-deficient, as a means of compensating for the loss of oxidative phosphorylation by increasing the production of acetyl-CoA (Liu and Butow, 1999). Peroxisome proliferation has also been observed in mammals in response to chemicals or during development and cell differentiation. The transcription factor PPARα, a member of the steroid-hormone receptor superfamily, mediates this induction in mammalian cells by turning on specific genes.

Morphological observations, particularly with yeast, have led to the view that peroxisomes arise from preexisting peroxisomes, and that this occurs by budding and fission of old peroxisomes (Lazarow and Fujiki, 1985). There are several reports in the literature that question this long-held view, and suggest that under certain conditions (e.g. peroxisome induction by external cues), the organelle can arise *de novo*.

Peroxisome biogenesis intermediates during induction of the organelle

The principal argument questioning the generation of new peroxisomes solely from old ones is that, in certain yeast and human cell lines deficient in specific peroxins, no peroxisome remnants were detectable (i.e. no pre-existing organelles), and yet the organelle was recovered (apparently by *de novo* biogenesis) upon genetic complementation with the appropriate gene (South and Gould, 1999; Waterham et al., 1993).

An alternative reason for questioning this hypothesis is that some peroxins have been localized to distinct compartments that are distinguishable from mature peroxisomes, and their function is critical for peroxisome proliferation (Faber et al., 1998; Titorenko et al., 2000). One study has characterized five subpopulations of peroxisomes (P_1–P_5) and shown that vesicle fusion and maturation between these subpopulations allows a systematic progression to yield mature peroxisomes (Titorenko et al., 2000). It was fairly clear from these studies that peroxisomes can be induced to proliferate in a manner where they do not arise only from preexisting mature peroxisomes. However, it remains possible that the old hypothesis of peroxisome formation may apply during constitutive division of peroxisomes (South and Gould, 1999).

Insights into the early steps of peroxisome induction have come from genetic and biochemical studies. Using various *pex* mutants of *P. pastoris*, it was found that a *pex3Δ* mutant had no detectable remnants (Hettema et al., 2000; Muntau et al., 2000; Wiemer et al., 1996), *pex19Δ* cells had very small (early) pre-peroxisomes (Snyder et al., 1999a), and other *pex* mutants had intermediate size (late) pre-peroxisomes (Snyder et al., 1999a). It was suggested that these pre-peroxisomes were intermediates in peroxisome biogenesis (Subramani et al., 2000).

Elegant biochemical studies in *Yarrowia lipolytica* have shown the existence of even more intermediates (P_1–P_5) in addition to mature peroxisomes (P_6) (Titorenko et al., 2000). These subpopulations are distinguishable by the densities and by their content of proteins. The P_1 and P_2 populations first fuse in a Pex1p/Pex6p and ATP-dependent manner to yield P_3, which then matures, as shown by pulse-chase experiments, to P_4, then P_5 and eventually to mature peroxisomes P_6 (Figure 12.4). These data clearly contradict the view that peroxisomes arise from pre-existing mature peroxisomes.

Even in *pex* mutants that contain no detectable peroxisomes, it is impossible to say whether complementation by the missing gene generates new peroxisomes in a strictly *de novo* fashion, or whether some non-peroxisomal

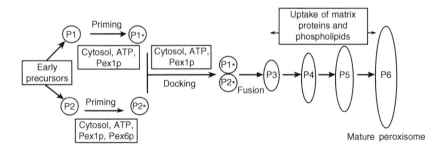

Figure 12.4 Proposed scheme for the generation of mature peroxisomes in *Y. lipolytica* from other biogenesis intermediates (reproduced with permission from Titorenko et al., 2000, *J Cell Biol* **150**: 881–886, by copyright permission of The Rockefeller University Press). The early precursors have been shown to fuse after priming and docking steps requiring the presence of the AAA-family ATPases, Pex1p and Pex6p. Fusion of P1* and P2* yields P3, which then matures via P4 and P5, to yield mature peroxisomes, P6.

compartment is recruited for this purpose. There are proposals of an involvement of the endoplasmic reticulum in peroxisome production (Titorenko and Rachubinski, 1998).

Summary: *Peroxisome inheritance and biogenesis intermediates*

- Peroxisomes are associated with microtubules in mammalian cells. They are distributed to daughter cells in a stochastic manner during cell division.
- Peroxisomes proliferate (i.e. divide) both constitutively and in response to metabolic cues.
- The prevailing view that all new peroxisomes arise by budding and fission of preexisting peroxisomes may apply only to the constitutive division of peroxisomes.
- There is increasing evidence that a number of intermediates of different size, composition and density are involved in peroxisome biogenesis, especially when the organelle divides in response to environmental cues.
- Peroxisomal enzymes and metabolic pathways can influence the proliferation of the organelle via metabolites that might modulate transcription factors.
- The origin of the lipids and membranes for the biogenesis intermediates has not been defined clearly, but the involvement of the endoplasmic reticulum has been suggested. Resolution of this question is needed before one can answer whether peroxisomes truly arise *de novo*.

PEROXISOMES AND HUMAN DISEASE

In the early 1960s, Hans Zellweger and his colleagues described the cerebrohepatorenal syndrome, later named Zellweger syndrome (Subramani et al., 2000). The role of peroxisomes in this disease became apparent through the efforts of Sidney Goldfischer. In the last four decades, two broad classes of disorders involving peroxisomes have been recognized: (a) those affecting peroxisomal metabolic pathways, and (b) peroxisome import/biogenesis disorders. The former are caused by single gene mutations affecting the localization or activity of individual peroxisomal enzymes and do not involve mutations in *PEX* genes (Wanders and Tager, 1998). Diseases in the latter group result from mutations in *PEX1, PEX2, PEX3, PEX5, PEX6, PEX7, PEX10, PEX12, PEX13, PEX16* and *PEX19* (Fujiki, 2000; Gould and Valle, 2000). They include diseases such as Zellweger syndrome, infantile Refsum disease, neonatal adrenoleukodystrophy and rhizomelic chondrodysplasia punctata (RCDP). The first three of these diseases can be caused by mutations of varying severity in the same or different *PEX* gene (except *PEX7*), while RCDP is caused by mutations in the *PEX7* gene (Fujiki, 2000; Gould and Valle, 2000).

Several important advances contributed to the definition of the genes involved in the human peroxisome biogenesis disorders. Although the patients invariably succumb to these diseases before the age of 10, cells from the patients can be grown in the laboratory and used in genetic complementation tests involving fused cells to see if they recover peroxisomes or metabolic activities that reside in this organelle. This led to the establishment of over a dozen complementation groups that were defective (a) only in the PTS1 pathway (e.g. certain *PEX5* mutations), (b) only the PTS2 pathway (e.g. *PEX7* mutations), (c) in both PTS1 and PTS2 pathways (e.g. mutations in the genes encoding the docking or translocation subcomplex components), or (d) in PMP biogenesis (e.g. *PEX3*, *PEX16* or *PEX19* mutations) (Gould and Valle, 2000; Slawecki et al., 1995).

The attention of scientists in the field turned from impaired peroxiomes more specifically to the possibility of biogenesis defects when it was discovered that certain patients with Zellweger syndrome had membrane remnants or 'ghosts' that could be identified using indirect immunofluorescence with antibodies against PMPs, but most matrix proteins such as catalase and thiolase were mislocalized to the cytosol (Santos et al., 1988). This was followed by a direct demonstration of a matrix import defect in these cells when it was found that firefly luciferase, the peroxisomal protein that was used to elucidate the PTS1 sequence, was targeted to peroxisomes upon microinjection into normal human fibroblasts, but was cytosolic in cells from a Zellweger syndrome patient containing the peroxisome remnants (Walton et al., 1992). A more systematic analysis of the peroxisomal matrix and membrane protein import pathways in all the human complementation groups showed very clearly the same classes of biogenesis phenotypes observed in the yeast *pex* mutants (Slawecki et al., 1995). The peroxisome remnants seen in at least one *PEX5*-deficient CHO mutant cell line have been recently shown to be true intermediates in peroxisome biogenesis (Yamasaki et al., 1999).

Another advance was the generation and characterization of CHO cell mutants affected in *PEX* genes, which in fact paved the way for the isolation of the first mammalian *PEX* gene (*PEX2*) by the use of functional complementation (Tsukamoto et al., 1991), and the subsequent cloning of the homologous human gene. However, the single most important factor that led to the identification of the human genes involved in these disorders was the fact that the process of peroxisome biogenesis and the *PEX* genes are conserved in evolution. In other words, the knowledge of the yeast genes led quickly to the homologous human counterparts. As the yeast genes were discovered by genetic complementation of yeast *pex* mutants, human homologs were found in databases being generated by the Human Genome Project. The human candidate *PEX* genes identified initially by homology to yeast *PEX* genes were cloned and introduced into the human or CHO cell lines corresponding to different complementation groups, leading to the definitive identification of the genes and mutations responsible for most of the inherited human peroxisome biogenesis disorders (Fujiki, 2000; Gould and Valle, 2000).

Summary: *Peroxisomes and human disease*

- There are at least 17 human disorders affecting peroxisomes. These are caused by mutations either in *PEX* genes or in genes, other than those encoding peroxins, for various peroxisomal enzymes.
- The conservation of peroxisome biogenesis in evolution has accelerated the discovery of the genes and mutations causing most of the known human peroxisome biogenesis disorders.

ACKNOWLEDGMENTS

This work was supported by grant NIH DK41737 to S.S. We wish to thank past members of the laboratory for their scientific contributions.

REFERENCES

* Albertini, M., Rehling, P., Erdmann, R. et al. (1997) Pex14p, a peroxisomal membrane protein binding both receptors of the two PTS-dependent import pathways. *Cell* **89**: 83–92.

Baerends, R.J., Faber, K.N., Kram, A.M. et al. (2000) A stretch of positively charged amino acids at the N terminus of Hansenula polymorpha Pex3p is involved in incorporation of the protein into the peroxisomal membrane. *J Biol Chem* **275**: 9986–9995.

Chang, C.C., South, S., Warren, D. et al. (1999a) Metabolic control of peroxisome abundance. *J Cell Sci* **112**: 1579–1590.

Chang, C.C., Warren, D.S., Sacksteder, K.A. et al. (1999b) PEX12 interacts with PEX5 and PEX10 and acts downstream of receptor docking in peroxisomal matrix protein import. *J Cell Biol* **147**: 761–774.

De Duve, C. and Baudhuin, P. (1966) Peroxisomes (microbodies and related particles). *Physiol Rev* **46**: 323–357.

Diestelkotter, P. and Just, W.W. (1993) In vitro insertion of the 22-kD peroxisomal membrane protein into isolated rat liver peroxisomes. *J Cell Biol* **123**: 1717–1725.

* Dodt, G. and Gould, S.J. (1996) Multiple PEX genes are required for proper subcellular distribution and stability of Pex5p, the PTS1 receptor: evidence that PTS1 protein import is mediated by a cycling receptor. *J Cell Biol* **135**: 1763–1774.

* Further reading suggested by the author

* Dunn, W.A., Jr. (1994) Autophagy and related mechanisms of lysosome-mediated protein degradation. *Trends Cell Biol* **4**: 139–143.

Dyer, J.M., McNew, J.A. and Goodman, J.M. (1996) The sorting sequence of the peroxisomal integral membrane protein PMP47 is contained within a short hydrophilic loop. *J Cell Biol* **133**: 269–280.

Eitzen, G.A., Szilard, R.K. and Rachubinski, R.A. (1997) Enlarged peroxisomes are present in oleic acid-grown *Yarrowia lipolytica* overexpressing the PEX16 gene encoding an intraperoxisomal peripheral membrane peroxin. *J Cell Biol* **137**: 1265–1278.

Elgersma, Y., Kwast, L., Klein, A. et al. (1996) The SH3 domain of the peroxisomal membrane protein Pex13p functions as a docking site for Pex5p, a mobile receptor for peroxisomal proteins. *J Cell Biol* **135**: 97–109.

Erdmann, R. and Blobel, G. (1995) Giant peroxisomes in oleic acid-induced *Saccharomyces cerevisiae* lacking the peroxisomal membrane protein Pmp27p. *J Cell Biol* **128**: 509–523.

Erdmann, R. and Blobel, G. (1996) Identification of Pex13p a peroxisomal membrane receptor for the PTS1 recognition factor. *J Cell Biol* **135**: 111–121.

Faber, K.N., Heyman, J.A. and Subramani, S. (1998) Two AAA family peroxins, PpPex1p and PpPex6p, interact with each other in an ATP-dependent manner and are associated with different subcellular membranous structures distinct from peroxisomes. *Mol Cell Biol* **18**: 936–943.

Fan, C.Y., Pan, J., Usuda, N. et al. (1998) Steatohepatitis, spontaneous peroxisome proliferation and liver tumors in mice lacking peroxisomal fatty acyl-CoA oxidase. Implications for peroxisome proliferator-activated receptor alpha natural ligand metabolism. *J Biol Chem* **273**: 15639–15645.

* Fujiki, Y. (2000) Peroxisome biogenesis and peroxisome biogenesis disorders. *FEBS Lett* **476**: 42–46.

* Glover, J.R., Andrews, D.W. and Rachubinski, R.A. (1994) *Saccharomyces cerevisiae* peroxisomal thiolase is imported as a dimer. *Proc Natl Acad Sci USA* **91**: 10541–10545.

* Gould, S.J. and Valle, D. (2000) Peroxisome biogenesis disorders: genetics and cell biology. *Trends Genet* **16**: 340–345.

Gould, S.J., Keller, G.-A. and Subramani, S. (1987) Identification of a peroxisomal targeting signal at the carboxy terminus of firefly luciferase. *J Cell Biol* **105**: 2923–2931.

Gould, S.J., Keller, G.A., Hosken, N. et al. (1989) A conserved tripeptide sorts proteins to peroxisomes. *J Cell Biol* **108**: 1657–1664.

Gould, S.J., Kalish, J.E., Morrell, J.C. et al. (1996) Pex13p is an SH3 protein of the peroxisome membrane and a docking factor for the predominantly cytoplasmic PTS1 receptor. *J Cell Biol* **135**: 85–95.

Hausler, T., Stierhof, Y.-D., Wirtz, E. et al. (1996) Import of DHFR hybrid protein into glycosomes *in vivo* is not inhibited by the folate-analogue aminopterin. *J Cell Biol* **132**: 311–324.

Hettema, E.H., Ruigrok, C.C.M., Koerkamp, M.G. et al. (1998) The cytosolic DnaJ-like protein djp1p is involved specifically in peroxisomal protein import. *J Cell Biol* **142**: 421–434.

* Further reading suggested by the author

Hettema, E.H., Girzalsky, W., van Den Berg, M. et al. (2000) *Saccharomyces cerevisiae* Pex3p and Pex19p are required for proper localization and stability of peroxisomal membrane proteins. *EMBO J* **19**: 223–233.

Huhse, B., Rehling, P., Albertini, M. et al. (1998) Pex17p of *Saccharomyces cerevisiae* is a novel peroxin and component of the peroxisomal protein translocation machinery. *J Cell Biol* **140**: 49–60.

Imanaka, T., Shiina, Y., Takano, T. et al. (1996) Insertion of the 70-kDa peroxisomal membrane protein into peroxisomal membranes *in vivo* and *in vitro*. *J Biol Chem* **271**: 3706–3713.

Johnson, M.A., Snyder, W.B., Cereghino, J.L. et al. (2001) *Pichia pastoris* Pex14p, a phosphorylated peroxisomal membrane protein, is part of a PTS-receptor docking complex and interacts with many peroxins. *Yeast* **18**: 621–641.

Just, W.W. and Diestelkotter, P. (1996) Protein insertion into the peroxisomal membrane. *Ann NY Acad Sci* **804**: 60–75.

Karpichev, I.V. and Small, G.M. (1998) Global regulatory functions of Oaf1p and Pip2p (Oaf2p), transcription factors that regulate genes encoding peroxisomal proteins in *Saccharomyces cerevisiae*. *Mol Cell Biol* **18**: 6560–6570.

Kunau, W.H., Beyer, A., Franken, T. et al. (1993) Two complementary approaches to study peroxisome biogenesis in *Saccharomyces cerevisiae*: forward and reversed genetics. *Biochimie* **75**: 209–224.

* Lazarow, P.B. and Fujiki, Y. (1985) Biogenesis of peroxisomes. *Annu Rev Cell Biol* **1**: 489–530.

Lee, M.S., Mullen, R.T. and Trelease, R.N. (1997) Oilseed isocitrate lyases lacking their essential type 1 peroxisomal targeting signal are piggybacked to glyoxysomes. *Plant Cell* **9**: 185–197.

Legakis, J.E. and Terlecky, S.R. (2001) PTS2 Protein import into mammalian peroxisomes. *Traffic* **2**: 252–260.

Liu, Z. and Butow, R.A. (1999) A transcriptional switch in the expression of yeast tricarboxylic acid cycle genes in response to a reduction or loss of respiratory function. *Mol Cell Biol* **19**: 6720–6728.

Marshall, P.A., Krimkevich, Y.I., Lark, R.H. et al. (1995) Pmp27 promotes peroxisomal proliferation. *J Cell Biol* **129**: 345–355.

Marshall, P.A., Dyer, J.M., Quick, M.E. et al. (1996) Redox-sensitive homodimerization of Pex11p: a proposed mechanism to regulate peroxisomal division. *J Cell Biol* **135**: 123–137.

Marzioch, M., Erdmann, R., Veenhuis, M. et al. (1994) PAS7 encodes a novel yeast member of the WD-40 protein family essential for import of 3-oxoacyl-CoA thiolase, a PTS2-containing protein, into peroxisomes. *EMBO J* **13**: 4908–4918.

McCollum, D., Monosov, E. and Subramani, S. (1993) The pas8 mutant of Pichia pastoris exhibits the peroxisomal protein import deficiencies of Zellweger syndrome cells – the PAS8 protein binds to the COOH-terminal tripeptide peroxisomal targeting signal, and is a member of the TPR protein family. *J Cell Biol* **121**: 761–774.

* McNew, J.A. and Goodman, J.M. (1994) An oligomeric protein is imported into peroxisomes *in vivo*. *J Cell Biol* **127**: 1245–1257.

* Further reading suggested by the author

Muntau, A.C., Mayerhofer, P.U., Paton, B.C. et al. (2000) Defective peroxisome membrane synthesis due to mutations in human PEX3 causes Zellweger syndrome, complementation group G. *Am J Hum Genet* **67**: 967–975.

Nakagawa, T., Imanaka, T., Morita, M. et al. (2000) Peroxisomal membrane protein Pmp47 is essential in the metabolism of middle-chain fatty acid in yeast peroxisomes and Is associated with peroxisome proliferation. *J Biol Chem* **275**: 3455–3461.

Okumoto, K., Abe, I. and Fujiki, Y. (2000) Molecular anatomy of the peroxin Pex12p: ring finger domain is essential for Pex12p function and interacts with the peroxisome-targeting signal type 1-receptor Pex5p and a ring peroxin, Pex10p. *J Biol Chem* **275**: 25700–25710.

Osumi, T., Tsukamoto, T., Hata, S. et al. (1991) Amino-terminal presequence of the precursor of peroxisomal 3-ketoacyl-CoA thiolase is a cleavable signal peptide for peroxisomal targeting. *Biochem Biophys Res Commun* **181**: 947–954.

Otera, H., Harano, T., Honsho, M. et al. (2000) The mammalian peroxin Pex5pL, the longer isoform of the mobile peroxisome targeting signal (PTS) type 1 transporter, translocates the Pex7p.PTS2 protein complex into peroxisomes via its initial docking site, Pex14p. *J Biol Chem* **275**: 21703–21714.

Pause, B., Diestelkotter, P., Heid, H. et al. (1997) Cytosolic factors mediate protein insertion into the peroxisomal membrane. *FEBS Lett* **414**: 95–98.

Preisig-Muller, R., Muster, G. and Kindl, H. (1994) Heat shock enhances the amount of prenylated Dnaj protein at membranes of glyoxysomes. *Eur J Biochem* **219**: 57–63.

Purdue, P.E., Yang, X. and Lazarow, P.B. (1998) Pex18p and Pex21p, a novel pair of related peroxins essential for peroxisomal targeting by the PTS2 pathway. *J Cell Biol* **143**: 1859–1869.

Rachubinski, R.A. and Subramani, S. (1995) How proteins penetrate peroxisomes. *Cell* **83**: 525–528.

Rapp, S., Saffrich, R., Anton, M. et al. (1996) Microtubule-based peroxisome movement. *J Cell Sci* **109**: 837–849.

Rehling, P., Marzioch, M., Niesen, F. et al. (1996) The import receptor for the peroxisomal targeting signal 2 (PTS2) in *Saccharomyces cerevisiae* is encoded by the PAS7 gene. *EMBO J* **15**: 2901–2913.

Rehling, P., Skaletz-Rorowski, A., Girzalsky, W. et al. (2000) Pex8p, an intraperoxisomal peroxin of *Saccharomyces cerevisiae* required for protein transport into peroxisomes binds the PTS1 receptor Pex5p. *J Biol Chem* **275**: 3593–3602.

Rhodin, J. (1954) Correlation of ultrastructural organization and function in normal and experimentally changed proximal convoluted tubule cells of the mouse kidney. Karolinska Institute, Stockholm, Sweden.

Sacksteder, K.A., Jones, J.M., South, S.T. et al. (2000) PEX19 binds multiple peroxisomal membrane proteins, is predominantly cytoplasmic, and is required for peroxisome membrane synthesis. *J Cell Biol* **148**: 931–944.

Sakai, Y., Marshall, P.A., Saiganji, A. et al. (1995) The *Candida boidinii* peroxisomal membrane protein Pmp30 has a role in peroxisomal proliferation and is functionally homologous to Pmp27 from *Saccharomyces cerevisiae*. *J Bacteriol* **177**: 6773–6781.

Sakai, Y., Yurimoto, H., Matsuo, H. et al. (1998) Regulation of peroxisomal proteins and organelle proliferation by multiple carbon sources in the methylotrophic yeast, *Candida boidinii*. *Yeast* **14**: 1175–1187.

Santos, M.J., Imanaka, T., Shio, H. et al. (1988) Peroxisomal membrane ghosts in Zellweger syndrome – aberrant organelle assembly. *Science* **239**: 1536–1538.

Schrader, M., Burkhardt, J.K., Baumgart, E. et al. (1996) Interaction of microtubules with peroxisomes. Tubular and spherical peroxisomes in HepG2 cells and their alteration induced by microtubule-active drugs. *Eur J Cell Biol* **69**: 24–35.

Schrader, M., Reuber, B.E., Morrell, J.C. et al. (1998) Expression of PEX11β mediates peroxisome proliferation in the absence of extracellular stimuli. *J Biol Chem* **273**: 29607–29614.

Slawecki, M.L., Dodt, G., Steinberg, S. et al. (1995) Identification of three distinct peroxisomal protein import defects in patients with peroxisome biogenesis disorders. *J Cell Sci* **108**: 1817–1829.

Small, G.M., Karpichev, I.V. and Luo, Y. (1997) Regulation of peroxisomal fatty acyl-CoA oxidase in the yeast *Saccharomyces cerevisiae. Adv Exp Med Biol* **422**: 157–166.

Smith, J.J. and Rachubinski, R.A. (2001) A role for the peroxin Pex8p in Pex20p-dependent thiolase import into peroxisomes of the yeast *Yarrowia lipolytica. J Biol Chem* **276**: 1618–1625.

Snyder, W.B., Faber, K.N., Wenzel, T.J. et al. (1999a) Pex19p interacts with Pex3p and Pex10p and is essential for peroxisome biogenesis in *Pichia pastoris. Mol Biol Cell* **10**: 1745–1761.

Snyder, W.B., Koller, A., Choy, A.J. et al. (1999b) Pex17p is required for import of both peroxisome membrane and lumenal proteins and interacts with Pex19p and the peroxisome targeting signal-receptor docking complex in *Pichia pastoris. Mol Biol Cell* **10**: 4005–4019.

Snyder, W.B., Koller, A., Choy, A.J. et al. (2000) The peroxin Pex19p interacts with multiple, integral membrane proteins at the peroxisomal membrane. *J Cell Biol* **149**: 1171–1178.

South, S.T. and Gould, S.J. (1999) Peroxisome synthesis in the absence of preexisting peroxisomes. *J Cell Biol* **144**: 255–266.

Subramani, S. (1993) Protein import into peroxisomes and biogenesis of the organelle. *Annu Rev Cell Biol* **9**: 445–478.

Subramani, S., Koller, A. and Snyder, W.B. (2000) Import of peroxisomal matrix and membrane proteins. *Annu Rev Biochem* **69**: 399–418.

Swinkels, B.W., Gould, S.J., Bodnar, A.G. et al. (1991) A novel, cleavable peroxisomal targeting signal at the amino-terminus of the rat 3-ketoacyl-CoA thiolase. *EMBO J* **10**: 3255–3262.

Szilard, R.K., Titorenko, V.I., Veenhuis, M. et al. (1995) Pay32p of the yeast *Yarrowia lipolytica* is an intraperoxisomal component of the matrix protein translocation machinery. *J Cell Biol* **131**: 1453–1469.

Terlecky, S.R., Nuttley, W.M., McCollum, D. et al. (1995) The *Pichia pastoris* peroxisomal protein PAS8p is the receptor for the C-terminal tripeptide peroxisomal targeting signal. *EMBO J* **14**: 3627–3634.

Terlecky, S.R., Legakis, J.E., Hueni, S.E. et al. (2001) Quantitative analysis of peroxisomal protein import *in vitro. Exp Cell Res* **263**: 98–106.

Titorenko, V.I. and Rachubinski, R.A. (1998) The endoplasmic reticulum plays an essential role in peroxisome biogenesis. *Trends Biochem Sci* **23**: 231–233.

Titorenko, V.I., Smith, J.J., Szilard, R.K. et al. (1998) Pex20p of the yeast *Yarrowia lipolytica* is required for the oligomerization of thiolase in the cytosol and for its targeting to the peroxisome. *J Cell Biol* **142**: 403–420.

* Titorenko, V.I., Chan, H. and Rachubinski, R.A. (2000) Fusion of small peroxisomal vesicles *in vitro* reconstructs an early step in the *in vivo* multistep peroxisome assembly pathway of *Yarrowia lipolytica. J Cell Biol* **148**: 29–44.

Tsukamoto, T., Miura, S. and Fujiki, Y. (1991) Restoration by a 35K membrane protein of peroxisome assembly in a peroxisome-deficient mammalian cell mutant. *Nature* **350**: 77–81.

Tuttle, D.L. and Dunn, W.A. (1995) Divergent modes of autophagy in the methylotrophic yeast *Pichia pastoris. J Cell Sci* **108**: 25–35.

Urquhart, A.J., Kennedy, D., Gould, S.J. et al. (2000) Interaction of Pex5p, the type 1 peroxisome targeting signal receptor, with the peroxisomal membrane proteins Pex14p and Pex13p. *J Biol Chem* **275**: 4127–4136.

van der Klei, I.J., Hilbrands, R.E., Kiel, J.A. et al. (1998) The ubiquitin-conjugating enzyme Pex4p of *Hansenula polymorpha* is required for efficient functioning of the PTS1 import machinery. *EMBO J* **17**: 3608–3618.

van Roermund, C.W., Tabak, H.F., van Den Berg, M. et al. (2000) Pex11p plays a primary role in medium-chain fatty acid oxidation, a process that affects peroxisome number and size in *Saccharomyces cerevisiae. J Cell Biol* **150**: 489–498.

Walton, P.A., Gould, S.J., Feramisco, J.R. et al. (1992) Transport of microinjected proteins into peroxisomes of mammalian cells: inability of Zellweger cell lines to import proteins with the SKL tripeptide peroxisomal targeting signal. *Mol Cell Biol* **12**: 531–541.

Walton, P.A., Wendland, M., Subramani, S. et al. (1994) Involvement of 70-kDa heat-shock proteins in peroxisomal import. *J Cell Biol* **125**: 1037–1046.

* Walton, P.A., Hill, P.E. and Subramani, S. (1995) Import of stably folded proteins into peroxisomes. *Mol Biol Cell* **6**: 675–683.

Wanders, R.J. and Tager, J.M. (1998) Lipid metabolism in peroxisomes in relation to human disease. *Mol Aspects Med* **19**: 69–154.

Waterham, H.R., Titorenko, V.I., Swaving, G.J. et al. (1993) Peroxisomes in the methylotrophic yeast *Hansenula polymorpha* do not necessarily derive from pre-existing organelles. *EMBO J* **12**: 4785–4794.

Wiemer, E.A. and Subramani, S. (1994) Protein import deficiencies in human peroxisomal disorders. *Mol Genet Med* **4**: 119–152.

Wiemer, E.A.C., Luers, G., Faber, K.N. et al. (1996) Isolation and characterization of Pas2p, a peroxisomal membrane protein essential for peroxisome biogenesis in the methylotrophic yeast *Pichia pastoris. J Biol Chem* **271**: 18973–18980.

Wiemer, E.A.C., Wenzel, T., Deerinck, T.J. et al. (1997) Visualization of the peroxisomal compartment in living cells: dynamic behavior and association with microtubules. *J Cell Biol* **136**: 71–80.

Wimmer, B., Lottspeich, F., van der Klei, I. et al. (1997) The glyoxysomal and plastid molecular chaperones (70-kDa heat shock protein) of watermelon cotyledons are encoded by a single gene. *Proc Natl Acad Sci USA* **94**: 13624–13629.

---

* Further reading suggested by the author

Yamasaki, M., Hashiguchi, N., Fujiwara, C. et al. (1999) Formation of peroxisomes from peroxisomal ghosts in a peroxisome-deficient mammalian cell mutant upon complementation by protein microinjection. *J Biol Chem* **274**: 35293–35296.

Zhang, J.W. and Lazarow, P.B. (1995) PEB1 (PAS7) in *Saccharomyces cerevisiae* encodes a hydrophilic, intra-peroxisomal protein that is a member of the WD repeat family and is essential for the import of thiolase into peroxisomes. *J Cell Biol* **129**: 65–80.

Zhang, J.W. and Lazarow, P.B. (1996) Peb1p (Pas7p) is an intraperoxisomal receptor for the NH_2-terminal, type 2, peroxisomal targeting signal of thiolase: Peb1p itself is targeted to peroxisomes by an NH_2-terminal peptide. *J Cell Biol* **132**: 325–334.

13

NUCLEOCYTOPLASMIC TRANSPORT

DIRK GÖRLICH AND STEFAN JÄKEL

The cell nucleus is the most prominent cellular organelle and the defining feature of the eukaryotic branch of life. It is surrounded by the nuclear envelope (NE), which separates the nuclear from the cytoplasmic compartment, uncouples protein synthesis (translation) from transcription and RNA processing and thereby also necessitates nucleocytoplasmic transport of RNAs and proteins. This transport proceeds through nuclear pore complexes (NPCs) and normally requires nuclear transport receptors that confer specificity to the transport processes. The transport machinery employs roughly 100 different, often highly abundant proteins and thus utilizes considerable cellular resources. However, these expenses clearly pay off, as indicated by the fact that only eukaryotes evolved into complex, multicellular organisms.

One can think of several reasons why such cellular complexity requires a cell nucleus. First, the containment of the genome within a specialized organelle certainly improves genetic stability and is probably a key factor that allows eukaryotes to handle 1000-times larger genomes than prokaryotes. Second, this compartmentation permits regulation of key cellular events at a level unavailable to prokaryotes, e.g. by controlling the access of transcriptional regulators to chromatin. And finally, there is the composition of typical eukaryotic genes from exons and introns, which requires the primary transcript to be spliced before translation may occur. The translation of unspliced pre-mRNAs typically yields mutilated, non-functional and potentially even dominant-negative protein fragments and so a number of mechanisms act in concert to make this an unlikely event. The NE is certainly instrumental for this purpose by preventing access of the cytoplasmic translation machinery to nascent transcripts and splicing intermediates in the nucleus.

Protein Targeting, Transport & Translocation
ISBN 0-12-200731-X

GENERAL ASPECTS OF NUCLEAR TRANSPORT

Before we discuss components and mechanisms of nuclear transport in detail, we would like to summarize a number of general characteristics and also point to parallels and differences between nuclear transport and some other cellular targeting pathways. Nuclear transport is a tremendous activity (for further reviews see Görlich and Kutay, 1999; Mattaj and Englmeier, 1998; Nakielny and Dreyfuss, 1999). Every nuclear protein, such as histones, polymerases, transcription factors etc., originates from the cytoplasm and needs to be imported. Considering that $\approx22\%$ of the $>35\,000$ human proteins are nuclear (Cokol et al., 2000), this amounts to a very large number of import substrates. mRNA and transfer RNA, on the other hand, are synthesized in the nucleus and need to be exported to the cytoplasm, where they function in translation. The biogenesis of ribosomes even involves multiple crossings of the NE: ribosomal proteins are first imported from the cytoplasm, assemble in the nucleolus with rRNA and finally are re-exported as ribosomal subunits to the cytoplasm. The biogenesis of the signal recognition particle (SRP) appears to occur in an analogous manner to ribosomes.

NPCs allow passage of material in essentially two modes: passive diffusion and facilitated translocation. Passive diffusion occurs with 'inert objects' that show no specific interaction with the NPCs. It is fast for small molecules, such as metabolites, but becomes restricted and inefficient as the diffusing objects approach a size-limit of $20–40\,kDa$ (Bonner, 1978; Paine et al., 1975). In contrast, facilitated translocation allows passage of even very large particles, extreme examples being ribosomal subunits (1.4 and 2.8 MDa), Balbiani ring particles (which are giant mRNPs of $>10\,MDa$, see Daneholt, 1997) or viral particles (Whittaker and Helenius, 1998). It is often coupled to an input of metabolic energy, which in turn permits active transport against a gradient of chemical activity.

Facilitated transport through NPCs does not occur at random, but instead in a signal- and receptor-mediated fashion. Many nuclear import and export signals have been identified to date; they often comprise just short peptides, but sometimes also large protein domains. This diversity of signals is recognized by a plethora of cognate nuclear transport receptors which fall into several categories. First there are importin β (Impβ)-type nuclear transport receptors, which account for the bulk of protein import and export, as well as for the export of tRNA, UsnRNA (uridylate-rich small nuclear RNA) and apparently also of SRP and ribosomal subunits (see below). There are 14 such receptors in the yeast *Saccharomyces cerevisiae* and probably 22 in human. The second type of receptor is NTF2 (nuclear transport factor 2). It mediates import of the GTPase Ran, whose nuclear localization is, in turn, required to make Impβ-type receptors function (see below). And finally, there is mRNA export which utilizes its own set of export mediators such as Mex67p (see also below).

Nuclear transport receptors bind cargo molecules on one side of the NE, translocate through NPCs, release their cargo and finally return to the original compartment to mediate a next round of transport. The employment of such shuttling receptors is distinctly different from import into the rough endoplasmic reticulum (rER), mitochondria or plastids (see Chapters 5, 10 and 11, respectively) and for a long time appeared to be a unique feature of the nuclear transport system. Very recent data, however, indicate that peroxisomal protein import might also rely on such shuttling receptors (Dammai and Subramani, 2001; Chapter 12, this volume).

Import signals that direct proteins from the cytoplasm into the rER, mitochondria or chloroplasts are normally removed during the transport event. In contrast, nuclear transport signals are not cleaved, for good reasons. First, nuclear import signals are often part of functional domains, such as RNA- or DNA-binding motifs. Secondly, many proteins constantly circulate between nucleus and cytoplasm and therefore need to be repeatedly imported into the nucleus. In addition, the open mitosis in higher eukaryotes results (once per cell cycle) in the mixing of the nuclear and cytoplasmic compartments. After mitosis, nuclear proteins must be re-imported into the newly formed nuclei, which is only possible if these proteins retained their import signals.

NPCs can transport native, fully folded proteins and even multi-subunit complexes or ribonucleoprotein particles (RNPs, such as ribosomal subunits) up to a diameter of ≈25 nm (Feldherr and Akin, 1990). Considering that the translocating species normally also includes bound transport receptors, the effective diameter of the translocating species might even amount to 40 nm (Ribbeck and Görlich, 2001). This transport of native proteins and large protein assemblies through the NPCs is fundamentally different from protein import into mitochondria, chloroplasts, or the rER where proteins cross the membrane only one by one and in a fully unfolded state (see Chapters 5, 10 and 11). However, the ability of transporting native proteins is also common to import into peroxisomes (Chapter 12), all forms of vesicular transport (Chapter 16) and transport through plasmodesmata (Zambryski and Crawford, 2000).

NUCLEAR PORE COMPLEXES

NPCs were first noticed in the early 1950s on electron micrographs and from very early on they were suspected to constitute the sites of nucleocytoplasmic exchange (for a review of this earlier work see Maul, 1977). This was eventually proven by experiments that combined microinjections of colloidal gold particles into amphibian oocytes with electron microscopy and these experiments indeed caught some gold particles in the state of NPC passage (Feldherr, 1962). Colloidal gold can easily be coated with a variety of proteins and more sophisticated versions of these experiments demonstrated that NPCs are highly selective for the uptake of nuclear (signal-bearing) proteins (Dingwall et al., 1982; Feldherr et al., 1984).

The number of NPCs per cell depends on the demand for nuclear transport and varies greatly with cell size and synthetic and proliferative activity. There are 200 NPCs in a yeast cell (Maul, 1977; Rout and Blobel, 1993), approximately 2000–4000 in a proliferating human cell (Maul et al., 1972) and ≈50 million in a mature *Xenopus* oocyte (Cordes et al., 1995).

NPCs are giant molecular machines with a mass of 125 MDa in higher eukaryotes (Reichelt et al., 1990). Their morphology has been studied at considerable detail by electron microscopy (Akey and Radermacher, 1993; Fahrenkrog et al., 2001; Pante and Aebi, 1996) and a schematic view is pictured in Figure 13.1. NPCs are characterized by an 8-fold rotational symmetry. Apart from the largely membrane-imbedded central core structure,

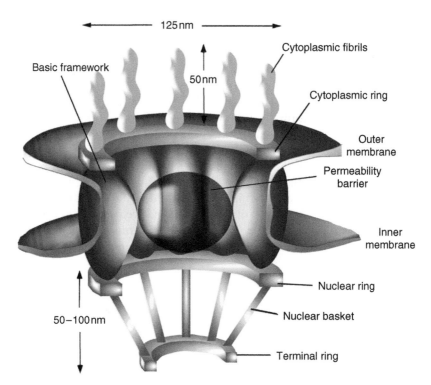

Figure 13.1 Cut-open model of a nuclear pore complex (NPC) embedded in the nuclear envelope.

NPCs also contain cytoplasmic and nuclear extensions that form cytoplasmic filaments and nuclear baskets, respectively. The inner and the outer nuclear membrane join at nuclear pore complexes, leaving a large central channel. Topologically, this is an important detail implying that nuclear transport does not proceed through a lipid bilayer, but so to say 'along' these membranes. The ≈40 nm central channel is not fully open, but filled by some permeability barrier. This barrier restricts the flux of inert macromolecules, but allows passage of nuclear transport receptors and receptor–cargo complexes. We will come back to this issue after having discussed the nuclear transport receptors in more detail.

NPCs are proteinaceous structures whose protein constituents are collectively referred to as nucleoporins. There are 30 different nucleoporins in yeast (Rout et al., 2000) and probably 30–50 in higher eukaryotes (Vasu and Forbes, 2001). For reasons of symmetry, the number of copies of a given nucleoporin per pore should be 8 or multiples thereof. Assuming a mass of 120 kDa for an average nucleoporin (Vasu and Forbes, 2001), one can calculate that higher-eukaryotic NPCs must be composed of ≈1000 individual polypeptides.

NPCs are not only giant by mass but also very respectable in terms of capacity. A single NPC can apparently accommodate a mass-flux of macromolecules of up to 80 MDa per second (Ribbeck and Görlich, 2001), which is roughly 5 orders of magnitude higher than the translocation capacity of the translocon in the rER membrane. The demand for nuclear transport capacity is of similar scale. For a proliferating HeLa cell, it can be estimated that each of their ≈3000 NPCs must transport 10–20 MDa material per second in order to supply the nucleus with enzymes, histones, ribosomal proteins and the cytoplasm with ribosomes, mRNA and tRNA (see Ribbeck and Görlich, 2001).

IMPORTIN β-LIKE NUCLEAR TRANSPORT RECEPTORS

Many of the fundamental insights into the nuclear transport machinery were gained through studying the import of proteins that carry a so-called classical nuclear localization signal (NLS). The first proof for the existence of the classical NLS was obtained through an elegant study of nuclear accumulation of nucleoplasmin (Dingwall et al., 1982). Nucleoplasmin is a pentameric nuclear protein from *Xenopus laevis* oocytes and consists of a protease-resistant core domain and a 'tail'. Intact nucleoplasmin rapidly enters the nucleus after being injected into the cytoplasm. When all the 'tails' were removed from the pentamer by protease treatment, the residual 'core' remained pentameric but failed to enter the nucleus. In contrast, the detached tails showed rapid nuclear accumulation, indicating that the tails contain some signal for nuclear targeting. Shortly after, the NLS from the SV40 large T-antigen could be precisely delineated (Kalderon et al., 1984). The import signals from nucleoplasmin and the SV40 large T-antigen represent the prototypes of the classical NLS (Kalderon et al., 1984; Makkerh et al., 1996; Robbins et al., 1991) and are recognized through the importin α/β (Impα/β) complex (see below). Besides this, many more nuclear signals exist that confer import by other receptors (see also below). Apart from a few exceptions, they usually coincide with the most basic region of a protein and are often part of RNA- or DNA-binding domains.

A great advance towards the identification of mediators of NLS-dependent nuclear protein import has been an *in vitro* assay based on permeabilized mammalian cells (Adam et al., 1990). The selective permeabilization of the cholesterol-rich plasma membrane with digitonin has two crucial consequences. First, a fluorescent import substrate can be introduced into the cells and its uptake followed by fluorescence microscopy. Second, the cells are depleted of their soluble contents. The observation that import required the re-addition of cytosol or cytosolic fractions (Adam et al., 1990; Moore and Blobel, 1992) provided an assay for essential, soluble transport factors. This approach resulted in the purification, molecular cloning and functional characterization of four key players in the classical, NLS-dependent nuclear import pathway, namely, the NLS-receptor importin α (Impα) (Adam and Adam, 1994; Görlich et al., 1994), Impβ (Chi et al., 1995; Görlich et al., 1995a; Imamoto et al., 1995a), Ran (Melchior et al., 1993; Moore and Blobel, 1993) and NTF2 (Moore and Blobel, 1994; Paschal and Gerace, 1995). Their function will be discussed below in detail. These factors have been given different names. Importins, for example, have occasionally also been called karyopherins and NTF2 is sometimes called p10. To avoid confusion, we will stick here to the importin and NTF2 nomenclature.

Impβ mediates the facilitated translocation of the NLS–Impα/β complex through NPCs (Görlich et al., 1996a; Weis et al., 1996) and therefore represents the actual import receptor in the classical import pathway. However, it cannot bind the classical NLS directly but instead through Impα, which functions as an import adaptor (Adam and Adam, 1994; Görlich et al., 1995a; Imamoto et al., 1995a; Weis et al., 1995). It has to be mentioned that Impβ can also directly bind and import cargo molecules, such as certain ribosomal proteins (Jäkel and Görlich, 1998). The use of an adaptor for NLS import complicates the transport scheme considerably and for the sake of simplicity, we first consider mechanistic aspects of the more simple cases, before we later return to the adaptor problem.

Transport cycles of Impβ-type transport receptors

Impβ is the prototype of an entire class of nuclear transport receptors (Fornerod et al., 1997b; Görlich et al., 1997) that comprises import mediators (importins) as well as exportins. These receptors recognize and bind cargo molecules, confer facilitated translocation through NPCs and thereby mediate cargo transport across the nuclear envelope (Figure 13.2). They constantly circulate between nucleus and cytoplasm and the question of how a receptor that moves in- and outward can mediate uni-directional import or export had been a major issue in the field.

An importin, for example, must bind its cargoes initially in the cytoplasm, translocate into the nucleus, release the cargo there and finally return to the cytoplasm in order to accomplish another round of import. This scenario predicts that cargo loading and release are regulated in a compartment-specific manner and that importins (and exportins) can 'somehow sense' a nuclear or a cytoplasmic environment. The RanGTP gradient model provides a plausible explanation for how this can be accomplished (Görlich et al., 1996b; Izaurralde et al., 1997; Nachury and Weis, 1999). Ran is a small GTPase that switches between a GDP- and GTP-bound form (Bischoff and Ponstingl, 1991b). The regulators of Ran's nucleotide-bound state are localized to opposite sides of the NE. The GTPase-activating protein (RanGAP, Bischoff et al., 1994) is excluded from the nucleus and depletes RanGTP from the cytoplasm, whereas the nucleotide exchange factor (called RCC1 or RanGEF, Bischoff and Ponstingl, 1991a) is nuclear, generating RanGTP in the nucleus. The expected result is a RanGTP gradient across the NE with a high RanGTP concentration in the nucleus and low levels in the cytoplasm. Strikingly, importins and exportins are RanGTP-binding proteins that respond to the RanGTP gradient by loading and unloading their cargo in the appropriate compartment.

Importins bind cargo molecules initially in the cytoplasm, release them upon binding to RanGTP in the nucleus (Chi et al., 1996; Görlich et al., 1996b; Izaurralde et al., 1997; Rexach and Blobel, 1995; Siomi et al., 1997) and return to the cytoplasm as RanGTP complexes without their cargo (Hieda et al., 1999; Izaurralde et al., 1997). To allow binding and import of another import substrate, RanGTP then needs to be removed from the importins. This happens by hydrolysis of the Ran-bound GTP, which in turn is triggered by RanGAP and either RanBP1 or RanBP2 as a cofactor (Bischoff and Görlich, 1997; Floer et al., 1997; Lounsbury and Macara, 1997).

An exportin, on the other hand, binds its substrates preferentially in the nucleus, forming a trimeric complex with RanGTP (Fornerod et al., 1997a; Kutay et al., 1997). The trimeric complex is then transferred to the cytoplasm where it is disassembled and the Ran-bound GTP is hydrolyzed. The substrate-free and Ran-free exportin can then re-enter the nucleus and bind and export the next cargo molecule (see Figure 13.2).

Importins and exportins constantly export RanGTP from the nucleus, which implies that the RanGTP gradient would soon collapse unless the nuclear RanGTP pool was efficiently replenished. This is accomplished by NTF2 (Ribbeck et al., 1998; Smith et al., 1998) and the exclusively

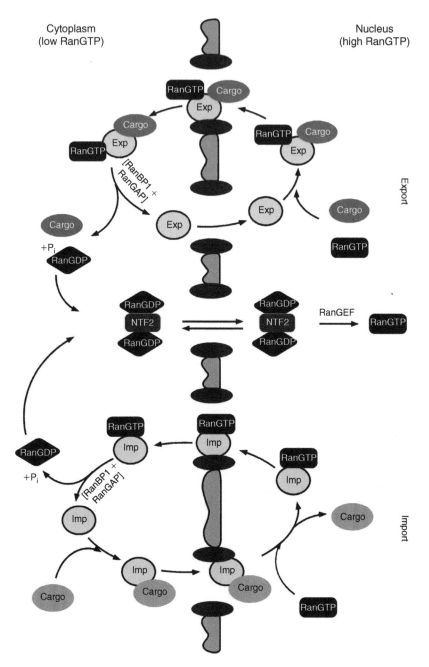

Figure 13.2   Transport cycles of importins (Imp) and exportins (Exp) and their coordination by the RanGTPase system.

nuclear-localized nucleotide exchange factor RCC1 (Bischoff and Ponstingl, 1991a). NTF2 mediates nuclear import of RanGDP, while the RCC1 recharges Ran with GTP in the nucleus.

Importins and exportins can accumulate cargoes against a gradient of chemical activity, which is an energy-consuming task. The RanGTPase system hydrolyzes per transport cycle one GTP molecule to GDP (see Figure 13.2) and, remarkably, this cytoplasmic GTP hydrolysis event represents the sole input of metabolic energy. This is an important point and implies that the facilitated translocation through NPCs per se is not directly coupled to nucleotide hydrolysis (see Englmeier et al., 1999; Kose et al., 1997; Nachury and Weis, 1999; Nakielny and Dreyfuss, 1998; Ribbeck et al., 1998, 1999; Schwoebel et al., 1998 and below). It also implies that nucleocytoplasmic exchange represents a quite economical mode of transport: the cleavage of a single high-energy phosphate bond normally suffices to actively transport one protein molecule across the NE. For comparison, more than 1000 high-energy phosphate bonds must be hydrolyzed to synthesize a 30 kDa protein from amino acids.

Transport events mediated by Impβ-type transport receptors

Having discussed some mechanistic aspects of receptor function, we will now describe several Impβ-type receptors in more detail. These receptors all interact with NPCs and specifically bind RanGTP. They are usually large (90–140 kDa) and acidic proteins. The overall sequence similarity between the various transport receptors is low and, in many cases, restricted to the N-terminal RanGTP-binding motif (Fornerod et al., 1997c; Görlich et al., 1997). This can at least in part be explained by the fact that these receptors bind very different cargoes, such as basic import signals in the case of Impβ (Görlich et al., 1996a; Henderson and Percipalle, 1997; Jäkel and Görlich, 1998; Weis et al., 1996), tRNA in the case of exportin-t (Arts et al., 1998a; Kutay et al., 1998), or a leucine-rich NES in the case of CRM1 (Fischer et al., 1995; Fornerod et al., 1997a; Stade et al., 1997; Wen et al., 1995). The RanGTP-binding motif can thus be considered as a diagnostic feature of Impβ-related transport receptors and in fact it allowed the identification of most of the 14 family members from the yeast *S. cerevisiae* (Fornerod et al., 1997c; Görlich et al., 1997; Wozniak et al., 1998). Higher eukaryotes employ an even larger number of transport receptors, at least 22 in the case of mammals (see Table 13.1 and E. Hartmann and D. Görlich, unpublished).

Importin β

Impβ is the prototypic import receptor (Chi et al., 1995; Görlich et al., 1995a; Imamoto et al., 1995a). It is essential in the yeast *S. cerevisiae* (Iovine

Table 13.1 Vertebrate nuclear transport receptors of the importin β-like family, their adaptors and examples of their cargo. Eleven receptors are designated importins, four are exportins. Importin 13 can apparently function as importin and exportin. Combinatorial usage of adaptors or co-receptors contributes to the broad substrate diversity of importin β and CRM1. The function of the following family members still remains to be elucidated: transportin 2 (Siomi et al., 1997), RanBP16 and 17 (Kutay et al., 2000), as well as RanBP4, 6, 21 and 22 (unpublished)

Receptor	Adaptor/ co-receptor	Examples of cargo	Reference
Importins			
Importin β		HIV Rev, HIV tat	Henderson and Percipalle, 1997; Truant and Cullen, 1999
		Ribosomal proteins	Jäkel and Görlich, 1998
	Importin αs	Classical NLS cargoes	See text
	Snurportin 1	m₃G capped UsnRNPs	Huber et al., 1998
	XRIP	Replication protein A	Jullien et al., 1999
	Importin 7	Linker histones	Jäkel et al., 1999
Importin 5		Ribosomal proteins	Deane et al., 1997; Jäkel and Görlich, 1998
Importin 7		Ribosomal proteins	Jäkel and Görlich, 1998
Importin 8		SRP19	Dean et al., 2001
Importin 9a+b		Ribosomal proteins	Unpublished
Importin 11		UbcM2	Plafker and Macara, 2000
Importin 13		hUbc9, RBM8/Y14 (and export of eIF1A)	Mingot et al., 2001
Transportin		hnRNP proteins A1, F (B, D, E)	Fridell et al., 1997; Pollard et al., 1996; Siomi et al., 1997
		Ribosomal proteins	Jäkel and Görlich, 1998
		TAP/NXF1	Bear et al., 1999; Truant et al., 1999
Transportin SR1+2		SR proteins	Kataoka et al., 1999; Lai et al., 2000
Exportins			
CRM1		Leucine-rich NES cargo	See text
	PHAX+CBC	m⁷G capped UsnRNAs	Ohno et al., 2000
	HIV Rev	RRE containing RNAs	Fischer et al., 1995; Malim et al., 1989, 1991
CAS		Importin αs	Kutay et al., 1997
Exportin-t		tRNAs	Arts et al., 1998a; Kutay et al., 1998
Exportin-4		eIF-5A	Lipowsky et al., 2000

et al., 1995; Koepp et al., 1996) and apparently conserved in all eukaryotes. The crystal structures of Impβ complexed with either RanGTP, the Impβ-binding (IBB-) domain from Impα, or with Phe-rich nucleoporin repeats have been solved (Bayliss et al., 2000; Cingolani et al., 1999; Vetter et al., 1999). The structures show Impβ as a purely α-helical protein consisting of tandemly

repeated motifs in which two α-helices are separated by either a flexible linker or a third short α-helix.

In the simplest case, Impβ binds and imports its cargo molecules directly, examples being the HIV Rev protein and some ribosomal proteins (Henderson and Percipalle, 1997; Jäkel and Görlich, 1998). However, Impβ from higher eukaryotes has a remarkable combinatorial flexibility and can combine with a variety of other factors to expand its substrate specificity. For example, it forms a dimer with another nuclear import receptor, importin 7 (Imp7), to import the linker histone H1 (Jäkel et al., 1999). Once the trimeric Impβ–Imp7–H1 complex has reached the nuclei, it is disassembled into its constituents by RanGTP-binding to both Impβ and Imp7. Both Impβ and Imp7 are autonomous transport receptors and can therefore return to the cytoplasm without the help of any *trans*-acting factor.

Impβ can also form complexes with the already mentioned adaptor molecules. Adaptors are involved in substrate recognition, but incapable of autonomous, facilitated translocation through NPCs. The best characterized example for such an adaptor is Impα (Adam and Adam, 1994; Görlich et al., 1994; Imamoto et al., 1995b). As already mentioned before, Impα recognizes the so-called classical NLS, which in turn is probably the most common nuclear import signal. While the yeast *S. cerevisiae* has only a single Impα species (known also as Srp1p or KAP60), mammals have at least six (see Köhler et al., 1999). There is significant functional overlap between the various Impα forms. However, some NLS substrate also have a great preference for one of the isoforms. All Impαs have an N-terminal binding site for Impβ, the IBB domain (Görlich et al., 1996a; Weis et al., 1996). The IBB domain itself constitutes an extremely potent nuclear import signal. However, it is also considerably larger (41 residues) and more basic than the classical NLS.

NLS–Impα–Impβ complexes form in the cytoplasm and as soon as they have entered nuclei, the complex is dissociated by RanGTP-binding to Impβ. While the Impβ–RanGTP complex can rapidly return to the cytoplasm on its own (Görlich et al., 1995b; Izaurralde et al., 1997), Impα cannot and needs the help of a *trans*-acting exportin. This exportin is called CAS (Cse1p in yeast) and forms trimeric RanGTP–CAS–Impα complexes (Kutay et al., 1997). The complex formation is highly cooperative, which ensures that CAS interacts with Impα only in the presence of RanGTP, i.e. in a nuclear environment. The complex can translocate through NPCs and becomes disassembled in the cytoplasm under GTP hydrolysis. Impα can then re-combine with Impβ to import another NLS protein, while CAS will re-enter nuclei to export another Impα molecule. A remarkable property of CAS is its preference for NLS-free Impα (Kutay et al., 1997; Solsbacher et al., 1998), which ensures that Impα exits nuclei only without the cargo it just carried in. This is obviously crucial to ensure a uni-directional NLS import cycle.

Snurportin 1 is a second type of Impβ-specific import adaptor (Huber et al., 1998) and involved in the biogenesis of the spliceosomal UsnRNPs U1, U2, U3, U4 and U5 (reviewed in Will and Lührmann, 2001). These pre-UsnRNA are initially synthesized as 7-methylguanosine (m^7G)-capped RNA polymerase II transcripts and exported for maturation to the cytoplasm. There, they assemble with the Sm proteins and their m^7cap becomes hyper-methylated to a 2,2,7-trimethylguanosine (m_3G) cap. The m_3G cap in turn is the signal for re-import into the nucleus and recognized by snurportin 1. Snurportin 1 also binds Impβ through an IBB domain and thereby triggers nuclear import of the mature UsnRNPs. Unlike Impα, snurportin 1 is not exported by CAS but instead by the exportin CRM1 (Paraskeva et al., 1999).

XRIPα is a third type of import adaptor (Jullien et al., 1999) and accounts for the import of the replication protein A (RPA) in *Xenopus*. So far nothing is known about its re-export to the cytoplasm.

Interestingly, the Impα/β dimer not only functions during interphase in nuclear import, but also has a function in mitosis, where it helps to orient the mitotic spindle. This, however, leads outside the focus of this chapter and we would refer the reader to some of the excellent original papers on that issue (Gruss et al., 2001; Nachury et al., 2001; Wiese et al., 2001).

IMPORT OF RIBOSOMAL PROTEINS

The biogenesis of ribosomes is a very complex process that involves both nuclear import and export events. Ribosomal proteins are first imported from the cytoplasm. Once in the nucleus, the ribosomal proteins assemble with rRNA in the nucleolus to form ribosomal subunits which are then finally re-exported to the cytoplasm. Ribosome biogenesis impressively demonstrates that nuclear transport is a major activity. For example, a HeLa cell contains 10 million ribosomes. It duplicates its contents and divides every 24 hours. This means that a total of ≈15 000 ribosomal subunits must be exported and 600 000 molecules of ribosomal proteins be imported every minute. Ribosomal proteins constitute thus an extremely abundant class of import substrates.

The import of so far only very few ribosomal proteins has been studied in detail. However, human rpL23a and its yeast homolog L25 might serve as a paradigm. L23a is not imported by the classical Impα/β pathway. Instead at least four distinct transport receptors, namely Impβ, transportin, importin 5 and importin 7, can directly bind and import rpL23a (Jäkel and Görlich, 1998). Likewise, yeast rpL25 can be imported by at least two factors, Yrb4p/KAP123 and Pse1p/KAP121 (Rout et al., 1997; Schlenstedt et al., 1997). This redundancy demonstrates an apparently common principle in nuclear import, namely that some substrates can 'choose' between several different carriers (see also Mosammaparast et al., 2001).

Ribosomal proteins are evolutionarily more ancient than are nuclear transport receptors and probably constituted one of the first import substrates of the putative progenitor of nowadays Impβ-like import receptors. While these receptors diversified in evolution, they probably acquired additional, specialized binding sites such as that for Impα in the case of Impβ, or that for the M9 domain (the import signal of the hnRNP A1 protein) in the case of transportin; but obviously they also maintained their capacity to bind and import ribosomal proteins. Transportin is indeed a good example to illustrate this: it binds its two types of import substrates (M9 domain and L23a) through distinct and non-overlapping binding sites (Jäkel and Görlich, 1998; Pollard et al., 1996). Extrapolating this to other nuclear transport receptors, we can expect to find an even greater number of import or export signals than there are nuclear transport receptors.

EXPORTIN-t

Eukaryotic tRNAs are initially synthesized as pre-tRNAs in the nucleus, processed to mature tRNA and exported to the cytoplasm. There, they participate in cycles of aminoacylation, binding to the elongation factor eEF1A, and function in translation. The nuclear export of tRNA is mediated by exportin-t in higher eukaryotes or by its ortholog Los1p in the yeast *S. cerevisiae*. Exportin-t functions according to the exportin paradigm described earlier and is so far the only nuclear transport receptor known to bind an RNA directly.

The maturation of pre-tRNAs occurs in the nucleus and includes trimming of the 5′ and 3′ ends, modification of a number of nucleosides, the post-transcriptional addition of the 3′ CCA end to which the amino acid is later attached, and in some cases also the removal of a small intron (for review see Wolin and Matera, 1999). Only mature tRNAs are finally exported to the cytoplasm (see for example Melton et al., 1980). It is quite remarkable that exportin-t preferentially binds and exports mature tRNAs which contain correctly processed 3′ and 5′ends and the appropriate nucleoside modifications (Arts et al., 1998b; Kutay et al., 1998; Lipowsky et al., 1999). Exportin-t mediated export thus constitutes a proof-reading or quality-control mechanism that coordinates RNA processing with export and thereby ensures that only functional tRNAs arrive in the cytoplasm.

CRM1

CAS and exportin-t are each specialized on the export of a single class of substrates, namely Impα and tRNA, respectively. In contrast, CRM1 (also called exportin 1) exports a very broad range of substrates, proteins as well as RNAs (Fornerod et al., 1997a; Ossareh-Nazari et al., 1997; Stade et al., 1997; Wolff et al., 1997). CRM1 appears conserved and essential in all eukaryotes and is the cellular target of the cytotoxic drug leptomycin B (LMB; Hamamoto et al., 1983). LMB covalently modifies a single cysteine in CRM1 and thereby selectively inactivates this receptor (Kudo et al., 1999; Neville and Rosbash, 1999). As LMB is membrane-permeable, it can be easily applied to a variety of cell types and used to test for an involvement of CRM1 in a given transport process.

CRM1 directly binds and exports proteins with a so-called leucine-rich nuclear export signal (NES; Bogerd et al., 1996; Fischer et al., 1995; Wen et al., 1995), examples being the protein kinase inhibitor (PKI) (see below), the tumor suppressor p53 and numerous transcription factors (Freedman and Levine, 1998; Kehlenbach et al., 1998). Like Impβ, CRM1 can also use adaptor molecules to expand its substrate specificity and in the following, we will describe a number of illustrative examples. Our first example is the HIV Rev protein, which plays a critical role during the replication of HIV (reviewed in Pollard and Malim, 1998). HIV-1, like other retroviruses, uses nuclear host enzymes for replication and produces proteins from several alternatively spliced mRNAs. Late in infection, the full-length genomic RNA must be exported from the nucleus in order to be packaged into viral particles. The problem is the presence of introns within this genomic RNA which would normally retain the RNA in the nucleus. One intron in the unspliced HIV-1 RNAs therefore contains the RRE (Rev responsive element), to which several copies of the export adaptor Rev bind. Rev, in turn, recruits CRM1 through its 'activation domain' (which is a leucine-rich NES), thereby allowing CRM1-mediated export of the unspliced RNA (Fischer et al., 1995; Malim et al., 1991; Wolff et al., 1997). Rev is then returned by Impβ to the nucleus. Here it is crucial that Rev cannot bind RNA and Impβ at the same time (Henderson and Percipalle, 1997), which ensures that only Rev is imported, while the RNA stays in the cytoplasm. In the nucleus, the Rev protein is dissociated by RanGTP from Impβ and can then bind and export a further RRE-containing RNA.

Our second example concerns the already mentioned export of the m^7G-capped pre-UsnRNAs. The monomethyl cap structure of the UsnRNAs serves in this case as the export signal (Fischer and Lührmann, 1990; Hamm and Mattaj, 1990) and recruits a complicated adaptor system. The cap structure is primarily recognized by the nuclear CAP-binding complex (CBC; Izaurralde et al., 1995) which also binds PHAX (Ohno et al., 2000). PHAX, in turn contains a leucine-rich NES that finally recruits CRM1 and thereby allows export.

Our third example is the export of the large (60S) ribosomal subunit to the cytoplasm. The export adaptor in this case is called NMD3. It binds to the ribosomes via the ribosomal protein L10 and also recruits CRM1 for export (Gadal et al., 2001; Ho et al., 2000). NMD3 and CRM1 are clearly essential for 60S export (at least in the yeast *S. cerevisiae*). However, it is still unclear whether the recruitment of a single exportin molecule is sufficient to mediate NPC passage of such a large particle. Alternatively, additional adaptor molecules might be involved. Export of the SRP is apparently also mediated by CRM1 (Ciufo and Brown, 2000).

TRANSPORT RECEPTORS THAT FUNCTION BOTH IN IMPORT AND IN EXPORT

A standard importin carries cargo only into the nucleus, but exits nuclei cargo-free. Exportins operate exactly the opposite way. The vast majority of Impβ-type transport receptors indeed appear to function either as importins or as exportins. However, recently two exceptions have been described. The yeast exportin Msn5p is on the one hand specialized on the export of phosphorylated transcription factors (Kaffman et al., 1998a, see below). However, it can also import RPA in yeast (Yoshida and Blobel, 2001). The second example is the human importin 13 (Mingot et al., 2001). It functions primarily in import and mediates nuclear uptake of ribosomal proteins and hUBC9. However, it can also export the translation initiation factor eIF1A and thereby helps to confine the translation machinery to the cytoplasm.

mRNA EXPORT

mRNAs are initially transcribed as precursors (pre-mRNAs, hnRNA) that need to be processed, assembled into ribonucleoprotein particles (RNPs) and finally exported to the cytoplasm. The maturation of the pre-mRNA includes a 5′ addition of a m^{7}G cap structure, removal of introns by splicing and 3′ poly-adenylation, while the assembly into RNPs involves a recruitment of numerous (pre-) mRNA binding proteins (such as hnRNP or SR proteins) and begins already during transcription. It is generally accepted that these mRNA-binding proteins play a critical role in all aspects of mRNA maturation and export.

mRNA maturation normally needs to be completed before export can occur. This order of events certainly makes sense as it avoids a cytoplasmic accumulation of immature mRNA. It is mainly the retention of intron-containing RNA by the splicing machinery that prevents a cytoplasmic appearance of unspliced mRNA (Chang and Sharp, 1989; Hamm and Mattaj, 1990; Legrain and Rosbash, 1989). However, cells do not entirely rely on this mechanism. A second line of defense represents the so-called nonsense-mediated decay (NMD), which rapidly degrades incorrectly spliced mRNA that have escaped the aforementioned retention and made it to the cytoplasm (Hentze and Kulozik, 1999; Lykke-Andersen, 2001).

The protein composition of a given RNP is not fixed, but changes as the (pre-) mRNA passes through the splicing, export and translation machineries. As a result of the splicing reaction, mRNAs recruit a specific set of mRNA-binding proteins and ultimately also export mediators (Kataoka et al., 2000; Le Hir et al., 2000; Strasser and Hurt, 2000; Zhou et al., 2000). The probably best characterized of these export mediators are the Mex67p–Mtr2p complex in yeast and its higher eukaryotic counterpart, the TAP–p15 complex (Grüter et al., 1998; Katahira et al., 1999; Segref et al., 1997) (note, TAP is also called NXF1 and has nothing to do with the TAP transporter in the rER membrane). The yeast Mex67 and vertebrate NXF1 complexes are essential for mRNA export, they bind to NPCs and facilitate the NPC passage of the mRNA.

Mex67/NXF1 and also Mtr2 and p15 are unrelated to Impβ-type nuclear transport receptors (Conti and Izaurralde, 2001). Accordingly, mRNA export per se occurs largely independently of the RanGTPase system. However, Ran-binding, Impβ-type receptors have an indirect role in mRNA export, because they recycle many of the (pre-) mRNA binding proteins and export factors back to the nucleus.

If mRNA-export is per se not directly coupled to the RanGTPase system, how can directionality be achieved in this case? The dead-box-protein DBP5 is an excellent candidate for that function (Schmitt et al., 1999; Snay-Hodge et al., 1998; Tseng et al., 1998). It is an ATP-driven RNA-helicase located at the cytoplasmic filaments of the NPC. It is believed that DBP5 removes export mediators from the mRNAs, making the NPC passage irreversible. As mentioned before, the released exporters are then rapidly re-imported and thus removed from the equilibrium. One could even imagine that the DBP5-mediated unwinding of the RNA exerts a pulling force that helps the mRNA out of the nucleus.

REGULATED NUCLEAR LOCALIZATION

Gene expression or cell cycle progression are regulated at many levels. One of these levels is regulated nuclear transport by which, for example, the access of key regulators to their nuclear targets can be accurately controlled (for reviews see Kaffman and O'Shea, 1999; Komeili and O'Shea, 2000; Takizawa and Morgan, 2000; Hoppe et al., 2001; Patil and Walter, 2001). The great number of so far described regulated nuclear transport events cannot be covered within a single chapter and we will therefore describe only a selection of illustrative cases.

Our first example is the yeast transcription factor Pho4p which becomes activated upon phosphate starvation and induces genes that ultimately improve phosphate utilization (for review see Kaffman and O'Shea, 1999). When yeast cells are grown in phosphate-rich medium, Pho4p is entirely cytoplasmic and thus physically separated from its target genes. Under these conditions, the Pho85–Pho80 kinase complex phosphorylates multiple sites in Pho4p (Kaffman et al., 1994), which inactivates the Pse1p-dependent NLS and thereby prevents nuclear import (Kaffman et al., 1998b). The phosphorylation also activates an Msn5p-dependent nuclear export signal and thereby promotes rapid nuclear export (Kaffman et al., 1998a). Phosphate starvation causes rapid dephosphorylation of Pho4p, which induces import, blocks export and thereby shifts Pho4 to the nucleus, where phosphate starvation genes can now be activated. The dual regulation of Pho4p import and export allows the system to respond rapidly and in both directions. An induction of the system by regulated nuclear import can obviously be much faster than a *de novo* synthesis of Pho4p that would include transcription, mRNA processing and export, as well as translation, and that would also appear costly when cellular resources are limited. Phosphorylation of Pho4p also reduces its capacity of transcriptional activation, but it is probably the combination of this regulation level with regulated nuclear import and export that allows a very tight control (Kaffman and O'Shea, 1999).

Pho4p is not the only Msn5-specific export substrate. This exportin also exports phosphorylated forms of the Mig1p glucose repressor (DeVit and Johnston, 1999), Rtg3p (which regulates nitrogen utilization; Komeili et al., 2000) and Far1p (which regulates the pheromone response; Blondel et al., 1999) and thus controls several cellular processes. This might also explain why it is a phosphorylation of the cargo, and not a modification of the receptor, that regulates the transport event. Switching on or off the entire export pathway would have pleiotropic effects and indiscriminately affect each of these signal transduction pathways.

Our next example is protein kinase A (PKA). It is involved in several signal transduction pathways and phosphorylates cytoplasmic as well as nuclear proteins. The inactive form of PKA consists of two catalytic and two (inhibitory) regulatory subunits (reviewed in Taylor et al., 1990). This complex is too large to diffuse into nuclei and is additionally retained in the cytoplasm by tethering to cytoplasmic structures (Meinkoth et al., 1993; Feliciello et al., 2001). Binding of cAMP to the regulatory subunits liberates the catalytic ones (reviewed in Taylor et al., 1990). The free kinase subunit is then small enough to enter nuclei by passive diffusion (Nigg et al., 1985) and phosphorylates nuclear targets such as the CREB protein (cyclic AMP response element binding protein; reviewed in Montminy, 1997). As cAMP levels decline, the (cytoplasmic) regulatory subunits rebind the catalytic ones and thereby rapidly quench PKA activity in the cytoplasm. The PKA inhibitor (PKI) is critical for the inactivation of the nuclear PKA pool. It binds and inhibits the catalytic subunits (Meinkoth et al., 1993). In addition, it provides an NES for rapid CRM1-mediated retrieval to the cytoplasm (Wen et al., 1995), which in turn allows a consequent silencing of the signal.

The sterol response element binding protein (SREBP) exemplifies another principle for regulated import of a transcription factor. SREBP initially resides as an integral membrane protein in the ER membrane and becomes proteolytically processed upon cholesterol depletion (Wang et al., 1994). The released soluble cytoplasmic domain is imported into the nucleus by Impβ (Nagoshi et al., 1999) and once in the nucleus, it activates genes required for cholesterol uptake or synthesis (Brown and Goldstein, 1999). The proteolytic activation of SREBP is obviously irreversible and so the transcriptional response is apparently only attenuated by the high turnover rate of the liberated transcription factor (Wang et al., 1994).

Regulated nuclear transport controls many more than just the processes described here. There are, for example, the ligand-induced import of nuclear hormone receptors (Picard and Yamamoto, 1987), regulated transport of the tumor suppressor p53 (Stommel et al., 1999), of the NFAT and Yap1 transcription factors (Kehlenbach et al., 1998; Yan et al., 1998), of cell cycle regulators such as CDC6 (Petersen et al., 1999) or the cyclin B–Cdc2 complex (Hagting et al., 1998; Yang et al., 1998) . In conclusion, there can be no doubt that eukaryotes make good use of the great regulatory potential offered by the nuclear envelope.

NUCLEAR PORE FUNCTION

We have so far discussed that a cargo, be it a protein, an mRNA, tRNA or even a ribosome, can traverse NPCs by facilitated translocation, provided appropriate transport receptors have been recruited. We will now turn to the question as to why these receptors are capable of facilitated translocation through NPCs while 'normal' proteins are not. This is the central, but largely unresolved problem in the nuclear transport field. However, a number of pieces to the puzzle have already been identified and so we will try to put them together.

NPC passage is not directly coupled to NTP hydrolysis or any other irreversible step and thus represents some kind of diffusion (see above). It proceeds through the central NPC channel, which has a diameter of roughly 40 nm. This channel cannot be fully open, but instead must be filled by some permeability barrier. This barrier is selective in terms of size: proteins smaller than 10–20 kDa pass nearly freely, while bovine serum albumin (BSA), for example, (68 kDa) remains essentially excluded (Bonner, 1978; Paine et al., 1975). However, size is not the only selectivity criterion. The import receptor transportin, for example, is larger ($\approx$100 kDa) than BSA and yet traverses NPCs >500 times faster (Ribbeck and Görlich, 2001). One can now ask why the translocation of transportin is so fast. This question can, however, also be rephrased to why NPC passage of BSA is so slow. The mechanism of facilitated translocation can obviously not be separated from the question for the nature of the permeability barrier; the two reflect one and the same problem.

What makes nuclear transport receptors so special is apparently their ability to interact with the so-called phenylalanine-rich repeats (see Bayliss et al., 2000 and references therein). These repeats can be considered as a diagnostic feature of nucleoporins and are characterized by short clusters of hydrophobic residues separated by very hydrophilic spacers (Rout and Wente, 1994).

Several lines of evidence support the assumption that these Phe-rich repeats might be major and functionally relevant constituents of the permeability barrier. First, monoclonal antibodies recognizing such repeats or the lectin wheat germ agglutinin, which binds sugars within the repeat regions, also stain the central channel (Akey and Goldfarb, 1989). Secondly, the repeats are estimated to be present in more than 1000 copies per NPC (Bayliss et al., 1999) and would thus be sufficiently abundant to constitute the principal structural element of the permeability barrier. Finally and most importantly, point mutations in NTF2 or Impβ, which impair their interaction with repeat domains, also compromise the facilitated translocation (Bayliss et al., 1999, 2000).

How can the interaction between the translocating species and the repeats facilitate NPC passage? A simple binding cannot explain the phenomenon and instead should cause a retention of the transport receptors and a delay of their passage. One possible explanation might be given by the selective phase model (Ribbeck and Görlich, 2001). In this model, the permeability barrier within the central channel is created by mutual attractions between the hydrophobic, Phe-rich clusters of nucleoporin repeats. This should result in a meshwork that restricts the flow of inert molecules. Such attraction would ensure the structural integrity of the permeability barrier, while the presence of the hydrophilic spacers between the Phe-rich clusters would prevent a collapse of the structure. It is easy to imagine how such interactions could create a sieve-like structure that allows passage of small molecules but restricts the flow of larger ones. Translocating material, however, can be incorporated into the meshwork, because it is able to interact with the Phe-rich clusters and thus take part in the mutual attraction between the repeats. The translocating species could thus selectively partition into the permeability barrier and use this 'selective solvation' to cross this permeability barrier at a high rate. In other words, the plug would seal around the translocating species and remain a barrier for inert molecules even when large objects pass.

ACKNOWLEDGMENTS

We thank Katharina Ribbeck for supplying Figure 13.1 and critical reading of the manuscript. Research in the authors' laboratory has been supported by the Deutsche Forschungsgemeinschaft.

REFERENCES

Adam, E.J.H. and Adam, S.A. (1994) Identification of cytosolic factors required for nuclear location sequence-mediated binding to the nuclear envelope. *J Cell Biol* **125**: 547–555.

Adam, S.A., Marr, R.S. and Gerace, L. (1990) Nuclear protein import in permeabilized mammalian cells requires soluble cytoplasmic factors. *J Cell Biol* **111**: 807–816.

Akey, C.W. and Goldfarb, D.S. (1989) Protein import through the nuclear pore complex is a multistep process. *J Cell Biol* **109**: 971–982.

Akey, C.W. and Radermacher, M. (1993) Architecture of the *Xenopus* nuclear pore complex revealed by three-dimensional cryo-electron microscopy. *J Cell Biol* **122**: 1–19.

Arts, G.J., Fornerod, M. and Mattaj, I.W. (1998a) Identification of a nuclear export receptor for tRNA. *Curr Biol* **8**: 305–314.

Arts, G.J., Kuersten, S., Romby, P., Ehresmann, B. and Mattaj, I.W. (1998b) The role of exportin-t in selective nuclear export of mature tRNAs. *EMBO J* **17**: 7430–7441.

Bayliss, R., Ribbeck, K., Akin, D. et al. (1999) Interaction between NTF2 and xFxFG-containing nucleoporins is required to mediate nuclear import of RanGDP. *J Mol Biol* **293**: 579–593.

Bayliss, R., Littlewood, T. and Stewart, M. (2000) Structural basis for the interaction between FxFG nucleoporin repeats and importin-beta in nuclear trafficking. *Cell* **102**: 99–108.

Bear, J., Tan, W., Zolotukhin, A.S. et al. (1999) Identification of novel import and export signals of human TAP, the protein that binds to the constitutive transport element of the type D retrovirus mRNAs. *Mol Cell Biol* **19**: 6306–6317.

Bischoff, F.R. and Görlich, D. (1997) RanBP1 is crucial for the release of RanGTP from importin beta-related nuclear transport factors. *FEBS Lett* **419**: 249–254.

Bischoff, F.R. and Ponstingl, H. (1991a) Catalysis of guanine nucleotide exchange on Ran by the mitotic regulator RCC1. *Nature* **354**: 80–82.

Bischoff, F.R. and Ponstingl, H. (1991b) Mitotic regulator protein RCC1 is complexed with a nuclear ras-related polypeptide. *Proc Natl Acad Sci USA* **88**: 10830–10834.

Bischoff, F.R., Klebe, C., Kretschmer, J., Wittinghofer, A. and Ponstingl, H. (1994) RanGAP1 induces GTPase activity of nuclear ras-related Ran. *Proc Natl Acad Sci USA* **91**: 2587–2591.

Blondel, M., Alepuz, P.M., Huang, L.S. et al. (1999) Nuclear export of Far1p in response to pheromones requires the export receptor Msn5p/Ste21p. *Genes Dev* **13**: 2284–2300.

Bogerd, H.P., Fridell, R.A., Benson, R.E., Hua, J. and Cullen, B.R. (1996) Protein sequence requirements for function of the human T-cell leukemia virus type 1 Rex nuclear export signal delineated by a novel *in vivo* randomization-selection assay. *Mol Cell Biol* **16**: 4207–4214.

Bonner, W.M. (1978) Protein migration and accumulation in nuclei. In: Busch, H. (ed.) *The Cell Nucleus*, Vol. 6, part C, pp. 97–148. New York: Academic Press.

Brown, M.S. and Goldstein, J.L. (1999) A proteolytic pathway that controls the cholesterol content of membranes, cells, and blood. *Proc Natl Acad Sci USA* **96**: 11041–11048.

Chang, D.D. and Sharp, P.A. (1989) Regulation by HIV Rev depends upon recognition of splice sites. *Cell* **59**: 789–795.

Chi, N.C., Adam, E.J. and Adam, S.A. (1995) Sequence and characterization of cytoplasmic nuclear protein import factor p97. *J Cell Biol* **130**: 265–274.

Chi, N.C., Adam, E.J.H., Visser, G.D. and Adam, S.A. (1996) RanBP1 stabilises the interaction of Ran with p97 in nuclear protein import. *J Cell Biol* **135**: 559–569.

Cingolani, G., Petosa, C., Weis, K. and Muller, C.W. (1999) Structure of importin-beta bound to the IBB domain of importin-alpha. *Nature* **399**: 221–229.

Ciufo, L.F. and Brown, J.D. (2000) Nuclear export of yeast signal recognition particle lacking Srp54p by the Xpo1p/Crm1p NES-dependent pathway. *Curr Biol* **10**: 1256–1264.

Cokol, M., Nair, R. and Rost, B. (2000) Finding nuclear localization signals. *EMBO Rep* **1**: 411–415.

Conti, E. and Izaurralde, E. (2001) Nucleocytoplasmic transport enters the atomic age. *Curr Opin Cell Biol* **13**: 310–319.

Cordes, V.C., Reidenbach, S. and Franke, W.W. (1995) High content of a nuclear pore complex protein in cytoplasmic annulate lamellae of *Xenopus* oocytes. *Eur J Cell Biol* **68**: 240–255.

Dammai, V. and Subramani, S. (2001) The human peroxisomal targeting signal receptor, Pex5p, is translocated into the peroxisomal matrix and recycled to the cytosol. *Cell* **105**: 187–196.

Daneholt, B. (1997) A look at messenger RNP moving through the nuclear pore. *Cell* **88**: 585–588.

Deane, R., Schafer, W., Zimmermann, H.P. et al. (1997) Ran-binding protein 5 (RanBP5) is related to the nuclear transport factor importin-beta but interacts differently with RanBP1. *Mol Cell Biol* **17**: 5087–5096.

Dean, K.A., von Ahsen, O., Görlich, D. and Fried, H.M. (2001) Signal recognition particle protein 19 is imported into the nucleus by importin 8 (RanBP8) and transportin. *J Cell Sci* **114**: 3479–3485.

DeVit, M.J. and Johnston, M. (1999) The nuclear exportin Msn5 is required for nuclear export of the Mig1 glucose repressor of *Saccharomyces cerevisiae*. *Curr Biol* **9**: 1231–1241.

Dingwall, C., Sharnick, S.V. and Laskey, R.A. (1982) A polypeptide domain that specifies migration of nucleoplasmin into the nucleus. *Cell* **30**: 449–458.

Englmeier, L., Olivo, J.C. and Mattaj, I.W. (1999) Receptor-mediated substrate translocation through the nuclear pore complex without nucleotide triphosphate hydrolysis. *Curr Biol* **9**: 30–41.

Fahrenkrog, B., Stoffler, D. and Aebi, U. (2001) Nuclear pore complex architecture and functional dynamics. *Curr Top Microbiol Immunol* **259**: 95–117.

Feldherr, C.M. (1962) The nuclear annuli as pathways for nucleocytoplasmic exchanges. *J Cell Biol* **14**: 65–72.

Feldherr, C.M. and Akin, D. (1990) EM visualization of nucleocytoplasmic transport processes. *Electron Microsp Rev* **3**: 73–86.

Feldherr, C.M., Kallenbach, E. and Schultz, N. (1984) Movement of a karyophilic protein through the nuclear pores of oocytes. *J Cell Biol* **99**: 2216–2222.

Feliciello, A., Gottesman, M.E. and Avvedimento, E.V. (2001) The biological functions of A-kinase anchor proteins. *J Mol Biol* **308**: 99–114.

Fischer, U. and Lührmann, R. (1990) An essential signaling role for the m3G cap in the transport of U1 snRNP to the nucleus. *Science* **249**: 786–790.

Fischer, U., Huber, J., Boelens, W.C., Mattaj, I.W. and Lührmann, R. (1995) The HIV-1 Rev activation domain is a nuclear export signal that accesses an export pathway used by specific cellular RNAs. *Cell* **82**: 475–483.

Floer, M., Blobel, G. and Rexach, M. (1997) Disassembly of RanGTP-karyopherin beta complex, an intermediate in nuclear protein import. *J Biol Chem* **272**: 19538–19546.

Fornerod, M., Ohno, M., Yoshida, M. and Mattaj, I.W. (1997a) Crm1 is an export receptor for leucine rich nuclear export signals. *Cell* **90**: 1051–1060.

Fornerod, M., van Baal, S., Valentine, V., Shapiro, D.N. and Grosveld, G. (1997b) Chromosomal localization of genes encoding CAN/Nup214-interacting proteins – human CRM1 localizes to 2p16, whereas Nup88 localizes to 17p13 and is physically linked to SF2p32. *Genomics* **42**: 538–540.

Fornerod, M., van Deursen, J., van Baal, S. et al. (1997c) The human homologue of yeast CRM1 is in a dynamic subcomplex with CAN/Nup214 and a novel nuclear pore component Nup88. *EMBO J* **16**: 807–816.

Freedman, D.A. and Levine, A.J. (1998) Nuclear export is required for degradation of endogenous p53 by MDM2 and human papillomavirus E6. *Mol Cell Biol* **18**: 7288–7293.

Fridell, R.A., Truant, R., Thorne, L., Benson, R.E. and Cullen, B.R. (1997) Nuclear import of hnRNP A1 is mediated by a novel cellular cofactor related to karyopherin-beta. *J Cell Sci* **110**: 1325–1331.

Gadal, O., Strauss, D., Kessl, J. et al. (2001) Nuclear export of 60s ribosomal subunits depends on Xpo1p and requires a nuclear export sequence-containing factor, Nmd3p, that associates with the large subunit protein Rpl10p. *Mol Cell Biol* **21**: 3405–3415.

Görlich, D. and Kutay, U. (1999) Transport between the cell nucleus and the cytoplasm. *Annu Rev Cell Dev Biol* **15**: 607–660.

Görlich, D., Prehn, S., Laskey, R.A. and Hartmann, E. (1994) Isolation of a protein that is essential for the first step of nuclear protein import. *Cell* **79**: 767–778.

Görlich, D., Kostka, S., Kraft, R. et al. (1995a) Two different subunits of importin cooperate to recognize nuclear localization signals and bind them to the nuclear envelope. *Curr Biol* **5**: 383–392.

Görlich, D., Vogel, F., Mills, A.D., Hartmann, E. and Laskey, R.A. (1995b) Distinct functions for the two importin subunits in nuclear protein import. *Nature* **377**: 246–248.

Görlich, D., Henklein, P., Laskey, R.A. and Hartmann, E. (1996a) A 41 amino acid motif in importin alpha confers binding to importin beta and hence transit into the nucleus. *EMBO J* **15**: 1810–1817.

Görlich, D., Pante, N., Kutay, U., Aebi, U. and Bischoff, F.R. (1996b) Identification of different roles for RanGDP and RanGTP in nuclear protein import. *EMBO J* **15**: 5584–5594.

Görlich, D., Dabrowski, M., Bischoff, F.R. et al. (1997) A novel class of RanGTP binding proteins. *J Cell Biol* **138**: 65–80.

Gruss, O.J., Carazo-Salas, R.E., Schatz, C.A. et al. (2001) Ran induces spindle assembly by reversing the inhibitory effect of importin alpha on TPX2 activity. *Cell* **104**: 83–93.

Grüter, P., Tabernero, C., von Kobbe, C. et al. (1998) TAP, the human homolog of Mex67p, mediates CTE-dependent RNA export from the nucleus. *Mol Cell* **1**: 649–659.

Hagting, A., Karlsson, C., Clute, P., Jackman, M. and Pines, J. (1998) MPF localization is controlled by nuclear export. *EMBO J* **17**: 4127–4138.

Hamamoto, T., Gunji, S., Tsuji, H. and Beppu, T. (1983) Leptomycins A and B, new antifungal antibiotics. I. Taxonomy of the producing strain and their fermentation, purification and characterization. *J Antibiot (Tokyo)* **36**: 639–645.

Hamm, J. and Mattaj, I.W. (1990) Monomethylated cap structures facilitate RNA export from the nucleus. *Cell* **63**: 109–118.

Henderson, B.R. and Percipalle, P. (1997) Interactions between HIV Rev and nuclear import and export factors: the Rev nuclear localisation signal mediates specific binding to human importin-beta. *J Mol Biol* **274**: 693–707.

Hentze, M.W. and Kulozik, A.E. (1999) A perfect message: RNA surveillance and nonsense-mediated decay. *Cell* **96**: 307–310.

Hieda, M., Tachibana, T., Yokoya, F. et al. (1999) A monoclonal antibody to the COOH-terminal acidic portion of Ran inhibits both the recycling of Ran and nuclear protein import in living cells. *J Cell Biol* **144**: 645–655.

Ho, J.H., Kallstrom, G. and Johnson, A.W. (2000) Nmd3p is a Crm1p-dependent adaptor protein for nuclear export of the large ribosomal subunit. *J Cell Biol* **151**: 1057–1066.

Hoppe, T., Rape, M. and Jentsch, S. (2001) Membrane-bound transcription factors: regulated release by RIP or RUP. *Curr Opin Cell Biol* **13**: 344–348.

Huber, J., Cronshagen, U., Kadokura, M. et al. (1998). Snurportin1, an m_3G-cap-specific nuclear import receptor with a novel domain structure. *EMBO J* **17**: 4114–4126.

Imamoto, N., Shimamoto, T., Kose, S. et al. (1995a) The nuclear pore targeting complex binds to nuclear pores after association with a karyophile. *FEBS Lett* **368**: 415–419.

Imamoto, N., Tachibana, T., Matsubae, M. and Yoneda, Y. (1995b) A karyophilic protein forms a stable complex with cytoplasmic components prior to nuclear pore binding. *J Biol Chem* **270**: 8559–8565.

Iovine, M.K., Watkins, J.L. and Wente, S.R. (1995) The GLFG repetitive region of the nucleoporin Nup116p interacts with Kap95p, an essential yeast nuclear import factor. *J Cell Biol* **131**: 1699–1713.

Izaurralde, E., Lewis, J., Gamberi, C. et al. (1995) A cap-binding protein complex mediating U snRNA export. *Nature* **376**: 709–712.

Izaurralde, E., Kutay, U., von Kobbe, C., Mattaji, I. and Görlich, D. (1997) The asymmetric distribution of the constituents of the Ran system is essential for transport into and out of the nucleus. *EMBO J* **16**: 6535–6547.

Jäkel, S. and Görlich, D. (1998) Importin beta, transportin, RanBP5 and RanBP7 mediate nuclear import of ribosomal proteins in mammalian cells. *EMBO J* **17**: 4491–4502.

Jäkel, S., Albig, W., Kutay, U. et al. (1999) The importin beta/importin 7 heterodimer is a functional nuclear import receptor for histone H1. *EMBO J* **18**: 2411–2423.

Jullien, D., Görlich, D., Laemmli, U. K. and Adachi, Y. (1999) Nuclear import of RPA in Xenopus egg extracts requires a novel protein XRIPalpha but not importin alpha. *EMBO J* **18**: 4348–4358.

Kaffman, A. and O'Shea, E.K. (1999) Regulation of nuclear localization: a key to a door. *Annu Rev Cell Dev Biol* **15**: 291–339.

Kaffman, A., Herskowitz, I., Tjian, R. and O'Shea, E.K. (1994) Phosphorylation of the transcription factor PHO4 by a cyclin-CDK complex, PHO80-PHO85. *Science* **263**: 1153–1156.

Kaffman, A., Rank, N.M., O'Neill, E.M., Huang, L.S. and O'Shea, E.K. (1998a) The receptor Msn5 exports the phosphorylated transcription factor Pho4 out of the nucleus. *Nature* **396**: 482–486.

Kaffman, A., Rank, N.M. and O'Shea, E.K. (1998b) Phosphorylation regulates association of the transcription factor Pho4 with its import receptor Pse1/Kap121. *Genes Dev* **12**: 2673–2683.

Kalderon, D., Richardson, W.D., Markham, A.F. and Smith, A.E. (1984) Sequence requirements for nuclear location of Simian Virus 40 large T antigen. *Nature* **311**: 33–38.

Katahira, J., Strasser, K., Podtelejnikov, A. et al. (1999) The Mex67p-mediated nuclear mRNA export pathway is conserved from yeast to human. *EMBO J* **18**: 2593–2609.

Kataoka, N., Bachorik, J.L. and Dreyfuss, G. (1999) Transportin-SR, a nuclear import receptor for SR proteins. *J Cell Biol* **145**: 1145–1152.

Kataoka, N., Yong, J., Kim, V.N. et al. (2000) Pre-mRNA splicing imprints mRNA in the nucleus with a novel RNA-binding protein that persists in the cytoplasm. *Mol Cell* **6**: 673–682.

Kehlenbach, R.H., Dickmanns, A. and Gerace, L. (1998) Nucleocytoplasmic shuttling factors including Ran and CRM1 mediate nuclear export of NFAT *in vitro*. *J Cell Biol* **141**: 863–874.

Koepp, D.M., Wong, D.H., Corbett, A.H. and Silver, P.A. (1996) Dynamic localization of the nuclear import receptor and its interactions with transport factors. *J Cell Biol* **133**: 1163–1176.

Köhler, M., Speck, C., Christiansen, M. et al. (1999) Evidence for distinct substrate specificities of importin alpha family members in nuclear protein import. *Mol Cell Biol* **19**: 7782–7791.

Komeili, A. and O'Shea, E.K. (2000) Nuclear transport and transcription. *Curr Opin Cell Biol* **12**: 355–360.

Komeili, A., Wedaman, K.P., O'Shea, E.K. and Powers, T. (2000) Mechanism of metabolic control. Target of rapamycin signaling links nitrogen quality to the activity of the Rtg1 and Rtg3 transcription factors. *J Cell Biol* **151**: 863–878.

Kose, S., Imamoto, N., Tachibana, T., Shimamoto, T. and Yoneda, Y. (1997) Ran-unassisted nuclear migration of a 97-kD component of nuclear pore-targeting complex. *J Cell Biol* **139**: 841–849.

Kudo, N., Matsumori, N., Taoka, H. et al. (1999) Leptomycin B inactivates CRM1/exportin 1 by covalent modification at a cysteine residue in the central conserved region. *Proc Natl Acad Sci USA* **96**: 9112–9117.

Kutay, U., Bischoff, F.R., Kostka, S., Kraft, R. and Görlich, D. (1997) Export of importin alpha from the nucleus is mediated by a specific nuclear transport factor. *Cell* **90**: 1061–1071.

Kutay, U., Lipowsky, G., Izaurralde, E. et al. (1998) Identification of a tRNA-specific nuclear export receptor. *Mol Cell* **1**: 359–369.

Kutay, U., Hartmann, E., Treichel, N. et al. (2000) Identification of two novel RanGTP-binding proteins belonging to the importin beta superfamily. *J Biol Chem* **275**: 40163–40168.

Lai, M.C., Lin, R.I., Huang, S.Y., Tsai, C.W. and Tarn, W.Y. (2000) A human importin-beta family protein, transportin-SR2, interacts with the phosphorylated RS domain of SR proteins. *J Biol Chem* **275**: 7950–7957.

Le Hir, H., Izaurralde, E., Maquat, L.E. and Moore, M.J. (2000) The spliceosome deposits multiple proteins 20–24 nucleotides upstream of mRNA exon–exon junctions. *EMBO J* **19**: 6860–6869.

Legrain, P. and Rosbash, M. (1989) Some *cis*- and *trans*-acting mutants for splicing target pre-mRNA to the cytoplasm. *Cell* **57**: 573–583.

Lipowsky, G., Bischoff, F.R., Izaurralde, E. et al. (1999) Coordination of tRNA nuclear export with processing of tRNA. *RNA* **5**: 539–549.

Lipowsky, G., Bischoff, F.R., Schwarzmaier, P. et al. (2000) Exportin 4: a mediator of a novel nuclear export pathway in higher eukaryotes. *EMBO J* **19**: 4362–4371.

Lounsbury, K.M. and Macara, I.G. (1997) Ran-binding protein 1 (RanBP1) forms a ternary complex with Ran and karyopherin beta and reduces Ran GTPase-activating protein (RanGAP) inhibition by karyopherin beta. *J Biol Chem* **272**: 551–555.

Lykke-Andersen, J. (2001) mRNA quality control: marking the message for life or death. *Curr Biol* **11**: R88–91.

Makkerh, J.P.S., Dingwall, C. and Laskey, R.A. (1996) Comparative mutagenesis of nuclear localization signals reveals the importance of neutral and acidic amino acids. *Curr Biol* **6**: 1025–1027.

Malim, M.H., Hauber, J., Le, S.Y., Maizel, J. and Cullen, B.R. (1989) The HIV-rev *trans*-activator acts through a structured target sequence to activate nuclear export of unspliced viral mRNA. *Nature* **338**: 254–257.

Malim, M.H., McCarn, D.F., Tiley, L.S. and Cullen, B.R. (1991). Mutational definition of the human immunodeficiency virus type 1 Rev activation domain. *J Virol* **65**: 4248–4254.

Mattaj, I.W. and Englmeier, L. (1998) Nucleocytoplasmic transport: the soluble phase. *Annu Rev Biochem* **67**: 265–306.

Maul, G.G. (1977) The nuclear and the cytoplasmic pore complex: structure, dynamics, distribution, and evolution. *Int Rev Cytol Suppl*: 75–186.

Maul, G.G., Maul, H.M., Scogna, J.E. et al. (1972) Time sequence of nuclear pore formation in phytohemagglutinin-stimulated lymphocytes and in HeLa cells during the cell cycle. *J Cell Biol* **55**: 433–447.

Meinkoth, J.L., Alberts, A.S., Went, W. et al. (1993) Signal transduction through the cAMP-dependent protein kinase. *Mol Cell Biochem* **127–128**: 179–186.

Melchior, F., Paschal, B., Evans, E. and Gerace, L. (1993) Inhibition of nuclear protein import by nonhydrolyzable analogs of GTP and identification of the small GTPase Ran/TC4 as an essential transport factor. *J Cell Biol* **123**: 1649–1659.

Melton, D.A., De Robertis, E.M. and Cortese, R. (1980) Order and intracellular location of the events involved in the maturation of a spliced tRNA. *Nature* **284**: 143–148.

Mingot, J.M., Kostka, S., Kraft, R., Hartmann, E. and Görlich, D. (2001) Importin 13: a novel mediator of nuclear import and export. *EMBO J* **20**: 3685–3694.

Montminy, M. (1997) Transcriptional regulation by cyclic AMP. *Annu Rev Biochem* **66**: 807–822.

Moore, M.S. and Blobel, G. (1992) The two steps of nuclear import, targeting to the nuclear envelope and translocation through the nuclear pore, require different cytosolic factors. *Cell* **69**: 939–950.

Moore, M.S. and Blobel, G. (1993) The GTP-binding protein Ran/TC4 is required for protein import into the nucleus. *Nature* **365**: 661–663.

Moore, M.S. and Blobel, G. (1994) Purification of a Ran-interacting protein that is required for protein import into the nucleus. *Proc Natl Acad Sci USA* **91**: 10212–10216.

Mosammaparast, N., Jackson, K.R., Guo, Y. et al. (2001) Nuclear import of histone H2A and H2B is mediated by a network of karyopherins. *J Cell Biol* **153**: 251–262.

Nachury, M.V. and Weis, K. (1999) The direction of transport through the nuclear pore can be inverted. *Proc Natl Acad Sci USA* **96**: 9622–9627.

Nachury, M.V., Maresca, T.J., Salmon, W.C. et al. (2001) Importin beta is a mitotic target of the small GTPase Ran in spindle assembly. *Cell* **104**: 95–106.

Nagoshi, E., Imamoto, N., Sato, R. and Yoneda, Y. (1999) Nuclear import of sterol regulatory element-binding protein-2, a basic helix-loop-helix-leucine zipper (bHLH-Zip)-containing transcription factor, occurs through the direct interaction of importin beta with HLH-Zip. *Mol Biol Cell* **10**: 2221–2233.

Nakielny, S. and Dreyfuss, G. (1998) Import and export of the nuclear protein import receptor transportin by a mechanism independent of GTP hydrolysis. *Curr Biol* **8**: 89–95.

Nakielny, S. and Dreyfuss, G. (1999) Transport of proteins and RNAs in and out of the nucleus. *Cell* **99**: 677–690.

Neville, M. and Rosbash, M. (1999) The NES-Crm1p export pathway is not a major mRNA export route in *Saccharomyces cerevisiae*. *EMBO J* **18**: 3746–3756.

Nigg, E.A., Hilz, H., Eppenberger, H.M. and Dutly, F. (1985) Rapid and reversible translocation of the catalytic subunit of cAMP-dependent protein kinase type II from the Golgi complex to the nucleus. *EMBO J* **4**: 2801–2806.

Ohno, M., Segref, A., Bachi, A., Wilm, M. and Mattaj, I.W. (2000) PHAX, a mediator of U snRNA nuclear export whose activity is regulated by phosphorylation. *Cell* **101**: 187–198.

Ossareh-Nazari, B., Bachelerie, F. and Dargemont, C. (1997) Evidence for a role of CRM1 in signal-mediated nuclear protein export. *Science* **278**: 141–144.

Paine, P.L., Moore, L.C. and Horowitz, S.B. (1975) Nuclear envelope permeability. *Nature* **254**: 109–114.

Palmeri, D. and Malim, M.H. (1999) Importin beta can mediate the nuclear import of an arginine-rich nuclear localization signal in the absence of importin alpha. *Mol Cell Biol* **19**: 1218–1225.

Pante, N. and Aebi, U. (1996) Molecular dissection of the nuclear pore complex. *Crit Rev Biochem Mol Biol* **31**: 153–199.

Paraskeva, E., Izaurralde, E., Bischoff, F.R. et al. (1999) CRM1-mediated recycling of snurportin 1 to the cytoplasm. *J Cell Biol* **145**: 255–264.

Paschal, B.M. and Gerace, L. (1995) Identification of NTF2, a cytosolic factor for nuclear import that interacts with nuclear pore protein p62. *J Cell Biol* **129**: 925–937.

Patil, C. and Walter, P. (2001) Intracellular signaling from the endoplasmic reticulum to the nucleus: the unfolded protein response in yeast and mammals. *Curr Opin Cell Biol* **13**: 349–355.

Petersen, B.O., Lukas, J., Sorensen, C.S., Bartek, J. and Helin, K. (1999) Phosphorylation of mammalian CDC6 by cyclin A/CDK2 regulates its subcellular localization. *EMBO J* **18**: 396–410.

Picard, D. and Yamamoto, K.R. (1987) Two signals mediate hormone-dependent nuclear localization of the glucocorticoid receptor. *EMBO J* **6**: 3333–3340.

Plafker, S.M. and Macara, I.G. (2000) Importin-11, a nuclear import receptor for the ubiquitin-conjugating enzyme, UbcM2. *EMBO J* **19**: 5502–5513.

Pollard, V.W., Michael, W.M., Nakielny, S. et al. (1996) A novel receptor-mediated nuclear protein import pathway. *Cell* **86**: 985–994.

Pollard, V.W. and Malim, M.H. (1998) The HIV-1 Rev protein. *Annu Rev Microbiol* **52**: 491–532.

Reichelt, R., Holzenburg, A., Buhle, E.L., Jr. et al. (1990) Correlation between structure and mass distribution of the nuclear pore complex and of distinct pore complex components. *J Cell Biol* **110**: 883–894.

Rexach, M. and Blobel, G. (1995) Protein import into nuclei: association and dissociation reactions involving transport substrate, transport factors, and nucleoporins. *Cell* **83**: 683–692.

Ribbeck, K. and Görlich, D. (2001) Kinetic analysis of translocation through nuclear pore complexes. *EMBO J* **20**: 1320–1330.

Ribbeck, K., Lipowsky, G., Kent, H.M., Stewart, M. and Görlich, D. (1998) NTF2 mediates nuclear import of Ran. *EMBO J* **17**: 6587–6598.

Ribbeck, K., Kutay, U., Paraskeva, E. and Görlich, D. (1999) The translocation of transportin-cargo complexes through nuclear pores is independent of both Ran and energy. *Curr Biol* **9**: 47–50.

Robbins, J., Dilworth, S.M., Laskey, R.A. and Dingwall, C. (1991) Two interdependent basic domains in nucleoplasmin nuclear targeting sequence: identification of a class of bipartite nuclear targeting sequence. *Cell* **64**: 615–623.

Rout, M.P. and Blobel, G. (1993) Isolation of the yeast nuclear pore complex. *J Cell Biol* **123**: 771–783.

Rout, M.P. and Wente, S.R. (1994) Pores for thought: nuclear pore complex proteins. *Trends Cell Biol* **4**: 357–364.

Rout, M.P., Blobel, G. and Aitchison, J.D. (1997) A distinct nuclear import pathway used by ribosomal proteins. *Cell* **89**: 715–725.

Rout, M.P., Aitchison, J.D., Suprapto, A. et al. (2000) The yeast nuclear pore complex: composition, architecture, and transport mechanism. *J Cell Biol* **148**: 635–651.

Schlenstedt, G., Smirnova, E., Deane, R. et al. (1997) Yrb4p, a yeast ran-GTP-binding protein involved in import of ribosomal protein L25 into the nucleus. *EMBO J* **16**: 6237–6249.

Schmitt, C., von Kobbe, C., Bachi, A. et al. (1999) Dbp5, a DEAD-box protein required for mRNA export, is recruited to the cytoplasmic fibrils of nuclear pore complex via a conserved interaction with CAN/Nup159p. *EMBO J* **18**: 4332–4347.

Schwoebel, E.D., Talcott, B., Cushman, I. and Moore, M.S. (1998) Ran-dependent signal-mediated nuclear import does not require GTP hydrolysis by Ran. *J Biol Chem* **273**: 35170–35175.

Segref, A., Sharma, K., Doye, V. et al. (1997) Mex67p, a novel factor for nuclear mRNA export, binds to both Poly(a)+ RNA and nuclear pores. *EMBO J* **16**: 3256–3271.

Siomi, M.C., Eder, P.S., Kataoka, N. et al. (1997) Transportin-mediated nuclear import of heterogeneous nuclear RNP proteins. *J Cell Biol* **138**: 1181–1192.

Smith, A., Brownawell, A. and Macara, I.G. (1998) Nuclear import of ran is mediated by the transport factor NTF2. *Curr Biol* **8**: 1403–1406.

Snay-Hodge, C.A., Colot, H.V., Goldstein, A.L. and Cole, C.N. (1998) Dbp5p/Rat8p is a yeast nuclear pore-associated DEAD-box protein essential for RNA export. *EMBO J* **17**: 2663–2676.

Solsbacher, J., Maurer, P., Bischoff, F.R. and Schlenstedt, G. (1998) Cse1p is involved in export of yeast importin alpha from the nucleus. *Mol Cell Biol* **18**: 6805–6815.

Stade, K., Ford, C.S., Guthrie, C. and Weis, K. (1997) Exportin 1 (Crm1p) is an essential nuclear export factor. *Cell* **90**: 1041–1050.

Stommel, J.M., Marchenko, N.D., Jimenez, G.S. et al. (1999) A leucine-rich nuclear export signal in the p53 tetramerization domain: regulation of subcellular localization and p53 activity by NES masking. *EMBO J* **18**: 1660–1672.

Strasser, K. and Hurt, E. (2000) Yra1p, a conserved nuclear RNA-binding protein, interacts directly with Mex67p and is required for mRNA export. *EMBO J* **19**: 410–420.

Takizawa, C.G. and Morgan, D.O. (2000) Control of mitosis by changes in the subcellular location of cyclin-B1-Cdk1 and Cdc25C. *Curr Opin Cell Biol* **12**: 658–665.

Taylor, S.S., Buechler, J.A. and Yonemoto, W. (1990) cAMP-dependent protein kinase: framework for a diverse family of regulatory enzymes. *Annu Rev Biochem* **59**: 971–1005.

Truant, R. and Cullen, B.R. (1999) The arginine-rich domains present in human immunodeficiency virus type 1 tat and rev function as direct importin beta-dependent nuclear localization signals. *Mol Cell Biol* **19**: 1210–1217.

Truant, R., Kang, Y. and Cullen, B.R. (1999) The human tap nuclear RNA export factor contains a novel transportin-dependent nuclear localization signal that lacks nuclear export signal function. *J Biol Chem* **274**: 32167–32171.

Tseng, S.S., Weaver, P.L., Liu, Y. et al. (1998) Dbp5p, a cytosolic RNA helicase, is required for poly(A)+ RNA export. *EMBO J* **17**: 2651–2662.

Vasu, S.K. and Forbes, D.J. (2001) Nuclear pores and nuclear assembly. *Curr Opin Cell Biol* **13**: 363–375.

Vetter, I.R., Arndt, A., Kutay, U., Görlich, D. and Wittinghofer, A. (1999) Structural view of the Ran-Importin beta interaction at 2.3 A resolution. *Cell* **97**: 635–646.

Wang, X., Sato, R., Brown, M.S., Hua, X. and Goldstein, J.L. (1994) SREBP-1, a membrane-bound transcription factor released by sterol-regulated proteolysis. *Cell* **77**: 53–62.

Weis, K., Mattaj, I.W. and Lamond, A.I. (1995) Identification of hSRP1α as a functional receptor for nuclear localization sequences. *Science* **268**: 1049–1053.

Weis, K., Ryder, U. and Lamond, A.I. (1996) The conserved amino-terminal domain of hSRP1 alpha is essential for nuclear protein import. *EMBO J* **15**: 1818–1825.

Wen, W., Meinkoth, J.L., Tsien, R.Y. and Taylor, S.S. (1995) Identification of a signal for rapid export of proteins from the nucleus. *Cell* **82**: 463–473.

Whittaker, G.R. and Helenius, A. (1998) Nuclear import and export of viruses and virus genomes. *Virology* **246**: 1–23.

Wiese, C., Wilde, A., Moore, M.S. et al. (2001) Role of importin-beta in coupling Ran to downstream targets in microtubule assembly. *Science* **291**: 653–656.

Will, C.L. and Lührmann, R. (2001) Spliceosomal UsnRNP biogenesis, structure and function. *Curr Opin Cell Biol* **13**: 290–301.

Wolff, B., Sanglier, J.J. and Wang, Y. (1997) Leptomycin B is an inhibitor of nuclear export: inhibition of nucleo-cytoplasmic translocation of the human immunodeficiency virus type 1 (HIV-1) Rev protein and Rev-dependent mRNA. *Chem Biol* **4**: 139–147.

Wolin, S.L. and Matera, G. (1999) The trials and travels of tRNA. *Genes Dev* **13**: 1–10.

Wozniak, R.W., Rout, M.P. and Aitchison, J.D. (1998) Karyopherins and kissing cousins. *Trends Cell Biol* **8**: 184–188.

Yan, C., Lee, L.H. and Davis, L.I. (1998) Crm1p mediates regulated nuclear export of a yeast AP-1-like transcription factor. *EMBO J* **17**: 7416–7429.

Yang, J., Bardes, E.S., Moore, J.D. et al. (1998) Control of cyclin B1 localization through regulated binding of the nuclear export factor CRM1. *Genes Dev* **12**: 2131–2143.

Yoshida, K. and Blobel, G. (2001) The karyopherin kap142p/msn5p mediates nuclear import and nuclear export of different cargo proteins. *J Cell Biol* **152**: 729–740.

Zambryski, P. and Crawford, K. (2000) Plasmodesmata: gatekeepers for cell-to-cell transport of developmental signals in plants. *Annu Rev Cell Dev Biol* **16**: 393–421.

Zhou, Z., Luo, M.J., Straesser, K. et al. (2000) The protein Aly links pre-messenger-RNA splicing to nuclear export in metazoans. *Nature* **407**: 401–405.

14

PROTEIN TRANSPORT TO THE YEAST VACUOLE

TODD R. GRAHAM AND STEVEN F. NOTHWEHR

INTRODUCTION

The study of protein transport to the yeast vacuole is providing a wealth of information on a remarkable variety of mechanisms used to sort and transport eukaryotic proteins. Proteins can gain entry into the vacuole by receptor-mediated endocytosis, by inheritance through the transfer of contents from the mother cell vacuole to the daughter (bud) vacuole, or by the related cytosol-to-vacuole transport (*cvt*) and autophagic (*apg*) pathways. It appears, however, that the majority of newly synthesized vacuolar proteins traverse the endoplasmic reticulum (ER) and Golgi complex along with secreted proteins before being sorted to the vacuole. Some proteins, such as alkaline phosphatase (ALP) and the vacuolar t-SNARE Vam3p, can be delivered directly from the Golgi complex to the vacuole while others, such as carboxypeptidase Y (CPY) and proteinase A, pass through a late endosome *en route* to the vacuole (Figure 14.1). The *cvt*, endocytic and inheritance pathways have recently been the subject of a few excellent reviews (Catlett and Weisman, 2000; D'Hondt et al., 2000; Klionsky and Emr, 2000) and so this chapter will focus on how proteins traveling the secretory pathway are sorted to the vacuole.

TRANSPORT AND MODIFICATION OF CPY THROUGH THE SECRETORY PATHWAY

CPY is perhaps the best-characterized vacuolar protein and like many other soluble vacuolar proteins, it is synthesized as a high molecular weight

Protein Targeting, Transport & Translocation
ISBN 0-12-200731-X

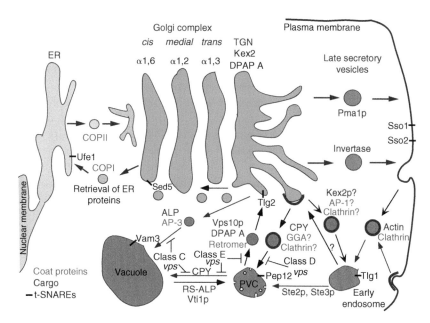

Figure 14.1   Protein transport through the secretory and endocytic pathways to the yeast vacuole. A few cargo proteins are listed in blue type with a larger list provided in Table 14.1. The ALP and CPY pathways define two major routes for delivery of newly synthesized proteins from the TGN to the vacuole, while the pheromone receptors, Ste2p and Ste3p, are internalized from the plasma membrane. CPY and the pheromone receptors pass through a late endosomal intermediate called the prevacuolar compartment (PVC) before arrival in the vacuole. Coat proteins that potentially mediate vesicular transport events between the TGN and endosomal/vacuolar system are listed in red type and are further defined in Tables 14.2 and 14.4. Genes encoding proteins implicated in fusion of transport intermediates with the PVC or vacuole are listed in Tables 14.3 and 14.6. The t-SNAREs implicated in these fusion steps label specific compartments in this membrane system. Note that the PVC internalizes membrane to produce a multivesicular body (MVB) and genes implicated in this process are listed in Table 14.5.

precursor with an N-terminal signal peptide that directs this protein across the membrane of the endoplasmic reticulum (ER). Through the action of ER resident enzymes, the cleavable signal peptide is removed, four N-linked oligosaccharides are added and CPY folds into a structure with the appropriate disulfide bonds. This 67 kDa core-glycosylated form of CPY is an inactive zymogen carrying an N-terminal propiece, and is designated the p1 or ER precursor form (reviewed in Klionsky et al., 1990).

The structure of the N-linked oligosaccharides added *en bloc* to glycoproteins in the yeast ER is the same as that found in most eukaryotic organisms ($GlcNAc_2Man_9Glc_3$). However, the subsequent trimming and elongation of these oligosaccharides are unique to the fungal system.

The three glucoses and a single mannose are removed in the ER before this structure is built up with additional mannose residues within the Golgi complex. Several secreted proteins, such as invertase and pro-α-factor, receive an extensive and heterogeneous outer chain composed of dozens of α1→6-, α1→2- and α1→3-linked mannose residues (reviewed in Dean, 1999). These modifications are added sequentially in functionally distinct Golgi compartments, which correspond respectively to the *cis-*, *medial-* and *trans-*Golgi cisternae. The *trans-*Golgi network (TGN) houses the Kex2p, Kex1p and dipeptidylaminopeptidase (DPAP A) proteases involved in the proteolytic cleavage of pro-α-factor and its subsequent processing to the mature α-factor peptide (Brigance et al., 2000).

CPY receives the same Golgi-specific carbohydrate modifications but in more modest portions. On average, CPY is modified with one α1→6-, one α1→2- and three α1→3-linked mannose residues on each N-linked oligosaccharide as it is transported through the Golgi complex (Ballou et al., 1990). This 69 kDa fully glycosylated CPY precursor is called the p2, or Golgi form of CPY. The N-linked oligosaccharides of CPY are also modified with phosphate but while mannose 6-phosphate is an important sorting determinant for lysosomal enzymes in animals, this modification does not contribute to the vacuolar sorting of CPY (reviewed in Klionsky et al., 1990).

COPII-coated transport vesicles mediate the export of p1 CPY from the ER and its subsequent import into the Golgi complex (reviewed in Barlowe, 2000). The *sec* mutants that carry temperature-sensitive forms of COPII components accumulate p1 CPY at the non-permissive temperature, demonstrating that transport to the Golgi complex is essential for further modification of this enzyme (Stevens et al., 1982). This anterograde protein transport pathway also appears to be coupled to the action of a COPI-dependent pathway *in vivo* since COPI mutants also accumulate p1 CPY. This is thought to reflect a requirement for COPI-coated vesicles to retrieve cargo receptors from the Golgi complex back to the ER for use in packaging CPY and other proteins into COPII vesicles (reviewed in Gaynor et al., 1998). The putative cargo receptor(s) that mediates exit of CPY from the ER is yet to be discovered, and so whether COPI plays a direct or indirect role in the anterograde transport of CPY, or other cargo proteins, has not been resolved. As described in Chapter 15, the mechanism of protein transport through the Golgi complex is also controversial.

CPY sorting from the secretory pathway

To identify the vacuolar localization signal within CPY that diverts this protein from the secretory pathway, Emr and colleagues expressed a set of fusion proteins between CPY and invertase in a strain carrying a deletion of the invertase gene (*SUC2*). This group demonstrated that a small segment of CPY, containing the signal peptide and the first 30 amino acids of the pro-piece, is sufficient to quantitatively deliver invertase to the vacuole. Taking a mutagenesis approach, Stevens and colleagues identified the same CPY segment and defined the critical amino acids, QRPL, lying just beyond the signal peptide cleavage site, that are necessary for sorting CPY to the vacuole. Mutation of this sorting signal causes secretion of p2 CPY (Klionsky and Emr, 1990; Conibear and Stevens, 1998).

Secreted and vacuolar proteins are equally affected in all of the *sec* mutants that exhibit a block in protein transport from the ER to the Golgi complex and through the Golgi complex. These pathways diverge within late Golgi compartments since *sec* mutations that block the fusion of late secretory vesicles with the plasma membrane have no effect on CPY transport to the vacuole (Stevens et al., 1982). To determine if vacuolar proteins enter the TGN *en route* to the vacuole, a Kex2 cleavage site was introduced into the junction between CPY and invertase in the fusion protein described above. This tripartite fusion protein received the full complement of outer chain mannose additions on invertase and was efficiently cleaved by Kex2p in the TGN before it could be sorted from the secretory pathway (Graham and Emr, 1991). These data suggest that CPY is sorted from secreted proteins within the TGN.

From the TGN, CPY is delivered first into a late endosome, also called the prevacuolar compartment (PVC), and subsequently to the vacuole, where the propiece of p2 CPY is removed by vacuolar proteases to produce the 61 kDa enzymatically active mature (mCPY) form. The proteolytic processing of CPY is dependent on the action of proteinase A (encoded by the *PEP4* gene), which controls the activation of several zymogens within the yeast vacuole, including proteinase B. It appears that cleavage of p2 CPY by both proteinase A and proteinase B results in the production of the final mCPY form (reviewed in Klionsky and Emr, 1990; Conibear and Stevens, 1998).

The presence of a sorting signal within the p2 CPY propiece implied that there must be other proteins responsible for recognizing the CPY sorting signal and delivering CPY to the vacuole. The Emr and Stevens groups predicted that mutations in these *trans*-acting proteins would have the same effect as mutating the CPY sorting signal: secretion of CPY. To identify the *trans*-acting machinery involved in this process, these groups devised clever genetic selection schemes to identify *vps* mutants exhibiting a defect in vacuolar protein sorting (see Historical Note 1). Others identified *vps* mutants in screens for mutants with defective vacuolar protease activity (*pep*) (Jones, 1977) or vacuolar morphology (*vam*) (Wada et al., 1988). What these investigators could not have predicted at the outset is that yeast can tolerate gross defects in vacuole biogenesis without loss of viability. This facilitated the mutational identification of approximately 55 *VPS* genes. All *vps* mutants secrete CPY but individual mutants have been assigned into six phenotypic classes (A–F) based on differences in the morphology of the vacuole and the localization of ALP and the vacuolar ATPase (Raymond et al., 1992).

Most of the *vps* mutants secrete the precursor form of several soluble vacuolar proteins (such as CPY, proteinase A and proteinase B). A notable exception is *vps10*, a mutant that secretes p2 CPY but exhibits a milder proteinase A sorting defect. Vps10p is an integral membrane resident of the TGN and is now known to be the CPY receptor (Marcusson et al., 1994). The proteins encoded by the remainder of the *VPS* genes function in many different protein trafficking steps between the Golgi complex, the endosomal system and the vacuole. The remainder of this chapter will explore these trafficking steps in greater detail.

Summary: *Transport and modification of CPY through the secretory pathway*

- CPY is synthesized as a high molecular weight precursor with an N-terminal signal sequence and propiece.
- Upon translocation into the ER, the signal peptide is cleaved and N-linked oligosaccharides are added to produce the p1 CPY precursor form.
- COPII-coated vesicles transport CPY from the ER to the Golgi complex, where additional glycosylation produces the p2 precursor form.
- A sorting signal (QRPL) in the CPY propiece is recognized by the Vps10 receptor within the TGN.
- CPY is transported from the TGN to a late endosome called the PVC before delivery to the vacuole. The propiece is proteolytically removed in the vacuole to produce the mature, enzymatically active form (mCPY).
- More than 50 different *vps* mutants have been isolated that exhibit a defect in protein transport between the TGN, PVC and vacuole. These mutants secrete p2 CPY.
- The *vps* mutants have been assigned to six phenotypic classes (A–F).

PROTEIN TRAFFICKING PATHWAYS EXITING THE TGN

A number of different transport vesicles, with different cargo and destinations, are produced from the TGN. In mammals, this compartment segregates lysosomal enzymes from proteins destined for the apical and basolateral surfaces of polarized epithelial cells, as well as proteins entering constitutive and regulated secretory pathways in endocrine cells. The yeast TGN can also produce two types of secretory vesicles that are distinguishable by their density and cargo content. The light vesicles carry the plasma membrane ATPase (Pma1p) and a β-glucanase (Bgl2p) while the more dense population carries invertase and acid phosphatase (Harsay and Bretscher, 1995). Golgi glycosyltransferases that traffic into the TGN appear to be sorted into a retrograde pathway for return to earlier Golgi compartments (Graham and Krasnov, 1995; Harris and Waters, 1996). The nature of the vesicles mediating this retrograde pathway is unknown but in the case of α1→3 mannosyltransferase (Mnn1p), the retrieval pathway requires a basal level of signaling through a MAP kinase cascade (Reynolds et al., 1998). There are also two pathways for sorting vacuolar proteins from the TGN: the CPY and the ALP pathways (reviewed in Odorizzi et al., 1998b). In addition, it appears that some proteins can cycle between the TGN and an early endosome in a pathway that might be distinct from the TGN to PVC route taken by CPY and its receptor (Holthuis et al., 1998; Black and Pelham, 2000). Thus, it is possible that three types of vesicles are produced from the yeast TGN that deliver proteins to the endosomal/vacuolar system although only the ALP pathway is well defined (Table 14.1).

Sorting of mammalian lysosomal enzymes into clathrin-coated vesicles

Historically, one of the most important paradigms in intracellular protein transport is the mannose 6-phosphate dependent sorting of lysosomal enzymes to the mammalian lysosome (see Historical Note 2). The mannose 6-phosphate recognition motif on N-glycans of lysosomal enzymes is generated in the early compartments of the Golgi complex by an enzyme that can recognize a tertiary structure common to many soluble lysosomal enzymes. Upon arrival in the TGN, lysosomal enzymes bind to the mannose 6-phosphate receptor and the complexes are packaged into clathrin-coated vesicles that bud from the TGN. Clathrin is composed of three heavy and three light chains that associate into a three-legged structure called a triskelion, which assembles with tetrameric adaptor protein (AP) complexes into a coat structure (Table 14.2). The small GTP binding protein ARF and the tetrameric adaptor protein AP-1, composed of two large (γ and β1) a medium (μ1) and a small (σ1) subunit, are required *in vitro* to recruit clathrin to the TGN. Packaging of the M6P receptor into clathrin-coated buds appears to be mediated by a sorting motif within the cytoplasmic tail of the M6P receptor that presumably interacts with AP-1 (Traub and Kornfeld, 1997).

After budding, the clathrin coat must be shed in order for the vesicle to fuse with the endosome. Uncoating of bovine brain clathrin-coated vesicles, primarily containing AP-2, requires the DnaJ protein auxilin and Hsc70 (the clathrin uncoating ATPase) (reviewed in Lemmon, 2001). The uncoated vesicles then fuse with a late endosome where the low pH environment

Table 14.1 Cargo proteins whose trafficking is commonly analyzed during study of the yeast TGN/endosomal system

Protein	Gene	Size*	Primary residence	Comment
α-factor	*MFA1*	165 aa	Secreted	Mating pheromone
A-ALP	*STE13-PHO8*	653 aa	TGN	DPAP A-ALP fusion; type II membrane protein
ALP	*PHO8*	566 aa	Vacuole	Alkaline phosphatase; type II membrane protein
CPS	*CPS1*	576 aa	Vacuole	Carboxypeptidase S; type II membrane protein
CPY	*PRC1*	532 aa	Vacuole	Carboxypeptidase Y; hydrophilic protein
DPAP A	*STE13*	931 aa	TGN	Dipeptidyl aminopeptidase A; type II membrane protein
Invertase	*SUC2*	532 aa	Secreted	Hyrolyzes sucrose; hydrophilic protein
Kex2p	*KEX2*	814 aa	TGN	Aspartyl endoprotease; type I membrane protein
Pma1p	*PMA1*	918 aa	PM	Proton ATPase, 10 transmembrane domains
PrA	*PEP4*	405 aa	Vacuole	Proteinase A; hydrophilic protein
RS-ALP	*pho8::rs*	569 aa	TGN	ALP with DPAP A FXFXD signal in cytosolic domain
Ste3p	*STE3*	470 aa	PM	**a**-factor receptor; 7 transmembrane domains
Ste2p	*STE2*	431 aa	PM	α-factor receptor; 7 transmembrane domains
Vps10p	*VPS10*	1579 aa	TGN/VPC	CPY receptor; type I membrane protein

*Size of precursor forms of proteins are given.

causes the dissociation of the lysosomal enzymes from the M6P receptor. The receptor is then returned to the TGN for further rounds of sorting.

Potential role for clathrin in CPY sorting at the TGN

In many ways the yeast CPY pathway seems to be analogous to the mammalian M6P receptor pathway. For example, Vps10p associates with CPY in the TGN and delivers CPY to a late endosome (PVC) where CPY dissociates and the receptor returns for further rounds of sorting. However, the

Table 14.2 Clathrin, adaptor proteins and other clathrin-associated proteins implicated in TGN to endosome or TGN to vacuole transport

Protein	Gene	Size	Comment
Clathrin			Involved in the internalization step of endocytosis and the retention of late Golgi proteins in the TGN. Potential role in protein transport from the TGN to endosomes.
heavy chain	*CHC1*	1653 aa	
light chain	*CLC1*	233 aa	
AP-1			Strains harboring deletions of AP-1 subunit genes do not exhibit any defects in growth or protein trafficking. However, these mutations accentuate the growth and TGN protein sorting defects of *chc1-ts* strains. When overexpressed, Apm2p can replace Apm1p to form an alternate AP-1 complex. Functional overlap with the Gga proteins.
β1-adaptin	*APL2*	726 aa	
γ-adaptin	*APL4*	832 aa	
μ1-adaptin	*APM1*	475 aa	
	APM2	605 aa	
σ1-adaptin	*APS1*	156 aa	
AP-2			No defects in growth or protein trafficking have been observed in strains harboring deletions of AP-2 subunit genes.
β2-adaptin	*APL1*	700 aa	
α-adaptin	*APL3*	1025 aa	
μ2-adaptin	*APM4*	491 aa	
σ2-adaptin	*APS2*	147 aa	
AP-3			The AP-3 complex is required for efficient transport of ALP and Vam3p from the TGN to the vacuole.
β3-adaptin	*APL6*	809 aa	
δ-adaptin	*APL5*	932 aa	
μ3-adaptin	*APM3*	483 aa	
σ3-adaptin	*APS3*	194aa	
Gga1p	*GGA1*	557 aa	Binds ARF-GTP and clathrin. VHS domain at N-terminus and γ1 ear homology at C-terminus. Deletion of both GGA genes perturbs CPY transport. Functional overlap with AP-1
Gga2p	*GGA2*	585 aa	
Vps1p	*VPS1*	704 aa	Dynamin-related protein required for TGN to endosome and TGN to vacuole protein transport.
Swa2/Aux1p	*SWA2/ AUX1*	668 aa	Auxilin-related protein. Binds clathrin and facilitates disassembly of clathrin coats. Deletion mimics clathrin null.
Drs2p	*DRS2*	1355 aa	Integral membrane P-type ATPase of the TGN. Deletion perturbs clathrin function at the TGN.

role of clathrin in this pathway has been enigmatic. Disruption of the clathrin heavy chain (*CHC1*) or light chain (*CLC1*) genes causes a dramatic mislocalization of TGN proteins, such as Kex2p, to the plasma membrane and perturbs endocytosis. Surprisingly, these clathrin mutations do not appear to affect CPY transport to the vacuole (Payne et al., 1988; Payne and Schekman, 1989; Chu et al., 1996). These observations seemed to discount a role for clathrin in the CPY pathway but the characterization of certain conditional alleles of the clathrin heavy chain challenged this view. Using a reverse genetic approach, Seeger and Payne generated a temperature-sensitive (ts) allele of the clathrin heavy chain gene (*chc1-521*) and found that mutants harboring this allele missorted CPY during the first hour after shifting to the non-permissive temperature (37°C). These cells somehow adapted to the loss of clathrin function and were able to sort CPY normally after longer times of incubation at 37°C. The defects in endocytosis and TGN protein localization in the *chc1-521* mutant did not adapt after the temperature shift and so this adaptation response was specific to the CPY pathway (Seeger and Payne, 1992).

The phenotype of the *chc1-521* mutant implies that a cellular adaptation process might mask a CPY sorting defect in the *chc1* strain. However, other *chc1* conditional mutants fail to display a CPY sorting defect or this adaptation process. Lemmon and colleagues produced a *chc1* ts allele (*chc1-57*) by deleting the C-terminal 57 amino acids from the heavy chain. Mutants carrying this allele exhibited defects in TGN protein localization and endocytosis at 37°C but sorted CPY normally at this temperature (Lemmon et al., 1991). Chen and Graham isolated a *chc1* ts allele (*chc1-5/swa5-1*) in a genetic screen for mutations that are synthetically lethal with an *arf1* mutation. This genetic interaction between clathrin and ARF mutations is consistent with the proposed role of ARF in recruiting clathrin to the TGN. The *chc1-5* allele carried a frameshift mutation that caused the replacement of the last 43 amino acids of the heavy chain with 28 missense amino acids. Deletion of these 28 missense amino acids to produce the *chc1-43* mutant also resulted in a ts allele (Chen and Graham, 1998).

To compare the effect of these clathrin mutations on protein transport, a set of isogenic strains carrying the four different *chc1-ts* alleles was produced. Each *chc1-ts* mutant strain exhibited a comparable defect in TGN protein localization but remarkably, the effect on CPY transport was unique in each strain. The *chc1-5* mutant exhibited a partial defect in glycosylation such that a clearly resolved p2 CPY form was not produced. CPY was sorted normally in this mutant but the rate of transport to vacuole was slowed 3-fold and did not adapt upon longer incubation at 37°C. The *chc1-43* mutant displayed the glycosylation defect but the sorting and transport kinetics of CPY were normal. The *chc1-521* mutant exhibited the transient CPY missorting phenotype as previously reported with no defect in glycosylation, and CPY modification and sorting in the *chc1-57* mutant was indistinguishable from a wild-type strain (Chen and Graham, 1998).

This panoply of *chc1-ts* phenotypes, with regard to CPY transport only, presents a significant challenge to interpret what role, if any, clathrin plays in the CPY transport pathway. At one extreme, it is possible that clathrin plays a key role in the sorting and transport of CPY from the TGN to the endosome. In this case, the *chc1-521* mutation might delay the adaptation process long enough to allow observation of the CPY sorting defect, while the *chc1-43* and *chc1-57* strains may adapt immediately to the loss of clathrin function. The nature of this adaptation process remains obscure. Does it reflect diversion of CPY into a second, clathrin-independent route to the vacuole or is it possible that other proteins can replace the function of clathrin in forming these transport vesicles? At the other extreme, it is possible that clathrin is not directly participating in the CPY transport pathway but certain mutant forms of clathrin can interfere with this pathway. This could occur directly by inappropriate association with transport vesicles or indirectly by perturbing the normal organization of the TGN in an allele-specific fashion. In this case, the adaptation of the *chc1-521* mutant might represent a gradual diminution of this interference. In support of a direct role for clathrin in CPY transport, it was recently reported that clathrin-coated vesicles immuno-isolated from yeast carry the Vps10 CPY receptor (Deloche et al., 2001). This vesicle preparation also contained AP-1 although it is not known if the same clathrin-coated vesicle contained Vps10p, CPY and AP-1.

Clathrin-associated proteins

The *arf1* synthetic lethal screen identified two other mutants, *swa2/aux1* and *swa3/drs2*, that appear to perturb clathrin function at the TGN. Swa2/Aux1 is an auxilin-like protein with a conserved DnaJ motif at the C-terminus. The *swa2/aux1* mutant exhibits a modest accumulation of clathrin-coated vesicles and most of the clathrin in this mutant is assembled with very few free triskelia left in the cytosol. Otherwise, the *swa2/aux1* strain exhibits the same TGN protein mislocalization and endocytosis defects exhibited by the *chc1* strain. Like *chc1-5/swa5-1*, CPY sorting is relatively normal in the *swa2/aux1* mutant but the kinetics of transport to the vacuole is 2–3-fold slower than normal (Chen and Graham, 1998; Gall et al., 2000). Swa3/Drs2p is an integral membrane P-type ATPase that localizes to the TGN. The *drs2* mutation shows a specific synthetic lethal relationship to *arf1* and *chc1-ts* alleles and causes TGN defects that mimic the clathrin mutants. Strains harboring *drs2* are cold-sensitive for growth and appear to have a defect in budding clathrin-coated vesicles at the non-permissive temperature. The effect of *drs2* on CPY transport is similar to that described for *swa2/aux1* and *chc1-5* (Chen et al., 1999).

The protein with a clathrin association that exerts the greatest influence on CPY transport is Vps1p, a large GTPase in the dynamin family; *vps1* mutants mislocalize TGN proteins to the cell surface and missort both CPY and ALP (Wilsbach and Payne, 1993; Nothwehr et al., 1995). Therefore, it appears that Vps1p is required for all protein transport pathways from the

TGN to the endosome. Additional work is required to determine if Vps1p functions to pinch off transport vesicles budding from the TGN in the same manner that dynamin is proposed to pinch off endocytic vesicles budding from the plasma membrane.

The effect of AP mutations in yeast is even more surprising than the clathrin mutant phenotypes. The expectation was that AP-1 mutants would exhibit TGN-specific defects comparable to the clathrin mutants, and the AP-2 mutants would exhibit a specific defect in the internalization step of endocytosis. The Payne and Lemmon laboratories have produced strains carrying multiple deletions of AP-1 and AP-2 subunits that appear perfectly normal (Huang et al., 1999; Yeung et al., 1999). Even a strain carrying deletions of all large AP-1, AP-2, AP-3 subunit genes and the two AP180 genes (encoding a clathrin assembly protein that is not part of a tetrameric complex) failed to display any phenotypes in common with clathrin mutants. In fact, normal appearing clathrin-coated vesicles were purified from this AP-deficient strain (Huang et al., 1999). A function for AP-1 was revealed by deletion of AP-1 subunit genes in a *chc1-ts* strain. The double mutant (*aps1 chc1-ts*) exhibited more severe defects in growth and TGN function at elevated temperatures than the *chc1-ts* single mutant (Rad et al., 1995; Stepp et al., 1995). The CPY sorting defect exhibited by *chc1-521*, however, was not accentuated by deletion of both *APS1* and *APS2* (the AP-2 small subunit gene). Surprisingly, deletion of *APS1* in the *chc1* background produced a double mutant that grows slower than *chc1* (Rad et al., 1995). This suggests that AP-1 can contribute to some sorting processes in the absence of clathrin, which is clearly the case for AP-3.

The ALP/AP-3 sorting pathway

A role for AP-3 in vacuolar protein transport was discovered through the analysis of ALP transport to the vacuole. Like CPY, ALP is also synthesized as a high molecular weight precursor but uses an uncleaved, signal-anchor domain for translocation into the ER. Thus, ALP is transported to the vacuole as a type II integral membrane protein where it undergoes a *PEP4*-dependent removal of a C-terminal extension (Klionsky et al., 1990). The ALP vacuolar sorting signal is present in its short N-terminal cytoplasmic tail and early on it was noted that ALP could be sorted to the vacuole in several *vps* mutants that missort most of CPY. This includes the endosomal t-SNARE Pep12p (Vps6p) and several other Vps proteins implicated in endosomal trafficking (Conibear and Stevens, 1998; Odorizzi et al., 1998b). Thus, ALP follows a pathway that is distinct from CPY and seems to move from the Golgi complex directly to the vacuole, bypassing the endosome in the process.

A genetic screen for factors specifically involved in ALP transport identified the two large subunits of yeast AP-3 (β and δ). In fact, deletion of any of the four AP-3 subunit genes results in the accumulation of ALP in cytoplasmic vesicles and tubules that are distinct from the normal-appearing vacuoles in these mutants. These mutants also disrupt transport of the vacuolar t-SNARE Vam3p to the vacuole but do not affect CPY transport. In contrast, ALP transport to the vacuole is normal in strains carrying deletions of AP-1 or AP-2 subunit genes (Conibear and Stevens, 1998; Odorizzi et al., 1998b). In addition, ALP transport is normal in a *chc1* strain and in the *chc1-521* and *chc1-5* mutants that perturb CPY transport (Seeger and Payne, 1992; Chen and Graham, 1998). These data indicate that AP-3 can function normally in the absence of clathrin and it is possible that another coat-like protein, such as Vps41p, replaces clathrin in this transport pathway (Rehling et al., 1999). But it should be noted that these data do not prove an absence of clathrin interaction with AP-3 during vesicle formation.

The AP-3 protein transport pathway appears to be well conserved among eukaryotes and several coat and eye color mutations in mice and *Drosophila* respectively have been mapped to AP-3 subunit genes. While the yeast ALP and CPY pathways converge on the same vacuole, the AP-3 coat seems to have been co-opted in metazoans for the production of melanosomes and platelet dense granules, which are specialized lysosomal compartments. A neuron-specific AP-3 complex involved in synaptic vesicle biosynthesis has also been found (reviewed in Odorizzi et al., 1998b). Whether or not AP-3 functions independently of clathrin in mammalian cells is unresolved since AP-3 can bind clathrin *in vitro* and in some cell types there is good co-localization of AP-3 with clathrin on endosomal membranes (Dell'Angelica et al., 1998). It is not clear, however, if this interaction between clathrin and AP-3 is functionally significant *in vivo*. In fact, in other cell types there is very little co-localization between clathrin and AP-3 (Simpson et al., 1996, 1997).

The GGA proteins may mediate a TGN to PVC protein transport pathway

Recently, a new family of potential coat proteins called GGAs (for Golgi localized, gamma ear-containing, ARF-binding proteins) have been implicated in protein trafficking from the TGN (Boman et al., 2000; Dell'Angelica et al., 2000; Hirst et al., 2000). As the name indicates, these proteins have a region of homology to the $\gamma 1$ subunit ear appendage of AP-1, directly bind ARF and localize to the TGN in mammalian cells. The GGAs are not part of a tetrameric adaptor complex but like other coat proteins, the GGAs dissociate from the Golgi complex of cells treated with brefeldin A (Boman et al., 2000; Dell'Angelica et al., 2000; Hirst et al., 2000), an inhibitor of the ARF guanine nucleotide exchange factor (Peyroche et al., 1999).

The yeast GGAs are encoded by the *GGA1* and *GGA2* genes, which can both be deleted without affecting viability or growth rate. However, these strains secrete some p2 CPY and exhibit a slower transport rate for delivery of the remaining p2 CPY to the vacuole (Boman et al., 2000; Dell'Angelica et al., 2000; Hirst et al., 2000). Recently, it has been reported that the yeast GGA proteins bind clathrin and co-fractionate with clathrin-coated vesicles. In addition, cells harboring deletions of both *GGA* genes and an AP-1 subunit gene exhibit a severe growth defect, suggesting a functional redundancy between the GGAs and AP-1. Consistent with this interpretation, deletion of *GGA2* and *APL2* (β1 subunit) causes a defect in carboxypeptidase S (CPS) transport to the PVC that is not observed in the single mutants (Costaguta et al., 2001).

Black and Pelham have found that the GGAs are required for sorting the t-SNARE Pep12p from the TGN to the PVC. Pep12, which normally marks the PVC, is mislocalized to early endosomes of the *gga1 gga2* mutant where it co-fractionates with Tlg1p, the early endosome t-SNARE. This Pep12p pathway also appears to require clathrin function, again suggesting the possibility that clathrin/GGA coated vesicles mediate this TGN to PVC transport pathway (Black and Pelham, 2000). While more work is needed to define the function of the GGA proteins, it appears that they act in budding clathrin-coated vesicles from the TGN that carry CPY, CPS and Pep12p to the PVC. It has been suggested that clathrin and AP-1 mediates the clathrin-dependent cycling of Kex2p (and other TGN proteins) to the early endosome (Black and Pelham, 2000). Thus, the redundancy between AP-1 and the GGA proteins may reflect a need of the cell to maintain one of these two TGN to endosome pathways. It is possible that CPY normally uses the GGA pathway but can be diverted into the AP-1 pathway in the absence of the GGA proteins.

Targeting transport vesicles containing CPY to the PVC

The fusion of transport vesicles with target organelles throughout the secretory and endocytic pathways requires a common set of proteins, often with distinct family members acting at different transport steps. These include t-SNARE/v-SNARE pairs, Rab proteins, Sec1-related proteins and the NSF/SNAP complex. Several class D *VPS* genes encode proteins involved in TGN-derived vesicle targeting and fusion with the PVC (Table 14.3), which is consistent with the observation that several class D *vps* mutants accumulate 40–60 nm transport vesicles. Pep12p is the resident t-SNARE of the PVC and it appears that Vti1p is the v-SNARE marking the TGN-derived transport vesicles. Vps21p and Vps9p are the respective Rab protein and cognate GTP/GDP exchange factor (Hama et al., 1999) while Vps45p is the Sec1-related protein operating at this step (reviewed in Conibear and Stevens, 1998).

Phosphatidylinositol-(3)-phosphate (PtdIns(3)P) has been implicated in this pathway by the discovery that the class D *VPS34* gene encodes a

Table 14.3   *VPS* genes implicated in TGN-to-PVC transport

Gene	Mutant class	Size of product	Comment
VPS3	D	1011 aa	Hydrophilic, mostly cytoplasmic
VPS6/PEP12	D	288 aa	t-SNARE marking the PVC
VPS8	A	1176 aa	Hydrophilic, membrane association requires Vps21, RING finger domain
VPS9	D	451 aa	GDP/GTP exchange factor for Vps21p, RABEX5 homolog
VPS15	D	1454 aa	Protein kinase, associates with and activates Vps34p
VPS19/ VAC1/PEP7	D	515 aa	FYVE finger domain binds PtdIns(3)P, Binds Vps21p and Vps45p, EEA1 homolog
VPS21	D	210 aa	Rab GTPase, binds Vps8p and Vps9p
VPS34	D	875 aa	PtdIns 3-kinase, binds Vps15p
VPS45	D	577 aa	Sec1-like protein, binds Pep12-Vti1 SNARE complex
VTI1	N/A	217 aa	Synaptobrevin member (v-SNARE)

phosphatidylinositol 3-kinase. Vps34p associates with a protein kinase, Vps15p, which is necessary for the activity of Vps34p and its peripheral association with TGN membranes. PtdIns(3)P plays an important role in vesicle fusion with the PVC and the membrane dynamics of this organelle through interaction with effector proteins containing a structural motif called the FYVE domain (named after the first letters of the first four proteins found to contain this domain) that binds directly to PtdIns(3)P. Effector proteins include a yeast homolog of the mammalian early endosome autoantigen 1 (EEA1) called Vac1p/Vps19p, which interacts with both Vps21p and Vps45p. Thus, Vac1p/Vps19p appears to couple Vps34p function (the production of PtdIns(3)P) to the machinery required for vesicle docking and fusion with the PVC (reviewed in Odorizzi et al., 2000).

Summary: *Protein trafficking pathways exiting the TGN*

- The yeast TGN may produce as many as five different transport vesicles carrying cargo proteins to different destinations (Figure 14.1).
- Mammalian lysosomal enzymes bound to the mannose 6-phosphate receptor appear to be packaged into clathrin-coated vesicles for delivery to an endosome.

- It is possible that CPY bound to its receptor, Vps10p, is packaged into clathrin-coated vesicles at the TGN. However, yeast can somehow adapt to the loss of clathrin and sort CPY normally.
- The dynamin-related protein Vps1p appears necessary for all TGN to endosome/vacuole protein transport pathways.
- Tetrameric adaptor proteins (AP-1, AP-2 and AP-3) do not appear to contribute significantly to CPY transport (but see below).
- However, the AP-3 complex is necessary for transporting ALP from the TGN directly to the vacuole.
- The GGA proteins are not part of a tetrameric adaptor complex but share a region of homology with the γ1 subunit of AP-1 and are functionally redundant with AP-1. In addition, the GGAs bind ARF and clathrin.
- Deletion of the GGA genes causes a partial CPY missorting phenotype suggesting a GGA-dependent pathway for CPY delivery to the PVC. It is possible that some CPY is diverted through an AP-1 pathway in the absence of the GGAs.
- The class D Vps proteins and PtdIns(3)P are required for targeting and fusion of TGN-derived vesicles with the PVC. The class D proteins include components of vesicle fusion machinery such as SNARE and Rab proteins (Table 14.3).

ENDOSOME-TO-TGN TRANSPORT

The yeast prevacuolar/late endosomal compartment (PVC) is the site of convergence of the endocytic pathway and the biosynthetic pathway that mediates delivery of newly synthesized proteins to the vacuole. For example, proteins originating from both the endocytic and biosynthetic pathways accumulate in the PVC in class E *vps* mutant cells that exhibit blocks in both PVC-to-vacuole anterograde transport and PVC-to-TGN retrograde transport (Raymond et al., 1992). Membrane proteins that are delivered to the PVC can either be transported to the vacuole or be retrieved back to the TGN. Recent studies have shed light on the molecular details of the endosome-to-TGN retrieval pathway.

The initial finding that CPY is sorted into the vacuolar pathway via a TGN integral membrane protein called Vps10p suggested that Vps10p may cycle between the TGN and endosomes similarly to the mannose 6-phosphate receptor (Marcusson et al., 1994). This idea was supported by the finding that CPY was synthesized at a rate 20-fold higher than Vps10p but bound to Vps10p with a 1:1 stoichiometry (Cooper and Stevens, 1996). This imbalance suggested that each Vps10p molecule must undergo multiple rounds of sorting. In wild-type cells Vps10p was predominantly localized to the

TGN but in class E *vps* mutant cells Vps10p accumulated in the PVC. Finally, mutation of aromatic amino acid-based sequences (e.g. $YSSL_{1495}$) within the Vps10p cytosolic domain caused mislocalization to the vacuole (Cooper and Stevens, 1996). Taken together these results strongly suggested that Vps10p cycles between the TGN and PVC in fulfillment of its role as the CPY sorting receptor.

Retrieval signals on proteins that cycle between the TGN and endosomes

The identification and study of yeast TGN resident membrane proteins (Kex2p, Kex1p, DPAP A, and A-ALP) provided critical evidence for this bi-directional pathway between the TGN and PVC. DPAP A, Kex2p, and Kex1p are required for proteolytic processing of the mating pheromone α-factor as it is transported through the TGN. A-ALP is a model TGN membrane protein consisting of the cytosolic domain of DPAP A fused to the transmembrane and lumenal domains of the vacuolar membrane protein ALP (Nothwehr et al., 1993). The cytosolic domain of DPAP A is necessary and sufficient for TGN localization; thus A-ALP is very efficiently targeted to the TGN. However, under conditions where A-ALP fails to be localized to the TGN it is delivered to the vacuole, whereupon the C-terminal propeptide of the ALP lumenal domain is removed by vacuolar proteases. This processing event can be followed by both SDS gel mobility and ALP activity, since A-ALP is inactive in the unprocessed form and is activated by removal of the propeptide.

Like Vps10p, each of these resident TGN proteins has a large lumenal domain, a single transmembrane domain, and a cytosolic domain of ~90–160 amino acids. Both DPAP A and Kex2p contain aromatic amino acid-based TGN localization signals, $FXFXD_{89}$ and FXY_{713}, respectively, where X is a non-critical amino acid (Wilcox et al., 1992; Nothwehr et al., 1993). Mutation of these TGN localization signals causes A-ALP and Kex2p mislocalization to the vacuole. To address whether the FXFXD signal was sufficient for TGN localization, a 10 amino acid sequence containing FXFXD was transplanted into the cytosolic domain of ALP so that it replaced the vacuolar targeting signal of ALP. The resulting protein, called RS-ALP, was localized to the TGN, albeit less efficiently than A-ALP. Thus the FXFXD signal is sufficient to confer TGN localization. As is the case with Vps10p, A-ALP accumulates in the PVC in a class E *vps* mutant indicating that A-ALP (and most likely Kex2p) visits this organelle as part of its normal trafficking itinerary (Raymond et al., 1992; Nothwehr et al., 1999).

Several lines of evidence indicate that the aromatic amino acid based sorting motifs on A-ALP, Kex2p and Vps10p are involved in retrieval of these proteins from the PVC back to the TGN. For example, mutant forms of DPAP A and Kex2p lacking retrieval signals are delivered to the vacuole independent of late secretory pathway functions (Wilcox et al., 1992;

Nothwehr et al., 1993). In addition, when TGN-localized RS-ALP was analyzed much of it was found to lack its propeptide (Bryant and Stevens, 1997). This result suggested RS-ALP had cycled back to the TGN after having reached a post-TGN compartment containing vacuolar proteolytic activity (e.g. the PVC or vacuole). RS-ALP containing a mutated FXFXD signal was localized to the vacuole, consistent with the idea that the aromatic signal was required for the retrieval event. In addition, mutant forms of A-ALP and Kex2p lacking retrieval signals (Kex2-Y_{713}A and A(F_{85}A, F_{87}A)-ALP) were found to be transported to the PVC at a similar rate as the corresponding wild-type proteins, ruling out a role for the aromatic signals in a static retention mechanism (Brickner and Fuller, 1997; Bryant and Stevens, 1997). Finally, in the class E mutant *vps27*, A-ALP was found to redistribute from the PVC back to the TGN upon induction of synthesis of wild-type Vps27p, whereas A(F_{85}A, F_{87}A)-ALP redistributed from the PVC to the vacuole under the same conditions (Bryant and Stevens, 1997). These results support a model in which TGN membrane proteins at some frequency are transported to the PVC, either directly or via an early endosome. Once at the PVC these proteins engage a sorting apparatus that recognizes their aromatic signals and packages them into vesicles for delivery back to the TGN.

Mutants defective in retrieval of cycling proteins

Early on, it was noted that the retrieval signals on yeast TGN proteins resembled aromatic sorting signals in animal cells consisting of YXXΦ (Y is tyrosine, X is any amino acid, and Φ is a residue with a bulky hydrophobic side chain). The YXXΦ signal has been shown to interact with the μ subunit of adaptor complexes present in clathrin vesicle coats (Heilker et al., 1999). This similarity, coupled with the observation that clathrin heavy chain yeast mutants mislocalized Kex2p and DPAP A to the cell surface, suggested that clathrin may be involved in the retrieval step. However, more recent data would appear to rule this out. The effect of inactivation of clathrin heavy chain on Kex2p and DPAP A localization appears to be exerted at the TGN since this effect occurs independent of the retrieval signal (Redding et al., 1996). Also, adaptor gene disruptions do not appear to cause a defect in Kex2p localization, although in some cases synthetic effects with clathrin heavy chain mutations have been observed. Finally, a different protein complex, the retromer, was found to be intimately involved in the retrieval step (see below).

Genetic screens have led to the identification of several proteins that function in PVC-to-Golgi retrieval of TGN proteins (Table 14.4). While class E Vps proteins are necessary for retrieval, this requirement is probably indirect since these proteins are also required for endosome-to-vacuole transport and endosome membrane dynamics (see below). An early clue in

Table 14.4 Genes implicated in PVC-to-TGN transport

Gene	Mutant class	Size of product	Comment
GRD19	A	162 aa	Hydrophilic, endosome-localized, PX domain
RIC1	B	1056 aa	Hydrophilic
VPS5	B	675 aa	Subunit of endosome-localized retromer complex, PX domain
VPS17	B	551 aa	Subunit of endosome-localized retromer complex, PX domain
VPS26	F	379 aa	Subunit of endosome-localized retromer complex
VPS29	A	282 aa	Subunit of endosome-localized retromer complex
VPS30	A	557 aa	Hydrophilic, associates with Vps34p/Vps15p complex
VPS35	A	944 aa	Subunit of endosome-localized retromer complex
VPS52	B	641 aa	Hydrophilic, TGN-localized, associates with Vps53p and Vps54p
VPS53	B	822 aa	Hydrophilic, TGN-localized, associates with Vps52p and Vps54p
VPS54	B	889 aa	Hydrophilic, TGN-localized, associates with Vps52p and Vps53p
VTI1	N/A	217 aa	Synaptobrevin member (v-SNARE)
YPT6	B	215 aa	Small GTPase, Rab6-like

the search for genes specifically involved in PVC-to-TGN retrieval came from the finding that a loss of CPY receptor function caused very specific effects on the CPY sorting pathway. Although *vps10* mutants exhibited a dramatic affect on sorting of CPY, other vacuolar hydrolases were delivered to the vacuole with near normal efficiency (Marcusson et al., 1994). Furthermore, no aberrant vacuolar morphologies were observed in *vps10* mutants in contrast to most other *vps* mutants. Interestingly, the *vps29*, *vps30* and *vps35* mutants were found to exhibit very similar phenotypes, suggesting that they may have specific roles in the trafficking/localization of Vps10p (Seaman et al., 1997). Subcellular fractionation experiments demonstrated that Vps10p was indeed mislocalized to the vacuole in each of these three mutant strains. Vps10p reached the vacuole in these strains in a manner dependent on the PVC t-SNARE Pep12p but independent of late secretory pathway functions. Kex2p and A-ALP were also mislocalized to the vacuole in a *vps35* mutant strain. Vps35p was shown to be required for redistribution of A-ALP from the PVC to the TGN in a *vps27* strain upon induction of Vps27p synthesis (Nothwehr et al., 1999).

The retromer complex

Biochemical and cell biological analyses strongly suggested that Vps35p and Vps29p were members of a protein complex at the PVC involved in retrieval of Golgi membrane proteins (Seaman et al., 1998). Vps35p, Vps29p, and Vps26p were shown to physically interact into a membrane-associated complex. A somewhat looser association was detected between this subcomplex and a subcomplex consisting of Vps5p and Vps17p. The full complex, termed the retromer, appears to associate with vesicular and endosomal membranes. The roles of each subcomplex appear to be distinct. For example, Vps5p has been observed to self assemble into very large homo-oligomers *in vitro* and thus could be a structural component of a vesicle coat. Although they are defective in localization of TGN membrane proteins, *vps5* and *vps17* mutants exhibit a wider range of mutant phenotypes than *vps35* and *vps29* mutants such as defects in vacuolar morphology (Horazdovsky et al., 1997; Nothwehr and Hindes, 1997). Taken together these observations led Emr and colleagues to propose that the Vps5p/ Vps17p subcomplex plays a structural role in vesicle formation while the Vps35p/Vps29p/Vps26p subcomplex is involved in sorting of cargo into the forming vesicles. Thus in the absence of Vps5p or Vps17p function no retrograde vesicles would form, causing a broader set of phenotypes than a loss of Vps29p or Vps35p function, which may only affect cargo sorting.

Which of the retromer subunits is responsible for recognition of the aromatic amino acid-based retrieval signals on TGN membrane proteins? A clue to this issue derived from a detailed mutagenic analysis of the *VPS35* gene (Nothwehr et al., 1999). *VPS35* null alleles are severely defective for retrieval of both A-ALP and Vps10p. One mutant allele, called *vps35-101*, obtained in a genetic screen was defective for A-ALP retrieval from the PVC but was normal for retrieval of Vps10p. Other alleles were obtained that exhibited the opposite cargo specificity, i.e. they exhibited defects in A-ALP retrieval but were normal for Vps10p. The two classes of mutations segregated to different regions of the 944 amino acid Vps35p sequence. These data suggested the possibility that Vps35p associates with sorting signals on A-ALP and Vps10p but does so using somewhat different structural features. Subsequent biochemical analysis demonstrated that Vps35p does indeed physically associate with the cytosolic domain of A-ALP in a retrieval signal-dependent manner (Nothwehr et al., 2000). Furthermore, a screen for suppressor mutations within the cytosolic domains of A-ALP and Vps10p yielded several mutations very near to the retrieval signals that suppressed cargo-specific mutations in Vps35p. These data strongly suggest that Vps35p directly interacts with the retrieval signal domains on these two cargo proteins and recruits these proteins into forming vesicles at the PVC membrane.

Another protein that might collaborate with Vps35p in retrieval is Grd19p (Voos and Stevens, 1998). Mutations in Grd19p cause defects in retrieval of

A-ALP and Kex2p but much weaker defects in retrieval of Vps10p. Grd19p binds to the DPAP A cytosolic domain expressed from *E. coli* and thus it is possible that Grd19p may also be a member of the retromer complex. Grd19p may be an accessory protein that is needed in combination with Vps35p for sorting of a subset of cargo proteins. Interestingly, both Grd19p and Vps5p contain NADPH oxidase p40 (PX) domains which have been found in a family of proteins involved in protein sorting – the sorting nexins. In animal cells, sorting nexin family members interact with the receptors for insulin, platelet-derived growth factor, and epidermal growth factor. Recently one of the mammalian sorting nexins, SNX15, was shown to be involved in trafficking through the endocytic pathway (Barr et al., 2000; Phillips et al., 2001).

The sequences of the retromer protein subunits are consistent with a peripheral association with membranes but otherwise give few clues to their function. However, the retromer subunits are highly conserved through eukaryotes. Furthermore, the human orthologs of yeast Vps26p, Vps29p, and Vps35p assemble into multimeric complexes (Haft et al., 2000). A retromer-mediated retrieval mechanism thus appears to be conserved from yeast to humans; however, the cargo that utilizes such a mechanism in higher eukaryotes is unknown.

Targeting and fusion of PVC-derived vesicles with the TGN

The mechanism by which PVC-derived vesicles are targeted to and fuse with the Golgi apparatus is not well understood. As is the case with other well-studied vesicle transport steps these vesicles are assumed to carry a v-SNARE that associates with a Golgi t-SNARE during vesicle fusion. A good candidate v-SNARE for this step is Vti1p (Fischer von Mollard et al., 1997). Temperature-sensitive *VTI1* alleles have been identified that are defective in TGN-to-PVC transport. Both genetic and physical interactions between Vti1p and the PVC t-SNARE Pep12p have been detected; thus it seems clear that Vti1p acts as a v-SNARE in the TGN-to-PVC transport step. Another *VTI1* temperature-sensitive allele is also defective in trafficking to the *cis*-Golgi compartment. Accordingly, physical interactions with the *cis*-Golgi t-SNARE Sed5p have been detected as well as interactions with the t-SNAREs Tlg1p and Tlg2p. Tlg1p and Tlg2p appear to localize to the TGN and early endosomal compartments. Therefore, it seems likely that Vti1p is recycled from the PVC back to the TGN on vesicle carriers whose formation is mediated by the retromer. These vesicles could either fuse with the *cis*-Golgi via a Vti1p/Sed5p interaction or directly with the TGN via a Vti1p/Tlg2p (or Tlg1p) interaction. The SNARE-mediated vesicle fusion event also appears to require Ypt6p, a small GTPase related to rab6, and Ric1p (Bensen et al., 2001).

Other components of the machinery for PVC-vesicle targeting and fusion with the Golgi apparatus are a protein complex consisting of Vps52p,

Vps53p, and Vps54p (Conibear and Stevens, 2000). In yeast cells lacking function of this complex TGN membrane proteins fail to be retrieved back to the TGN efficiently and are found in both vesicle-like structures and to a limited extent in the vacuole. These observations suggest that vesicles form from the PVC in cells lacking function of Vps52p/Vps53p/Vps54p complex but are unable to fuse with the Golgi apparatus. Such a fusion block could indirectly affect retrieval from the PVC if machinery necessary for retrieval were not able to be recycled back to the PVC. The Vps52p/Vps53p/Vps54p complex does not appear to localize to the PVC but localizes in part to the TGN. These data suggest that the destination for PVC-derived vesicles may in fact be the TGN rather than the *cis*-Golgi but additional work will be needed to resolve these issues fully.

Summary: *Endosome-to-TGN transport*

- Vps10p and TGN proteins such as Kex2 and DPAP A cycle between the TGN and endosomes.
- Class E *vps* mutants block PVC to vacuole and PVC to TGN transport causing accumulation of Vps10p and TGN proteins in an enlarged PVC.
- TGN proteins that cycle through the endosome have an aromatic amino acid-based retrieval signal (e.g. FXFXD) in their cytosolic tails.
- Several *vps* mutants exhibit a specific defect in the retrieval pathway (Table 14.4).
- Some of these *VPS* genes encode subunits of the retromer, which may be a vesicle coat and is composed of Vps26p/29p/35p and Vps5p/17p sub-complexes.
- The Vps35p subunit recognizes the retrieval signals on cycling proteins. Two distinct regions of Vps35p bind the DPAP A and Vps10p retrieval signals.
- Grd19p may also contribute to the recognition of the DPAP A retrieval signal.
- PVC-derived vesicles appear to carry the Vti1p v-SNARE and may be targeted to either the *cis*-Golgi or TGN.
- The Vps52p/53p/54p complex contributes to vesicle fusion at the TGN.

SORTING OF PROTEINS INTO THE LUMEN OF LATE ENDOSOMES: THE MULTIVESICULAR BODY PATHWAY

The yeast vacuole and mammalian lysosome are the sites of degradation of a variety of macromolecules including lipids and membrane protein receptors from the plasma membrane. In addition, the endocytic trafficking of receptors from the plasma membrane to the lysosome as a means for down-regulating them has been well documented. On the other hand, certain proteins and lipids delivered to the vacuole from the biosynthetic pathway are spared from degradation. These observations imply the existence of a mechanism that allows plasma-membrane-derived lipids and proteins to be selectively degraded in the vacuole while certain membrane components originating from the biosynthetic pathway are not. Cells appear to have solved this problem by sorting lipids and proteins present on the endosomal membrane into vesicles that undergo inverted budding and fission to become lumenal vesicles. The resulting mature endosome, called a multi-vesicular body (MVB), can then directly fuse with the vacuole. In this way the vesicle-associated material to be degraded becomes more accessible to the hydrolytic enzymes of the vacuolar lumen.

The MVB pathway was originally revealed by the analysis of the endocytic trafficking of receptors in mammalian cells (Gruenberg and Maxfield, 1995; Futter et al., 1996). For example, the epidermal growth factor (EGF) receptor undergoes clathrin-coated pit internalization from the plasma membrane and is transported to early endosomes. The endosome can then undergo a maturation process, which involves recycling of certain cargo back to the plasma membrane and sorting of other cargo into vesicles which bud into the lumenal space. The resulting MVB eventually fuses directly with the lysosome.

Recent work on yeast MVBs has shed new light on the issues of how MVB formation is regulated and the basis of cargo selectivity. Membrane proteins that reach the yeast vacuole appear to fall into two main classes: those that localize to the outer, limiting vacuolar membrane and those that localize to the vacuolar lumen. In addition, morphological studies of yeast endosomal compartments have suggested the presence of endosomes resembling MVBs (Prescianotto-Baschong and Riezman, 1998).

The trafficking of the type II integral membrane protein carboxypeptidase S (CPS) has been used as a model protein to understand the pathway by which membrane proteins reach the vacuolar lumen (Odorizzi et al., 1998a). When a fusion protein consisting of green fluorescent protein (GFP) fused to the cytosolic N-terminus of CPS was expressed in yeast, essentially all fluorescence was found exclusively in the vacuolar lumen. Analysis of vacuoles from strains lacking vacuolar hydrolytic activity revealed that GFP-CPS associated with 40–50 nm vesicles in the vacuolar lumen. These data indicated that the cytosolic as well as the other domains of CPS reached the lumen and were suggestive of the existence of an MVB-like pathway in yeast. Furthermore, a fusion protein containing GFP fused to the mating pheromone receptor Ste2p was also transported to the vacuolar lumen from the plasma membrane. Both CPS and Ste2p are known to reach the vacuole via the PVC.

Interestingly, ALP, a protein that bypasses the PVC while it is being transported from the TGN to the vacuole (see above), localizes to the outer vacuolar membrane. In addition, a GFP-CPS fusion containing the ALP vacuolar targeting signal is delivered to the outer vacuolar membrane. Passage through the PVC is therefore necessary for targeting to the vacuolar lumen. These data suggest that prior to fusion with the vacuole, the PVC matures into an MVB in which CPS is present on the lumenal vesicles. However, not all proteins that reach the vacuole via the PVC are lumenal, suggesting that proteins like CPS and Ste2p may be actively sorted into such vesicles. More clues to regulation of this sorting pathway has been derived from study of phosphoinositides and class E *vps* mutants.

Phosphoinositides may regulate MVB formation

Wurmser and Emr noticed that cellular levels of PtdIns(3)P dropped quickly upon inactivation of Vps34p (the PtdIns-3 kinase). Surprisingly, they found that lumenal vacuolar hydrolase activity was required for PtdIns(3)P turnover and mutations that blocked PVC to vacuole transport also stabilized PtdIns(3)P. In addition, mutants with stabilized PtdIns(3)P accumulated small vesicles within the vacuole or PVC. Thus, it was proposed that PtdIns(3)P is sorted into vesicles that bud into the PVC lumen to generate an MVB and these vesicles are degraded in the vacuole lumen. It is likely that PtdIns(3)P plays a direct role in the MVB pathway through interaction with the Vps27p and Fab1p effector proteins (Wurmser and Emr, 1998). Vps27p has an FYVE domain and is a class E peripheral membrane protein required for retrograde vesicle formation from the PVC and for PVC-to-vacuolar transport. Interestingly, Fab1p, which phosphorylates PtdIns(3)P at the 5 position, is an FYVE-domain-containing protein that appears to associate with the PVC (Odorizzi et al., 1998a). The FYVE and enzymatic domains appear to be structurally distinct since they map to opposite ends of the polypeptide primary sequence. Fab1p is the sole enzyme in the cell capable of synthesizing PtdIns(3,5)P$_2$ from PtdIns(3)P. Yeast strains deleted for *FAB1* do not exhibit vacuolar protein sorting defects but do exhibit extremely large vacuoles. These data suggested a defect in vacuolar membrane homeostasis and possibly a role in the MVB pathway.

Fab1p and other genes known to be involved in PVC function were evaluated for a role in MVB formation (Odorizzi et al., 1998a) (Table 14.5). Mutations in any of the 13 class E *vps* genes, including *VPS27*, results in accumulation of cargo in an aberrant endosome-like structure termed the class E compartment (Raymond et al., 1992). Several of the class E Vps proteins are known to associate directly with the PVC including Vps27p, Vps4p, Vps24p and Vps32. Mutations in class E *vps* genes cause a portion of the GFP-CPS to accumulate in the class E compartment. Interestingly, of the GFP-CPS that is transported to the vacuole, a significant pool is found on the outer membrane rather than in the lumen. These data suggest that formation of the lumenal vesicles within the PVC is partially defective in class E mutants.

Table14.5   Genes implicated in endosome membrane dynamics

Gene	Mutant class	Size of product	Comment
FAB1	N/A	2278 aa	PtdIns(3)P 5-kinase, endosome-localized, FYVE domain
VPS2	E		Not cloned
VPS4	E	437 aa	Hydrophilic, endosome-localized, AAA ATPase domain
VPS15	D	1454 aa	Ser/Thr protein kinase, associates with Vps34p
VPS20	E	221 aa	Hydrophilic, predicted coiled-coil domain
VPS22	E		Not cloned
VPS23	E	296 aa	Hydrophilic, endosome-localized, tumor susceptibility gene homolog
VPS24	E	224 aa	Hydrophilic, endosome-localized, predicted coiled-coil domain
VPS25	E	202 aa	Hydrophilic
VPS27	E	622 aa	Hydrophilic, endosome-localized, FYVE domain
VPS28	E	242 aa	Hydrophilic
VPS31	E		Not cloned
VPS32	E	240 aa	Hydrophilic, endosome-localized, predicted coiled-coil domain
VPS34	D	875 aa	PtdIns 3-kinase; associates with Vps15p
VPS36	E	566 aa	Hydrophilic
VPS37	E		Not cloned

A more impressive defect in sorting GFP-CPS to the vacuole lumen was observed in *fab1* mutants since endosome to vacuole transport is unaffected in these cells. In mutants carrying a *FAB1* deletion (*fab1*) or a point mutation that inactivates the Fab1p kinase activity, almost all of the GFP-CPS that was delivered to the vacuole was found on the outer membrane. Mutations in Fab1p ($G_{2042}V$, $G_{2045}V$) that lead to a 10-fold reduction in the cellular level of PtdIns(3,5)P_2 still sorted GFP-CPS into the vacuolar lumen. These results indicate that PtdIns(3,5)P_2 is necessary for MVB formation but that a relatively low level of this lipid is sufficient. It is tempting to speculate that the increased vacuole size in *fab1* mutants is caused by an inability of membrane to be targeted to the lumen of the PVC and vacuole. However, cells expressing Fab1p with the $G_{2042}V$, $G_{2045}V$ mutations exhibited the large vacuole phenotype even though CPS sorting to the lumen was normal. Thus the increase in vacuole size is not a simple matter of a defect in MVB formation. PtdIns(3,5)P_2 appears also to have a role for

MVB formation in mammalian cells since prevention of the production of its PtdIns(3)P precursor using the drug wortmannin blocked MVB formation in antigen-presenting cells (Fernandez-Borja et al., 1999).

In summary, Fab1p appears to play a pivotal role in dampening PtdIns(3)P signaling pathways and activating PtdIns(3,5)P_2 pathways. These two processes could play a role in maturation of PVCs into MVBs. According to this model (Odorizzi et al., 2000), a 'young' PVC would contain plenty of PtdIns(3)P enabling it to accept vesicular traffic from the TGN. TGN resident proteins would undergo retromer-mediated retrieval back to the TGN while at the same time the level of proteins in the PVC requiring delivery to the vacuole would increase. Over time, the action of Fab1p would convert more and more PtdIns(3)P to PtdIns(3,5)P_2, which would cause vesicle invagination. Class E proteins presumably play a role in the vesicle formation process, as well as unidentified effector proteins that would bind to PtdIns(3,5)P_2. Certain cargo such as CPS and specific lipids (possibly PtdIns(3)P itself) would then be sorted into the forming vesicles. In animal cells the lipid lysobisphosphatidic acid is also specifically sorted into the internal membranes of MVBs. Other vacuolar proteins such as Vph1p and DPAP B would instead remain on the outer membrane of the maturing endosome. Upon completion of the PVC-to-MVB maturation process the MVB would fuse with the vacuole (see below) and release its lumenal contents into the vacuolar lumen. The protein and lipid components of the vesicle would then be degraded by vacuolar hydrolases. Future progress in this field will involve identification of the PtdIns(3,5)P_2 effector proteins as well as characterizing the sorting signals on proteins such as CPS that are sorted into the vesicles of MVBs.

Summary: *Sorting of proteins into the lumen of late endosomes: the MVB pathway*

- The limiting membrane of the PVC invaginates to produce internal vesicles and generate a MVB.
- The integral membrane proteins CPS and Ste2p are sorted into the internal vesicles of the MVB.
- PtdIns(3)P also appears to be sorted into internal vesicles.
- Class E *vps* mutants exhibit a defect in sorting CPS to internal vesicles and accumulate an aberrant endosomal structure called the class E compartment.
- PtdIns(3)P appears to play an active role in the MVB pathway through interaction with the FYVE domain of effector proteins (Vps27p and Fab1p).
- Fab1p is a PtdIns(3)-5-kinase and *fab1* mutants exhibit a defect in the MVB pathway and enlarged vacuoles.
- Fab1p may dampen a PtdIns(3)P-dependent pathway and turn on a PtdIns(3,5)P-dependent pathway through unknown effector proteins.

TRANSPORT BETWEEN THE ENDOSOME AND VACUOLE

The final membrane trafficking event in the biosynthetic and endocytic pathways leading to the vacuole is the process by which membrane and proteins from the PVC are delivered to the vacuole. Although much is now known about the machinery mediating this process, the mechanism has been controversial (Luzio et al., 2000). According to one model, vesicles form from the PVC and fuse with the vacuolar membrane. However, a vesicle transport model is difficult to reconcile with the observation that CPS and other proteins are sorted into the lumen of MVBs on vesicles that are subsequently delivered to the lumen of the vacuole. In animal cells, late endosomes appear to fuse directly with the lysosome resulting in a hybrid organelle. Molecular degradation and recycling of membrane components back to endosomes causes the hybrid organelle to mature into a lysosome. These observations suggest that in yeast the PVC/MVB probably also fuses directly with the vacuole, releasing the lumenal contents of the MVB into the lumen of the vacuole. However, it is possible that a vesicle transport pathway could coexist with a direct fusion pathway.

Membrane fusion with the vacuole

In recent years much new information on the machinery that mediates endosome–vacuole fusion has been obtained (Table 14.6). Many of the class B and C *vps* mutants as well as the *vam* mutants exhibit severe vacuolar morphology defects and accumulation of vesicles and endosomal membranes. Most of these mutants exhibit either kinetic delays in or complete blocks of vacuolar processing of CPY and ALP. Thus these mutants block trafficking to the vacuole at a late stage, at or after the point of intersection of the CPY and ALP pathways. Many of these mutants also block homotypic fusion of vacuolar membranes; thus biochemical analysis of this fusion event has been highly informative for understanding endosome–vacuole fusion.

Fusion of endosomes with the vacuole involves a specialized, SNARE-based membrane fusion machinery. *VAM3* encodes a vacuole-localized t-SNARE required for fusion of multiple membrane intermediates with the vacuole (Darsow et al., 1997; Wada et al., 1997). A rapid loss of Vam3p function causes an immediate block in maturation of multiple hydrolases including CPY, proteinase A, CPS, ALP and aminopeptidase I (a marker of the cytosol-to-vacuole transport pathway). Vam3p associates with another vacuolar t-SNARE, Vam7p, a SNAP-25-related protein. Genetic and biochemical experiments indicate that during membrane fusion a ternary complex is formed between the two t-SNAREs on the vacuolar membrane and a v-SNARE, Vti1p, on the endosomal membrane. Mutations in any of these

Table 14.6   Genes implicated in endosome-to-vacuole transport

Gene	Mutant class	Size of product	Comment
SEC18	N/A	758 aa	Hydrophilic; yeast homolog of mammalian NSF
SEC17	N/A	292 aa	Hydrophilic; yeast homolog of mammalian α-SNAP
VAM3	B	283 aa	Syntaxin member (t-SNARE) on vacuolar membrane
VAM7	B	316 aa	SNAP-25-like; t-SNARE domain; PX domain
VPS11	C	1029 aa	Hydrophilic, vacuole-localized, RING-H2 finger
VPS16	C	798 aa	Hydrophilic, vacuole-localized
VPS18	C	918 aa	Hydrophilic, vacuole-localized, RING-H2 and zinc finger domains
VPS33	C	691 aa	Sec1-like protein, ATP-binding domain
VPS39	B	1049 aa	Hydrophilic, vacuole-localized
VPS41	B	992 aa	Hydrophilic, AP-3 assembly role, vacuole-localized
VTI1	N/A	217 aa	Synaptobrevin member (v-SNARE)
YPT7	B	208 aa	Small GTPase, Rab7-like

three SNARE proteins exhibit *in vivo* phenotypes associated with a loss of endosome–vacuole fusion and cause blocks in an *in vitro* homotypic vacuole fusion assay.

As is the case in other membrane fusion systems, other molecules aid SNAREs during membrane fusion of yeast endosomes with the vacuole. A Rab protein, Ypt7p, is required for docking/fusion of membrane intermediates with the vacuole and for homotypic vacuole fusion (Wichmann et al., 1992; Wada et al., 1997). In addition, the NSF homolog Sec18p and the SNAP homolog Sec17p are involved in dissociating SNARE complexes after membrane fusion has been completed (Ungermann et al., 1998).

The C–Vps complex

Sec1 homologs have also been implicated as essential factors in membrane fusion. For example, Sec1p in yeast and n-Sec1/munc18 in neuronal cells are required for the fusion of exocytic vesicles with the plasma membrane. In yeast endosome–vacuole fusion the Sec1 homolog Vps33p is required (Darsow et al., 1997). Vps33p is a member of a complex consisting of at least six polypeptides (Vps11p, Vps16p, Vps18p, Vps33p, Vps39p and Vps41p) called the C–Vps complex (Wurmser et al., 2000) or alternatively HOPS, for *ho*motypic fusion and vacuole *p*rotein *s*orting complex (Seals et al., 2000).

Loss of function of Vps11p, Vps16p, Vps18p or Vps33p causes a class C *vps* phenotype; i.e. cells lack a distinguishable vacuole and exhibit cargo trafficking defects similar to that described above for *vam3* mutants. Loss of function of Vps39p and Vps41p causes a somewhat less severe vacuolar morphology defect (class B phenotype) (Raymond et al., 1992). In addition, Vps41p collaborates with the AP-3 complex at the TGN in the vesicle-mediated sorting of cargo into the ALP pathway. Vps11p, Vps16p and Vps18p appear to act as a subcomplex that can recruit the Sec1 homolog Vps33p to the vacuole. Interestingly, the C–Vps complex is conserved in mammalian cells, *Drosophila* and *Caenorhabditis elegans* (Sevrioukov et al., 1999).

The C–Vps complex appears to have multiple roles in mediating endosome–vacuole fusion. This complex interacts genetically and biochemically with Vam3p but apparently not when it is complexed with Vti1p and Vam7p (Sato et al., 2000). The interaction of the C–Vps complex with Vam3p might prevent it from forming non-productive associations with Vti1p present on the vacuolar membrane. Inactivation of the C–Vps complex using an antibody that recognized epitope-tagged Vps18p blocked membrane fusion *in vitro*. Mutations in the class C *VPS* genes reduced stability of the Vam3-Vti1-Vam7 SNARE complex. These results are consistent with the behavior of Sec1 homologs in other systems that have been observed to associate with the t-SNARE alone. These observations suggest a model in which the C–Vps complex associates with unpaired Vam3p to mediate productive pairing with Vam7p and endosome-localized Vti1p leading to fusion. However, another study suggests instead that the C–Vps complex associates with the assembled Vam3–Vti1–Vam7 complex and this association is disrupted when the SNARE complex is dissociated (Price et al., 2000). Thus, while the details of the interaction of the C–Vps complex with the SNAREs are controversial, it seems clear that the interaction in some way mediates SNARE function.

Recent data indicating an interaction between the C–Vps complex and the Rab Ypt7p suggest yet another aspect to its function (Seals et al., 2000; Wurmser et al., 2000). The C–Vps complex, through the Vps39p subunit, directly binds to the GDP-bound or nucleotide-free forms of Ypt7p and stimulates nucleotide exchange. In addition, the C–Vps complex appears to function as a downstream effector of the GTP-bound form of Ypt7p. The C–Vps complex as well as the Rab Ypt7p can associate with the endosomal membrane as well as the vacuolar membrane. In several systems Rab proteins and their effectors have been shown to mediate a SNARE-independent initial docking of membranes, termed tethering. Tethering is thought to occur prior to SNARE pairing. It is tempting to speculate that the presence of Ypt7:GTP and the C–Vps complex on both membranes reflects a type of symmetrical bridge, or tether, that could mediate the initial association between the two membranes. In this regard, the C–Vps complex could perform a tethering function analogous to the TRAPP complex for fusion of vesicles with the *cis*-Golgi and the exocyst complex for exocytic vesicle fusion to the plasma membrane.

A vacuole to PVC pathway

As stated above, after fusion of the animal cell endosome with the lysosome to generate a hybrid organelle, a maturation process occurs that involves recycling of membrane proteins back to endosomes. A recent study suggests that a retrograde pathway for vacuole-to-endosome trafficking may exist in yeast (Bryant et al., 1998). In this study the trafficking of a mutant form of ALP containing the FXFXD signal for PVC retrieval in its cytosolic domain was studied. The mutant protein (RS-ALP) localizes predominantly to the TGN in wild-type cells. As expected, RS-ALP localizes to the PVC in class E cells, which accumulate an exaggerated PVC that is blocked in recycling to the TGN. RS-ALP is able to reach the PVC in class E cells when the CPY pathway out of the TGN is blocked in a class D/class E (*vps45 vps27*) double mutant. However, RS-ALP cannot reach the class E compartment in an AP-3 defective class E double mutant. Therefore, RS-ALP must reach the class E compartment by first taking the ALP pathway to the vacuole followed by retrograde transport from the vacuole to the class E compartment. Interestingly, the v-SNARE Vti1p also appears to be recycled via the retrograde vacuole-to-PVC pathway. Whether yeast vacuoles actually engage in a maturation process from a hybrid organelle to a mature vacuole, or whether all vacuolar structures in the yeast cell are structurally and functionally equivalent, has yet to be determined.

Summary: *Transport between the endosome and vacuole*

- After maturation into an MVB, the PVC probably fuses directly with the vacuole, thus releasing the internal vesicles into the lumen of the vacuole for degradation.
- SNARE-dependent fusion machinery and the C–Vps protein complex (Vps11,16,18,33,39,41p) mediate membrane fusion at this step.
- The C–Vps complex may serve a tethering role to mediate membrane association prior to SNARE pairing.
- Certain proteins, such as the v-SNARE Vti1p, can be recycled from the vacuole membrane back to the endosome.

HISTORICAL NOTES

Historical Note 1

In the 1970s and 1980s yeast genetic screens were carried out that led to identification of mutants defective in vacuolar biogenesis. The first of these screens was designed to identify mutants with reduced ability to cleave the chymotrypsin substrate *N*-acetyl-DL-phenylalanine β-napthyl ester (Jones, 1977). Many of the resulting *pep* mutants were defective in genes encoding vacuolar proteases. It was later recognized that this screen also identified genes that are important for vacuolar biogenesis. As a result, such mutants were deficient in vacuolar protease activity. In the mid 1980s two screens were carried out that were based on the assumption that defects in sorting of vacuolar hydrolase CPY would cause it to be aberrantly secreted. The *vpt* mutants (for vacuolar targeting defective) were isolated by screening for yeast mutants that secreted a CPY-invertase reporter protein (Bankaitis et al., 1986; Robinson et al., 1988). In the second approach, the *vpl* (for vacuolar protein localization) mutants were identified in a leucine auxotrophic background by selecting for mutants that exhibited CPY activity at the cell surface (Rothman and Stevens, 1986; Rothman et al., 1989). The extracellular CPY activity enabled the mutants to cleave a CBZ-L-phenylalanine-L-leucine substrate, thereby liberating leucine which they could use as their sole source of leucine. Considerable genetic overlap was observed between the *pep*, *vpt* and *vpl* mutants and, in all, these mutants defined over 40 complementation groups. Currently, the number of genes required for sorting of vacuolar hydrolases has grown to over 50 and the genes are now collectively called *vps* for (vacuolar protein sorting defective).

Historical Note 2

The discovery of the mannose 6-phosphate recognition determinant on lysosomal enzymes derived from the study of human diseases caused by a deficiency of one or more lysosomal enzymes. In the early 1970s, the Neufeld laboratory discovered that fibroblasts from patients genetically deficient for a particular lysosomal enzyme could be 'corrected' by factors secreted from normal fibroblasts (Neufeld and Cantz, 1971). In these experiments, the normal lysosomal enzyme was taken up from the medium by receptor-mediated endocytosis and delivered to the mutant lysosome. The initial evidence that most lysosomal enzymes carry a common recognition component that mediates endocytosis came from studies of I-cell fibroblasts, a disease manifested by a deficiency of multiple lysosomal enzymes. These fibroblasts synthesized the enzymes normally but secreted them instead of sorting them to the lysosome. The secreted enzymes could not be taken up by normal fibroblasts although the I-cell fibroblasts could endocytose exogenously supplied enzymes. Thus, Neufeld proposed that the I-cell defect was in the production of a single recognition determinant common to many different lysosomal enzymes (Hickman and Neufeld, 1972). Preliminary experiments suggested that the recognition determinant was at least partly composed of carbohydrate and a screen for chemical inhibitors of enzyme endocytosis nicely confirmed this. Kaplan, Achord and Sly, and Sando and Neufeld independently reported that mannose 6-phosphate was a potent competitive inhibitor (Kaplan et al., 1977; Sando and Neufeld, 1977). It was later demonstrated that N-linked oliogosaccharides on lysosomal enzymes, but not other glycoproteins, carried mannose 6-phosphate (von Figura and Hasilik, 1986).

REFERENCES

Ballou, L., Hernandez, L.M., Alvarado, E. and Ballou, C.E. (1990) Revision of the oligosaccharide structures of yeast carboxypeptidase Y. *Proc Natl Acad Sci USA* **87**: 3368–3372.

Bankaitis, V.A., Johnson, L.M. and Emr, S.D. (1986) Isolation of yeast mutants defective in protein targeting to the vacuole. *Proc Natl Acad Sci USA* **83**: 9075–9079.

Barlowe, C. (2000) Traffic COPs of the early secretory pathway. *Traffic* **1**: 371–377.

Barr, V.A., Phillips, S.A., Taylor, S.I. and Haft, C.R. (2000) Overexpression of a novel sorting nexin, SNX15, affects endosome morphology and protein trafficking. *Traffic* **1**: 904–916.

Bensen, E.S., Yeung, B.G. and Payne, G.S. (2001) Ric1p and the Ypt6p GTPase function in a common pathway required for localization of *trans*-Golgi network membrane proteins. *Mol Biol Cell* **12**: 13–26.

Black, M.W. and Pelham, H.R. (2000) A selective transport route from Golgi to late endosomes that requires the yeast GGA proteins. *J Cell Biol* **151**: 587–600.

Boman, A.L., Zhang, C., Zhu, X. and Kahn, R.A. (2000) A family of ADP-ribosylation factor effectors that can alter membrane transport through the *trans*-Golgi. *Mol Biol Cell* **11**: 1241–1255.

Brickner, J.H. and Fuller, R.S. (1997) *SOI1* encodes a novel, conserved protein that promotes TGN-endosomal cycling of Kex2p and other membrane proteins by modulating the function of two TGN localization signals. *J Cell Biol* **139**: 23–36.

Brigance, W.T., Barlowe, C. and Graham, T.R. (2000) Organization of the yeast Golgi complex into at least four functionally distinct compartments. *Mol Biol Cell* **11**: 171–182.

Bryant, N.J. and Stevens, T.H. (1997) Two separate signals act independently to localize a yeast late Golgi membrane protein through a combination of retrieval and retention. *J Cell Biol* **136**: 287–297.

Bryant, N.J., Piper, R.C., Weisman, L.S. and Stevens, T.H. (1998) Retrograde traffic out of the yeast vacuole to the TGN occurs via the prevacuolar/endosomal compartment. *J Cell Biol* **142**: 651–663.

Catlett, N.L. and Weisman, L.S. (2000) Divide and multiply: organelle partitioning in yeast. *Curr Opin Cell Biol* **12**: 509–516.

Chen, C.Y. and Graham, T.R. (1998) An *arf1* synthetic lethal screen identifies a new clathrin heavy chain conditional allele that perturbs vacuolar protein transport. *Genetics* **150**: 577–589.

Chen, C.Y., Ingram, M.F., Rosal, P.H. and Graham, T.R. (1999) Role for Drs2p, a P-type ATPase and potential aminophospholipid translocase, in yeast late Golgi function. *J Cell Biol* **147**: 1223–1236.

Chu, D.S., Pishvaee, B. and Payne, G.S. (1996) The light chain subunit is required for clathrin function in *Saccharomyces cerevisiae*. *J Biol Chem* **271**: 33123–33130.

Conibear, E. and Stevens, T.H. (1998) Multiple sorting pathways between the late Golgi and the vacuole in yeast. *Biochim Biophys Acta* **1404**: 211–230.

Conibear, E. and Stevens, T.H. (2000) Vps52p, Vps53p, and Vps54p form a novel multisubunit complex required for protein sorting at the yeast late Golgi. *Mol Biol Cell* **11**: 305–323.

Cooper, A.A. and Stevens, T.H. (1996) Vps10p cycles between the late-Golgi and prevacuolar compartments in its function as the sorting receptor for multiple yeast vacuolar hydrolases. *J Cell Biol* **133**: 529–541.

Costaguta, G., Stefan, C.J., Benson, E.S., Emr, S.D. and Payne, G.S. (2001) Yeast Gga coat proteins function with clathrin in Golgi to endosome transport. *Mol Biol Cell* **12**: 1885–1896.

Darsow, T., Rieder, S.E. and Emr, S.D. (1997) A multispecificity syntaxin homologue, Vam3p, essential for autophagic and biosynthetic protein transport to the vacuole. *J Cell Biol* **138**: 517–529.

Dean, N. (1999) Asparagine-linked glycosylation in the yeast Golgi. *Biochim Biophys Acta* **1426**: 309–322.

Dell'Angelica, E.C., Klumperman, J., Stoorvogel, W. and Bonifacino, J.S. (1998) Association of the AP-3 adaptor complex with clathrin. *Science* **280**: 431–434.

Dell'Angelica, E.C., Puertollano, R., Mullins, C. et al. (2000) GGAs: a family of ADP ribosylation factor-binding proteins related to adaptors and associated with the Golgi complex. *J Cell Biol* **149**: 81–94.

Deloche, O., Yeung, B.G., Payne, G.S. and Schekman, R. (2001) Vps10p transport from the *trans*-Golgi network to the endosome is mediated by clathrin-coated vesicles. *Mol Biol Cell* **12**: 475–485.

D'Hondt, K., Heese-Peck, A. and Riezman, H. (2000) Protein and lipid requirements for endocytosis. *Annu Rev Genet* **34**: 255–295.

Fernandez-Borja, M., Wubbolts, R., Calafat, J. et al. (1999) Multivesicular body morphogenesis requires phosphatidyl-inositol 3-kinase activity. *Curr Biol* **9**: 55–58.

Fischer von Mollard, G., Nothwehr, S.F. and Stevens, T.H. (1997) The yeast v-SNARE Vti1p mediates two vesicle transport pathways through interactions with the t-SNAREs Sed5p and Pep12p. *J Cell Biol* **137**: 1511–1524.

Futter, C.E., Pearse, A., Hewlett, L.J. and Hopkins, C.R. (1996) Multivesicular endosomes containing internalized EGF-EGF receptor complexes mature and then fuse directly with lysosomes. *J Cell Biol* **132**: 1011–1023.

Gall, W.E., Higginbotham, M.A., Chen, C. et al. (2000) The auxilin-like phosphoprotein Swa2p is required for clathrin function in yeast. *Curr Biol* **10**: 1349–1358.

Gaynor, E.C., Graham, T.R. and Emr, S.D. (1998) COPI in ER/Golgi and intra-Golgi transport: do yeast COPI mutants point the way? *Biochim Biophys Acta* **1404**: 33–51.

Graham, T.R. and Emr, S.D. (1991) Compartmental organization of Golgi-specific protein modification and vacuolar protein sorting events defined in a *sec18*(NSF) mutant. *J Cell Biol* **114**: 207–218.

Graham, T.R. and Krasnov, V.A. (1995) Sorting of yeast alpha-1,3-mannosyltransferase is mediated by a lumenal domain interaction, and a transmembrane domain signal that can confer clathrin-dependant Golgi localization to a secreted protein. *Mol Biol Cell* **6**: 809–824.

Gruenberg, J. and Maxfield, F.R. (1995) Membrane transport in the endocytic pathway. *Curr Opin Cell Biol* **7**: 552–563.

Haft, C.R., de la Luz Sierra, M., Bafford, R. et al. (2000) Human orthologs of yeast vacuolar protein sorting proteins Vps26, 29, and 35: assembly into multimeric complexes. *Mol Biol Cell* **11**: 4105–4116.

Hama, H., Tall, G.G. and Horazdovsky, B.F. (1999) Vps9p is a guanine nucleotide exchange factor involved in vesicle-mediated vacuolar protein transport. *J Biol Chem* **274**: 15284–15291.

Harris, S.L. and Waters, M.G. (1996) Localization of a yeast early Golgi mannosyl-transferase, Och1p, involves retrograde transport. *J Cell Biol* **132**: 985–998.

Harsay, E. and Bretscher, A. (1995) Parallel secretory pathways to the cell surface in yeast. *J Cell Biol* **131**: 297–310.

Heilker, R., Spiess, M. and Crottet, P. (1999) Recognition of sorting signals by clathrin adaptors. *Bioessays* **21**: 558–567.

Hickman, S. and Neufeld, E.F. (1972) A hypothesis for I-cell disease: defective hydrolases that do not enter lysosomes. *Biochem Biophys Res Commun* **49**: 992–999.

Hirst, J., Lui, W.W., Bright, N.A. et al. (2000) A family of proteins with gamma-adaptin and VHS domains that facilitate trafficking between the *trans*-Golgi network and the vacuole/lysosome. *J Cell Biol* **149**: 67–80.

Holthuis, J.C.M., Nichols, B.J., Dhruvakumar, S. and Pelham, H.R.B. (1998) Two syntaxin homologues in the TGN/endosomal system of yeast. *EMBO J* **17**: 113–126.

Horazdovsky, B.F., Davies, B.A., Seaman, M.N. et al. (1997) A sorting nexin-1 homologue, Vps5p, forms a complex with Vps17p and is required for recycling the vacuolar protein-sorting receptor. *Mol Biol Cell* **8**: 1529–1541.

Huang, K.M., D'Hondt, K., Riezman, H. and Lemmon, S.K. (1999) Clathrin functions in the absence of heterotetrameric adaptors and AP180-related proteins in yeast. *EMBO J* **18**: 3897–3908.

Jones, E.W. (1977) Proteinase mutants of *Saccharomyces cerevisiae*. *Genetics* **85**: 23–33.

Kaplan, A., Achord, D.T. and Sly, W.S. (1977) Phosphohexosyl components of a lysosomal enzyme are recognized by pinocytosis receptors on human fibroblasts. *Proc Natl Acad Sci USA* **74**: 2026–2030.

Klinosky, D.J. and Emr, S.D. (1990) A new class of lysosomal/vacuolar protein sorting signals. *J Biol Chem* **265**: 5349–5352.

Klionsky, D.J. and Emr, S.D. (2000) Autophagy as a regulated pathway of cellular degradation. *Science* **290**: 1717–1721.

Klionsky, D.J., Herman, P.K. and Emr, S.D. (1990) The fungal vacuole: Composition, function, and biogenesis. *Microbiol Rev* **54**: 266–292.

Lemmon, S.K. (2001) Clathrin uncoating: Auxilin comes to life. *Curr Biol* **11**: R49–52.

Lemmon, S.K., Pellicena-Palle, A., Conley, K. and Freund, C.L. (1991) Sequence of the clathrin heavy chain from *Saccharomyces cerevisiae* and requirement of the COOH terminus for clathrin function. *J Cell Biol* **112**: 65–80.

Luzio, J.P., Rous, B.A., Bright, N.A. et al. (2000) Lysosome-endosome fusion and lysosome biogenesis. *J Cell Sci* **113**: 1515–1524.

Marcusson, E.G., Horazdovsky, B.F., Cereghino, J.L., Gharakhanian, E. and Emr, S.D. (1994) The sorting receptor for yeast carboxypeptidase Y is encoded by the *VPS10* gene. *Cell* **77**: 579–586.

Neufeld, E.F. and Cantz, M.J. (1971) Corrective factors for inborn errors of mucopolysaccharide metabolism. *Ann N Y Acad Sci* **179**: 580–587.

Nothwehr, S.F. and Hindes, A.E. (1997) The yeast VPS5/GRD2 gene encodes a sorting nexin-1-like protein required for localizing membrane proteins to the late Golgi. *J Cell Sci* **110**: 1063–1072.

Nothwehr, S.F., Roberts, C.J. and Stevens, T.H. (1993) Membrane protein retention in the yeast Golgi apparatus: dipeptidyl aminopeptidase A is retained by a cytoplasmic signal containing aromatic residues. *J Cell Biol* **121**: 1197–1209.

Nothwehr, S.F., Conibear, E. and Stevens, T.H. (1995) Golgi and vacuolar membrane proteins reach the vacuole in *vps1* mutant yeast cells via the plasma membrane. *J Cell Biol* **129**: 35–46.

Nothwehr, S.F., Bruinsma, P. and Strawn, L.A. (1999) Distinct domains within Vps35p mediate the retrieval of two different cargo proteins from the yeast prevacuolar/endosomal compartment. *Mol Biol Cell* **10**: 875–890.

Nothwehr, S.F., Ha, S.A. and Bruinsma, P. (2000) Sorting of yeast membrane proteins into an endosome-to-Golgi pathway involves direct interaction of their cytosolic domains with Vps35p. *J Cell Biol* **151**: 297–310.

Odorizzi, G., Babst, M. and Emr, S.D. (1998a) Fab1p PtdIns(3)P 5-kinase function essential for protein sorting in the multivesicular body. *Cell* **95**: 847–858.

Odorizzi, G., Cowles, C.R. and Emr, S.D. (1998b) The AP-3 complex: a coat of many colours. *Trends Cell Biol* **8**: 282–288.

Odorizzi, G., Babst, M. and Emr, S.D. (2000) Phosphoinositide signaling and the regulation of membrane trafficking in yeast. *Trends Biochem Sci* **25**: 229–235.

Payne, G.S. and Schekman, R. (1989) Clathrin: a role in the intracellular retention of a Golgi membrane protein. *Science* **245**: 1358–1365.

Payne, G.S., Baker, D., van Tuinen, E. and Schekman, R. (1988) Protein transport to the vacuole and receptor-mediated endocytosis by clathrin heavy chain-deficient yeast. *J Cell Biol* **106**: 1453–1461.

Peyroche, A., Antonny, B., Robineau, S. et al. (1999) Brefeldin A acts to stabilize an abortive ARF-GDP-Sec7 domain protein complex: involvement of specific residues of the Sec7 domain. *Mol Cell* **3**: 275–285.

Phillips, S.A., Barr, V.A., Haft, D.H., Taylor, S.I. and Renfrew Haft, C. (2001) Identification and characterization of SNX15, a novel sorting nexin involved in protein trafficking. *J Biol Chem* **276**: 5074–5084.

Prescianotto-Baschong, C. and Riezman, H. (1998) Morphology of the yeast endocytic pathway. *Mol Biol Cell* **9**: 173–189.

Price, A., Seals, D., Wickner, W. and Ungermann, C. (2000) The docking stage of yeast vacuole fusion requires the transfer of proteins from a *cis*-SNARE complex to a Rab/Ypt protein. *J Cell Biol* **148**: 1231–1238.

Rad, M.R., Phan, H.L., Kirchrath, L. et al. (1995) *Saccharomyces cerevisiae* Apl2p, a homologue of the mammalian clathrin AP beta subunit, plays a role in clathrin-dependent Golgi functions. *J Cell Sci* **108**: 1605–1615.

Raymond, C.K., Howald-Stevenson, I., Vater, C.A. and Stevens, T.H. (1992) Morphological classification of the yeast vacuolar protein sorting mutants: evidence for a prevacuolar compartment in class E *vps* mutants. *Mol Biol Cell* **3**: 1389–1402.

Redding, K., Seeger, M., Payne, G.S. and Fuller, R.S. (1996) The effects of clathrin inactivation on localization of Kex2 protease are independent of the TGN localization signal in the cytosolic tail of Kex2p. *Mol Biol Cell* **7**: 1667–1677.

Rehling, P., Darsow, T., Katzmann, D.J. and Emr, S.D. (1999) Formation of AP-3 transport intermediates requires Vps41 function. *Nat Cell Biol* **1**: 346–353.

Reynolds, T.B., Hopkins, B.D., Lyons, M.R. and Graham, T.R. (1998) The high osmolarity glycerol response (HOG) MAP kinase pathway controls localization of a yeast Golgi glycosyltransferase. *J Cell Biol* **143**: 935–946.

Robinson, J.S., Klionsky, D.J., Banta, L.M. and Emr, S.D. (1988) Protein sorting in *Saccharomyces cerevisiae*: Isolation of mutants defective in the delivery and processing of multiple vacuolar hydrolases. *Mol Cell Biol* **8**: 4936–4948.

Rothman, J.H. and Stevens, T.H. (1986) Protein sorting in yeast: mutants defective in vacuole biogenesis mislocalization vacuolar proteins into the late secretory pathway. *Cell* **47**: 1041–1051.

Rothman, J.H., Howald, I. and Stevens, T.H. (1989) Characterization of genes required for protein sorting and vacuolar function in the yeast *Saccharomyces cerevisiae*. *EMBO J* **8**: 2057–2065.

Sando, G.N. and Neufeld, E.F. (1977) Recognition and receptor-mediated uptake of a lysosomal enzyme, alpha-1-iduronidase, by cultured human fibroblasts. *Cell* **12**: 619–627.

Sato, T.K., Rehling, P., Peterson, M.R. and Emr, S.D. (2000) Class C Vps protein complex regulates vacuolar SNARE pairing and is required for vesicle docking/fusion. *Mol Cell* **6**: 661–671.

Seals, D.F., Eitzen, G., Margolis, N., Wickner, W.T. and Price, A. (2000) A Ypt/Rab effector complex containing the Sec1 homolog Vps33p is required for homotypic vacuole fusion. *Proc Natl Acad Sci USA* **97**: 9402–9407.

Seaman, M.N., Marcusson, E.G., Cereghino, J.L. and Emr, S.D. (1997) Endosome to Golgi retrieval of the vacuolar protein sorting receptor, Vps10p, requires the function of the VPS29, VPS30, and VPS35 gene products. *J Cell Biol* **137**: 79–92.

Seaman, M.N., McCaffery, J.M. and Emr, S.D. (1998) A membrane coat complex essential for endosome-to-Golgi retrograde transport in yeast. *J Cell Biol* **142**: 665–681.

Seeger, M. and Payne, G.S. (1992) A role for clathrin in the sorting of vacuolar proteins in the Golgi complex of yeast. *EMBO J* **11**: 2811–2818.

Sevrioukov, E.A., He, J.P., Moghrabi, N., Sunio, A. and Kramer, H. (1999) A role for the deep orange and carnation eye color genes in lysosomal delivery in *Drosophila*. *Mol Cell* **4**: 479–486.

Simpson, F., Bright, N.A., West, M.A. et al. (1996) A novel adaptor-related protein complex. *J Cell Biol* **133**: 749–760.

Simpson, F., Peden, A.A., Christopoulou, L. and Robinson, M.S. (1997) Characterization of the adaptor-related protein complex, AP-3. *J Cell Biol* **137**: 835–845.

Stepp, J.D., Pellicena-Palle, A., Hamilton, S., Kirchhausen, T. and Lemmon, S.K. (1995) A late Golgi sorting function for *Saccharomyces cerevisiae* Apm1p, but not for Apm2p, a second yeast clathrin AP medium chain-related protein. *Mol Biol Cell* **6**: 41–58.

Stevens, T., Esmon, B. and Schekman, R. (1982) Early stages in the yeast secretory pathway are required for transport of carboxypeptidase Y to the vacuole. *Cell* **30**: 439–448.

Traub, L.M. and Kornfeld, S. (1997) The *trans*-Golgi network: a late secretory sorting station. *Curr Opin Cell Biol* **9**: 527–533.

Ungermann, C., Nichols, B.J., Pelham, H.R. and Wickner, W. (1998) A vacuolar v-t-SNARE complex, the predominant form in vivo and on isolated vacuoles, is disassembled and activated for docking and fusion. *J Cell Biol* **140**: 61–69.

von Figura, K. and Hasilik, A. (1986) Lysosomal enzymes and their receptors. *Ann Rev Biochem* **55**: 167–193.

Voos, W. and Stevens, T.H. (1998) Retrieval of resident late-Golgi membrane proteins from the prevacuolar compartment of *Saccharomyces cerevisiae* is dependent on the function of Grd19p. *J Cell Biol* **140**: 577–590.

Wada, Y., Ohsumi, Y. and Anraku, Y. (1988) Isolation of vacuole-morphological mutants of the yeast, *Saccharomyces cerevisiae. Cell Struct Funct* **13**: 608.

Wada, Y., Nakamura, N., Ohsumi, Y. and Hirata, A. (1997) Vam3p, a new member of syntaxin related protein, is required for vacuolar assembly in the yeast *Saccharomyces cerevisiae. J Cell Sci* **110**: 1299–1306.

Wichmann, H., Hengst, L. and Gallwitz, D. (1992) Endocytosis in yeast: evidence for the involvement of a small GTP- binding protein (Ypt7p). *Cell* **71**: 1131–1142.

Wilcox, C.A., Redding, K.A., Wright, R. and Fuller, R.S. (1992) Mutation of a tyrosine signal in the cytosolic tail of yeast Kex2 protease disrupts Golgi retention and results in default transport to the vacuole. *Mol Biol Cell* **3**: 1353–1371.

Wilsbach, K. and Payne, G.S. (1993) Vps1p, a member of the dynamin GTPase family, is necessary for Golgi membrane protein retention in *Saccharomyces cerevisiae. EMBO J* **12**: 3049–3059.

Wurmser, A.E. and Emr, S.D. (1998) Phosphoinositide signaling and turnover: PtdIns(3)P, a regulator of membrane traffic, is transported to the vacuole and degraded by a process that requires lumenal vacuolar hydrolase activities. *EMBO J* **17**: 4930–4942.

Wurmser, A.E., Sato, T.K. and Emr, S.D. (2000) New component of the vacuolar class C–Vps complex couples nucleotide exchange on the Ypt7 GTPase to SNARE-dependent docking and fusion. *J Cell Biol* **151**: 551–562.

Yeung, B.G., Phan, H.L. and Payne, G.S. (1999) Adaptor complex-independent clathrin function in yeast. *Mol Biol Cell* **10**: 3643–3659.

15

THE SECRETORY PATHWAY

BENJAMIN S. GLICK

INTRODUCTION

The secretory pathway has two main functions. First, it transports newly synthesized proteins and lipids from the endoplasmic reticulum (ER) to the Golgi apparatus and then on to their final destinations. Second, it modifies many of these proteins and lipids during the transport process.

The organelles of the secretory and endocytic pathways collectively form the endomembrane system, which has the distinctive property that all of its compartments communicate with one another and with the extracellular environment by means of transport vesicles (see Chapter 16). Vesicular transport has implications for organelle identity. The simplest view is that the compartments of the endomembrane system are all stable entities that exchange material via transport vesicles. An alternative view, which is now gaining ground, is that vesicular transport can result in the formation and disappearance of compartments (Helenius et al., 1983; Glick and Malhotra, 1998). As diagrammed in Figure 15.1, transport vesicles may be able to fuse homotypically to generate a compartment *de novo*. Further rounds of vesicular transport would alter the composition of this new compartment in a process known as maturation. Finally, an existing compartment could disappear by fragmenting into transport vesicles. In this model, some compartments may exist only as transitory intermediates. Transport through the secretory pathway probably involves both vesicular transport between stable compartments and the vesicle-driven formation, maturation and disappearance of transitory compartments.

Protein Targeting, Transport & Translocation
ISBN 0-12-200731-X

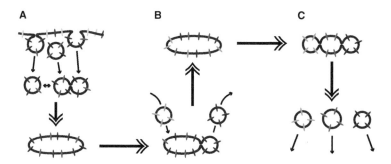

Figure 15.1   Life cycle of a hypothetical compartment of the endomembrane system. **A**, Formation – transport vesicles bud from a parental compartment and then fuse with one another to generate a new compartment. **B**, Maturation – this new compartment matures by receiving vesicles containing one set of components while exporting vesicles containing a different set of components. **C**, Disappearance – ultimately, the compartment disappears by fragmenting into various types of transport vesicles. Reprinted from *Cell*, **95**, Glick, B.S. and Malhotra, V. The curious status of the Golgi apparatus, pp. 883–889, Copyright 1998, with permission from Elsevier Science.

EXPORT FROM THE ER

Other chapters in this volume describe how newly synthesized proteins are translocated into or across the membrane of the ER, and how these proteins fold, oligomerize, and acquire disulfide bonds and N-linked oligosaccharide chains (Chapters 5, 6 and 7). A sophisticated quality control machinery in the ER lumen ensures that proteins are correctly folded and assembled (Chapter 9). Once a newly synthesized protein is released by the quality control machinery, it becomes eligible for entry into the secretory pathway.

The first step in the secretory pathway is the packaging of newly synthesized proteins into ER-derived COPII transport vesicles (Chapter 16). In most cell types, COPII vesicles bud from specialized ribosome-free ER subdomains that are termed transitional ER (tER) sites or ER exit sites. The number of tER sites in a cell varies from 2 to 5 in the budding yeast *Pichia pastoris* (Rossanese et al., 1999) to several hundred in a typical mammalian cell (Bannykh and Balch, 1997). These tER sites are long-lived, relatively immobile structures about 0.5 μm in diameter (Bannykh and Balch, 1997; Stephens et al., 2000; Hammond and Glick, 2000). Although the COPII assembly pathway is being characterized in detail (Barlowe, 2000), virtually nothing is known about how the budding of COPII vesicles is restricted to tER sites. One possibility is that the protein components that define tER sites spontaneously self-associate to form specialized patches in the ER membrane.

After COPII vesicles pinch off from the ER membrane and shed their coats, these vesicles apparently fuse with one another to generate the

'ER–Golgi intermediate compartment', or ERGIC (Bannykh and Balch, 1997; Klumperman, 2000). This homotypic fusion event is poorly understood, but it presumably involves the standard cellular fusion machinery (Chapter 16). Because the ERGIC has a complex topology, it is also referred to as vesicular-tubular clusters, or VTCs (Bannykh and Balch, 1997). The ERGIC is a site of protein sorting. Proteins such as cargo receptors (Herrmann et al., 1999) are recycled from the ERGIC to the ER for another round of action, whereas secretory cargo proteins remain in the ERGIC for transport to the Golgi. This sorting is carried out, at least in part, by COPI vesicles that bud from ERGIC elements and selectively recycle certain proteins to the ER (Klumperman, 2000; Barlowe, 2000). COPI-mediated recycling leads to a net increase in the concentration of secretory cargo proteins in the ERGIC.

In many unicellular eukaryotes, Golgi stacks are immediately adjacent to tER sites, and the ERGIC is presumably located between these two compartments (Becker and Melkonian, 1996; Rossanese et al., 1999). ERGIC elements in these cells are difficult to distinguish from the early Golgi. This distinction is clearer in vertebrate cells, in which the Golgi forms a ribbon near the nucleus whereas many of the tER sites are located in the peripheral cytoplasm (Figure 15.2). ERGIC elements that are generated near peripheral tER sites translocate along microtubules to the juxtanuclear Golgi (see below).

The ERGIC represents the first post-ER compartment in the secretory pathway, and serves as a carrier for delivering secretory cargo molecules to the Golgi. As described below, the conceptual framework for understanding transport to and through the Golgi stack depends upon the underlying model for how the Golgi operates.

Summary: *Export from the ER*

- Newly synthesized proteins exit the ER in COPII transport vesicles, which form at ribosome-free ER subdomains termed transitional ER sites.
- COPII vesicles are thought to shed their coats and then fuse homotypically to form the ER–Golgi intermediate compartment (ERGIC). The ERGIC produces COPI transport vesicles, which recycle proteins such as cargo receptors to the ER.

TRANSPORT THROUGH THE GOLGI

The Golgi is an elaborate structure that consists of multiple disk-shaped membranes called cisternae. In most cell types these cisternae form ordered

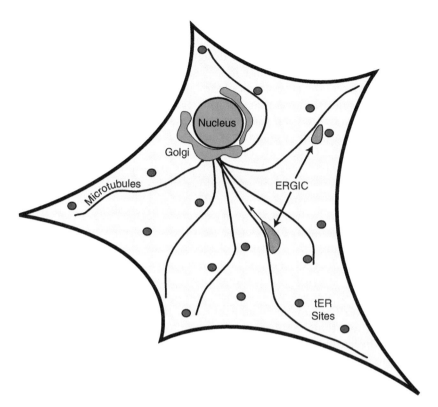

Figure 15.2   Relationship between tER sites, ERGIC elements and the Golgi in vertebrate cells. A typical vertebrate cell contains several hundred tER sites that are distributed throughout the ER network. ERGIC elements form near tER sites, and then translocate along microtubules toward the centrosome. This microtubule-dependent translocation generates a ribbon of interconnected Golgi stacks near the nucleus.

stacks. However, the morphology of the Golgi apparatus varies considerably in different organisms (Mollenhauer and Morré, 1991). A typical vertebrate cell contains several hundred Golgi stacks, with the cisternae being linked by lateral tubular connections to form a ribbon near the nucleus (Rambourg and Clermont, 1997). A typical plant cell also contains several hundred Golgi stacks, but they are scattered throughout the cytoplasm as individual units (Griffing, 1991). Many protist cells contain only one or a few Golgi stacks (Becker and Melkonian, 1996). The number of cisternae per stack varies from about four in certain yeasts (Rossanese et al., 1999) to about 30 in some algae (Becker and Melkonian, 1996). Golgi cisternae are not organized into stacks in some organisms, most notably the budding yeast *Saccharomyces cerevisiae*, in which individual cisternae are distributed throughout the cytoplasm (Preuss et al., 1992). Any models for Golgi function must take into account this remarkable diversity in morphology.

The different Golgi cisternae are designated *cis, medial* or *trans* to indicate the sequence of their function in the secretory pathway. Secretory cargo molecules arrive at the *cis* cisterna of the Golgi and then move successively to the *medial* and *trans* cisternae. During this time, the carbohydrate side chains on newly synthesized proteins and glycolipids are extensively modified by a series of resident Golgi enzymes (Mellman and Simons, 1992). At the exit face of the Golgi, often referred to as the *trans*-Golgi network or TGN, secretory cargo molecules are sorted into various types of transport carriers for delivery to either the endosomal/lysosomal/vacuolar system (Chapter 14) or the plasma membrane (Keller and Simons, 1997).

The mechanism by which secretory cargo molecules move through the Golgi has been debated for over 40 years. Two extreme possibilities are illustrated in Figure 15.3 (Glick, 2000). According to the stable compartments model, each cisterna of the Golgi is a permanent entity that contains a fixed complement of resident Golgi enzymes, and secretory cargo molecules travel from one cisterna to the next in transport vesicles (Figure 15.3A). Incoming ERGIC elements would fuse with the *cis* cisterna. Small cargo molecules would then be transported from one cisterna to the next in COPI vesicles, whereas larger cargoes would be transported in 'megavesicles' (Rothman and Wieland, 1996; Pelham and Rothman, 2000). At the TGN, the various cargo molecules would be segregated into different classes of vesicles for delivery to endosomes or the plasma membrane. In this view, the resident Golgi enzymes remain in the cisternae while the secretory cargo molecules are packaged into transport vesicles. An alternative model called 'cisternal maturation' is shown in Figure 15.3B. According to this model, each Golgi cisterna is a transitory compartment (Glick and Malhotra, 1998). A new cisterna would form at the *cis* face of the stack by the homotypic fusion of ERGIC elements. This cisterna would then progressively move through the stack toward the *trans* face, maturing in the process. Maturation would be mediated by retrograde-directed COPI vesicles: a given cisterna would export one set of Golgi enzymes in vesicles targeted to younger cisternae, while receiving a different set of Golgi enzymes in vesicles derived from older cisternae. Finally, at the TGN stage, the terminally mature cisterna would disappear by fragmentation into various classes of transport carriers. In this view, secretory cargo molecules remain in the maturing cisternae while the resident Golgi enzymes are packaged into transport vesicles.

No consensus has yet been reached about which of these two models is more accurate, and many researchers believe that intra-Golgi transport involves a combination of the two mechanisms shown in Figure 15.3. It has been argued that cisternal maturation occurs too slowly to account for the observed rates of transport through the Golgi stack (Pelham and Rothman,

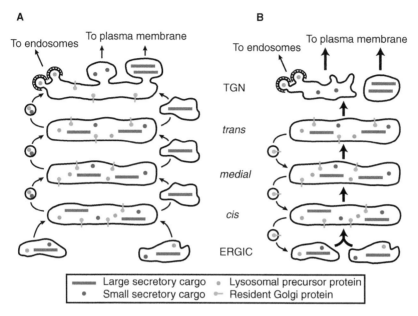

A

To endosomes   To plasma membrane

B

To endosomes   To plasma membrane

TGN

trans

medial

cis

ERGIC

——— Large secretory cargo   • Lysosomal precursor protein
• Small secretory cargo   •— Resident Golgi protein

Figure 15.3   Two models for transport through the Golgi stack: stable compartments (**A**) and cisternal maturation (**B**). Thin arrows indicate vesicle-mediated transport whereas thick arrows indicate maturation events. Note that some researchers have postulated hybrid models that incorporate both of the mechanisms depicted. (**A**) Vesicle-mediated transport between stable Golgi cisternae. ERGIC elements fuse with a preexisting *cis* cisterna. Different secretory cargo molecules then transit from one Golgi cisterna to the next in COPI vesicles and/or megavesicles. At the TGN, cargo molecules are sorted either into clathrin-coated vesicles for delivery to endosomes, or into secretory vesicles for delivery to the plasma membrane. (**B**) Maturation of transitory Golgi cisternae. ERGIC elements fuse with one another to generate a new *cis* cisterna. Secretory cargo molecules remain in this cisterna, which progressively matures due to the COPI-mediated recycling of resident Golgi proteins. At the TGN stage, clathrin-coated vesicles pinch off to carry certain cargo molecules to endosomes, and the remaining cargo molecules are segregated into cisternal remnants that become secretory carriers. Reprinted from *Current Opinion in Cell Biology*, **12**, Glick, B.S., Organization of the Golgi apparatus, pp. 450–456, Copyright 2000, with permission from Excerpta Medica Inc.

2000), but experimental support for this notion is lacking. Perhaps the most significant open question concerns the contents and directionality of COPI vesicles (Chapter 16). According to the stable compartments model, COPI vesicles should contain secretory cargo molecules and should move in the forward direction, whereas according to the maturation model, COPI vesicles should contain resident Golgi proteins and should move in the retrograde direction. Evidence in favor of both interpretations has been presented (Orci et al., 2000a; Lanoix et al., 1999). This analysis is complicated by the finding that Golgi cisternae produce more than one class of COPI vesicle (Orci et al., 2000b). It is not even clear whether COPI vesicles move in a directed fashion through the Golgi stack, or whether they simply fuse with random cisternae to allow cargo proteins to sample the entire Golgi (Orci et al., 2000b). Obviously, more work is needed to address these fundamental questions about how the Golgi operates.

Equally mysterious are the mechanisms that control Golgi organization. The stacking of Golgi cisternae can potentially be explained by the maturation model: if a new cisterna forms near a tER site, and then another cisterna forms behind the first, and so on, these cisternae should pile up to form a stack (Glick, 2000). This mechanism may be used by organisms such as *Pichia pastoris* that contain relatively simple Golgi stacks located next to tER sites (Rossanese et al., 1999). The situation is more complex in vertebrate and plant cells, which employ stacking proteins to align neighboring cisternae (Linstedt, 1999; Ladinsky et al., 1999). Another prominent and widely conserved aspect of Golgi organization is the tubulation of cisternal rims (Mollenhauer and Morré, 1991). Vertebrate Golgi cisternae are connected laterally by tubules, which probably arise when the homotypic fusion of ERGIC elements generates an extended *cis* cisterna (see above). In addition, non-connecting tubular projections are found in both the linked Golgi stacks of vertebrate cells and the separate Golgi stacks of other cell types (Mollenhauer and Morré, 1991; Ladinsky et al., 1999). The generation of these tubular projections seems to involve reactions of lipid metabolism (de Figueiredo et al., 1999; Weigert et al., 1999). While the function of Golgi tubules is unknown, this phenomenon probably reflects a more general partitioning of Golgi cisternae into biochemically distinct subdomains (Weidman et al., 1993).

Why is the Golgi divided into multiple cisternae? This question is relevant because in principle, the carbohydrate processing and protein sorting reactions that take place in the Golgi could all occur in a single compartment. Two explanations have been proposed. The 'distillation tower' model postulates that each cisterna selectively extracts escaped ER resident proteins for recycling to the ER, resulting in a progressive refinement of the cargo that is moving forward through the Golgi (Rothman and Wieland, 1996). This recycling of ER proteins is indeed an important process (Pelham, 1995), but the distillation hypothesis does not readily account for the large number of Golgi cisternae seen in many cells, because ER export is quite selective and the ER resident proteins that do escape rarely travel beyond the *cis*-most portion of the Golgi. The 'delay timer' model views the Golgi as an assembly line, and postulates that the biosynthetic events occurring in the Golgi require a certain amount of time that varies for different cell types (Glick and Malhotra, 1998). Thus, cells that produce more complex secretory products should contain more cisternae per stack, so that the secretory products will spend more total time in the Golgi. In support of this idea, there seems to be a rough correlation between the number of cisternae per Golgi stack and the complexity of the secretory products being synthesized by the cell (Becker and Melkonian, 1996).

Summary: *Transport through the Golgi*

- The Golgi consists of multiple cisternal compartments, which often display tubular projections. In most eukaryotes the cisternae form ordered stacks that may be either separate or interconnected. However, some budding yeasts contain non-stacked Golgi cisternae.
- Secretory cargo molecules traverse the Golgi in the order *cis* → *medial* → *trans* → *trans*-Golgi network (TGN). The compartmental nature of the Golgi facilitates the multistep processing of cargo molecules.
- The movement of cargo molecules through the Golgi may involve vesicular transport or cisternal maturation, or some combination of the two. According to the vesicular transport model, Golgi compartments are stable entities that exchange material by means of vesicles. According to the cisternal maturation model, Golgi compartments are transitory structures that form *de novo*, mature, and ultimately disappear.

EXPORT FROM THE TGN

The TGN is the site of sorting during exit from the Golgi stack (Keller and Simons, 1997). High-resolution electron microscopy indicates that the Golgi cisternae undergo an abrupt functional switch from producing COPI vesicles in the *cis* through *trans* cisternae to producing clathrin-coated vesicles in the TGN (Ladinsky et al., 1999). These clathrin-coated vesicles transport material to the endosomal/lysosomal/vacuolar system (Chapter 14). In addition, the TGN generates secretory carriers, which include constitutive secretory vesicles and regulated secretory granules. These sorting events involve the segregation of different cargo proteins and lipids into distinct subdomains of the TGN (Keller and Simons, 1997; Brown et al., 2001). The process of sorting at the TGN is elaborate and poorly understood, but it seems to require the recruitment from the cytosol of various scaffolding and membrane remodeling proteins (Stow and Heimann, 1998; Liljedahl et al., 2001; Brown et al., 2001).

The two models shown in Figure 15.3 imply different mechanisms for the formation of secretory carriers. In the stable compartments model (Figure 15.3A), secretory carriers bud repeatedly from the same TGN compartment. In the maturation model (Figure 15.3B), secretory carriers form by the fission and terminal maturation of the TGN. Consistent with this latter model, the production of secretory carriers does not seem to involve classical coat-driven budding events. Instead, regulated secretory granules undergo a progressive maturation and condensation (Arvan and Castle, 1998), and so-called secretory vesicles are actually large and pleiomorphic structures that appear to be remnants of TGN cisternae (Polishchuk et al., 2000). Thus, sorting at the TGN probably involves a combination of coated vesicle budding, membrane subdomain formation, and compartmental maturation.

The delivery of secretory carriers to the cell surface is facilitated by cytoskeletal elements (see below). Polarized cells contain multiple classes of secretory vesicles that are delivered to one or more distinct plasma membrane domains (Keller and Simons, 1997; Pruyne and Bretscher, 2000). In budding yeasts, an oligomeric tethering complex called the exocyst cooperates with a small GTPase to target secretory vesicles to sites of polarized growth (Guo et al., 1999). The mammalian version of the exocyst targets secretory vesicles to the basolateral membrane domain of polarized epithelial cells (Mostov et al., 2000). As with other vesicular transport events in the cell (Chapter 16), the tethering of secretory vesicles to the plasma membrane is followed by a protein-catalyzed fusion reaction. Soluble secretory cargo molecules are then released from the cell while membrane-associated cargo molecules become incorporated into the plasma membrane.

Summary: *Export from the TGN*

- At the TGN, different cargo molecules are sorted into distinct transport carriers, including clathrin-coated vesicles and secretory carriers. The formation of secretory carriers might involve either budding from a stable TGN compartment or else terminal maturation of TGN cisternae.
- Specific proteins mediate the plasma membrane targeting and subsequent fusion of secretory carriers.

PROTEIN LOCALIZATION TO THE ER AND GOLGI

Some of the proteins that are initially translocated into or across the ER membrane are not destined for secretion or for delivery to the endosomal/lysosomal/vacuolar system, but instead reside permanently in the ER. For example, the lumen of the ER contains a high concentration of soluble chaperone proteins (Chapters 7 and 9). These chaperones are largely excluded from budding COPII vesicles (Barlowe, 2000). The chaperone molecules that do leak out of the ER are captured by a specific Golgi-localized transmembrane receptor, which recognizes the tetrapeptide KDEL or a related sequence at the very C-terminus of the chaperone (Pelham,

1995). The complex between the chaperone and the KDEL receptor is packaged into COPI vesicles for recycling to the ER (Aoe et al., 1998). A similar type of retrieval mechanism recycles escaped ER membrane proteins that contain a C-terminal K(X)KXX peptide (X = any amino acid) at their cytosolic C-terminus (Pelham, 1995). This K(X)KXX signal binds directly to the COPI coat (Cosson and Letourneur, 1997). In addition, some escaped ER membrane proteins are recycled by the Golgi-localized protein Rer1, which binds both to COPI and to the transmembrane sequences of these ER proteins (Sato et al., 2001). In all of these cases, COPI-dependent retrieval serves merely to recycle the molecules that occasionally 'leak' out of the ER. The primary mechanism that localizes proteins to the ER is retention by means of exclusion from COPII vesicles.

Certain other proteins that enter the secretory pathway leave the ER but then take up permanent residence in the Golgi. Membrane proteins of the TGN are recycled from endosomes and/or the plasma membrane, and this recycling is triggered by signals in the cytosolic tails of the TGN proteins (Gleeson, 1998). Thus, TGN proteins are localized primarily or exclusively by retrieval. The localization process is less well understood for resident proteins of the *cis*, *medial* and *trans* Golgi. Most resident Golgi enzymes are type II membrane proteins with short N-terminal cytosolic tails, and the localization signals for these proteins are contained within the transmembrane domains and flanking regions (Gleeson, 1998). However, it is still unclear whether these localization sequences are retention signals or retrieval signals. According to the stable compartments model (Figure 15.3A), resident Golgi enzymes should be retained in the cisternae. In this case, the localization signals would somehow exclude the Golgi enzymes from forward-directed COPI vesicles. According to the maturation model (Figure 15.3B), resident Golgi enzymes should be continually retrieved from maturing cisternae. In this case, the localization signals would somehow promote the packaging of Golgi enzymes into retrograde-directed COPI vesicles. Either of these models might invoke the recognition of Golgi localization signals by specific partner proteins, or the partitioning of resident Golgi enzymes into specific lipid subdomains, or both mechanisms (Munro, 1998).

As this discussion illustrates, virtually every protein of the endomembrane system spends time in two or more compartments, and the composition of some compartments can evolve, so the 'residence' of proteins in a given secretory compartment is a rather fuzzy concept. In particular, the maturation model implies that the entire Golgi is constantly in flux. This concept is difficult because we tend to think of cells as machines; but whereas humans build machines from stable, well-defined components, cells seem to employ some components that continually change their identities. If we accept this idea that some compartments of the endomembrane system are constantly turning over, then the challenge is to provide a rigorous, quantitative description of trafficking events. One promising approach is kinetic modeling. For example, fluorescence photobleaching methods yield rate constants for the movement of a protein between different compartments, and these numerical values can be incorporated into models that have predictive and explanatory power (Hirschberg et al., 1998). Another example concerns the distribution of resident Golgi enzymes within the organelle. It has been known for many years that various Golgi enzymes are concentrated in either *cis*, *medial* or *trans* cisternae (Farquhar and Hauri, 1997), but accumulating evidence indicates that resident Golgi enzymes move between the cisternae and that a given enzyme species is actually present in multiple cisternae (Harris and Waters, 1996; Rabouille et al., 1995). These observations fit with the maturation model, which predicts that Golgi enzymes continually recycle as the cisternae mature (Figure 15.3B). However, it is not immediately obvious how a maturation mechanism could produce a polarized steady-state concentration of different Golgi enzymes in specific cisternae. To explain this Golgi polarity, it was proposed that the resident Golgi enzymes compete with one another for packaging into retrograde-directed COPI vesicles (Glick et al., 1997). The strongest competitors should end up concentrated in *cis* cisternae, whereas weaker competitors would be swept forward to *medial* or *trans* cisternae. This idea was tested by a numerical computer simulation, which confirmed that a competition mechanism can potentially account for the observed patterns of Golgi enzyme distribution (Figure 15.4). Of course, this result does not prove that competitive maturation actually operates in the Golgi, and later simulations have shown that other mechanisms could also account for Golgi polarity (Weiss and Nilsson, 2000). These preliminary efforts suggest that quantitative kinetic modeling will be a useful tool for understanding the organization of the secretory pathway.

Summary: *Protein localization to the ER and Golgi*

- Resident ER proteins are largely excluded from COPII transport vesicles, but occasionally escape and are retrieved by COPI vesicles. Typical retrieval signals for resident ER proteins include C-terminal KDEL and K(X)KXX peptides.
- Resident TGN proteins are localized by retrieval from later compartments.
- Type II resident Golgi enzymes contain localization signals within their transmembrane domains and flanking regions. It is still unknown whether these localization signals mediate retention or retrieval.
- Most proteins of the endomembrane system continually cycle between multiple compartments.

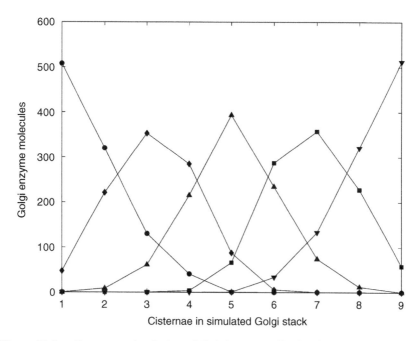

Figure 15.4   Computer simulation of Golgi enzyme distributions in a competitive maturation model. The model incorporates five hypothetical Golgi enzymes (marked by ●, ♦, ▲, ■, ▼) in a stack of nine cisternae. As each cisterna matures, its resident Golgi enzymes compete with one another in a stochastic fashion for packaging into retrograde COPI vesicles, which deliver their contents to the adjacent younger cisterna. Each enzyme in the series ●, ♦, ▲, ■, ▼ is approximately 12-fold more likely than its predecessor to be incorporated into a given COPI vesicle. As soon as the *trans*-most cisterna is entirely depleted of Golgi enzymes, the remaining cisternae advance by one step and a new *cis* cisterna is initiated. After the system has reached a steady state, the average number of enzyme molecules per cisterna is recorded. The result is that the five Golgi enzymes show partially overlapping peaks of distribution across the stack. This pattern resembles the experimentally observed distributions of resident enzymes in actual Golgi stacks. Reproduced with permission from Glick et al. (1997).

ROLE OF THE CYTOSKELETON IN THE FUNCTION AND ORGANIZATION OF THE SECRETORY PATHWAY

The cytoskeleton seems to be dispensable for basic secretory functions, including the selective packaging of material into coated vesicles, the targeting of vesicles to the appropriate compartments, and the fusion of vesicles with their target membranes (Chapter 16). However, all eukaryotic cells take advantage of the cytoskeleton to facilitate certain steps along the secretory pathway. The steps that are facilitated and the cytoskeletal elements that are employed differ in various cell types.

In vertebrate cells, microtubules play an important role in secretion and Golgi organization. The vertebrate Golgi also interacts with actin (Valderrama et al., 1998), certain spectrin isoforms (De Matteis and Morrow, 2000) and intermediate filaments (Gao and Sztul, 2001), but the interactions with microtubules seem to be the most significant. ERGIC elements that are generated near peripheral tER sites translocate along microtubules to the juxtanuclear Golgi ribbon (see Figure 15.2) in a process that requires the cytoplasmic dynein/dynactin motor protein complex (Burkhardt, 1998). The translocation of ERGIC elements toward the centrosome is responsible for assembling the Golgi ribbon. Protein recycling from the Golgi to the ER is also microtubule-dependent, and is mediated by motor proteins of the kinesin superfamily (Lippincott-Schwartz, 1998). Finally, kinesin-driven transport along microtubules delivers secretory carriers from the TGN to the plasma membrane (Lippincott-Schwartz, 1998). This microtubule-dependent transport of secretory carriers is probably facilitated by the association of the Golgi ribbon with the centrosome.

In some eukaryotes, microtubules appear to play no role in the secretory pathway. For example, microtubules in budding yeasts partition the nuclear material during cell division but do not influence the dynamics of cytoplasmic organelles (Botstein et al., 1997). Thus, when the yeast *Pichia pastoris* is treated with nocodazole to depolymerize microtubules, the organization of the tER–Golgi system is unaltered (Rossanese et al., 1999) (Figure 15.5). By contrast, nocodazole treatment of vertebrate cells induces a fragmentation of the Golgi ribbon into multiple 'ministacks' (Kreis et al., 1997). These ministacks are located near tER sites (Cole et al., 1996; Hammond and Glick, 2000). In other words, a nocodazole-treated vertebrate cell resembles an enlarged *Pichia pastoris* cell with regard to tER and Golgi organization (Figure 15.5). This similarity suggests that the association of Golgi stacks with tER sites is an evolutionarily conserved phenomenon, and that vertebrate cells have superimposed an additional layer of complexity by transporting Golgi elements along microtubules.

How does the vertebrate Golgi ribbon break down after addition of nocodazole? This question is still being debated. For many years it was assumed that after microtubules are disrupted, the preexisting Golgi ribbon fragments into smaller units that gradually diffuse throughout the cytoplasm. This fragmentation model continues to receive experimental support, and it fits with the view of the Golgi as a stable, independent organelle (Shima et al., 1998; Pelletier et al., 2000). Recently, several groups have proposed that blocking the microtubule-dependent transport of ERGIC elements to the Golgi ribbon leads to the formation of new Golgi stacks next to tER sites, with the concomitant shrinking of the preexisting Golgi ribbon (Cole et al., 1996; Storrie et al., 1998). This *de novo* formation model fits

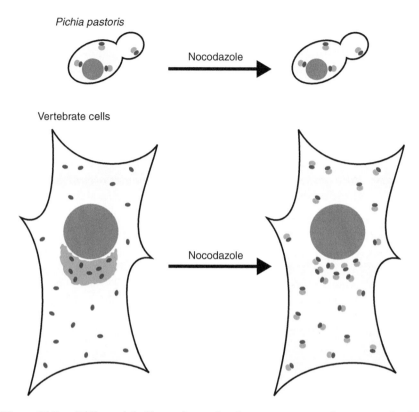

Figure 15.5   Differential effects of nocodazole on secretory pathway organization in two representative cell types. tER sites are shown in red and Golgi elements are shown in green. In the budding yeast *Pichia pastoris*, individual Golgi stacks are located next to tER sites. Disrupting yeast microtubules with nocodazole does not alter the tER–Golgi system. In vertebrate cells, tER sites are present throughout the cytoplasm whereas the Golgi forms a ribbon near the nucleus. Disrupting vertebrate microtubules with nocodazole blocks assembly of the Golgi ribbon and leads to the appearance of scattered Golgi stacks that are located next to tER sites. Thus, a nocodazole-treated vertebrate cell resembles a giant *Pichia pastoris* cell with regard to organization of the tER–Golgi system, suggesting that the close relationship between the tER and the Golgi has been evolutionarily conserved.

with the view of the Golgi as a dynamic outgrowth of the ER. Therefore, studies of Golgi–cytoskeleton interactions are shedding light on basic questions of Golgi identity.

Budding yeast and plant cells use actin filaments to position Golgi elements (Rossanese et al., 2001; Nebenfuhr and Staehelin, 2001). Moreover, budding yeasts rely on actomyosin-driven transport to deliver secretory carriers from the Golgi to the appropriate sites on the plasma membrane

(Pruyne and Bretscher, 2000). When actin function is disrupted in *Saccharomyces cerevisiae*, secretory carriers are no longer targeted to sites of polarized growth, but instead fuse with the entire plasma membrane. A similar effect is seen in polarized vertebrate epithelial cells, which require microtubule function for the selective delivery of secretory carriers to the distinct plasma membrane domains (Lippincott-Schwartz, 1998). Thus, cytoskeletal elements provide spatial cues that help to refine the specificity of membrane trafficking events.

Summary: *Role of the cytoskeleton in the function and organization of the secretory pathway*

- The cytoskeleton is not essential for the basic functioning of the secretory pathway, but helps to direct transport carriers to the appropriate locations.
- In vertebrate cells, microtubules guide ERGIC elements to the juxtanuclear Golgi and guide secretory carriers from the TGN to the plasma membrane. Disrupting microtubules causes the vertebrate Golgi ribbon to fragment into 'ministacks'.
- In plants and budding yeasts, actin filaments influence secretion and Golgi organization.

PERSPECTIVES

The initial elucidation of the secretory pathway was achieved using light and electron microscopy (Berger, 1997). During the past two decades, genetic and biochemical studies have extended this early descriptive work to provide detailed molecular information about vesicular transport (Duden and Schekman, 1997; Farquhar and Hauri, 1997). Now microscopy is once again coming to the fore as investigators reexamine the organization and dynamics of the secretory compartments. Video fluorescence microscopy is highlighting the tremendous plasticity of these compartments, while three-dimensional electron microscope tomography is imaging organelles at unprecedented resolution (Lippincott-Schwartz et al., 2001; Ladinsky et al., 1999).

Every advance in this field emphasizes how much remains to be learned. A quick tour through the secretory pathway illustrates this point. For example, how are tER sites generated and maintained? Do COPII vesicles fuse homotypically to generate ERGIC elements, and if so, how is this process regulated? Similarly, do ERGIC elements fuse to form a new *cis* cisterna, and if so, how does the cell determine when cisternal formation is complete? How do secretory cargo molecules transit through the Golgi: by vesicular transport, or cisternal maturation, or both processes? What are the contents of COPI vesicles, and how are these vesicles directed to their target compartments? What regulatory mechanisms ensure that forward ER-to-Golgi traffic is balanced by retrograde Golgi-to-ER traffic? Why is the Golgi divided into multiple cisternae, and how does a cell regulate the size and number of Golgi cisternae? How are different classes of cargo molecules sorted into distinct subdomains of the TGN? Do secretory carriers bud from a stable TGN, or are they the remnants of terminally mature TGN cisternae? If the reader is intrigued by these questions, please feel free to join the effort – we need all the help we can get.

REFERENCES

Aoe, T., Lee, A.J., van Donselaar, E., Peters, P.J. and Hsu, V.W. (1998) Modulation of intracellular transport by transported proteins: insights from regulation of COPI-mediated transport. *Proc Natl Acad Sci USA* **95**: 1624–1629.

Arvan, P. and Castle, D. (1998) Sorting and storage during secretory granule biogenesis: looking backward and looking forward. *Biochem J* **332**: 593–610.

Bannykh, S.I. and Balch, W.E. (1997) Membrane dynamics at the endoplasmic reticulum–Golgi interface. *J Cell Biol* **138**: 1–4.

Barlowe, C. (2000) Traffic COPs of the early secretory pathway. *Traffic* **1**: 371–377.

Becker, B. and Melkonian, M. (1996) The secretory pathway of protists: spatial and functional organization and evolution. *Microbiol Rev* **60**: 697–721.

Berger, E.G. (1997) The Golgi apparatus: from discovery to contemporary studies. In: Berger, E.G. and Roth, J. (eds) *The Golgi Apparatus*, pp. 1–35. Basel: Birkhäuser Verlag.

Botstein, D., Amberg, D., Mulholland, J., Huffaker, T. et al. (1997) The yeast cytoskeleton. In: Pringle, J.R., Broach, J.R. and Jones, E.W. (eds) *The Molecular and Cellular Biology of the Yeast Saccharomyces*, pp. 1–90. Cold Spring Harbor, NY: Cold Spring Harbor Laboratory Press.

Brown, D.L., Heimann, K., Lock, J. et al. (2001) The GRIP domain is a specific targeting sequence for a population of trans-Golgi network derived tubulo-vesicular carriers. *Traffic* **2**: 336–344.

Burkhardt, J.K. (1998) The role of microtubule-based motor proteins in maintaining the structure and function of the Golgi complex. *Biochim Biophys Acta* **1404**: 113–126.

Cole, N.B., Sciaky, N., Marotta, A., Song, J. and Lippincott-Schwartz, J. (1996) Golgi dispersal during microtubule disruption: regeneration of Golgi stacks at peripheral endoplasmic reticulum exit sites. *Mol Biol Cell* **7**: 631–650.

Cosson, P. and Letourneur, F. (1997) Coatomer (COPI)-coated vesicles: role in intracellular transport and protein sorting. *Curr Opin Cell Biol* **9**: 484–487.

de Figueiredo, P., Polizotto, R.S., Drecktrah, D. and Brown, W.J. (1999) Membrane tubule-mediated reassembly and maintenance of the Golgi complex is disrupted by phospholipase A_2 antagonists. *Mol Biol Cell* **10**: 1763–1782.

De Matteis, M.A. and Morrow, J.S. (2000) Spectrin tethers and mesh in the biosynthetic pathway. *J Cell Sci* **113**: 2331–2343.

Duden, R. and Schekman, R. (1997) Insights into Golgi function through mutants in yeast and animal cells. In: Berger, E.G. and Roth, J. (eds) *The Golgi Apparatus*, pp. 219–246. Basel: Birkhäuser Verlag.

Farquhar, M.G. and Hauri, H.-P. (1997) Protein sorting and vesicular traffic in the Golgi apparatus. In: Berger, E.G. and Roth, J. (eds) *The Golgi Apparatus*, pp. 63–129. Basel: Birkhäuser Verlag.

Gao, Y. and Sztul, E. (2001) A novel interaction of the Golgi complex with the vimentin intermediate filament cytoskeleton. *J Cell Biol* **152**: 877–894.

Gleeson, P.A. (1998) Targeting of proteins to the Golgi apparatus. *Histochem Cell Biol* **109**: 517–532.

Glick, B.S. (2000) Organization of the Golgi apparatus. *Curr Opin Cell Biol* **12**: 450–456.

Glick, B.S. and Malhotra, V. (1998) The curious status of the Golgi apparatus. *Cell* **95**: 883–889.

Glick, B.S., Elston, T. and Oster, G. (1997) A cisternal maturation mechanism can explain the asymmetry of the Golgi stack. *FEBS Lett* **414**: 177–181.

Griffing, L.R. (1991) Comparisons of Golgi structure and dynamics in plant and animal cells. *J Electron Microsc Tech* **17**: 179–199.

Guo, W., Roth, D., Walch-Solimena, C. and Novick, P. (1999) The exocyst is an effector for Sec4p, targeting secretory vesicles to sites of exocytosis. *EMBO J* **18**: 1071–1080.

Hammond, A.T. and Glick, B.S. (2000) Dynamics of transitional endoplasmic reticulum sites in vertebrate cells. *Mol Biol Cell* **11**: 3013–3030.

Harris, S.L. and Waters, M.G. (1996) Localization of a yeast early Golgi mannosyltransferase, Och1p, involves retrograde transport. *J Cell Biol* **132**: 985–998.

Helenius, A., Mellman, I., Wall, D. and Hubbard, A. (1983) Endosomes. *Trends Biochem Sci* **8**: 245–250.

Herrmann, J.M., Malkus, P. and Schekman, R. (1999) Out of the ER–outfitters, escorts and guides. *Trends Cell Biol* **9**: 5–7.

Hirschberg, K., Miller, C.M., Ellenberg, J. et al. (1998) Kinetic analysis of secretory protein traffic and characterization of Golgi to plasma membrane transport intermediates in living cells. *J Cell Biol* **143**: 1485–1503.

Keller, P. and Simons, K. (1997) Post-Golgi biosynthetic trafficking. *J Cell Sci* **110**: 3001–3009.

Klumperman, J. (2000) Transport between ER and Golgi. *Curr Opin Cell Biol* **12**: 445–449.

Kreis, T.E., Goodson, H.V., Perez, F. and Rönnholm, R. (1997) Golgi apparatus–cytoskeleton interactions. In: Berger, E.G. and Roth, J. (eds) *The Golgi Apparatus*, pp. 179–193. Basel: Birkhäuser Verlag.

Ladinsky, M.S., Mastronarde, D.N., McIntosh, J.R., Howell, K.E. and Staehelin, L.A. (1999) Golgi structure in three dimensions: functional insights from the normal rat kidney cell. *J Cell Biol* **144**: 1135–1149.

Lanoix, J., Ouwendijk, J., Lin, C.C. et al. (1999) GTP hydrolysis by arf-1 mediates sorting and concentration of Golgi resident enzymes into functional COPI vesicles. *EMBO J* **18**: 4935–4948.

Liljedahl, M., Maeda, Y., Colanzi, A. et al. (2001) Protein kinase D regulates the fission of cell surface destined transport carriers from the trans–Golgi network. *Cell* **104**: 409–420.

Linstedt, A.D. (1999) Stacking the cisternae. *Curr Biol* **9**: R893–R896.

Lippincott-Schwartz, J. (1998) Cytoskeletal proteins and Golgi dynamics. *Curr Opin Cell Biol* **10**: 52–59.

Lippincott-Schwartz, J., Snapp, E. and Kenworthy, A. (2001) Studying protein dynamics in living cells. *Nat Rev Mol Cell Biol* **2**: 444–456.

Mellman, I. and Simons, K. (1992) The Golgi complex: in vitro veritas? *Cell* **68**: 829–840.

Mollenhauer, H.H. and Morré, D.J. (1991) Perspectives on Golgi apparatus form and function. *J Electron Microsc Tech* **17**: 2–14.

Mostov, K.E., Verges, M. and Altschuler, Y. (2000) Membrane traffic in polarized epithelial cells. *Curr Opin Cell Biol* **12**: 483–490.

Munro, S. (1998) Localization of proteins to the Golgi apparatus. *Trends Cell Biol* **8**: 11–15.

Nebenfuhr, A. and Staehelin, L.A. (2001) Mobile factories: Golgi dynamics in plant cells. *Trends Plant Sci* **6**: 160–167.

Orci, L., Amherdt, M., Ravazzola, M., Perrelet, A. and Rothman, J.E. (2000a) Exclusion of Golgi residents from transport vesicles budding from Golgi cisternae in intact cells. *J Cell Biol* **150**: 1263–1270.

Orci, L., Ravazzola, M., Volchuk, A. et al. (2000b) Anterograde flow of cargo across the Golgi stack potentially mediated via bidirectional 'percolating' COPI vesicles. *Proc Natl Acad Sci USA* **97**: 10400–10405.

Pelham, H. (1995) Sorting and retrieval between the endoplasmic reticulum and Golgi apparatus. *Curr Opin Cell Biol* **7**: 530–535.

Pelham, H.R. and Rothman, J.E. (2000) The debate about transport in the Golgi – two sides of the same coin? *Cell* **102**: 713–719.

Pelletier, L., Jokitalo, E. and Warren, G. (2000) The effect of Golgi depletion on exocytic transport. *Nat Cell Biol* **2**: 840–846.

Polishchuk, R.S., Polishchuk, E.V., Marra, P. et al. (2000) Correlative light-electron microscopy reveals the tubular-saccular ultrastructure of carriers operating between the Golgi apparatus and plasma membrane. *J Cell Biol* **148**: 45–58.

Preuss, D., Mulholland, J., Franzusoff, A., Segev, N. and Botstein, D. (1992) Characterization of the *Saccharomyces* Golgi complex through the cell cycle by immunoelectron microscopy. *Mol Biol Cell* **3**: 789–803.

Pruyne, D. and Bretscher, A. (2000) Polarization of cell growth in yeast. II. The role of the cortical actin cytoskeleton. *J Cell Sci* **113**: 571–585.

Rabouille, C., Hui, N., Hunte, F. et al. (1995) Mapping the distribution of Golgi enzymes involved in the construction of complex oligosaccharides. *J Cell Sci* **108**: 1617–1627.

Rambourg, A. and Clermont, Y. (1997) Three-dimensional structure of the Golgi apparatus in mammalian cells. In: Berger, E.G. and Roth, J. (eds) *The Golgi Apparatus*, pp. 37–61. Basel: Birkhäuser Verlag.

Rossanese, O.W., Soderholm, J., Bevis, B.J. et al. (1999) Golgi structure correlates with transitional endoplasmic reticulum organization in *Pichia pastoris* and *Saccharomyces cerevisiae*. *J Cell Biol* **145**: 69–81.

Rossanese, O.W., Reinke, C.A., Bevis, B.J. et al. (2001) A role for actin, Cdc1p and Myo2p in the inheritance of late Golgi elements in *Saccharomyces cerevisiae*. *J Cell Biol* **153**: 47–61.

Rothman, J.E. and Wieland, F.T. (1996) Protein sorting by transport vesicles. *Science* **272**: 227–234.

Sato, K., Sato, M. and Nakano, A. (2001) Rer1p, a retrieval receptor for endoplasmic reticulum membrane proteins, is dynamically localized to the Golgi apparatus by coatomer. *J Cell Biol* **152**: 935–944.

Shima, D.T., Cabrera-Poch, N., Pepperkok, R. and Warren, G. (1998) An ordered inheritance strategy for the Golgi apparatus: visualization of mitotic disassembly reveals a role for the mitotic spindle. *J Cell Biol* **141**: 955–966.

Stephens, D.J., Lin-Marq, N., Pagano, A., Pepperkok, R. and Paccaud, J.-P. (2000) COPI-coated ER-to-Golgi transport complexes segregate from COPII in close proximity to ER exit sites. *J Cell Sci* **113**: 2177–2185.

Storrie, B., White, J., Röttger, S. et al. (1998) Recycling of Golgi-resident glycosyl-transferases through the ER reveals a novel pathway and provides an explanation for nocodazole-induced Golgi scattering. *J Cell Biol* **143**: 1505–1521.

Stow, J.L. and Heimann, K. (1998) Vesicle budding on Golgi membranes: regulation by G proteins and myosin motors. *Biochim Biophys Acta* **1404**: 161–171.

Valderrama, F., Babia, T., Ayala, I. et al. (1998) Actin microfilaments are essential for the cytological positioning and morphology of the Golgi complex. *Eur J Cell Biol* **76**: 9–17.

Weidman, P., Roth, R. and Heuser, J. (1993) Golgi membrane dynamics imaged by freeze-etch electron microscopy: views of different membrane coatings involved in tubulation versus vesiculation. *Cell* **75**: 123–133.

Weigert, R., Silletta, M.G., Spanò, S. et al. (1999) CtBP/BARS induces fission of Golgi membranes by acylating lysophosphatidic acid. *Nature* **402**: 429–433.

Weiss, M. and Nilsson, T. (2000) Protein sorting in the Golgi apparatus: a consequence of maturation and triggered sorting. *FEBS Lett* **486**: 2–9.

16

VESICULAR TRANSPORT

JOACHIM OSTERMANN, TOBIAS STAUBER AND
TOMMY NILSSON

INTRODUCTION

By revealing the importance of organelles such as the endoplasmic reticulum
(ER) and the Golgi apparatus in protein secretion at the ultrastructural
level in the 1950s and 1960s, Palade and others laid the ground for our current understanding of intracellular transport. We now know that newly synthesized proteins destined for the plasma membrane are imported into the
ER where they fold, oligomerize, receive N-linked oligosaccharides, and are
then checked by the quality control machinery before export from the transitional ER through vesicular intermediates (Figure 16.1). These vesicular
intermediates uncoat and fuse to form a pleiotropic membrane structure
composed of highly fenestrated tubular membranes termed ER-to-Golgi
intermediate carriers (ERGICs) which subsequently move inwards to the
central Golgi apparatus from 100 or so different peripheral sites (Saraste
and Svensson, 1991; Scales et al., 1997; Presley et al., 1997). Transported in a
dynein and dynactin dependent manner on microtubules, the numerous
ERGICs dock at the *cis* face of the Golgi stack where they flatten out and
connect laterally to form a new *cis* cisterna (Burkhardt et al., 1997;
Ladinsky et al., 1999; Marsh et al., 2001). This newly formed cisterna will
then serve as a template onto which incoming ERGICs dock, flatten out,
and fuse and by doing so, displace the previous cisterna in the *trans* direction. This process is continuously repeated, pushing previously formed
cisternae with their cargo forward towards the *trans* face where cisternae
shed vesicles and tubules containing newly synthesized proteins for further
transport to the plasma membrane (see Chapter 15). What remains of the
trans cisterna is then somehow consumed, possibly by fusing with portions

Protein Targeting, Transport & Translocation
ISBN 0-12-200731-X

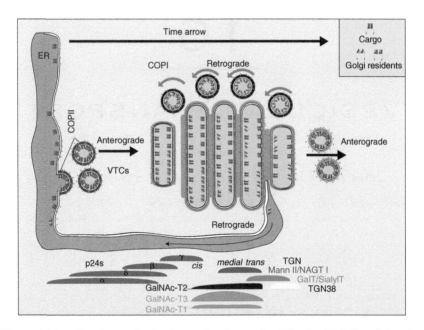

Figure 16.1 Newly synthesized proteins leave the ER via COPII vesicles that bud off from a hundred or so peripheral ER exit sites. This is followed by transport to the central and juxta-nuclear collection of laterally connected Golgi stacks termed the Golgi apparatus. Here, newly synthesized proteins receive modifications such as glycosylation while moving from the *cis* to the *trans* side. Anterograde transport is balanced by retrograde transport of Golgi resident proteins and lipids to keep the Golgi stack in a dynamic equilibrium. Cisternal carriers assemble through lateral fusion of uncoated, COPI and COPII vesicles. Over time, cisternal carriers receive resident proteins such as glycosylation enzymes acting on the cargo. After a brief period of time, resident proteins are sorted into COPI vesicles which deliver these to an earlier cisternal carrier. At the *trans* side, cisternae disassemble, releasing cargo for the plasma membrane or the endocytic pathway. The remaining *trans* cisterna is then consumed by the ER giving rise to COPI-independent but Rab6-dependent recycling of some Golgi resident proteins and a direct access to the ER for some protein toxins such as Shiga toxin and Shiga-like toxins (see Storrie et al., 2000, for further details). Note the gradient-like distribution of each resident protein.

of the ER known to pervade the *trans* part of the Golgi stack (see Hermo et al., 1991; Ladinsky et al., 1999; Marsh et al., 2001; Storrie et al., 2000).

At first glance, this model of the secretory pathway looks relatively simple. Membranes form, move forward and are then consumed upon release of its cargo (as indeed suggested by morphologists during the late 1950s). But though simplicity is part of the story, there exists an intricate machinery to ensure that modifying enzymes and structural components of the pathway move in a counter-current manner so that these are maintained in the pathway and not lost to the plasma membrane. Furthermore, as some enzymes need to act early whereas others act late, enzymes are kept in discrete but overlapping recycling loops that intersect the newly synthesized proteins at various stages of the cisternal progression process (Figure 16.1). This requires a recycling machinery that can recognize the constituents of the pathway and sort these away from newly synthesized proteins. The initial elucidation of a molecular machinery enabling this process is the combined outcome of two groups, the first using a genetic approach (Novick et al., 1980) and the second, biochemical fractionation and *in vitro* complementation assays (Fries and Rothman, 1980; Balch et al., 1984). Their work and those of others has defined key molecular components and vesicular transport intermediates that make up the transport and recycling machinery of the secretory pathway, the main topic of this chapter. Lately, use of green fluorescence protein (GFP) has also provided useful information, the most important being the visualization of ERGICs moving from the periphery to the central Golgi. In discussing the role of transport vesicles which lie at the heart of the secretory pathway, we will focus mostly on biochemical *in vitro* studies but also review, to some extent, important findings derived from other approaches.

TRANSPORT VESICLES – ARGUMENTS AGAINST THEIR EXISTENCE?

The existence of transport vesicles as transport intermediates in the secretory pathway has never been universally accepted. The most recent criticism is based on the failure to observe transport vesicles at all transport steps between organelles in the living cell. By constructing hybrid molecules between GFP and anterograde transport markers such as the G-protein of vesicular stomatitis virus (VSV-G) or resident proteins of the pathway known to recycle, transport can be observed in the living cell. The absence of detectable vesicles at the light microscopy level is, however, not an argument to discount their existence. Transport vesicles are supported by a large body of evidence derived from morphological, molecular, genetic as well as biochemical studies. Denying their existence has led to widespread speculation on other, hypothetical mechanisms of transport. However, in the absence of any direct evidence for such non-vesicular transport mechanisms, the dictum ascribed to the Western philosopher William of Occam (1287–1349), '*entia non sunt multiplicanda sine necessitate*' (also known as Occam's razor), applies. It is also easy to explain why transport vesicles have not yet been visualized in the living cell. When using GFP, long-range movements are predominantly highlighted. Vesicles moving between adjacent Golgi cisternae or leaving the ER are always in close proximity to larger and much brighter (GFP fluorescent) membrane structures and are therefore impossible to discern. Moreover, recycling vesicles (see below) transporting material from the Golgi to the ER do not have to travel from the central Golgi apparatus to peripheral ER exit sites. Rather, these may dock and fuse directly with the ER since the ER is present throughout the cell and, therefore, is always in close proximity to the Golgi apparatus. High resolution electron microscopy (EM) followed by three-dimensional (3D) reconstruction has consistently demonstrated the existence of numerous vesicles adjacent to Golgi stacks (see, for example, Rambourg et al., 1981). To avoid fixation artifacts whereby hypothetical tubules could fragment into uniformly sized vesicles (a process that is in itself hypothetical), EM and 3D high resolution tomography has been carried out on samples that were quick-frozen to preserve their ultrastructure. In agreement with earlier morphological work, multiple vesicles as well as tubules were seen in close proximity to Golgi cisternae always tethered to membranes or to each other (Ladinsky et al., 1999; Marsh et al., 2001).

COPI AND COPII – A ROLE FOR COAT PROTEINS IN CARGO SELECTION

The principle of cargo selection through vesicle coat was established by Brown, Goldstein and Pearse in the late 1970s. Studying receptor-mediated endocytosis, they showed that cargo selection and vesicle formation went hand in hand. Typically, ligand binding to surface receptors rapidly leads to sorting and concentration into clathrin-coated pits which then invaginate to form clathrin-coated vesicles. Clathrin, a cytosolic coat-forming protein, binds to cytoplasmic domains of surface receptors through linker proteins termed adaptor proteins. Homology exists between adaptor proteins and some coat proteins (COPs) found in the secretory pathway, making it likely that these have evolved from common ancestorial genes and that they share common functions (Scales et al., 2000). Two different COPs mediate the vesicular transport steps in the secretory pathway, COPI and COPII. Whereas COPI vesicles are mostly in close proximity to the Golgi stack from where they bud (Orci et al., 1986), COPII vesicles are always observed at the ER (Barlowe et al., 1994). Both vesicle coats differ in terms of their compositions as well as morphologically from the clathrin/adaptor coat in that they appear more fuzzy or irregular (Figure 16.2).

COPI

The COPI coat is composed of two main components: coatomer, which consists of seven subunits, α, β, β', γ, δ, ε, and ζ (Waters et al., 1991), and a

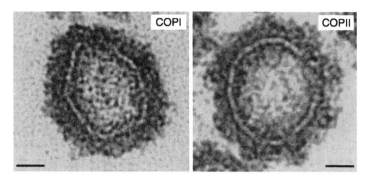

Figure 16.2   COPI and COPII transport vesicles purified from the yeast endoplasmic reticulum (Bar = 20 nm) (see Schekman and Orci, 1996). Micrographs courtesy of Professor Lelio Orci, University of Geneva, Switzerland.

small GTPase, Arf-1 (Serafini et al., 1991). Assembly of coatomer takes place in the cytosol, forming a stable complex that is recruited onto the membrane with the help of Arf-1. Under certain *in vitro* conditions, coatomer can be disassembled into smaller complexes or individual subunits. This has enabled the mapping of molecular interactions, both between subunits as well as to Arf-1 and other effectors (Faulstich et al., 1996; Lowe and Kreis, 1996; Goldberg, 1999; Eugster et al., 2000; Szafer et al., 2000). The ability of COPI to interact with cytoplasmic domains of resident membrane-bound proteins was discovered by Cosson and Letourneur (1994) when screening for molecular components interacting with K(X)KXX, a cytosolic recycling motif found in several resident proteins of the ER and the ER-to-Golgi interface (Nilsson et al., 1989; Jackson et al., 1990, 1993). Subsequent experiments involving yeast genetics confirmed that this interaction was indeed a relevant physiological event (Letourneur et al., 1994). This work remains important for two reasons. First, it showed that COPI shares similarities in *modus operandi* with the clathrin/adaptor coats in terms of cargo selection. Second, the link between K(X)KXX and COPI provided firm evidence that COPI vesicles are involved in recycling in the secretory pathway.

In order for coatomer to bind to membranes, Arf-1 is needed. This small GTPase is recruited from the cytosol in its GDP state onto the membrane, where a weak association to lipids is enabled through a 17 amino acid amphipathic helix containing a myristic acid at its amino terminus. Exchange of GDP to GTP is then required for the much stronger and productive membrane association and this is catalyzed by a guanine exchange factor (GEF) which binds tightly to Arf-1^{GDP} (Goldberg, 1998; Mossessova et al., 1998; Cherfils et al., 1998). The nucleotide exchange results in a major conformational change in Arf-1 causing hydrophobic residues to be exposed in the

N-terminal helix and enabling a tight association with the membrane. The hunt for Arf-1 GEF proved to be an interesting example of the synergy between genetics and biochemistry. It resulted in the successful identification of Arf-1 GEF in both yeast and mammalian cells. Sec7, the first Arf GEF, was isolated in one of the original screens for secretion mutants by Novick, Field and Schekman (Novick et al., 1980) but no GEF function was attributed to Sec7 at this stage. Almost 10 years later, studies were conducted on the fungal metabolite brefeldin A (BFA), which causes loss of coatomer from Golgi membranes, tubulation and subsequent fusion with the ER. Fractionation experiments identified a BFA-sensitive GEF activity on Golgi membranes (Donaldson et al., 1992; Helms and Rothman, 1992) but no molecular identification was made at that stage. Subsequently, p200/GEP was partly purified by Moss and Vaughan from mammalian cells (Morinaga et al., 1996) and in yeast, Jackson identified Gea1 and Gea2 as Arf-1 GEFs (Peyroche et al., 1996) by screening for suppression of dominant Arf-2 (functionally equivalent to Arf-1) mutants that are poor GTP binders. Cloning of p200/GEP (also termed BIG1) was completed in 1997 (Morinaga et al., 1997) and as with Gea1 and Gea2, was shown to be BFA sensitive. In parallel, ARNO and later ARNO3 and cytohesin-1 were identified as Arf GEFs in mammalian cells though their molecular weights were predicted to be 47 kDa, much smaller than Gea1, Gea2 or p200/GEP. These small molecular weight GEFs are not inhibited by BFA. Nevertheless, using ARNO, Chardin and co-workers established that it is the Sec7 domain (a common domain shared by all Arf GEFs) that catalyzes the nucleotide exchange on Arf (Chardin et al., 1996). Today, after nearly 20 years, a large family of Arf-GEFs has emerged whose members exert their enzymatic roles across the cell through their Sec7 domains.

With Arf-1^{GTP} tightly bound to the membrane, coatomer can now be recruited. This binding is greatly enhanced in the presence of resident proteins which display K(X)KXX-like motifs (Bremser et al., 1999). Thus, one could imagine that when resident proteins are present, coatomer will bind sufficiently for vesicles to form. But this process does not appear to be as straightforward as it seems. For resident proteins to be properly sorted into COPI vesicles, GTP hydrolysis by Arf-1 is required (Lanoix et al., 1999). But GTP hydrolysis by Arf-1 causes release of coatomer from the membrane and uncoating of COPI vesicles. We showed that addition of GTPγS (a non-hydrolyzable analog of GTP) or a GTP-restricted mutant of Arf-1 blocks sorting of Golgi resident proteins into COPI vesicles as well as subsequent uncoating of formed vesicles. The need for GTP hydrolysis therefore suggests that for proper sorting to take place, coatomer needs to cycle on and off the membrane. A possible scenario for how this can be envisaged is outlined in Figure 16.3 (based on Lanoix et al., 1999, 2001; Springer et al., 1999; Goldberg, 2000). A GTPase activating protein (GAP) is required for Arf-1 mediated GTP hydrolysis (Randazzo and Kahn, 1994; Makler

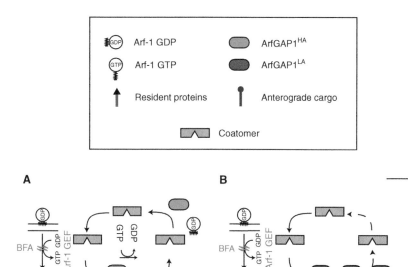

Golgi cisternae

Figure 16.3 On the left, coatomer binds to Arf-1^{GTP} at the cytosolic face of the cisternal membrane (**A**). GTP hydrolysis by Arf-1 stimulated by GAP (denoted by HA for high activity) removes coatomer, which is then free to rebind. This cycle is repeated but is gradually slowed down as resident proteins accumulate (**B**). The attraction of resident proteins is through the ability to interact with coatomer directly via their cytoplasmic domains (as suggested by Dominguez et al., 1998 and Lanoix et al., 1999). When resident proteins interact with coatomer (**B**) and GAP, the ability of GAP to stimulate GTP hydrolysis decreases (denoted by LA for low activity), allowing coatomer to remain sufficiently for the vesicle bud to form (Springer et al., 1999; Goldberg, 2000; Lanoix et al., 2001).

et al., 1995). Without GAP, the rate of GTP hydrolysis by Arf-1 is 1000-fold lower. The ability of GAP to stimulate Arf-1 mediated GTP hydrolysis is modulated by the presence of resident proteins in such a way that the rate of hydrolysis drops in response to the increase of resident proteins. Only when a sufficient amount of resident proteins is present will coatomer remain on the membrane long enough for the vesicle to form. This gives rise to a higher likelihood for coatomer to polymerize laterally thus forming patches in the membrane where resident proteins bound by coatomer complexes are concentrated. In this scenario, addition of GTPγS will stabilize coatomer on the membrane as Arf-1 will be locked in its GTP-like conformation. This negates the need for resident proteins to downregulate GAP activity thus enabling production of coated vesicles but with far fewer resident proteins.

COPII

A requirement for GTP hydrolysis for concentrating newly synthesized proteins into COPII vesicles has not yet been demonstrated. For quite some time it was thought that once cleared of the ER quality control machinery, newly synthesized proteins entered COPII vesicles by default without any need for concentration. This and other observations led to the hypothesis of 'bulk flow' negating the need for sorting and concentration in the secretory pathway (Wieland et al., 1987). Contrary to predictions, newly synthesized proteins were later found to be concentrated in COPII vesicles (Mizuno and Singer, 1993; Balch et al., 1994), suggesting an underlying active sorting process. As with COPI, direct binding between cargo and coat has been demonstrated.

The protein coat of COPII vesicles is composed of two heterodimers, Sec23/24 and Sec13/31 as well as a small GTPase, Sar1p (for review, see Barlowe, 2000). A membrane-bound GEF, Sec12, is needed for the GDP-to-GTP exchange of Sar1p. Sar1p^{GTP} is required for Sec23/24 and subsequently, Sec13/31 recruitment to form the protein coat. As with Arf-1, Sar1p requires a GAP for GTP hydrolysis and this activity is provided by Sec23 (Yoshihisa et al., 1993). The COPII coat can be assembled onto the membrane in at least two ways. Sed4, a membrane-bound component homologous to Sec12, but functionally distinct, interacts with Sec16, a peripheral protein tightly associated with the ER membrane, and serves to recruit the COPII coat (Espenshade et al., 1995; Gimeno et al., 1995). Cytoplasmic domains of ERGIC53, p24 proteins, Bet1p and Bos1p mediate a direct binding of Sec23/24 and/or Sar1p. These are all membrane proteins residing at the interface between the ER and the *cis* Golgi (Kappeler et al., 1997; Dominguez et al., 1998; Springer and Schekman, 1998). For membrane-bound cargo, Sar1p has been shown to interact directly with the cytoplasmic domain of VSV-G (Aridor et al., 2001) and for soluble cargo, selection and packaging is thought to be mediated by membrane-bound cargo receptors such as ERGIC53 and p24 proteins (Kuehn et al., 1998; Appenzeller et al., 1999; Muniz et al., 2000). Several isoforms of COPII subunits have been identified. In yeast, Sec24 isoforms have been shown to direct particular cargo into different subpopulations of vesicles (Pagano et al., 1999; Peng et al., 2000; Shimoni et al, 2000; Kurihara et al., 2000). This suggests that different subpopulations of COPII vesicles exist, each tailored to the particular needs of different cargo. Some proteins also need special chaperones for proper incorporation into COPII vesicles. For example, amino acid permeases in yeast require Shr3p, an integral membrane protein of the ER, for packaging into COPII vesicles and transport out of the ER (Kuehn et al., 1996). Deletion of Shr3p results in the accumulation of permeases in the ER and its packaging chaperone-like activity is thought to bind to permeases and to recruit COPII components for vesicle formation. Yet, Shr3p does not enter the vesicle but remains behind in the ER. How this is achieved is not yet understood.

Summary: *COPI and COPII – A role for coat proteins in cargo selection*

- Cargo selection is in part mediated by cytosolic coat proteins.
- Both COPI and COPII components interact directly with cytoplasmic motifs present in cargo proteins.
- Whereas COPI vesicles mediate recycling, COPII vesicles mediate ER to Golgi transport.
- Coat assembly is mediated by small GTPases (e.g. Arf-1 and Sar1) when in their GTP bound state. Coversion from GDP to GTP bound state is mediated by GEFs.
- Coat disassembly is mediated by GAPs. These act on the small GTPases, allowing them to hydrolyze their bound GTP.
- Sorting of Golgi resident proteins into COPI vesicles requires GTP hydrolysis by Arf-1. Thus, Arf-1 has a dual role in sorting and vesicle formation.

THE ROLE OF p24 PROTEINS IN VESICLE FORMATION

We have so far underscored the role of cargo molecules in vesicle formation through their ability to interact directly with coat proteins. But some proteins can bind coat proteins better than others, prompting the idea that they perhaps serve as coat receptors. SNARE proteins as well as ERGIC53 and p24 proteins have been shown to bind COPII, strongly suggesting that they could perform a coat receptor function. For the p24 proteins, this idea has evolved further into a proposed function where different members could play a role in the biogenesis of two types of COPI vesicles, recycling COPI vesicles and anterograde COPI vesicles, the latter transporting newly synthesized proteins forward. Such proposed functions are, however, unlikely to be essential as deletion of all eight p24 proteins in yeast did not inhibit protein transport (Springer et al., 2000). Nevertheless, the p24 proteins are interesting as they bind both COPI as well as COPII efficiently, perhaps in order to ensure that they are actively recycled between the ER and the *cis* face of the Golgi stack. In mammalian cells, five members have so far been characterized: p23 ($p24\delta_1$), p24 ($p24\beta_1$), gp25 ($p24\alpha_2$), p26 ($p24\gamma_4$) and gp27 ($p24\gamma_3$), and have all been found to reside between the ER and the Golgi. Some members have been shown to bind COPI directly via their K(X) KXX-like motifs (the β-subunit of coatomer) as well as COPII (Sec23/ 24) via their F/YXXXXF/Y motif (can also bind to the COP I δ-subunit if coatomer is artificially disassembled prior to binding). What role do the p24 proteins then play? One role could be to ensure that the ER quality control operates with high fidelity. A mutant in a p24 protein in *Caenorhabditis elegans* causes an increase in mutations in cell surface receptors, suggesting that the quality control machinery has been compromised (Wen and Greenwald, 1999). A role for p24 proteins in ER exit comes from the observation that p24 proteins participate in the formation of ER exit sites. Antibodies to the cytoplasmic domain of gp25 (p24α_2) inhibited the formation of ER exit sites, *in vitro*, a process that also requires COPI (Lavoie et al., 1999). Given their abundance, it is possible that p24 proteins through their ability to bind coat proteins help to ensure that newly synthesized proteins released from the quality control machinery are segregated and concentrated before final exit out of the ER. If p24 proteins are deleted, this process would be hampered and slowed down with the result that mutated and misfolded proteins can escape.

So what happens at the ER exit site from where COPII vesicles bud and ERGICs form? The form and shape of ER exit sites suggests that they are composed of highly tubulated and fenestrated membrane patches onto which both COPII and COPI can cycle to sort and segregate newly synthesized proteins into ERGICs (Stephens et al., 2000; Aridor et al., 2001). It is not yet fully understood where and how COPII vesicles function in this process. Either they bud, uncoat and fuse to form a pleiomorphic structure which, together with COPI, evolves into an ERGIC or COPII plays an active role in the initial stages of cargo sorting and formation of ERGICs by causing tubulation of ER membranes, as suggested by recent studies by Balch and co-workers (Aridor et al., 2001). Future work will have to show exactly where COPII vesicles exert their function in the transition of newly synthesized proteins from the ER into ERGICs. Once formed and detached from the ER exit site, ERGICs move inwards towards the central Golgi. During this transport process, COPI associates and helps to segregate resident proteins of the pathway away from newly synthesized proteins, leading to the appearance of subdomains within the ERGICs so that newly synthesized proteins are projected towards the Golgi stack (Shima et al., 1999). Whether or not COPI vesicles form during the transit of ERGICs is unclear but the reshaping of the structure is also accompanied by a marked increase in the concentration of newly synthesized proteins per membrane surface (Martinez-Menarguez et al., 1999). Thus, COPI plays a central role in segregating resident proteins away from newly synthesized proteins already in the initial stages of protein transport following the action of COPII.

Summary: *The role of p24 proteins in vesicle formation*

- p24 proteins are small but abundant type I transmembrane proteins that shuttle between the ER and the *cis* face of the Golgi apparatus.

- p24 proteins and other recycling proteins can recruit both COPI and COPII components.
- p24 proteins are non-essential proteins in yeast. Thus, they cannot perform essential functions such as coat and cargo receptors. Their function remains to be determined.

TETHERING FACTORS, RAB PROTEINS AND SNARE PROTEINS

Based on high resolution tomography studies, transport vesicles are mostly seen in close proximity to membranes to which they are tethered or sometimes away from cisternae but then always tethered to each other (Ladinsky et al., 1999; Marsh et al., 2001). Vesicles attach themselves to each other or to cisternal membranes via specialized proteins collectively termed 'tethering proteins'. Usually extended coiled-coil proteins, these are recruited from the cytosol with the help of Rab proteins, small GTPases which are required for optimal docking and fusion of transport vesicles and for fusion of membrane structures to each other (for a recent review, see Zerial and McBride, 2001). A number of tethering proteins have now been identified, some operating between the ER and the *cis* side of the Golgi, others in the middle of the stack or at the *trans* side of the stack. A large number of these belong to the family of golgins whose members include 230/245/256, golgin-97, GM130/golgin-95, golgin-160/MEA-2/GCP170 and giantin/macrogolgin. Golgins are recruited through their shared GRIP domain (Kjer-Nielsen et al., 1999; Barr, 1999; Munro and Nichols, 1999) onto the membrane in a Rab^{GTP} dependent manner. Similarly, Rab1 has been shown to specifically recruit p115 and GM130, which function at the ER-to-Golgi interface in stacking of cisternal membranes (Allan et al., 2000; Moyer et al., 2001; Weide et al., 2001). In yeast, ypt1 is activated through association with a protein complex termed TRAPP (for review, see Guo et al., 2000) consisting of multiple proteins such as bet3 and bet5, which form a high molecular weight complex that serves as a docking post for incoming vesicles/membranes. In addition, Sec34/35 and p115 are known to assist vesicle docking and in mammalian cells, p115 has been shown to be recruited onto ER exit sites in a $Rab1^{GTP}$ specific manner (Allan et al., 2000). It is not clear at which stages the different tethering proteins act and what their exact roles are but sufficient evidence exists to support a role in the tethering/docking stage of both COPI and COPII transport vesicles and presumably, also for incoming ERGICs helping these to align themselves to the *cis* side of the Golgi stack in order to create a new cisterna.

Intrinsically linked to the tethering process is the function exerted by a class of proteins termed SNAREs, lately the centre of much attention.

These membrane proteins were functionally identified as membrane receptors for soluble NSF attachment proteins (SNAPs) linking these functionally to membrane fusion. The first SNAREs (*SNAP Re*ceptors) to be identified were the two presynaptic proteins syntaxin1 and SNAP-25 (synaptosome associated protein of 25 kDa, not to be confused with soluble NSF attachment proteins) present at the plasma membrane and synaptobrevin/VAMP on synaptic vesicles (for review, see Mayer, 1999).

Based on their localization, SNAREs were initially classified as v-SNAREs (on the *v*esicle membrane) and t-SNAREs (on the *t*arget membrane) but as the family of SNARE proteins grew, this nomenclature appeared increasingly impractical. This is perhaps most evident upon aligning different SNARE proteins to determine the degree of homology. No clear correlation is seen between their v- or t-membrane localization and relative homology. Instead, a more unambiguous terminology has been adopted whereby SNARE proteins are classified on the basis of their structure as Q- or R-SNAREs (for a review, see Jahn and Südhof, 1999). All SNARE proteins contain what is termed a SNARE motif, a conserved domain of about 60 amino acids. Most SNAREs (like syntaxins, synaptobrevins, Bet1p and Bos1p) have a single SNARE motif flanked by a variable N-terminal sequence and a C-terminal transmembrane domain but some, such as SNAP-25 and related SNAREs, have two motifs. Other SNAREs with one SNARE motif may lack a transmembrane region (such as Vam7p) but are probably membrane anchored by a lipid-modified C-terminus. SNAP-25 is attached to the membrane by palmitic acid attached to two cysteines between the two SNARE motifs. SNAREs form a very stable complex through interactions between their respective motifs. The crystal structure of the core complex of the synaptic trimeric complex shows that four SNARE motifs (two from SNAP-25 and one each from syntaxin and synaptobrevin), form a parallel 12 nm long twisted four-helical bundle (Sutton et al., 1998). Most interactions within this coiled-coil structure are hydrophobic, as in other helical bundles. However, in the centre, flanked by leucine-zippers, one positively charged arginine from synaptobrevin and three polar glutamines, one from syntaxin and one each from both SNARE motifs of SNAP-25, form an ionic layer. At this central position within the SNARE motif all SNAREs have either a glutamine (Q-SNAREs) or an arginine (R-SNAREs). This 3Q:1R ratio in the SNARE complex has been shown to be important for SNARE function *in vivo*, by studying the effect of single amino acid mutations in yeast (Ossig et al., 2000; Katz and Brennwald, 2000).

CD-spectroscopy shows that SNARE proteins undergo immense conformational changes upon assembly into a complex (Fasshauser et al., 1997). If not incorporated into a complex the SNARE motifs appear unstructured. Only when assembled into a complex do they display a helical conformation. This structure is energetically favored so that SNAREs assemble into a complex as soon as they are solubilized in non-denaturing detergents

(Otto et al., 1997). Since the SNARE complex is very stable (Fasshauser et al., 1997; Hayashi et al., 1994) the complex needs to be disassembled before (Mayer et al., 1996) and after (Swanton et al., 2000) fusion. This is required so that SNAREs now interacting in *cis* can be made available for another round of *trans* interactions, a postulated prerequisite for membrane fusion. Complex dissociation is mediated by NSF (*N*-ethylmaleimide sensitive factor, termed Sec18p in yeast) in an ATP hydrolysis dependent manner (Söllner et al., 1993). To enable this, three α-SNAP molecules (termed Sec17p in yeast) need to bind to the trimeric SNARE complex (termed the 7S complex). α-SNAP then recruits NSF in the form of a hexamer to form a 20S complex (Söllner et al., 1993). Within the last few years, electron microscopy (Hanson et al., 1997) and crystallography (Lenzen et al., 1998; Yu et al., 1998) has provided insight into the composition and disassembly of the 20S complex. The NSF hexamer forms two rings around the SNARE core complex at its side distal from the membrane anchors, the more distal consisting of the D2-, the other of the D1-domains. The N-domains interact with α-SNAPs which are positioned proximal to the membrane anchors of the SNAREs. ATP hydrolysis by the D1-domains (the D2-domains has only a weak hydrolysis activity and is responsible for NSF oligomerization (Nagiec et al., 1995)) leads to a conformational change within the hexamer resulting in disassembly of the 20S as well as disassembly of the 7S complex.

It is clear that NSF and α-SNAP are involved in most intracellular transport events (Rothman, 1994). However, in some cases, other AAA proteins (ATPases associated with various cellular activities) such as p97 (in yeast Cdc48p) are required. This holds for homotypic fusion events like postmitotic reassembly of Golgi cisternae (Rabouille et al., 1995; Acharya et al., 1995) and the nuclear envelope (Latterich et al., 1995). It is speculated that p97 with its cofactor p47 might be specialized in disassembling Q-Q-SNARE complexes that are thought to form during these events (Rabouille et al., 1998). In the final fusion event, the formation of a *trans* complex of SNAREs bridging both membranes is suggested to pull the membranes to within 4 nm of each other (Jahn and Südhof, 1999). Indeed, Weber et al. (1998) could fuse artificial liposomes containing SNAREs and this, together with the finding that such fusion depends on how the SNAREs are anchored to the membranes (McNew et al., 2000a), provide suggestive evidence for a SNAREpin scenario which, albeit extremely slowly, can drive fusion *in vitro*.

Is this all there is to SNAREs? A crucial point of the SNARE hypothesis formulated by Rothman and co-workers in the early 1990s was that SNARE proteins specifically target transport vesicles to the right membrane with which they are to fuse. They postulated that only 'cognate' SNARE forms stable complexes in only one particular trafficking step. However, in 1994 Scheller and co-workers (Calakos et al., 1994) showed that synaptobrevin 2 binds to different syntaxins. The yeast SNAREs Sed5p and Vti1p seem to function in more than one transport step (for a review see Nichols and Pelham, 1998). *In vitro*, complexes can only be formed with four different SNAREs, one R- and three Q-SNAREs. One of the Q-SNAREs has to belong to the syntaxin subfamily and the two others, to subfamilies homologous to the first and the second SNARE motif of SNAP-25, respectively. Substitution of particular SNAREs within these defined subfamilies can occur without influencing complex formation (Fasshauer et al., 1999). Rothman and co-workers investigated the specificity of SNAREs in liposome fusion using different combinations (McNew et al., 2000b) and demonstrated that cognate SNAREs result in better fusion than non-cognate ones. Unfortunately, they did not test whether SNAREs of the same subfamily (see above) could substitute for each other.

There exist great question marks regarding the SNAREpin hypothesis. Data from other systems, such as *in vitro* assembly of post-mitotic Golgi membranes using an ATPase-deficient mutant of NSF which is unable to disassemble the SNARE complex, mediated fusion comparable to that observed with wild-type (wt) NSF, arguing that neither pre- nor post-disassembly of SNAREs is a prerequisite for fusion (Müller et al., 1999). This is further supported by the observation that following docking, the SNARE *trans* complex can be disassembled upon addition of excess NSF and α-SNAP without affecting the fusion rate (Ungermann, 1998). Moreover, using a liposome-based assay not unlike the one deployed by Rothman and co-workers, we could show that both NSF and p97, when together with their corresponding co-factors, mediate fast and efficient fusion in the absence of SNAREs (Otter-Nilsson, 1999). This with at least 100-fold faster kinetics and a 100-fold fewer proteins per fusion event as compared to the SNARE based liposome assay. Mayer and co-workers (Peters et al., 2001) recently provided evidence for a very different mechanism for fusion. Their biochemical data suggest that the initial fusion pore is not lipidic (as it is in the case of viral fusion) but proteinaceous. According to their model, a protein channel is formed when two V0 hexamers, one each from both vacuole membranes, bind head-to-head in a process that requires Ypt7-GTP and calmodulin. This *trans*-interaction requires a prior formation of *trans* SNARE complexes to bring membranes in close proximity but once the V0 *trans* complex is formed, SNARE complexes are no longer required to initiate fusion. Upon signaling by calcium-bound calmodulin the V0 hexamers segregate whereby lipids can invade the space to form an aqueous fusion pore that can then be expanded. In conclusion, though the SNAREpin hypothesis has broad support in the field, there exists little if any direct evidence for their ability to mediate the actual fusion event. Clearly, more direct experimental tests of this hypothesis are needed before attempting to close this chapter.

Summary: *Tethering factors, Rab proteins and SNARE proteins*

- Tethering factors are recruited by Rab proteins when in their GTP bound state.
- Tethering factors are required for vesicle docking.
- SNARE proteins are believed to mediate docking and/or fusion.
- AAA proteins such as NSF and p97 act on SNARE proteins to disassemble both *cis* and *trans* SNARE complexes.
- Proteins other than SNARE proteins have been shown to mediate fusion, *in vitro*.

RECONSTITUTION OF VESICULAR TRANSPORT, *IN VITRO*

A large portion of what we know today about vesicular transport has been derived from *in vitro* assays. Several assays have been established for both COPI and COPII vesicle transport. The most famous one is the *in vitro* assay established by Rothman and co-workers (Fries and Rothman, 1980; Balch et al., 1984). Using Golgi membranes isolated from Chinese hamster ovary (CHO) wt cells and a mutant CHO cell line deficient in the enzymatic activity of N-acetylglucosaminyltransferase I (GlcNAc-T1) infected with VSV, they showed that VSV-G could receive GlcNAc in a manner which required membrane fusion, such as fusion of transport intermediates, which were later identified as COPI vesicles. At that time, much attention was given to the idea that COPI vesicles transported VSV-G rather than resident proteins including GlcNAc-T1. Nevertheless, the *in vitro* transport assay proved highly successful and enabled the functional identification of several transport factors which today form parts of the essential machinery for protein transport (for review, see Rothman and Wieland, 1996). But lately, this transport assay has been subjected to alternative interpretations where it has been suggested that rather than measuring transport by a vesicular transport intermediate, the assay was registering homotypic fusion between the two populations of CHO membranes (Happe and Weidman, 1998). Though homotypic fusion occurs, for example in the *in vitro* fusion of mitotic Golgi fragments, it requires much higher membrane concentrations. Furthermore, a kinetic analysis of the vesicle fusion reaction suggests that it is unlikely that the much larger and less quickly diffusing Golgi cisternae could fuse under the highly diluted conditions of this *in vitro* assay (Ostermann, 2001).

Today, several lines of evidence support the view that the transport assay by Rothman does measure vesicular transport but with several major caveats. Rather than monitoring the transfer of VSV-G via COPI vesicles, the transport assay is measuring the transfer of GlcNAc-T1 from the wt membranes to the mutant membranes, allowing VSV-G to be glycosylated. Furthermore, cellular homogenates contain small pleiomorphic tubular vesicular structures which efficiently confer GlcNAc-T1 activity to Golgi membranes (Love et al., 1998). As they preferably fused with early compartments, it is likely that these represent retrograde transport vesicles, such as COPI vesicles. These can also be formed, *in vitro*, from rat liver Golgi membranes. Here, they were shown to contain resident proteins which had been concentrated before vesicle formation in an Arf-1 GTP hydrolysis-dependent manner (Lanoix et al., 1999). Docking and fusion of such vesicles to Golgi membranes shows that only one vesicle per cisterna is required for full glycosylation of VSV-G, underscoring the efficiency whereby COPI vesicles transport glycosylation enzymes (Ostermann, 2001). Using this assay, it should be possible to address the role of tethering proteins and other factors such as SNARE proteins in targeting and docking of transport vesicles leading up to final fusion.

MODELING AND SIMULATION

Important insights into the mechanism of a complex reaction can also be obtained by studying the kinetics of the process. This refers to quantitative measurements of the time-dependence of the generation of an assay signal. This is then compared with a mathematical model that describes the process and this model may contain adjustable parameters, such as rate constants, or probabilities with which certain events occur. But the success of any model comes from predictions that can be verified experimentally. Also good agreement must exist between the model and data obtained experimentally.

The quality of the overall model can be ascertained by comparing the predicted data, or the curve shape predicted by a particular model, after an empirical adjustment of the free parameters with the actually measured data. A systematic deviation of the data from the model, such as when all data points in the first half of the experiment fall below the values predicted by the model and all data in the second half are above their predicted values, suggests a failure of the model to account for the full complexity of the real process. Random deviation from the model suggests that the model is sufficient to describe the process at the level at which it is measurable with the existing experimental tools.

Using live cell video microscopy, it is now possible to quantitatively measure protein transport from the ER and through the secretory pathway. Lippincott-Schwartz and others have recently applied kinetic measurements to study the mechanism of secretory protein transport (Hirschberg et al., 1998). They find that a surprisingly simple process seems to account for the overall kinetics with which secretory proteins travel from the ER through the Golgi apparatus to the plasma membrane. A simple model of two first-order steps, one for exit from the ER and one for exit from the Golgi, is sufficient to explain the data. This, they argue, shows that the secretory pathway is all interconnected, negating the need for vesicular transport at any stage. Instead, once released from the ER, newly synthesized proteins would 'diffuse' through the Golgi before being transported to the plasma membrane. In other words, the two exit rate constants observed can be seen as probabilities with which a protein that is present in the ER or Golgi exits and progresses to the Golgi or plasma membrane, respectively. It is clear that such a simple model must underestimate the complexity that exists in the secretory pathway. Exit from the ER, for example, involves multiple steps such as folding (in the case of the misfolded mutant protein that was used in this study), assembly of transport intermediates at the ER, movement of transport intermediates from the cellular periphery to the Golgi, and fusion of these intermediates with the Golgi. Similarly, transport through the Golgi is likely to involve several distinct steps. As such, the resolution of the kinetic experiments was simply insufficient to detect the small effects on the overall kinetics that a combination of distinct steps would have as opposed to a single rate-determining step. It is tempting to dismiss this criticism and apply the simplest hypothesis consistent with the data approach. While the simplest interpretation of the kinetic data may indeed be that only two steps are needed to go from the ER to the Golgi, other data must be taken into account. As seen by the example of ER exit, it is clear that the simplest hypothesis emerging from the kinetic data is falsified when information from other experimental approaches is included. That the simplest hypothesis consistent with the kinetic data eliminates the known complexity merely demonstrates the limitation of the approach taken.

Building on this work, but with a different goal in mind, we have recently applied kinetics to study *in vitro* Golgi transport assays. For us, the goal is not so much to test different hypotheses about the overall mechanism of the process, but rather to use kinetics to improve the resolution with which the reaction can be studied and to increase the rigor with which experiments are quantitatively evaluated. We find that when isolated transport vesicles and Golgi membranes are incubated together, the time course of the assay signal that is generated by membrane fusion can be described by two rate constants. The first rate constant describes how fast vesicles and Golgi membranes come together to form a pre-fusion intermediate, which is most likely a docked vesicle that has not fused with the Golgi membranes. The second rate constant describes the subsequent mixing of bilayers. It is important to note that the rate constants that are measured by this approach describe the probability for an individual vesicle/Golgi pair to undergo docking or fusion at any time. They do not describe the time course of the molecular rearrangements of docking and fusion. Such kinetic information is only available from measurements of individual fusion events, which cannot currently be addressed by the existing experimental systems (Ostermann, 2001).

Modeling is also useful to demonstrate and identify critical parameters in complex systems. For this, coarse grain stochastic modeling has the advantage that multiple parameters can be grouped without knowing the exact details of each individual one. As a good example, Glick, Elston and Oster showed that the observed gradient-like distributions created by constant recycling via COPI vesicles in the maturing Golgi stack could be generated if different resident proteins were ascribed different degrees of competitiveness (Glick et al., 1997). A *cis* resident protein would in this model compete better than a *medial* or *trans* resident and therefore have a higher probability of being recycled earlier. This single parameter was shown to be sufficient to generate asymmetric gradient-like distributions across the stack. In a refined model, we introduced a milieu-induced trigger for sorting to create a more robust system where differences in competitiveness could be constrained within physiological parameters (Weiss and Nilsson, 2000). If this is how sorting occurs remains to be tested but importantly, these studies show that simple parameters can define models that predict the asymmetric distributions, offering clues to how the secretory pathway operates.

Summary: *Reconstitution of vesicular transport,* **in vitro;** *and modeling and simulation*

- Modeling and simulation can be used to constrain complex systems (e.g. *in vitro* transport, budding and docking/fusion assays) by mathematics.
- When put into a mathematical framework or model, predictions can be made which can then be tested experimentally. The experimental outcome is then used to refine the model.
- Modeling and simulation allows for the identification of rate limiting parameters. These are usually the ones of interest when dissecting mechanisms.

HISTORICAL NOTES

Historical Note 1

When screening for ER proteins associated with calnexin, Bergeron and co-workers cloned and sequenced a glycoprotein termed gp25L which was the first member of the p24 protein family to be identified (now termed $p24\alpha_1$)(Wada et al., 1991). This was followed by the identification of emp24 ($p24\beta_1$) in yeast through a genetic screen for proteins affecting endocytosis (Singer Kruger et al., 1993). A few years later, $p24\beta_1$ was mapped as an intergral protein of COPI and COPII vesicles in mammalian and yeast cells, respectively (Stamnes et al., 1995; Schimmöller et al., 1995). Deletion of p24β_1 in yeast caused a partial inhibition of ER export of selected cargo leading to the proposal that p24 proteins serve as cargo receptors. In parallel, p24 members were shown to directly bind coatomer of COPI and Sec23/24 of COPII via their cytoplasmic domains (Dominguez et al., 1998 and references therein) leading to the suggestion that p24 proteins serve as coat receptors needed for vesicle biogenesis. As deletion of all eight members in yeast resulted in viable cells with only slight defects in protein transport, this conclusively rules out any essential functions such as coat or cargo receptors (Springer et al., 2000). However, it is possible that p24 proteins act as modulators of cargo export and/or coat assembly but the reality is that the function of these small transmembrane proteins remains to be determined.

REFERENCES

Acharya, U., Jacobs, R., Peters, J.M. et al. (1995) The formation of Golgi stacks from vesiculated Golgi membranes requires two distinct fusion events. *Cell* **82**: 895–904.

Allan, B.B., Moyer, B.D. and Balch, W.E. (2000) Rab1 recruitment of p115 into a cis-SNARE complex: programming budding COPII vesicles for fusion. *Science* **289**: 444–448.

Appenzeller, C., Andersson, H., Kappeler, F. et al. (1999) The lectin ERGIC-53 is a cargo transport receptor for glycoproteins. *Nat Cell Biol* **1**: 330–334.

Aridor, M., Fish, K.N., Bannykh, S. et al. (2001) The Sar1 GTPase coordinates biosynthetic cargo selection with endoplasmic reticulum export site assembly. *J Cell Biol* **152**: 213–229.

Balch, W.E., Dunphy, W.G., Braell, W.A. et al. (1984) Reconstitution of the transport of protein between successive compartments of the Golgi measured by the coupled incorporation of *N*-acetylglucosamine. *Cell* **39**: 405–416.

Balch, W.E., McCaffery, J.M., Plutner, H. et al. (1994) Vesicular stomatitis virus glycoprotein is sorted and concentrated during export from the endoplasmic reticulum. *Cell* **76**: 841–852.

Barlowe, C. (2000) Traffic COPs of the early secretory pathway. *Traffic* **1**: 371–377.

Barlowe, C., Orci, L., Yeung, T. et al. (1994) COPII: a membrane coat formed by Sec proteins that drive vesicle budding from the endoplasmic reticulum. *Cell* **77**: 895–907.

Barr, F.A. (1999) A novel Rab6-interacting domain defines a family of Golgi-targeted coiled-coil proteins. *Curr Biol* **9**: 381–384.

Bremser, M., Nickel, W., Schweikert, M.H. et al. (1999) Coupling of coat assembly and vesicle budding to packaging of putative cargo receptors. *Cell* **96**: 495–506.

Burkhardt, J.K., Echeverri, C.J., Nilsson, T. et al. (1997) Overexpression of the dynamitin (p50) subunit of the dynactin complex disrupts dynein-dependent maintenance of membrane organelle distribution. *J Cell Biol* **139**: 469–484.

Calakos, N., Bennett, M.K., Peterson, K.E. et al. (1994) Protein-protein interactions contributing to the specificity of intracellular vesicular trafficking. *Science* **263**: 1146–1149.

Chardin, P., Paris, S., Antonny, B. et al. (1996) A human exchange factor for ARF contains Sec7- and pleckstrin-homology domains. *Nature* **384**: 481–484.

Cherfils, J., Menetrey, J., Mathieu, M. et al. (1998) Structure of the Sec7 domain of the Arf exchange factor ARNO. *Nature* **392**: 101–105.

Cosson, P. and Letourneur, F. (1994) Coatomer interaction with di-lysine endoplasmic reticulum retention motifs. *Science* **263**: 1629–1631.

Dominguez, M., Dejgaard, K., Füllekrug, J. et al. (1998) gp25L/emp24/p24 protein family members of the cis-Golgi network bind both COP I and II coatomer. *J Cell Biol* **140**: 751–765.

Donaldson, J.G., Finazzi, D. and Klausner, R.D. (1992) Brefeldin A inhibits Golgi membrane-catalysed exchange of guanine nucleotide onto ARF protein. *Nature* **360**: 350–352.

Espenshade, P., Gimeno, R.E., Holzmacher, E. et al. (1995) Yeast SEC16 gene encodes a multidomain vesicle coat protein that interacts with Sec23p. *J Cell Biol* **131**: 311–324.

Eugster, A., Frigerio, G., Dale, M. et al. (2000) COP I domains required for coatomer integrity, and novel interactions with ARF and ARF-GAP. *EMBO J* **19**: 3905–3917.

Fasshauer, D., Bruns, D., Shen, B. et al. (1997) A structural change occurs upon binding of syntaxin to SNAP-25. *J Biol Chem* **272**: 4582–4590.

Fasshauer, D., Antonin, W., Margittai, M. et al. (1999) Mixed and non-cognate SNARE complexes. Characterization of assembly and biophysical properties. *J Biol Chem* **274**: 15440–15446.

Faulstich, D., Auerbach, S., Orci, L. et al. (1996) Architecture of coatomer: molecular characterization of delta-COP and protein interactions within the complex. *J Cell Biol* **135**: 53–61.

Fries, E. and Rothman, J.E. (1980) Transport of vesicular stomatitis virus glycoprotein in a cell-free extract. *Proc Natl Acad Sci USA* **77**: 3870–3874.

Gimeno, R.E., Espenshade, P. and Kaiser, C.A. (1995) SED4 encodes a yeast endoplasmic reticulum protein that binds Sec16p and participates in vesicle formation. *J Cell Biol* **131**: 325–338.

Glick, B.S., Elston, T. and Oster, G. (1997) A cisternal maturation mechanism can explain the asymmetry of the Golgi stack. *FEBS Lett* **414**: 177–181.

Goldberg, J. (1998) Structural basis for activation of ARF GTPase: mechanisms of guanine nucleotide exchange and GTP-myristoyl switching. *Cell* **95**: 237–248.

Goldberg, J. (1999) Structural and functional analysis of the ARF1-ARFGAP complex reveals a role for coatomer in GTP hydrolysis. *Cell* **96**: 893–902.

Goldberg, J. (2000) Decoding of sorting signals by coatomer through a GTPase switch in the COPI coat complex. *Cell* **100**: 671–679.

Guo, W., Sacher, M., Barrowman, J. et al. (2000) Protein complexes in transport vesicle targeting. *Trends Cell Biol* **10**: 251–255.

Hanson, P.I., Roth, R., Morisaki, H. et al. (1997) Structure and conformational changes in NSF and its membrane receptor complexes visualized by quick-freeze/deep-etch electron microscopy. *Cell* **90**: 523–535.

Happe, S. and Weidman, P. (1998) Cell-free transport to distinct Golgi cisternae is compartment specific and ARF independent. *J Cell Biol* **140**: 511–523.

Hayashi, T., McMahon, H., Yamasaki, S. et al. (1994) Synaptic vesicle membrane fusion complex: action of clostridial neurotoxins on assembly. *EMBO J* **13**: 5051–5061.

Helms, J.B. and Rothman, J.E. (1992) Inhibition by brefeldin A of a Golgi membrane enzyme that catalyses exchange of guanine nucleotide bound to ARF. *Nature* **360**: 352–354.

Hermo, L., Green, H. and Clermont Y. (1991) Golgi apparatus of epithelial principal cells of the epididymal initial segment of the rat: structure, relationship with endoplasmic reticulum, and role in the formation of secretory vesicles. *Anat Rec* **229**: 159–176.

Hirschberg, K., Miller, C.M., Ellenberg, J. et al. (1998) Kinetic analysis of secretory protein traffic and characterization of golgi to plasma membrane transport intermediates in living cells. *J Cell Biol* **143**: 1485–1503.

Jackson, M.R., Nilsson, T. and Peterson, P.A. (1990) Identification of a consensus motif for retention of transmembrane proteins in the endoplasmic reticulum. *EMBO J* **9**: 3153–3162.

Jackson, M.R., Nilsson, T. and Peterson, P.A. (1993) Retrieval of transmembrane proteins to the endoplasmic reticulum. *J Cell Biol* **121**: 317–333.

Jahn, R. and Südhof, T.C. (1999) Membrane fusion and exocytosis. *Annu Rev Biochem* **68**: 863–911.

Kappeler, F., Klopfenstein, D.R., Foguet, M. et al. (1997) The recycling of ERGIC-53 in the early secretory pathway. ERGIC-53 carries a cytosolic endoplasmic reticulum-exit determinant interacting with COPII. *J Biol Chem* **272**: 31801–31808.

Katz, L. and Brennwald, P. (2000) Testing the 3Q:1R 'rule': mutational analysis of the ionic 'zero' layer in the yeast exocytic SNARE complex reveals no requirement for arginine. *Mol Biol Cell* **11**: 3849–3858.

Kjer-Nielsen, L., Teasdale, R.D., van Vliet, C. et al. (1999) A novel Golgi-localisation domain shared by a class of coiled-coil peripheral membrane proteins. *Curr Biol* **9**: 385–388.

Kuehn, M.J., Schekman, R. and Ljungdahl, P.O. (1996) Amino acid permeases require COPII components and the ER resident membrane protein Shr3p for packaging into transport vesicles in vitro. *J Cell Biol* **135**: 585–595.

Kuehn, M.J., Herrmann, J.M. and Schekman, R. (1998) COPII-cargo interactions direct protein sorting into ER-derived transport vesicles. *Nature* **391**: 187–190.

Kurihara, T., Hamamoto, S., Gimeno, R.E. et al. (2000) Sec24p and Iss1p function interchangeably in transport vesicle formation from the endoplasmic reticulum in *Saccharomyces cerevisiae*. *Mol Biol Cell* **11**: 983–998.

Ladinsky, M.S., Mastronarde, D.N., McIntosh, J.R. et al. (1999) Golgi structure in three dimensions: functional insights from the normal rat kidney cell. *J Cell Biol* **144**: 1135–1149.

Lanoix, J., Ouwendijk, J., Lin, C.C. et al. (1999) GTP hydrolysis by arf-1 mediates sorting and concentration of Golgi resident enzymes into functional COPI vesicles. *EMBO J* **18**: 4935–4948.

Lanoix, J., Ouwendijk, J., Stark, A., Szafer, E., Cassel, D., Dejgaard, K., Weiss, M. and Nilsson, T. (2001) Sorting of Golgi resident proteins into different sub populations of COPI vesicles: a role for ArfGAP1. *J Cell Biol* **155**: 1199–1212.

Latterich, M., Frohlich, K.U. and Schekman, R. (1995) Membrane fusion and the cell cycle: Cdc48p participates in the fusion of ER membranes. *Cell* **82**: 885–893.

Lavoie, C., Paiement, J., Dominguez, M. et al. (1999) Roles for alpha(2)p24 and COPI in endoplasmic reticulum cargo exit site formation. *J Cell Biol* **146**: 285–299.

Lenzen, C.U., Steinmann, D., Whiteheart, S.W. et al. (1998) Crystal structure of the hexamerization domain of N-ethylmaleimide-sensitive fusion protein. *Cell* **94**: 525–536.

Letourneur, F., Gaynor, E.C., Hennecke, S. et al. (1994) Coatomer is essential for retrieval of dilysine-tagged proteins to the endoplasmic reticulum. *Cell* **79**: 1199–1207.

Love, H.D., Lin, C.C., Short, C.S. et al. (1998) Isolation of functional Golgi-derived vesicles with a possible role in retrograde transport. *J Cell Biol* **140**: 541–551.

Lowe, M. and Kreis, T.E. (1996) In vivo assembly of coatomer, the COP-I coat precursor. *J Biol Chem* **271**: 30725–30730.

Makler, V., Cukierman, E., Rotman, M. et al. (1995) ADP-ribosylation factor-directed GTPase-activating protein. Purification and partial characterization. *J Biol Chem* **270**: 5232–5237.

Marsh, B.J., Mastronarde, D.N., Buttle, K.F. et al. (2001) Organellar relationships in the Golgi region of the pancreatic beta cell line, HIT-T15, visualized by high resolution electron tomography. *Proc Natl Acad Sci USA* **98**: 2399–2406.

Martinez-Menarguez, J.A., Geuze, H.J., Slot, J.W. et al. (1999) Vesicular tubular clusters between the ER and Golgi mediate concentration of soluble secretory proteins by exclusion from COPI-coated vesicles. *Cell* **98**: 81–90.

Mayer, A. (1999) Intracellular membrane fusion: SNAREs only? *Curr Opin Cell Biol* **11**: 447–452.

Mayer, A., Wickner, W. and Haas, A. (1996) Sec18p (NSF)-driven release of Sec17p (alpha-SNAP) can precede docking and fusion of yeast vacuoles. *Cell* **85**: 83–94.

McNew, J.A., Weber, T., Parlati, F. et al. (2000a) Close is not enough: SNARE-dependent membrane fusion requires an active mechanism that transduces force to membrane anchors. *J Cell Biol* **150**: 105–117.

McNew, J.A., Parlati, F., Fukuda, R. et al. (2000b) Compartmental specificity of cellular membrane fusion encoded in SNARE proteins. *Nature* **407**: 153–159.

Mizuno, M. and Singer, S.J. (1993) A soluble secretory protein is first concentrated in the endoplasmic reticulum before transfer to the Golgi apparatus. *Proc Natl Acad Sci USA* **90**: 5732–5736.

Morinaga, N., Tsai, S.C., Moss, J. et al. (1996) Isolation of a brefeldin A-inhibited guanine nucleotide-exchange protein for ADP ribosylation factor (ARF) 1 and ARF3 that contains a Sec7-like domain. *Proc Natl Acad Sci USA* **93**: 12856–12860.

Morinaga, N., Moss, J. and Vaughan, M. (1997) Cloning and expression of a cDNA encoding a bovine brain brefeldin A-sensitive guanine nucleotide-exchange protein for ADP-ribosylation factor. *Proc Natl Acad Sci USA* **94**: 12926–12931.

Mossessova, E., Gulbis, J.M. and Goldberg, J. (1998) Structure of the guanine nucleotide exchange factor Sec7 domain of human arno and analysis of the interaction with ARF GTPase. *Cell* **92**: 415–423.

Moyer, B.D., Allan, B.B. and Balch, W.E. (2001) Rab1 interaction with a GM130 effector complex regulates COPII vesicle *cis*-Golgi tethering. *Traffic* **2**: 268–276.

Müller, J.M., Rabouille, C., Newman, R. et al. (1999) An NSF function distinct from ATPase-dependent SNARE disassembly is essential for Golgi membrane fusion. *Nat Cell Biol* **1**: 335–340.

Muniz, M., Nuoffer, C., Hauri, H.P. et al. (2000) The Emp24 complex recruits a specific cargo molecule into endoplasmic reticulum-derived vesicles. *J Cell Biol* **148**: 925–930.

Munro, S. and Nichols, B.J. (1999) The GRIP domain – a novel Golgi-targeting domain found in several coiled-coil proteins. *Curr Biol* **9**: 377–380.

Nagiec, E.E., Bernstein, A. and Whiteheart, S.W. (1995) Each domain of the N-ethylmaleimide-sensitive fusion protein contributes to its transport activity. *J Biol Chem* **270**: 29182–29188.

Nichols, B.J. and Pelham, H.R. (1998) SNAREs and membrane fusion in the Golgi apparatus. *Biochim Biophys Acta* **1404**: 9–31.

Nilsson, T., Jackson, M. and Peterson, P.A. (1989) Short cytoplasmic sequences serve as retention signals for transmembrane proteins in the endoplasmic reticulum. *Cell* **58**: 707–718.

Novick, P., Field, C. and Schekman, R. (1980) Identification of 23 complementation groups required for post-translational events in the yeast secretory pathway. *Cell* **21**: 205–215.

Orci, L., Glick, B.S. and Rothman, J.E. (1986) A new type of coated vesicular carrier that appears not to contain clathrin: its possible role in protein transport within the Golgi stack. *Cell* **46**: 171–184.

Ossig, R., Schmitt, H.D., de Groot, B. et al. (2000) Exocytosis requires asymmetry in the central layer of the SNARE complex. *EMBO J* **19**: 6000–6010.

Ostermann, J. (2001) Stoichiometry and kinetics of transport vesicle fusion with Golgi membranes. *EMBO Rep* **2**: 324–329.

Otter-Nilsson, M., Hendriks, R., Pecheur-Huet, E.I. et al. (1999) Cytosolic ATPases, p97 and NSF, are sufficient to mediate rapid membrane fusion. *EMBO J* **18**: 2074–2083.

Otto, H., Hanson, P.I. and Jahn, R. (1997) Assembly and disassembly of a ternary complex of synaptobrevin, syntaxin, and SNAP-25 in the membrane of synaptic vesicles. *Proc Natl Acad Sci USA* **94**: 6197–6201.

Pagano, A., Letourneur, F., Garcia-Estefania, D. et al. (1999) Sec24 proteins and sorting at the endoplasmic reticulum. *J Biol Chem* **274**: 7833–7840.

Peng, R., De Antoni, A. and Gallwitz, D. (2000) Evidence for overlapping and distinct functions in protein transport of coat protein Sec24p family members. *J Biol Chem* **275**: 11521–11528.

Peters, C., Bayer, M.J., Buhler, S. et al. (2001) Trans-complex formation by proteolipid channels in the terminal phase of membrane fusion. *Nature* **409**: 581–588.

Peyroche, A., Paris, S. and Jackson, C.L. (1996) Nucleotide exchange on ARF mediated by yeast Gea1 protein. *Nature* **384**: 479–481.

Presley, J.F., Cole, N.B., Schroer, T.A. et al. (1997) ER-to-Golgi transport visualized in living cells. *Nature* **389**: 81–85.

Rabouille, C., Levine, T.P., Peters, J.M. et al. (1995) An NSF-like ATPase, p97, and NSF mediate cisternal regrowth from mitotic Golgi fragments. *Cell* **82**: 905–914.

Rabouille, C., Kondo, H., Newman, R. et al. (1998) Syntaxin 5 is a common component of the NSF- and p97-mediated reassembly pathways of Golgi cisternae from mitotic Golgi fragments in vitro. *Cell* **92**: 603–610.

Rambourg, A., Clermont, Y. and Hermo, L. (1981) Three-dimensional structure of the Golgi apparatus. *Methods Cell Biol* **23**: 155–166.

Randazzo, P.A. and Kahn, R.A. (1994) GTP hydrolysis by ADP-ribosylation factor is dependent on both an ADP-ribosylation factor GTPase-activating protein and acid phospholipids. *J Biol Chem* **269**: 10758–10763.

Rothman, J.E. (1994) Mechanisms of intracellular protein transport. *Nature* **372**: 55–63.

Rothman, J.E. and Wieland, F.T. (1996) Protein sorting by transport vesicles. *Science* **272**: 227–234.

Saraste, J. and Svensson, K. (1991) Distribution of the intermediate elements operating in ER to Golgi transport. *J Cell Sci* **100**: 415–430.

Scales, S.J., Pepperkok, R. and Kreis, T.E. (1997) Visualization of ER-to-Golgi transport in living cells reveals a sequential mode of action for COPII and COPI. *Cell* **90**: 1137–1148.

Scales, S.J., Gomez, M. and Kreis, T.E. (2000) Coat proteins regulating membrane traffic. *Int Rev Cytol* **195**: 67–144.

Schekman, R. and Orci, L. (1996) Coat proteins and vesicle budding. *Science* **271**: 1526–1533.

Schimmöller, F., Singer-Kruger, B., Schroder, S. et al. (1995) The absence of emp24p, a component of ER-derived COPII-coated vesicles, causes a defect in transport of selected proteins to the Golgi. *EMBO J* **14**: 1329–1339.

Serafini, T., Orci, L., Amherdt, M. et al. (1991) ADP-ribosylation factor is a subunit of the coat of Golgi-derived COP-coated vesicles: a novel role for a GTP-binding protein. *Cell* **67**: 239–253.

Shima, D.T., Scales, S.J., Kreis, T.E. et al. (1999) Segregation of COPI-rich and anterograde-cargo-rich domains in endoplasmic-reticulum-to-Golgi transport complexes. *Curr Biol* **9**: 821–824.

Shimoni, Y., Kurihara, T., Ravazzola, M. et al. (2000) Lst1p and Sec24p cooperate in sorting of the plasma membrane ATPase into COPII vesicles in *Saccharomyces cerevisiae*. *J Cell Biol* **151**: 973–984.

Singer-Kruger, B., Frank, R., Crausaz, F. et al. (1993) Partial purification and characterization of early and late endosomes from yeast. Identification of four novel proteins. *J Biol Chem* **268**: 14376–14386.

Söllner, T., Bennett, M.K., Whiteheart, S.W. et al. (1993) A protein assembly-disassembly pathway in vitro that may correspond to sequential steps of synaptic vesicle docking, activation, and fusion. *Cell* **75**: 409–418.

Springer, S. and Schekman, R. (1998) Nucleation of COPII vesicular coat complex by endoplasmic reticulum to Golgi vesicle SNAREs. *Science* **281**: 698–700.

Springer, S., Chen, E., Duden, R. et al. (2000) The p24 proteins are not essential for vesicular transport in *Saccharomyces cerevisiae*. *Proc Natl Acad Sci USA* **97**: 4034–4039.

Springer, S., Spang, A. and Schekman, R. (1999) A primer on vesicle budding. *Cell* **97**: 145–148.

Stamnes, M.A., Craighead, M.W., Hoe, M.H. et al. (1995) An integral membrane component of coatomer-coated transport vesicles defines a family of proteins involved in budding. *Proc Natl Acad Sci USA* **92**: 8011–8015.

Stephens, D.J., Lin-Marq, N., Pagano, A. et al. (2000) COPI-coated ER-to-Golgi transport complexes segregate from COPII in close proximity to ER exit sites. *J Cell Sci* **113**: 2177–2185.

Storrie, B., Pepperkok, R. and Nilsson, T. (2000) Breaking the COPI monopoly on Golgi recycling. *Trends Cell Biol* **10**: 385–390.

Sutton, R.B., Fasshauer, D., Jahn, R. et al. (1998) Crystal structure of a SNARE complex involved in synaptic exocytosis at 2.4 A resolution. *Nature* **395**: 347–353.

Swanton, E., Bishop, N., Sheehan, J. et al. (2000) Disassembly of membrane-associated NSF 20S complexes is slow relative to vesicle fusion and is Ca(2+)-independent. *J Cell Sci* **113**: 1783–1791.

Szafer, E., Pick, E., Rotman, M. et al. (2000) Role of coatomer and phospholipids in GTPase-activating protein-dependent hydrolysis of GTP by ADP-ribosylation factor-1. *J Biol Chem* **275**: 23615–23619.

Ungermann, C., Sato, K. and Wickner, W. (1998) Defining the functions of trans-SNARE pairs. *Nature* **396**: 543–548.

Wada, I., Rindress, D., Cameron, P.H. et al. (1991) SSR α and associated calnexin are major calcium binding proteins of the endoplasmic reticulum membrane. *J Biol Chem* **266**: 19599–19610.

Waters, M.G., Serafini, T. and Rothman, J.E. (1991) 'Coatomer': a cytosolic protein complex containing subunits of non-clathrin-coated Golgi transport vesicles. *Nature* **349**: 248–251.

Weber, T., Zemelman, B.V., McNew, J.A. et al. (1998) SNAREpins: minimal machinery for membrane fusion. *Cell* **92**: 759–772.

Weide, T., Bayer, M., Koster, M. et al. (2001) The Golgi matrix protein GM130: a specific interacting partner of the small GTPase rab1b. *EMBO Rep* **2**: 336–341.

Weiss, M. and Nilsson, T. (2000) Protein sorting in the Golgi apparatus: a consequence of maturation and triggered sorting. *FEBS Lett* **486**: 2–9.

Wen, C. and Greenwald, I. (1999) p24 proteins and quality control of LIN-12 and GLP-1 trafficking in *Caenorhabditis elegans*. *J Cell Biol* **145**: 1165–1175.

Wieland, F.T., Gleason, M.L., Serafini, T.A. et al. (1987) The rate of bulk flow from the endoplasmic reticulum to the cell surface. *Cell* **50**: 289–300.

Yoshihisa, T., Barlowe, C. and Schekman, R. (1993) Requirement for a GTPase-activating protein in vesicle budding from the endoplasmic reticulum. *Science* **259**: 1466–1468.

Yu, R.C., Hanson, P.I., Jahn, R. et al. (1998) Structure of the ATP-dependent oligomerization domain of N-ethylmaleimide sensitive factor complexed with ATP. *Nat Struct Biol* **5**: 803–811.

Zerial, M. and McBride, H. (2001) Rab proteins as membrane organizers. *Nat Rev Mol Cell Biol* **2**: 107–117.

17

CONCLUSION/PERSPECTIVE

ROSS E. DALBEY AND GUNNAR VON HEIJNE

The preceding chapters described how proteins are transported across the bacterial plasma membrane, the endoplasmic reticulum (ER) membrane, the chloroplast thylakoid membrane, and how proteins are imported into mitochondria, chloroplasts, peroxisomes, and the nucleus. The problem of how hydrophilic proteins traverse the hydrophobic membrane bilayer is in each of these membrane and organellar systems the same, but solved with diverse translocation machineries. In this concluding chapter, we compare the translocation machineries that are employed in the different membrane systems to highlight their similarities and unique properties.

The evolutionarily related Sec pathways in the ER, the bacterial plasma membrane, and the chloroplast thylakoid membrane have both conserved and distinct features (Figure 17.1). For example, the ER Sec61 translocon comprises the Sec61α, β and γ subunits; Sec61α and Secγ are homologous to bacterial and chloroplast SecY and SecE, respectively. The signal sequences that target proteins to the Sec61/SecYEG translocation machineries are conserved. Proteins are translocated by the Sec pathways in an 'unfolded state'. The thylakoid Sec system is more similar to the bacterial than to the eukaryotic one, as expected based on the fact that chloroplast originated from a bacterial ancestor. Both the bacterial and chloroplast SecYE translocation systems employ SecA, a molecular motor protein that uses ATP hydrolysis and cycles of membrane insertion and deinsertion to drive proteins across the membrane. Aside from these similarities, only bacteria employ SecB to target proteins to the membrane. In the ER system, the signal recognition particle (SRP) routes proteins to the membrane by its interaction with the signal sequence of the nascent exported protein and the membrane bound SRP receptor.

The Sec translocation machineries play a critical role in membrane protein insertion, in addition to protein translocation. The accessory membrane

Protein Targeting, Transport & Translocation
ISBN 0-12-200731-X

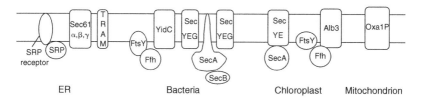

Figure 17.1 The Sec and Oxa1p translocation systems. *Bacteria*: Exported proteins are usually targeted to the membrane by SecB. Proteins are translocated across the membrane using SecA, and the heterotrimeric components SecYEG and SecDFYajC. Membrane protein insertion requires Ffh and FtsY for targeting, and the Sec components and YidC for insertion. Sec-independent proteins require YidC for membrane insertion. *Endoplasmic reticulum*: SRP and SRP receptor is necessary for targeting to the membrane. Co-translational translocation requires the Sec61α, β and γ components. TRAM is believed to assist in the movement of the membrane protein's transmembrane segment out of the aqueous Sec61 channel into the lipid bilayer. *Chloroplast*: Exported proteins are translocated across the thylakoid membrane using SecA and SecYE complex. Membrane proteins are targeted to the membrane using cpSRP and SRP43, and the SRP receptor called FtsY. Integration of some membrane proteins requires Alb3. *Mitochondrion*: Proteins are inserted into the inner membrane from the matrix side using Oxa1p.

component TRAM (translocating chain-associating membrane) assists in this process in the ER membrane (Figure 17.1). While TRAM's function is not yet completely defined, it is believed to cooperate with the Sec61 channel to move the membrane protein's hydrophobic segments out of the aqueous channel and into the lipid bilayer. A similar role seems to be performed by the YidC family of proteins in the bacterial and thylakoid systems (Figure 17.1). YidC may also function separate from the SecYE translocase for certain membrane proteins. Alb3, the chloroplast YidC homolog, is essential for the membrane integration of light harvesting chloroplast protein (LHCP). Whether Alb3 cooperates with SecYE remains an open question. In contrast, Oxa1p, the YidC homolog in *Saccharomyces cerevisiae* mitochondria, works in the absence of the Sec translocase because *S. cerevisiae* mitochondria lack the Sec components. Nevertheless, Oxa1p plays a critical function for proteins that are inserted into the mitochondrial inner membrane from the matrix compartment.

Quite different translocation pathways are utilized for translocation of folded proteins across membranes in chloroplasts, bacteria and peroxisomes (Figure 17.2). In chloroplast and bacteria, this is achieved by the twin arginine translocation (Tat) translocase, comprising the TatA, TatB, TatC and TatE proteins. The Tat system is present in both the bacterial plasma membrane and the chloroplast thylakoid membrane. Interestingly, in at least some cases, the protein must be in a folded state for Tat-dependent translocation to occur. Typically, the Tat pathway in bacteria functions to transport proteins containing metal ion cofactors that participate in oxidation/reduction reactions.

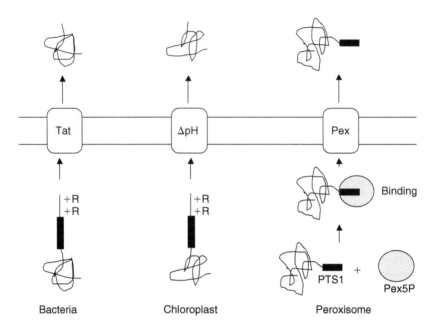

Figure 17.2   The Tat and peroxisomal pathway. Twin arginine transfer peptides target exported proteins into the Tat pathway in bacteria and chloroplasts. The peroxisomal transfer signal comprising the SKL (serine-lysine-leucine) motif targets most proteins to the peroxisome. This C-terminal signal is recognized by the PTS1 (peroxisome targeting signal 1) receptor called Pex5P. After docking to the membrane, the protein is translocated across the membrane by a protein complex comprising Pex10, Pex12 and Pex2. The mechanism by which the Tat and peroxisomal pathway exports folded proteins is not clear.

Export by this pathway requires a twin arginine signal peptide and the proton motive force. Analogous to the Tat pathway, import into the peroxisome can also tolerate folded proteins and even oligomers, although the peroxisomal translocon components have no discernable sequence similarity to the Tat components. In addition, export of folded proteins across membranes can be achieved by the Type II secretion system in Gram-negative eubacteria.

The Tom/Tim and Toc/Tic translocases located, respectively, in the outer and inner membranes of mitochondria and chloroplast, function to import proteins across the two envelope membranes of the organelles. While the mitochondrial and chloroplast machineries do not share sequence homology, they seem to function in an analogous way. The Tom/Tim complex translocates proteins directly from the cytoplasm into the mitochondrial matrix. The Tom complex contains receptors that recognize the signal sequences of mitochondrial imported proteins. Import into the organelle requires ATP hydrolysis and the membrane electrochemical potential across the inner membrane. The ATP-driven 'motor'-chaperone Hsp70 localized within the matrix promotes import by binding to both the incoming precursor protein and the Tim44 translocon protein. The Toc/Tic complex in chloroplast also uses the energy of ATP hydrolysis to catalyze import. The Toc receptors recognize the chloroplast transit peptide in a step regulated by GTP hydrolysis. This GTP regulated step distinguishes the chloroplast system from the mitochondrial one. After import into either organelle, an intra-organellar processing peptidase removes the transit peptide.

Quality control mechanisms ensure that proteins fold correctly after entering the lumen of the ER. For many proteins, correct folding in the ER involves the interaction with chaperones such as BIP, calnexin, calreticulin, processing by the signal peptidase complex, glycosylation by oligosaccharyl transferase, and introduction of disulfide bonds by protein disulfide isomerase. Chaperones and lectins play center stage by recognizing and retaining misfolded proteins in the ER. Failure to fold correctly often induces the unfolded protein response, which leads to increased amounts of chaperones in the ER. Misfolded proteins can be retro-translocated back into the cytoplasm by the Sec61 translocon where they are ubiquinated and finally destroyed by the proteosome.

Protein transport from the ER to the Golgi occurs by packaging of cargo proteins into COPII-coated vesicles. These vesicles then lose their COPII coats and fuse to form the intermediate compartment, which then forms the *cis* Golgi compartment. Movement through the Golgi stacks in the *cis* to *trans* direction may occur by anterograde vesicle movement. Alternatively, the *cis* Golgi stack matures into the *medial* stack, the *medial* stack matures into the *trans* cisternae, and so on. In this latter maturation model, retrograde movement of proteins via COPI-coated vesicles ensures that the membrane content of the cisternae is kept and that each compartment maintains its own unique composition of components. In the *trans* Golgi, proteins can be packaged into secretory vesicles. Proteins destined for the vacuole/lysosome are also packaged into vesicles, usually fusing with the late endosome, prior to making their way to the vacuole.

The transport of proteins in and out of the nucleus occurs via a large nuclear pore complex. This transport system is fundamentally different from those found in bacteria, ER, chloroplasts, mitochondria and peroxisomes. The nuclear pore is a huge complex that spans two unit membranes and allows free diffusion of small to medium-sized proteins. For larger proteins and protein complexes, nuclear protein import requires a nuclear localization signal (NLS) containing basic residues that is recognized by the importin α subunit of the importin α/β complex. After the cargo protein/ importin complex docks to the cytoplasmic surface of the nuclear pore complex, it is translocated to the nuclear side by a largely unknown mechanism. On the nuclear side, RanGTP binds to importin β, which results in the release of the exported protein from importin α. Export of proteins out of the nucleus is mediated by short, leucine-rich nuclear export signals.

A deeper understanding of the various translocation mechanisms will require answering many difficult questions. For instance, how is the SecAYEG complex regulated to promote membrane translocation? What is the structure of a polypeptide as it crosses the membrane? How do the accessory proteins YidC or TRAM facilitate the movement of hydrophobic regions in membrane proteins from within the Sec channel into the lipid bilayer? How do all the enzymes such as signal peptidase, oligosaccharyl transferase, chaperones, protein and disulfide isomerase function in concert to generate biologically active proteins in the ER? Regulation of transport out of the ER and into membrane vesicles must be coordinated by proteins that can sense whether the exported proteins are folded correctly. How is the recognition of improperly folded proteins in the ER coordinated with retro-translocation? Finally, how are large folded proteins transported across the bilayer using the Tat or peroxisomal translocases while maintaining the permeability barrier of the membrane?

The field of protein targeting has come a long way since the pioneering studies by Günter Blobel and collaborators in the early 1970s. Literally, thousands of postdoc-years have gone into figuring out the complex targeting pathways and translocation mechanisms that have evolved to ship proteins within and between cells. And yet, we cannot claim to have a full, atomic-level understanding of even a single translocation process – there is certainly more to come! And new books to be written …

ACKNOWLEDGMENT

R.E.D. would like to thank Minyong Chen, Fenglei Jiang, Liang Yi and Rong Xu who recently contributed to the work on the membrane biogenesis of proteins in bacteria.

INDEX

Page references in *italics* indicate tables and figures; those in **bold** indicate main discussions.

A-ALP *328*, 337–8, 339, 340–1
AAA proteins 389
ABC transporter 186, 188
Aβ₄₂ 168
'acid chain' hypothesis 223–4
acid phosphatase 327
actin 370, 371–2
activating transcription factors *see* ATF4;
    ATF6
acyl-CoA oxidase 282
adaptor molecules
    nuclear export 306–7
    nuclear import 298, *302*, 303–4
adaptor proteins (AP) 327, *329*, 332, 380
    see also AP-1; AP-2; AP-3
adenosine 5'-(β,γ-imidotriphosphate)
    (AMP-PNP) 59
ADP/ATP carrier (AAC) 229
adrenoleukodystrophy, neonatal 284–5
Albino3 (Alb3) 259, *260*, 405
alkaline phosphatase (ALP)
    bacterial 7, 8
    disulfide bonds 135
    fusion proteins *328*, 337–8
    transport to yeast vacuole 322, *323*, *328*,
        343–4
        clathrin-associated proteins and
            331–2
        sorting pathway **332–3**
    yeast vacuole to PVC transport 350
alpha factor (α-factor)
    pro- *see* pro-α-factor
    transport to vacuole *328*, 337
α helices, transmembrane *see*
        transmembrane (TM) helices
α1-antitrypsin 183–4, *192*, 198, 204
α1-proteinase inhibitor 189, *192*
Alzheimer's disease 168
amber codon technique 22
amino acid permeases 384
amino acids
    charge bias, in membrane proteins 115–16
    in transmembrane segments 109
aminoacyl-tRNA analogs 22–3, *23*, 102

amyloid precursor protein (APP) 168
amyloplasts 240
AP-1 327, *329*, 332, 333
    in CPY transport 331
    redundancy with GGA proteins 334
AP-2 *329*, 332, 333
AP-3 *329*, **332–3**
apolipoprotein B (apoB) 91–2, *192*, 204
apoptosis 4, 170, **172–3**
aquaporin2 *192*
Arabidopsis thaliana 62, 240
archaea 52, 60, *61*, **61–2**
ARF 327, 330, 333
Arf-1 381–3, *383*
Arf-2 382
Arf GEFs 381–2
ARNO 382
asialoglycoprotein 204
    receptor H1 116
ATF4 170
ATF6 165, *166*, 168–9, 170–1, 174
ATP
    in mitochondrial protein import 219,
        225–8
    in protein export in bacteria 56–7, **63–5**,
        66–7
    in protein import at chloroplast envelope
        246–7
    in SNARE complex disassembly 389
ATPase
    plasma membrane (Pma1p) 327, *328*
    SecA activity 57, 68
autophagic (*apg*) pathway 322
auxilin 327
azide 6, 65

Bacillus subtilis 47, 52
bacteria 1
    disulfide bond formation **134–43**
    genetic studies **12–15**
    in vitro studies **20**, 24
    in vivo studies **6–8**
    membrane protein insertion **107–24**
        assay 7–8

bacteria (*continued*)
  see also membrane proteins, insertion
      into bacterial membranes and ER
  protein export **47–68**
    energetics 63–7
    historical notes 67–8
    Tat pathway 62–3, 405–6, *406*
    translocase *see* translocase, bacterial
  Sec pathway *see* Sec pathway, bacterial
  signal peptides 36–8, 49–50, 54, 110
  subcellular fractionation 8
  toxins *192*, **196–7**
  see also Escherichia coli
bacteriorhodopsin *108*, 123, 124
Balbiani ring particles 294
Bet1p 384
bet3 387
bet5 387
β-barrels 107
Bgl2p 327
BIG1 382
binding chain hypothesis *220*, 223–4
biotechnology 4
BiP *see* Kar2p/BiP
Bos1p 384
bovine-pancreatic-trypsin inhibitor 245,
    257
bovine serum albumin (BSA) 310–11
bovine spongiform encephalopathy (BSE)
    199
brefeldin A (BFA) 333, 382
'Brownian ratchet' model, mitochondrial
    protein import 229
'bulk flow' hypothesis 384

C–Vps complex **348–9**
Caenorhabditis elegans 167, 170, 386
calcium pump 155, 187
calmodulin 38, 390
calnexin 84, *85*, 91, 185–6
calreticulin *85*, 91, 185–6
Candida boidinii 282
CAP-binding complex (CBC) *302*, 306
carbonate 65
carbonyl cyanide *p*-chlorophenylhydrazone
    (CCCP) 6
carboxypeptidase 9
carboxypeptidase S (CPS) *328*, 334, 343–6
carboxypeptidase Y (CPY) *328*, 351
  mature (mCPY) form 325
  mutated (CPY*), quality control 186–7,
      188–90, *192*
  p1 or ER precursor form 323
  p2 or Golgi form 324, 325
  receptor *see* Vps10p
  transport to yeast vacuole **322–6**, *323*,
      336–7
    role of clathrin in sorting at TGN
        328–32, 334

sorting from secretory pathway 324–6
targeting to PVC 334–5
carrier translocase *see* TIM22 complex
CAS *302*, 303
caspase-12 172–3
CD4 receptor *192*, 197, 204
CDC6 310
Cdc34 202
Cdc48p 389
cell biology techniques 5, **27–8**
cell cycle arrest, in UPR induction 171–2
cerebrohepatorenal syndrome 284–5
CF$_O$ assembly, subunit II 259–61
CFTR *see* cystic fibrosis transmembrane
        conductance regulator channel,
        protein-conducting *see* pore
        (channel), protein-conducting
        chaperones
  in bacteria 52, 53–4
  chloroplast preproteins 244–5
  cytosolic 52, 79, 219
  ER lumen 79, *85*, 91
  ER membrane 84, *85*
  in ER quality control 182, *183*, 184, 185
  localization to ER 366–7
  membrane protein insertion and 116–17
  mitochondrial 228, 230
  in peroxisomal protein import 277
  in protein folding 151, 153, 407
  for transport vesicle formation 384–5
  in unfolded protein response 158–9
charge difference rule 116
Chinese hamster ovary (CHO) cells 26–7,
        272, 277, 285–6, 391
CHIP 185
chloroplasts 1–2, 3, **240–61**
  general import pathway 241, *242*, 251
  isolation 11, *243*
  protein import at envelope membrane
      241–52, *242*
    cytosolic chaperones 244–5
    energetics 246–7
    experimental techniques *243*, 246
    'guidance complex' *242*, 245–6
    inner membrane machinery *see*
        Tic complex
    outer membrane machinery *see* Toc
        complex
    targeting sequences 40–1, **241–4**, 251,
        252
    versus peroxisomes 280
  protein insertion into envelope
      membranes 251–2
  stroma-targeting signals (transit peptides)
      41, 253
  thylakoid proteins *see* thylakoid proteins
cholera toxin *192*, 196–7
cholesteryl esters 91–2
choline 153

CHOP (C/EBP homologous protein-10) 155, 170, 172
chromoplasts 240
Cim/Com44 (now Tic40) 250–1
cisternae, Golgi see Golgi apparatus, cisternae
cisternal maturation model, Golgi 362–3, 363, 365–6, 367–8, 369
clathrin 327, 329, 380
    in ALP/AP-3 sorting pathway 333
    associated proteins 329, **331–2**
    in CPY sorting pathway **328–32**, 334
    in endosome-to-TGN retrograde transport 338
clathrin-coated vesicles **327–8**, 365
co-translational translocation
    bacteria **50–2**, 53
    ER 8–9, 78–9
        in vitro assay 17–18, 18
        in mammalian cells **88–90**, 89
        mechanisms 79, 80–2, 80
        membrane protein integration **92–4**, 93
        nascent chain processing and folding 89, **91–2**
        protein targeting **75–8**, 110
        regulation 96, 97
        translocon components 83–4, 85, 86
        in yeast 89, **90–1**
coat proteins (COPs)
    in cargo selection **380–5**
    role of p24 proteins 385–7
    see also COPI transport vesicles; COPII transport vesicles
coatomer 380–3, 383
'conservative sorting' model 40, 218
COPI transport vesicles 360, 378, 381, 407
    CPY transport to yeast vacuole 324
    escaped ER proteins 367
    formation **380–3**, 383
    Golgi interactions 362, 363–4, 363
    in vitro assays 391–2
    role of p24 proteins **385–7**, 394–5
    tethering 387–91
COPII transport vesicles 359–60, 378, 380, 381, 407
    CPY transport to yeast vacuole 324
    formation **384–5**
    in vitro assays 391–2
    role of p24 proteins **385–7**, 394–5
    tethering 387–91
CPY see carboxypeptidase Y
CREB (cyclic AMP response element binding protein) 309
Creutzfeldt–Jakob disease (CJD) 199
CRM1 301, 302, 304, **305–7**
Cse1p 303
Cue1p 188, 194
CXXC motifs 136, 137–8, 140, 144, 146–7
cyanobacteria 62, 249, 254

cyclin B–Cdc2 complex 310
cyclin D1 171
cystic fibrosis **198**, 204
cystic fibrosis transmembrane conductance regulator (CFTR) 192, 198, 204
    δF508 mutant 184, 185, 188, 190, 198
cytochrome c oxidase subunit IV 218
cytochrome b5 115
cytochrome bc_1 complex 39
cytochrome P-450 116
cytohesin-1 382
cytomegalovirus (HCMV) 190, 192, 197, 204
cytoplasmic membrane, bacteria 47
cytoskeleton
    in secretory pathway **369–72**
    see also microtubules
cytosol 47
cytosol-to-vacuole transport (cvt) pathway 322

DBP5 308
DDP1 protein 230
deafness dystonia syndrome 230
density gradient centrifugation 8, 11, 18–20, 19
Der1p 190
Der3p/Hrd1p 188, 190, 194
detergent-dilution method, rapid 24
diffusion, passive 294
digitonin 25, 298
dihydrofolate reductase 116, 218, 257
dipeptidylaminopeptidase (DPAP) A 324, 328, 337, 338
dipeptidylaminopeptidase (DPAP) B 346
disulfide bond formation 3, **131–47**
    in E. coli periplasm 134–43
        isomerization pathway 135, 139–43, 141
        oxidation/isomerization as futile cycle 143
        oxidative pathway 135–9, 139
    in eukaryotes 143–7
        PDI reoxidization 145, 146
        PDI structure 144–5
        role of PDI homologs 145–7
    mechanism 132–3, 134
    role in protein folding 131–2, 151
disulfide modifying proteins 183
dithiothreitol (DTT) 135
djp1p 277
DnaJ 277, 327
DnaK 53
    see also Hsp70
docking complex, peroxisomal membrane 272–4, 274
dolichol pathway 155
DPAP see dipeptidylaminopeptidase
Drosophila melanogaster 167, 170, 185
Drs2p 329, 331
drugs 6–7

DsbA 135–7, *135*, 139–40, *139*
  in futile cycle 143
  reoxidation 138
  versus PDI 144, 147
DsbB 137–8, *137*, *139*
DsbC *136*, 140–1, *141*
  dimerization 143, 146–7
  versus PDI 144
DsbC/DcbG-DsbD pathway 139–43, *142*
DsbD 140, 141–2, *142*
DsbG 140
dynein/dynactin 370

early endosome antigen 1 (EEA1) 335
early light-induced protein (Elip) 261
eIF2-α 169, 170
electron microscopy 27
electron transport chain 138, *139*
elongation arrest *76*, 77
emphysema, lung **198**
end mutants 16
endocytic pathway, yeast vacuole 322, *323*
endoglycosidase H (endo H) sensitivity 11
endomembrane system 358, *359*
  constant turnover 367–8
endoplasmic reticulum (ER) 3
  –Golgi intermediate compartment *see*
    ERGIC elements
  CPY transport to yeast vacuole 323–4
  disulfide bond formation 132, **143–7**
  enzyme regulation 193–4
  exit sites *see* transitional ER (tER) sites
  lumenal proteins
    quality control 188
    recycling of escaped 366–7
  membrane proteins
    insertion *see* membrane proteins,
      insertion into bacterial membranes
      and ER
    recycling of escaped 367
  methods of study
    in vitro protein export assay 17–18, *18*
    in vitro techniques 17–20, 24–5
    in vivo techniques 9–11, *10*
    isolation 11
  protein export 359–60, 377–9, *378*
    modeling and simulation 392–4
    see also secretory pathway; vesicular
      transport
  protein folding 89, **91–2**, 151–2, 182
  protein localization to **366–8**
  protein processing 3, 89, **91–2**, 182
  protein quality control 152, **180–204**
  protein sorting **74–102**
  retrotranslocation *see* retrotranslocation
  in secretory pathway 151–2, 181–2,
    359–60, 377–9
  signal peptides (sequences) 35–7, 76
    degradation 38

release from SRP 78
SRP-mediated recognition 76–7, *76*
targeting to membrane 50–1, **75–8**, 110
  flotation assay 18–20, *19*
translocation across membrane 75,
  78–100
  compared to other systems 404–5, *405*
  versus peroxisomes 280
translocon *see* translocon
unfolded protein response *see* unfolded
  protein response
endoplasmic reticulum (ER)-associated
  degradation (ERAD) **155–6**, 181,
  200–1
  components *183*
  in disease **196–200**
  historical notes 201–4
  machinery **187–93**, *193*
  substrates *192*
  targeting mechanisms 184–7
  translocon functions 89, **94–5**, 189–90
endosome
  early 334
  late
    in lysosomal enzyme sorting 327–8
    on way to yeast vacuole *see*
      prevacuolar compartment
    to TGN retrograde transport **336–42**
energetics
  mitochondrial protein import 225–8
  nuclear transport 301
  protein export in bacteria **63–7**
  protein import at chloroplast envelope
    246–7
  translocation across ER membrane **97–8**
epidermal growth factor (EGF) receptor
  343
epoxide hydrolase 118
ER *see* endoplasmic reticulum
ER stress response element (ERSE) 153
ERAD *see* endoplasmic reticulum
  (ER)-associated degradation
ERGIC elements 360, *361*, 377, 387
  cytoskeletal interactions *361*, 370
  Golgi interactions 362, *363*
ERGIC53 384, 385
Ero1 145, *146*, 152–3
ERp57 84, *85*, 185
ERp72 146, 152
Escherichia coli 47
  membrane protein insertion 111–13
  protein targeting to translocase 51, 52–3,
    258
  signal peptides 38, 49, 50
  subcellular fractionation 8
  Tat pathway 255–6, *256*, 258
  translocase 56
ESP1 146
etioplasts 240

EUG1 146, 152
eukaryotes
    cells 1–2, *2*
    disulfide bond formation **143–7**
    genetic studies **15–16**
    in vitro studies 17–20, 24–7
    in vivo studies **8–11**
    protein-conducting channel *61*, **61–2**
    signal peptides 36–7, 38
    single-spanning membrane proteins *114*,
        115
    subcellular fractionation 11
    see also mammalian cells; *Saccharomyces
        cerevisiae*; yeast
eukaryotic translation initiation factor 2
    (eIF2-α) 169, 170
evolutionary conservation
    heterotrimeric protein-conducting
        channel *61*, **61–2**
    PEX genes 285–6
    unfolded protein response 174
exocyst 366
exportin-1 *see* CRM1
exportin-4 *302*
exportin-t 301, *302*, **305**
exportins 299, *302*
    adaptor molecules 306
    function as importins 307
    transport cycles 299–301, *300*

Fab1p 344–6
Far1p 309
farnesylpyrophosphate 194
fatty acid oxidation 269, 281–2
ferredoxin (Fd) 244
Ffh 51, 77
Fkbp2 152
flavin adenine dinucleotide (FAD) 145,
    146
FlgI 135
flotation assay 18–20, *19*
fluorescence microscopy 27–8
fluorescence techniques **21–4**
folded proteins 405–6
    import into chloroplasts 245
    nuclear transport 295
    peroxisomal import 276, 280
    transport by Tat pathway **62–3**, 257–8
    see also misfolded/misassembled proteins;
        protein folding; unfolded proteins
folding, protein *see* protein folding
4.5S RNA 51
14-3-3 proteins 245
'frustrated' topologies 119
FtsH 58
FtsY 51, *53*, 258
    chloroplast homolog 259, *260*
Fur4p *192*
FYVE domain 335, 344

G-protein, vesicular stomatitis virus
    (VSV-G) 26–7, *26*
GADD153 *see* CHOP
β-galactosidase 12, 60
α-galactosidase A *192*
GCN2 169
Gcn2p 170
Gcn4p 170
Gea1 382
Gea2 382
GEFs *see* guanine exchange factors
general import pore (GIP) complex **222–3**,
    232
genetic studies 5, **12–17**
    in bacteria 12–15
    in eukaryotes 15–16
Gerstmann–Sträussler–Scheinker disease
    (GSS) 199
GGA proteins **333–4**
Gga1p *329*
Gga2p *329*
β-glucanase 327
glucose-6-phosphate dehydrogenase 8
glucose regulatory proteins (GRPs) 152
    GRP58 152
    GRP78 *see* Kar2p/BiP
    GRP94 151, 152
    GRP170 152
glucosidases I and II 185, 186
Glut1 119
glutathione 145
glycophorin A 122–3
glycoprotein glucosyltransferase 185
glycoproteins
    asparagine (Asn)-linked, processing state
        9–11
    formation *see* glycosylation
    quality control mechanism 185–6
glycosomes 269
glycosylation 8, 91
    membrane proteins 120
    in yeast 186, 323–4
glycosyltransferases, Golgi 327
glyoxysomes 269, 277
GM130 387
Golgi apparatus 2, 3
    cisternae 360–2, 364–5, 377–9
        cis, *medial* and *trans* 362
        'delay timer' model 364–5
        'distillation tower' model 364
        tubulation of rims 364
    constant turnover 367–8, *369*, 378–9, *378*
    CPY transport to yeast vacuole 324, 325
    cytoskeletal interactions *361*, 370–2, *371*
    in vivo study methods 9–11, *10*
    protein localization to **366–8**
    protein processing 324
    protein transport through **360–5**,
        377–8, *378*

Golgi apparatus (*continued*)
　　cisternal maturation model 362–3, *363*,
　　　364, 365–6, 367–8, *369*
　　modeling and simulation 392–4
　　reconstitution *in vitro* **6–7**, *26*, 393–4
　　stable compartments model 362–3, *363*,
　　　365, 367
　　relations with tER sites and ERGIC
　　　elements 360, *361*
　　see also trans-Golgi network
golgins 387
gp25 (p24α₂) 385, 386
gp25L 394
gp27 (p24γ₃) 385
Gram-negative bacteria 47, 51–2
Gram-positive bacteria 47, 52
Grd19p *339*, 340–1
green fluorescent protein (GFP) 28, 281,
　　343–4, 379–80
GRIP domain 387
GroEL 53
GRPs *see* glucose regulatory proteins
GTP
　　in chloroplast preprotein translocation
　　　247
　　in nuclear transport 301, 303
　　in SRP-mediated protein targeting 51, *53*,
　　　78
　　in Toc34 insertion into membrane 252
　　in transport vesicle formation 382–3, *383*,
　　　384
GTPase activating protein (GAP) 382–3,
　　383, 384
GTPases
　　in co-translational protein targeting 50–1
　　in transport vesicle formation 381–3, 384
　　see also Rab proteins
GTPγS 382, 383
guanine exchange factors (GEF) 334, 381–2,
　　384
　　see also RanGEF
'guidance complex' *242*, 245–6

h-βTrCP 197
HAC1 mRNA splicing 157, **159–60**, *161*
　　intracellular localization 164
　　mammalian Ire1p homologs 167
　　regulation by ER stress 163–5
　　ribonuclease responsible for **160–3**
Hac1p 157, 159–60, **163–5**, 201
HBsu 52
Hcf106 255, *256*
heat shock proteins (Hsps) 151
　　in peroxisomal protein import
　　　277, 281
　　see also Hsp40; Hsp70; Hsp100
heme-regulated inhibitor (HRI) 169
histidine 15–16
histidinol dehydrogenase (HD) 15–16

histone H1 303
HIV
　　CD4 receptor *192*, 197, 204
　　Rev protein *302*, 303, 306
HOPS (C–Vps complex) **348–9**
Hrd1p 188, 190, 194
Hrd3p 190, 194
Hsc70 185, 327
Hsp40 277, 281
Hsp70
　　in chloroplast preprotein translocation
　　　244, 245, 247
　　mitochondrial (mtHsp70) 225–9, *226*, *228*
　　in mitochondrial protein import 219
　　in peroxisomal protein import 277, 281
Hsp100 247
Htm1p 186, 189
human cytomegalovirus (HCMV) 190, *192*,
　　197, 204
human disease
　　ER-associated degradation (ERAD) and
　　　192, **196–200**
　　lysosomal enzyme deficiencies 351
　　peroxisomes and **284–6**
hydrophobic effects 111, 117–18, 121–3
hydroxymethylglutaryl-CoA reductase
　　(HMG-R) *192*, 193–4, 203

I-cell disease 351
IAP *see* Toc complex
IgE receptor 184, *192*
IgM 184, *192*, 203
immunofluorescence 28
immunoglobulin
　　light chains 35, 36
　　quality control 184, 185–6, *192*, 203
immunogold electron microscopy 27
immunology 4
immunoprecipitation assays 7–8
importin 5 *302*
importin 7 *302*, 303
importin 8 *302*
importin 9a+b *302*
importin 11 *302*
importin 13 *302*, 307
importin α 25, 38, 298, *302*, 303
importin β (Impβ) 25, 298–9, **301–4**, *302*
　　adaptor molecules 298, *302*, 303–4
　　nuclear pore passage 311
　　regulation of nuclear localization 310
　　transport events mediated by 303–4, 306
importin β-like nuclear transport receptors
　　294, **297–307**, *302*
　　adaptors/co-receptors 298, *302*, 303–4
　　with combined exportin/importin function
　　　307
　　in mRNA export 308
　　transport cycles 298–301
　　transport events mediated by 301–7

importins 298, 299, *302*
    adaptor molecules 298, *302*, 303–4
    function as exportins 307
    transport cycles 299–301, *300*
in vitro techniques 5, **17–27**
    nuclear transport assay 25, 298
    peroxisomal matrix protein import 277
    photocrosslinking and fluorescence
        methods 21–4
    reconstitution of Golgi transport 26–7
    reconstitution of protein translocation
        24–5
    translocation into lipid vesicles 25
    translocation into membrane vesicles
        17–20
in vivo techniques 5, **6–11**
    bacteria 6–8
    eukaryotes 8–11
inheritance
    peroxisome 281
    yeast vacuole proteins 322
inner membrane peptidase (Imp) 218
inositol 153
insomnia, fatal familial 199
insulin receptor *192*
intermediate filaments 370
invertase 9, *328*
    fusion proteins 16, 195–6, 324–5
    processing 324
    transport from Golgi 327
inverted membrane vesicles (INV or IMV)
        20
iodide ions 81
Ire1α 165, *166*, 167–8, 171
Ire1β 165, *166*, 167–8
    physiological responses 171, 172
Ire1p 157, 174, 200–1
    in apoptotic response 172
    kinase domain 163, 172
    mammalian homologs 165, **167–8**
    regulation of Hac1p production 159–60,
        164
    ribonuclease activity 160–3, *161*
    sensing of unfolded proteins **159–60**, *161*,
        182
    similarity to PERK 169

JNKs (c-jun amino-terminal kinases) 172

K(X)KXX motifs 367, 381, 385–6
KAP60 303
Kar2p/BiP 62, 84, *85*
    functions 88, *89*, 90–1, 121, 151
    as molecular ratchet 97, 100
    in protein quality control 186–7, 189
    in unfolded protein response 152, 159,
        161, 171, 182
karyopherins *see* importins
KDEL receptor 366–7

Kex1p 324, 337
Kex2p *328*
    clathrin-dependent transport 329–30, 334
    protein processing 324, 325
    PVC-to-TGN transport 337–8, 339, 341
kinesins 370
kinetic modeling 368

L10 306–7
L23a 304, 305
L25 304
lactose
    permease 123, 124
    sensitivity 12
LacZ 12–13, *13*, 15
LamB 13, *13*, 15, 67
λ-repressor protein 195–6
leader peptidase 109
    membrane insertion 111–12
    see also signal peptidase
leader sequence 109, 110
    see also signal peptides
lectin-like proteins 152, *183*, 185, 407
lepB gene *14*
leptomycin B (LMB) 306
Lhs1p 152
light-harvesting chlorophyll *a/b*-binding
        preprotein (preLHCP) 244
light-harvesting chlorophyll-binding protein
        (Lhcb1) 259, *260*, 261
light-harvesting chlorophyll-binding protein
        (Lhcb5) 261
'linear insertion model' 119
lipid metabolism, peroxisome function 269
lipid vesicles, protein translocation into
        25
lipoproteins
    bacterial, signal peptides **37**
    particle formation 92
Los1p 305
low-density lipoprotein receptor *192*
lspA gene *14*
luciferase, firefly 270, 285
lung emphysema **198**
Lys-tRNA 22–3, *23*, 102
lysosomal enzymes
    deficiencies causing disease 351
    sorting into clathrin-coated vesicles
        327–8
lysosome 2, 342–3, 347

M13 procoat protein 25, 259, *260*, 261
MalE-LacZ protein 12, 15
maltodextrin 13, *13*, 15, 67
maltose
    resistance 12, 15
    sensitivity 12, *13*
maltose binding protein (MBP), preprotein
        8, 12

mammalian cells
co-translational protein targeting 50–1, 110
co-translational translocation **88–90**, *89*
heterotrimeric protein-conducting pore
61, 62
nuclear transport assay 25, 298
PDI homologs 146–7
peroxisomes 276–7, 281
protein quality control 185–6, 190–1
protein translocation at ER membrane
62, 78–9
sorting of lysosomal enzymes 327–8
translocon 83–4, *85*, 86
unfolded protein response 152, **165–73**,
174
vesicular transport 382
mannose 6-phosphate (M6P) 327, 351
receptor 327–8
mannose trimming/removal 185–6, 324
α1,2-mannosidase 186
mannosidase I 185–6
α1–3 mannosyltransferase (Mnn1p) 327
Markov models, hidden 42
masking–unmasking signals 184–5
medium chain fatty acids (MCFA) 281–2
melanoma, malignant **198**
melanosomes 333
membrane potential (Δψ) 225, *228*, *230*, 231
membrane proteins **107–24**
insertion in chloroplast envelope 251–2
insertion into bacterial membranes and
ER **109–24**
anchoring/orienting transmembrane
helices 113–22
bundling of transmembrane helices
122–3
direct versus translocase-mediated
111–13, *112*
historical notes 124
targeting from cytosol 109–10
translocon-mediated 89, **92–4**, *93*
localization in ER and Golgi 367
multi-spanning
insertion into ER membrane 94,
119–22, *120*
structure 108–9, *108*
quality control 185, 187–8, 189
single-spanning *93*, 94, **113–19**
flanking charges 115–16
folding 116–17
hydrophobicity 117–18
structure 108
types 113–15, *114*
structure 107–9
tail-anchored 195, 220
transmembrane helices *see*
transmembrane (TM) helices
membranes, isolation methods 8, 11
menaquinone 138, *139*

methionine, radiolabeled ([^{35}S]-methionine) 6
methods **5–28**
cell biology techniques 27–8
genetic techniques 12–17
in vitro techniques 17–27
in vivo techniques 6–11
Mex67p-Mtr2p complex 308
Mge1 *226*, 228, *228*
MHC class I molecules 190, *192*, 197, 204
MHC complexes 38
microscopy 27–8, 372
microsomes 11, 17–20, 24
microtubules
peroxisome association 281
in secretory pathway *361*, 370–1
Mig1p 309
misfolded/misassembled proteins 152,
182–3
discovery and retention in ER 184–7
retrotranslocation and degradation 89,
94–5, **187–93**
see also unfolded protein response
mitochondria 1, 3, **214–33**
intermembrane space proteins
'conservative sorting' model 40, 218
'stop-transfer' model 40, 218
targeting sequences 40, 218
isolation 11
matrix proteins, targeting sequences
39–40, 218
Oxa1p 405, *405*
protein import 215, *216*
'Brownian ratchet or trapping' model
228–9
important discoveries *217*
inner membrane machinery 224–31
outer membrane machinery 219–24
'pulling' model 229
unanswered questions 233
versus peroxisomes 280
targeting sequences 39–40, **215–19**
translocases
biogenesis 232–3
inner membrane *see* TIM complexes
outer membrane *see* TOM complex
mitochondrial import stimulating factor
(MSF) 219
mitochondrial matrix targeting peptides
(mTP) 39
mitochondrial processing peptidase (MPP)
39, 218, *228*
mitosis
nuclear transport and 295, 304
peroxisome inheritance 281
Mnl1p 186, 189
MPD1 146
MPD2 146
mPTS *see* peroxisome targeting signals
(PTS), membrane

mRNA
  nuclear export 295, **307–8**
  truncated technology 21–4, *21*
mRNA-binding proteins 308
Msn5 309
Msn5p 307
Mtr10 308
multivesicular bodies (MVB) **342–6**
  regulation by phosphoinositides 344–6
  transport to vacuole from 347

N-acetylglucosamine-transferase I
  (GlcNAc-T1) 26, 391–2
N-end rule 202
N-glycan modifying proteins *183*
N-glycans 185
NADH oxidase 8
NADPH oxidase p40 (PX) domains 341
nascent chains
  integration into ER membrane 89, **92–4**,
    93
  processing and folding 89, **91–2**
  ribosome complex *see* ribosome–nascent
    chain (RNC) complex
  translocation *see* co-translational
    translocation; post-translational
    translocation
  transmembrane (TM) sequences 92–3,
    93, 94
  see also preproteins
nascent-polypeptide-associated complex
  (Nac) 22
NBD (7-nitrobenz-2-oxa-1,3-diazole) 23–4
neural networks 42
neuregulin precursor 117
neurodegenerative diseases 198–9
Neurospora crassa 215, 222
nexins, sorting 341
NFAT 310
NLS *see* nuclear localization signals
NMD3 306–7
nocodazole 370, *371*
nonsense-mediated decay (NMD) 307
Notch 168
NPCs *see* nuclear pore complexes
NSF (*N*-ethylmaleimide sensitive factor)
  389, 390
NTF2 294–5, 298, 311
  RanGTP nuclear import 299–301, *300*
nuclear envelope 293
nuclear export signal (NES), leucine-rich 306
nuclear hormone receptors 310
nuclear localization signals (NLS) 38
  adaptors 298
  classical 297–8
  Impα-Impβ complexes 303
  receptors *see* nuclear transport receptors
nuclear pore complexes (NPCs) 4, 293, 294,
  295–7

facilitated translocation through 310–11
  morphology 296–7, *296*
nuclear transport **293–311**, 407
  general aspects 294–5
  import signals 38, 295
  in vitro assay 25, 298
  mRNA export 307–8
  nuclear pore function 310–11
  regulation of nuclear localization 308–10
  ribosomal protein import 304–5
nuclear transport receptors 294–5,
  297–307
  mRNA transport 295
  nuclear pore passage 310–11
  shuttling 295, 298–301, *300*
  see also exportins; importin β-like nuclear
    transport receptors; importins; NTF2
nucleoplasmin 297–8
nucleoporins 297, 311
nucleus 1, 4, 293
NXF1 308

Oaf1 282
OEP *see* Toc complex
oligomerized proteins, peroxisomal import
  276, 280
oligopeptidase IV 38
oligopeptidase A 38
oligosaccharides, N-linked 9–11, *10*, 323–4
oligosaccharyltransferase (OST) 84, *85*, 86,
  91, 92
OmpA 7, 8, 121, 135
organelles, isolation methods 11
outer membrane, bacteria 47
outer membrane protein A *see* OmpA
Oxa1 proteins 233, 259
Oxa1p 405, *405*

p10 *see* NTF2
p15 308
p23 (p24δ₁) 385
p24 (p24β₁) 385, 394–5
p24 proteins 384, **385–7**, 394–5
p26 (p24γ₄) 385
p47 389
p53 tumor suppressor 306, 310
p97 389, 390
p115 387
p200/GEF 382
Pael receptor 199–200
parathyroid hormone 203
Parkin 199–200
Parkinsonism, juvenile 199–200
PDI *see* protein disulfide isomerase
PDI1 146
Pdr5, mutant (Pdr5*) 186, 187, 189, *192*
pep mutants 325, 350–1
Pep12p 332, 334, 339
peptidyl prolyl isomerase 152, *183*

periplasm 47
  disulfide bond formation 132, **134–43**
  isolation 8
PERK 165, *166*, **169**, 171–2, 174
peroxins 272, *273*, 281
  see also individual proteins
peroxisome targeting signals (PTS) 40,
      270–1, *271*
  membrane (mPTSs) 271, *271*, 279
    receptor 279
  'piggy back' entry of proteins lacking 276,
      280
  receptors 272–6, *274*
    cargo shuttling role 275–6
    docking at peroxisomal membrane
      272–4
    downstream peroxin interactions 275
    structure 272, *273*
    see also PTS1; PTS2
peroxisomes 2, 3, **268–86**, 406, *406*
  biogenesis 270–81
    intermediates 282–4
  discovery 268
  functions 268–70
  in human disease 284–6
  import of matrix proteins **272–8**, *274*
    docking of PTS receptor/cargo complex
      272–4
    in vitro systems 277
    other proteins involved 277–8
    PTS recognition in cytosol 272
    PTS1 and PTS2 pathway linkage in
      mammals 276–7
    PTS1 receptor shuttling 275–6
    subcellular locations of Pex5p 276
    targeting signals 40, 270–1, *271*
    translocation across membrane 276
  inheritance 281
  membrane protein (PMP) import **278–80**
    peroxins implicated 279–80
    targeting signals 40, 271, *271*, 279
  metabolic control of proliferation 281–2
  unique features of protein import 280–1
pertussis toxin *192*, 196
PEX genes 272, *273*
  complementation groups 285, 286
  disease-causing mutations 284–6
  evolutionary conservation 285–6
pex mutants 283–4
Pex2p *274*, 275, 279
Pex3p 272, *274*, 279, 280
Pex5p (PTS1 receptor) 272, *274*, 278
  cargo-shuttling role 275–6
  docking complex interaction 272–4
  downstream peroxin interactions 275
  long isoform (Pex5pL) 276–7
  PTS2 pathway interaction in mammals
      276–7
  short isoform (Pex5pS) 276–7

structure 248, 272, *273*
  subcellular locations 276
Pex7p (PTS2 receptor) 272, *274*, 275, 278
  docking complex interaction 272–4
  PTS1 pathway interaction in mammals
      276–7
  structure 272, *273*
Pex8p 278
Pex10p *274*, 275
Pex11p 281–2
Pex12p *274*, 275
Pex13p 272–4, *274*
Pex14p 272–4, *274*, 277
Pex16p 279–80
Pex17p 272, *274*, 279, 280
Pex18p 278
Pex19p *274*, 279
Pex20p 278
Pex21p 278
Pf3 coat protein 25, 111, 115–16
PHAX *302*, 306
pheromone receptors *323*
Pho4p 308–9
phosphate carrier 229
phosphatidylinositol-3-kinase (PtdIns-3
      kinase, Vps34p) 334–5, 344
phosphatidylinositol-(3,5)-diphosphate
      (PtdIns(3,5)P_2) 344, 345–6
phosphatidylinositol-(3)-phosphate
      (PtdIns(3)P) 334–5, 344, 346
phospholipids 49
phosphorylation
  chloroplast presequences 242–4, 245
  CPY in yeast 324
  regulation of nuclear localization 309
photocrosslinking studies **21–4**, *23*, 80–1, 83
photosystem II, subunits W, X and Y 259–60
'phylogenetic profiles', targeting sequences
      43
Pichia pastoris
  peroxisomes 275, 283
  secretory pathway 359, 364, 370, *371*
'piggy back' entry into peroxisomes 276, 280
Pip2 282
PKR 169
plants 62
  see also chloroplasts
plastids 240–1
platelet dense granules 333
Pma1p 327, *328*
PMP22 279
PMP47 282
PMP70 279
Pmr1p 187
polarized epithelial cells 372
polyubiquitination 181, 190
pore (channel), protein-conducting
  ER membrane *see* translocon
  heterotrimeric 61–2, *61*

peroxisome translocase 276
SecYEG complex 59–60
TIM complexes 225, 231
Toc complex 249
TOM complex 222–3
see also nuclear pore complexes
'positive-inside' rule 115
post-translational translocation
  bacteria *48*, 50, **52–4**, *55*
  ER 8–9, 79
    mechanism *89*, **90–1**
    regulation 96, 97
    targeting 79, 90
    translocon components 84, *85*
  mitochondria 219
potassium (K⁺) channel
  ATP-sensitive 184–5, *192*
  KcsA 123
PPARα 282
prepilin peptidase 38
preproteins
  import at chloroplast envelope
    membranes 241–52
  targeting *see* protein targeting
  see also nascent chains; post-translational
    translocation
presenilin-1 (PS1) 168
presequence translocase *see* TIM23 complex
prevacuolar (late endosomal) compartment
  (PVC) *323*
  CPY transport to 325, 334
  maturation into multivesicular bodies
    344–6
  protein sorting into lumen **342–6**
  targeting CPY-containing vesicles to
    334–5
  to TGN retrograde transport **336–42**
    genes implicated *339*
    mutant studies 338–9
    retrieval signals on cycling proteins
      337–8
    retromer complex 340–1
    vesicle targeting/fusion with TGN
      341–2
  transport to vacuole from **347–50**
    C–Vps complex 348–9
    membrane fusion with vacuole 347–8
    retrograde vacuole to PVC pathway
      350
prion protein *192*, 199
prl mutants 12–14, *14*, 57, 67
prlA gene *see secY* gene
pro-α-factor 9
  mutant, quality control 186, 187, 188–9,
    192
  processing in Golgi 324
prokaryotes *see* archaea; bacteria;
  cyanobacteria
proline 109

proteases
  ATF6 cleavage 168
  cleavage sites, engineered 8
  mapping studies 7, 9
  multicatalytic complex *see* proteasome
  protein processing 324, 325
  signal peptide degradation **38**
  yeast vacuole 350–1
proteasome
  20S 202
  26S 181, 202
  alternative pathway 194–6
  degradation of malfolded ER proteins
    188
  discovery 202–3
protein degradation 181, 201–2
  ER-associated *see* endoplasmic reticulum
    (ER)-associated degradation
  mammalian UPF effectors 167–9, 170–1
protein disulfide isomerase (PDI) 133,
    143–7
  cholera toxin interaction 197
  in co-translational translocation 84, *85*, 91
  homologs 145–7
  in protein quality control 187
  reoxidization 145, *146*
  structure 144–5
  in unfolded protein response 152
protein export 3
  in bacteria **47–68**
  ER *see* endoplasmic reticulum (ER),
    protein export
  quality control *see* quality control, protein
  TGN *see under trans*-Golgi network
protein folding 151–2, 182, 407
  chloroplast protein import and 244–5
  enzymes in ER 152–3
  in ER membrane *89*, **91–2**
  membrane protein insertion and 116–17,
    124
  mitochondrial protein import and 229
  role of disulfide bonds 131–2, 151
  see also folded proteins;
    misfolded/misassembled proteins;
    unfolded protein response
protein import 3
  see also translocation
protein kinase 242
protein kinase A (PKA) 309
protein kinase inhibitor (PKI) 306, 309
protein localization
  to Golgi and ER **366–8**
  prediction methods **41–3**
  within cells 2
  see also protein targeting
protein processing/modification 3, *89*, **91–2**,
    182
  in *trans*-Golgi network 324
  see also glycosylation

protein quality control *see* quality control, protein
protein sorting (trafficking) **74–102**
   historical notes 100–2
   principles 74–5
   targeting to ER membrane 75–8
   translocation across ER membrane 78–100
protein study methods *see* methods
protein targeting 3
   to bacterial translocase 48, 49–55
   to ER membrane 50–1, **75–8**, 110
   to mitochondria 215–19
   to peroxisomes *see under* peroxisomes
   sequences *see* targeting sequences
   to thylakoid membranes *see under* thylakoid proteins
proteinase A (PrA) 322, 325, *328*
proteinase B 325
proteoliposomes 24
proteolysis *see* protein degradation
proton motive force (PMF)
   in bacterial protein export 47, 63, *64*, **65–7**
   in thylakoid lumenal protein import 254–5
PsaN 254
PsbS 261
PsbW *260*
PSORT method 43
PTS *see* peroxisome targeting signals
PTS1 40, 270, *271*
   receptor *see* Pex5p
PTS2 40, 270–1, *271*
   receptor *see* Pex7p
'pulling' model, mitochondrial protein import 229
pulse-chase experiments 6, 10–11
puromycin 81
PVC *see* prevacuolar compartment

Q-SNAREs 388, 390
quality control, protein 152, **180–204**, 407
   alternative pathways 194–6
   disease and 196–200
   in enzyme regulation 193–4
   ERAD *see* endoplasmic reticulum (ER)-associated degradation
   essential nature 200–1
   historical notes 201–4
   mechanism 182–7
   system components 180–2, *183*
quinones 138, *139*

R-SNAREs 388, 390
Rab proteins 334, 348, 349, **387–91**
Rab1 387
Rad6 202
Ran 298, 299
RanBP1 299
RanBP2 299
RanGAP 299

RanGEF (or RCC1) 299, 301
RanGTP 299, 303
   gradient model 299–301, *300*
RanGTP-binding motif 301
RCC1 (or RanGEF) 299, 301
reconstitution
   Golgi transport *26*, **26–7**, 393–4
   protein translocation **24–5**
   vesicular transport **391–2**
redox potential 132–3, *134*, 140–1
redox proteins **62–3**, 258
Refsum disease, infantile 284–5
replication protein A (RPA) 304, 307
Rer1 367
retromer complex 338, **340–1**
retrotranslocation 189
   regulation 98
   toxins 196–7
   translocon functions *89*, **94–5**, 189–90
Rev protein, HIV *302*, 303, 306
reverse signal-anchor sequence *114*, 115
rhizomelic chondrodysplasia punctata (RCDP) 285–6
ribonucleoprotein particles (RNPs), nuclear export 295, 307–8
ribophorin I (RI332) 86, 191, *192*
ribosome–nascent chain (RNC) complex
   flotation assay 18–20, *19*
   photocrosslinking and fluorescence techniques 21–4, *21*
   SRP-mediated targeting 51, 52, *53*, **75–8**, *76*
   translocation across ER membrane **78–83**
   see also nascent chains; ribosomes
ribosomes 75
   biogenesis 294, 304
   cytoplasmic versus membrane-bound 101–2
   proteins, nuclear import **304–5**
   SecYEG association 56
   SRP binding 77
   stoichiometry 86
   subunits, nuclear transport 294, 304, 306–7
   translocon alignment/association 82, *82*, **96**
   transmembrane (TM) sequence recognition 92–3
   see also ribosome–nascent chain (RNC) complex
Ric1p *339*, 341
ricin A chain *192*, 196, 197
rlg1-100 allele 157, 162
RLG1 gene 157
Rlg1p *see* tRNA ligase
RNaseL 160–2
RNC complex *see* ribosome–nascent chain (RNC) complex
RND (resistance/nodulation/cell division) family 60–1
rpL23a 304

rpL25 304
RS-ALP *328*, 337–8, 350
Rtg3p 309
Rubisco activase 244

Saccharomyces cerevisiae
    alternative quality control pathway 195
    ERAD 186, 187–90, *192*, 200–1, 202–3
    genetic techniques 15–16, *16*
    Golgi apparatus 361–2
    in vivo techniques 8–11
    mitochondrial protein import 215, 219–24
    nuclear transport 301–2, 303, 304, 308–9
    peroxisomes 278, 281–2
    secretory pathway **322–51**, *323*, 372
    translocon 62
    unfolded protein response 152–3, **157–65**,
        161, 200–1
    vacuole *see* vacuole, yeast
    see also yeast
Sar1p 384
Sbh1p 62, 84, *85*, 90–1
SCAP (SREBP cleavage activating protein)
        168
scrapie 199
sec genes *14*, *16*
Sec mutants
    bacteria 6–7, 12–14, *13*
    yeast 9, 15–16, 324, 325
Sec pathway 404–5, *405*
    bacterial **48–62**, 254
        energetics 63–7
        historical notes 67–8
        signal peptides **36–7**, 49–50
    chloroplast 62, 254, *257*, 261, 404, *405*
    versus Tat pathway 257–8
    see also Sec61 complex; SecYEG complex
Sec1-related proteins 334, 348–9
Sec7 382
Sec11 9
Sec12 384
Sec13/31 384
Sec16 384
Sec17p 348, *348*, 389
Sec18p 348, *348*, 389
Sec23/24 384
Sec34/35 387
Sec61 complex
    evolutionary conservation *61*, 62
    functions 88–91, *89*
    in vitro studies 24–5
    mammalian 83–4, 86
    in membrane protein insertion *112*, 113
    pore 59, **86–7**
    in protein quality control 189–90
    regulation 98–9
    yeast 8–9, 84
SEC61 gene 16, 62
Sec61α 9, 62, 83, *85*, 86

Sec61β 62, 83–4, *85*
    in aqueous pore 86
    regulation 98
    stoichiometry 86
Sec61γ 62, 83–4, *85*, 86
Sec61p 9, 51, 62, 84, *85*
    function 90–1
    mutant, ERAD 186, 187–8, *192*
sec62 gene 16
Sec62p 62, 84, *85*, 90–1
sec63 gene 16
Sec63p 62, 84, *85*
    function 90–1, 97
    mammalian homolog 88
Sec71p 84, *85*, 90–1
Sec72p 84, *85*, 90–1
SecA **56–7**, 404
    chloroplast homolog 254
    energetics 64–5, *64*, 66–7
    evolutionary conservation 62
    in vitro studies 20
    in membrane protein insertion 111–12,
        113
    in protein targeting 53–4, *55*
    in protein translocation 56–7, 59–60, 62,
        68
    SecDFYajC interaction 60
secA gene 13, *14*, 56
    mutants 6, 12
    suppressor mutations 14–15
SecB
    absent chloroplast homolog 254
    in vitro studies 20
    protein targeting to translocase *48*, 52–4,
        55
secB gene 12, *14*
SecD 56, *58*, 60–1, 66
secD gene *14*, 60
SecDFYajC complex *58*, **60–1**, 113
SecE 56, 58
    evolutionary conservation 61–2, *61*
    mutants 6–7, 57–8
secE gene 13, *14*, 57–8
SecF 56, *58*, 60–1, 66
secF gene *14*, 60
SecG 24, 56, 58, 66
    in membrane protein insertion 111–12, 113
secG gene *14*
secretory carriers 365–6
secretory pathway 151–2, 181–2, **358–73**,
        377–9
    export from ER 359–60, 377–9, *378*
    export from TGN 365–6
    modeling and simulation 392–4
    perspectives 372–3
    protein localization to ER and Golgi 366–8
    role of cytoskeleton 369–72
    signal peptides *see* signal peptides
    transport through Golgi 360–5, *363*

secretory pathway (*continued*)
unfolded protein response (UPR) and
152–5, *154*
yeast vacuole **322–51**, *323*
see also Sec pathway; vesicular transport
secretory proteins
export in bacteria 48–68
misfolded/misassembled *see* misfolded/
misassembled proteins
quality control *see* quality control, protein
SecY 56, 57–8
evolutionary conservation 61–2, *61*
SecA interaction 57
secY (*prlA*) gene 13, *14*
mutants 57, 58, 65, 66–7, *67*
suppressor mutations 14–15
SecYE 58
in vitro studies 24
in membrane protein insertion 111–12, 113
SecYEG complex *48*, 56, **57–60**, *58*
chloroplast homolog 254, 259
pore formation 59–60
protein targeting *53*, 54, *55*, 258
purification and reconstitution 67–8
SecA interaction 56, 57, 59–60, 65
SecDFYajC association 60
see also translocase, bacterial
Sed4 384
Sed5p 390
Sed50 341
selective phase model, nuclear pore passage
311
Shiga toxin *192*, 196
Shr3p 384–5
shuttling
nuclear transport receptors 295, 298–301,
300
PTS receptor 275–6
signal-anchor sequence 92, 93–4
multi-spanning membrane proteins
119–20, *120*
reverse *114*, 115
single-spanning membrane proteins *114*,
115, 116–18
signal hypothesis 36, 101, 180
signal peptidase (SP) 109
bacteria 50
cleavage site 36–7, 50
ER membrane 84, *85*, 92
function 91
genes *14*
lipoprotein 37
stoichiometry 86
signal peptides (sequences) **35–8**, 109–10
bacterial lipoproteins 37
bacterial proteins 36–8, 49–50, 54, 110
chloroplast proteins *see* targeting
sequences, chloroplast
cleavage 8, 91, 109

degradation of cleaved 38
discovery 35–6, 101
ER *see* endoplasmic reticulum (ER),
signal peptides (sequences)
multi-spanning membrane proteins 94,
119–22, *120*
photocrosslinking studies 22
recognition 54, 76–7
Sec 36–7, 49–50
single-spanning membrane proteins 92,
93–4, 113–15, *114*, 116–18, *117*
Tat 38, 63
uncleaved 110
see also targeting sequences
signal recognition particle (SRP) 17–18,
50–2, 124, 404
discovery 101
evolutionary conservation 62
nuclear transport 294, 307
protein targeting 110
in bacteria *48*, 51–2, *53*, 54, 258
to chloroplast thylakoid membrane
258–9, *260*, 261
to ER membrane 76–8, *76*
signal peptide recognition/binding 54,
76–7, *76*
signal sequence release 78
SRP9/14 heterodimer 51–2, 77
SRP54 77
structure 51–2, 76–7
signal recognition particle (SRP) receptor
(SR) 24, 50–1
bacteria 51
discovery 17–18
ER membrane *76*, 78
regulation 98
SRα 51, 78
SRβ 51, 78
signal sequences *see* signal peptides
small subunit of Rubisco (SSU), preprotein
(preSSU) 244–5, 248
α-SNAP 389, 390
SNAP-25 388
SNAPs 388
SNARE motif 388
SNARE proteins **387–91**
complex formation 388–9
nomenclature 388
targeting and insertion into membrane 115
in transport vesicle formation 385
in transport to yeast vacuole 334, 341,
347–8
SNAREpin hypothesis 389–90
snurportin 1 *302*, 304
SNX15 341
sorting nexins 341
SP *see* signal peptidase
spectrin 370
spheroplasts 7

SR *see* signal recognition particle (SRP) receptor
SREBP (sterol-response element binding protein) 168, 310
SREBP cleavage activating protein (SCAP) 168
SRP *see* signal recognition particle
Srp1p 303
Ssh1p complex 62
Sss1p 62, 84, *85*, 90–1
stable compartments model, Golgi 362–3, *363*, 365, 367
Ste2p *328*, 343, 344
Ste3p *328*
sterol-response element binding protein (SREBP) 168, 310
'stop-transfer' model 40, 218
stop-transfer sequence
    multi-spanning membrane proteins 119, 121–2
    single-spanning membrane proteins 113, *114*
stress activated protein kinases (SAPKs) 172
subcellular fractionation 8, 11
sucrose density gradients 8, 11, 18–20, *19*
Sulfolobus acidocaldarius 115
suppressors of suppressors 14–15
SV40 large T-antigen 298
Swa2/Aux1p *329*, 331
Swa3/Drs2p 331
synaptic vesicles 333
synaptobrevin 2 390
synaptobrevin/VAMP 115, 388
Synechocystis 249, 254
syntaxins 388, 390
synToc75 249

T-cell receptor 184, *192*, 203
t-SNAREs 334, 341, 347–8, 388
tail-anchored membrane proteins 195, 220
TAP-p15 complex 308
targeting sequences 3, **35–43**, *37*
    chloroplast 40–1
        envelope membranes **241–4**, 251, 252
        stroma ('envelope transit' signals) 41, 253–4
        thylakoid proteins 41, 253–4, 255, 259–60
    ER membrane 76
    mitochondrial 39–40, **215–19**
    nuclear import 295
    peroxisomal 40, **270–1**, *271*, 279
    protein localization prediction **41–3**
    Tom and Tim proteins 232
    yeast vacuole proteins 324–6
    see also nuclear localization signals; signal peptides
TargetP program 42
tat genes 255–6, *256*
Tat pathway **62–3**, 258, 405–6, *406*

in chloroplast thylakoids 255–8, *256*, *257*
    signal peptide **38**, 63
TatA 255, *256*
TatB 255, *256*
TatC 255, *256*
TCP1 (T-complex protein 1) ring complex 279
techniques *see* methods
temperature-sensitive mutants 6–7, 9, 15, 68
tethering 349, 366, 387–91
    factors **387–91**
tetratrico-peptide repeat (TPR) motifs 222, 247–8
    peroxins 272, *273*
TGN *see trans*-Golgi network
Tha4 (TatA homolog) 255, *256*
thiol–disulfide exchange reaction 132–3
thiolase 278
thioredoxin 136, 141–2, *142*
    family 136, 144, 146
thylakoid lumen 253
thylakoid membranes 241, 253
thylakoid processing peptidase 41, 260, *260*
thylakoid proteins **252–61**
    historical note 261
    lumenal targeting **253–8**, *257*, 261
        Sec-independent pathway 254–8, 405–6, *406*
        Sec-related pathway 62, 254, 261, 404, *405*
    membrane targeting **258–61**
        SRP-dependent pathway 258–9, *260*
        SRP-independent/'spontaneous' pathway 259–61, *260*
    targeting sequences 41, 253–4
thyroglobulin 185–6, *192*
Tic complex 241, *242*, **249–51**, 406–7
    energetics of translocation 247
    experimental methods 246
    Toc complex interaction 250
Tic20 250, 251
Tic22 250, 251
Tic40 250–1
Tic55 250
Tic110 250
TIM complexes 215, **224–31**, 406–7
    TOM complex association 229
    see also TIM22 complex; TIM23 complex
Tim proteins 215, 224–5
    biogenesis 232–3
    tiny 229–30, *230*, 232
    see also individual proteins
Tim8 227, 230
Tim9 227, 229–30, *230*, 231
Tim10 *227*, 230, *230*, 231
Tim12 227, *230*, 231
Tim13 *227*, 230
Tim17 225, *226*, *228*, 231
Tim18 *227*, *230*, 231

Tim22 *227*, *230*, 231, 232
TIM22 complex (carrier translocase) 225,
    229–31, *230*
  biogenesis 232
  components *227*
Tim23 225, *226*, *228*, 231
  binding chain hypothesis 223, 224
TIM23 complex (presequence translocase)
    225–9, *228*
  biogenesis 232
  components 225, *226*
  TOM complex association 229
Tim44 225–9, *226*, *228*
Tim54 *227*, *230*, 231, 232
tiny Tim proteins 229–30, *230*, 232
Tlg1p 334, 341
Tlg2p 341
Toc complex 241, *242*, **247–9**, 406–7
  energetics of translocation 247
  experimental methods 246
  Tic complex interaction 250
Toc33 248–9
Toc34 247
  insertion into outer envelope membrane
    251–2
  presequence recognition 242–4
  receptor function 248–9
Toc36 (now Tic40) 250–1
Toc64 247–8
Toc75 245, 247, 249, 250, 252
Toc86 *see* Toc160
Toc120 248
Toc132 248
Toc160 (formerly Toc86) 244, 247, 248, 249
TOM complex 215, **219–24**, *220*, 406–7
  binding chain hypothesis *220*, 223–4
  biogenesis 232
  components *221*
  general import pore (GIP) complex 222–3
  import receptors 220–2
  TIM complex association 229
Tom proteins 215
  biogenesis 232–3
  see also individual proteins
Tom5 *221*, 222, 223, 224, 232
Tom6 *221*, 222, 223, 232
Tom7 *221*, 222, 223, 232
Tom20 39–40, 219, 220–2, *221*, 224
  targeting to mitochondria 232
  in Tom protein import 232
Tom22 220–2, *221*, 223
  binding chain hypothesis 223, 224
  in Tom protein import 232
Tom40 *221*, 222–3, 232
Tom70 39, 219, 220, *221*, 222, 248
toxins *192*, **196–7**
TRAF2 172
TRAM 25, 83, 84, *85*, 405
  in aqueous pore 86

regulation 98
stoichiometry 86
trans-Golgi network (TGN)
  CPY sorting within 325
  protein export from **365–6**
  protein export to yeast vacuole *323*, **326–36**
    ALP/AP-3 sorting pathway 332–3
    CPY sorting pathway 327, 328–31
    role of GGA proteins 333–4
    targeting vesicles to PVC 334–5, *335*
  protein localization to 367
  protein processing 324
  protein transport to 362, *363*
  retrograde transport from PVC *see*
    prevacuolar compartment (PVC), to
    TGN retrograde transport
transitional ER (tER, ER exit) sites 359–60,
    361, 370, *371*, 386
translation
  attenuation in UPR 170
  elongation arrest *76*, 77
  see also co-translational translocation
translocase
  bacterial *48*, **49–63**
    chloroplast homolog 254
    components 55–61
    evolutionary conservation 61–2, *61*
    -mediated membrane protein insertion
      111–13, *112*
    protein targeting 49–55, *53*, *55*
    Tat pathway 63
    see also SecA; SecYEG complex
  at inner membrane of chloroplasts *see* Tic
    complex
  at inner membrane of mitochondria *see*
    TIM complexes
  at outer membrane of chloroplasts *see* Toc
    complex
  at outer membrane of mitochondria *see*
    TOM complex
  peroxisome *274*, 275
translocation
  across ER membrane 75, **78–100**
  facilitated nuclear 294, 310–11
  into lipid vesicles **25**
  into membrane vesicles **17–20**
  mechanisms compared **404–8**, *405*, *406*
  photocrosslinking and fluorescence
    techniques **21–4**
  reconstitution **24–5**
  regulation **95–9**
  retrograde *see* retrotranslocation
  see also co-translational translocation;
    post-translational translocation;
    translocon
translocation-associated membrane protein
    see TRAM
translocon (ER) **79–100**, 181–2
  composition and structure **83–7**

accessory components and
    stoichiometry 84–6, *85*
primary components 83–4, *85*
properties of aqueous pore 86–7
evolutionary conservation 61–2, *61*
functions **87–95**, *89*
    in mammalian cells 88–90
    membrane protein integration *89*,
        92–4, *93*
    nascent chain processing and folding
        89, 91–2
    retrotranslocation and protein
        degradation *89*, 94–5, 189–90
    in yeast 90–1
historical notes 100–2
hypothesis and discovery **79–82**, 100–1
regulation **95–9**
    assembly, modification and turnover
        98–9
    directionality and energy requirements
        97–8
    dynamics 96–7
    ribosome alignment and coordination
        82, *82*, 96
transmembrane (TM) helices 107–9
    insertion into membrane 92–3, *93*, 94
    intra- and intermolecular bundling
        122–3
    membrane anchoring and orienting
        113–22
    uncleaved signal sequences 110
transport vesicles *see* vesicles, transport
transportin *302*, 305, 310
TRAPP 387
'trapping' model, mitochondrial protein
        import 229
trigger factor 54
trigger hypothesis 124
triglycerides 91–2
triskelion 327
tRNA ligase 157, **160–3**, *161*, 164
tRNAs, nuclear export 305
truncated mRNA technology 21–4, *21*
tryptophan 109
tunicamycin 9, 201
twin arginine translocation pathway *see* Tat
        pathway
type III secretion 36
tyrosinase *192*, 198
tyrosine 109

U-snRNPs 304, 306
UAS$_{ino}$ 153
Ubc1p 188
Ubc6p 187, 188, 191, 194–5
Ubc7p 188, 191, 194
ubiquinone 138, *139*
ubiquitin 181, 187, 201–2
    system 181, 187–9, 190–1, 202, 204

ubiquitin-activating enzyme (E1) 181,
        191
ubiquitin-conjugating enzyme
        (Ubc, E2) 181, 188, 191, 204
    yeast 187–8
ubiquitin ligase (E3) 181, 188, 190
ubiquitination 181
UDP-GlcNAc glycosyltransferase I 26–7
unfolded protein response (UPR) 95,
        151–75, 182, 200–1, 407
    in mammals 165–73, *166*, 174
        effectors of signaling pathway 167–9
        physiological responses 170–3
    output 152–6
        ER-associated degradation 155–6
        secretory pathway remodeling 152–5,
            154
    signaling pathway in yeast 157–65, *161*,
        174
        HAC1 mRNA splicing 159–63
        Hac1p translation in ER stress 163–5
        Ire1p-mediated sensing of unfolded
            proteins 158–9
        methods of identifying components
            157–8
    versus mammals 165, *166*, 174
unfolded protein response element (UPRE)
        153, *161*
unfolded proteins
    ER-associated degradation 155–6
    import into chloroplasts 244–5
    molecules sensing 158–9, 174
    Sec pathway transport 404
    see also folded proteins;
        misfolded/misassembled proteins;
        protein folding
uracil permease 195
urea 65, 80
urokinase 140

v-SNAREs 334, 341, 347–8, 388
V0 hexamers 390
Vac1p/Vps19p 335
vacuolar ATPase (V-ATPase) *192*, 195
vacuolar system 180, 201
vacuole, yeast **322–51**
    CPY transport via secretory pathway
        322–6
    endosome-to-TGN retrograde transport
        336–42
    endosome-to-vacuole transport 347–50
    historical notes 350–1
    multivesicular body pathway 342–6
    pathways exiting the TGN 326–36
    secretory and endocytic pathways *323*
vam mutants 325, 347
Vam3p 322, 333, 347, *348*, 349
Vam7p 347, *348*, 349
vasopressin precursor 118

vesicles
  clathrin-coated **327–8**, 365
  inverted membrane (INV or IMV) 20
  lipid, protein translocation into **25**
  membrane, protein translocation into
    17–20
  produced from TGN 326–7
  transport
    arguments for and against 379–80
    role of p24 proteins 385–7
    tethering 387–91
    see also COPI transport vesicles;
      COPII transport vesicles
vesicular stomatitis virus (VSV) G-protein
    (VSV-G) 26–7, *26*, 379, 384, 391–2
vesicular transport 3, 358, **377–95**, 407
  historical notes 394–5
  modeling and simulation 392–4
  reconstitution *in vitro* 391–2
  to yeast vacuole *323*
  see also vesicles, transport
vesicular-tubular clusters (VTCs) *see*
    ERGIC elements
viruses
  nuclear transport 294
  subversion of ERAD machinery *192*,
    197
  see also HIV; vesicular stomatitis virus
Vma22p 195
Von Willebrand factor *192*
Vph1p *192*, 195, 346
vpl mutants 16, 3553
VPS genes *16*, 325–6
  in multivesicular body (MVB) formation
    344, *345*
  in PVC-to-TGN transport 338–9, *339*,
    340, 341
  in PVC-to-vacuole transport 347, 348–9,
    348
  in TGN-to-PVC transport 334–5, *335*
vps mutants 16, 325–6, 336, 347, 351
Vps1p *329*, 331–2
Vps4p 344
Vps5p *339*, 340, 341
Vps6p (Pep12p) 332, 334, 339
Vps9p 334, *335*
Vps10p (CPY receptor) 195–6, 326, 328, *328*
  clathrin-mediated transport 331
  in PVC-to-TGN transport 336–7, 339
Vps11p 348–9, *348*
Vps15p 9, 335, *335*
Vps16p 348–9, *348*

Vps17p *339*, 340
Vps18p 348–9, *348*
Vps21p 334, 335, *335*
Vps24p 344
Vps26p *339*, 340, 341
Vps27p 338, 339, 344
Vps29p 339, *339*, 340, 341
Vps30p 339, *339*
Vps32p 344
Vps33p 348–9, *348*
Vps34p 334–5, *335*, 344
Vps35p 339, *339*, 340–1
Vps39p 348–9, *348*
Vps41p 333, 348–9, *348*
Vps45p 334, 335, *335*
Vps52p *339*, 341–2
Vps53p *339*, 342
Vps54p *339*, 342
vpt mutants 351
Vpu protein 197
Vti1p 334, *339*, 341, 347–8, *348*, 350, 390

WD40 repeats 272, *273*
Wilson disease 198–9
Wilson protein *192*, 198–9

Xenopus laevis 297–8, 304
XRIP *302*, 304

YajC 56, *58*, 60
Yap1 310
Yarrowia lipolytica 283, *283*
yeast
  co- and post-translational translocation
    79, *89*, **90–1**
  PDI homologs 145–6
  protein quality control 186, 187–90, 200–1
  protein targeting 110
  translocon *61*, 62, 84, *85*
  unfolded protein response 152–3, **157–65**,
    161, 174
  vacuole *see* vacuole, yeast
  vesicular transport 382
  see also Pichia pastoris; *Saccharomyces
    cerevisiae*
YidC 111, 112–13, *112*, 259, 405
ypt1 387
Ypt6p *339*, 341
Ypt7-GTP 390
Ypt7p 348, *348*, 349

Zellweger syndrome 284–5

Printed and bound by CPI Group (UK) Ltd, Croydon, CR0 4YY

03/10/2024

01040415-0012